Noise Pollution
and **its Control**

Contents at a Glance

Noise Pollution
and its Control

KJ Polak

CBS Publishers & Distributors Pvt Ltd

New Delhi • Bengaluru • Chennai • Kochi • Kolkata • Mumbai
Bhopal • Bhubaneswar • Hyderabad • Jharkhand • Nagpur • Patna • Pune • Uttarakhand • Dhaka (Bangladesh)

Noise Pollution
and its Control

ISBN: 978-93-88902-93-9

Copyright © Author and Publisher

First Edition: 2019

Published by Satish Kumar Jain and produced by Varun Jain for

CBS Publishers & Distributors Pvt Ltd
4819/XI Prahlad Street, 24 Ansari Road, Daryaganj, New Delhi 110 002, India.
Ph: 23289259, 23266861, 23266867   Fax: 011-23243014   Website: www.cbspd.com
e-mail: delhi@cbspd.com; cbspubs@airtelmail.in.
Corporate Office: 204 FIE, Industrial Area, Patparganj, Delhi 110 092
Ph: 4934 4934   Fax: 4934 4935   e-mail: publishing@cbspd.com; publicity@cbspd.com

Branches

- **Bengaluru:** Seema House 2975, 17th Cross, K.R. Road,
  Banasankari 2nd Stage, Bengaluru 560 070, Karnataka
  Ph: +91-80-26771678/79   Fax: +91-80-26771680   e-mail: bangalore@cbspd.com
- **Chennai:** 7, Subbaraya Street, Shenoy Nagar, Chennai 600 030, Tamil Nadu
  Ph: +91-44-26680620, 26681266   Fax: +91-44-42032115   e-mail: chennai@cbspd.com
- **Kochi:** 42/1325, 1326, Power House Road, Opp KSEB Power House,
  Ernakulam 682 018, Kochi, Kerala
  Ph: +91-484-4059061-65   Fax: +91-484-4059065   e-mail: kochi@cbspd.com
- **Kolkata:** 6/B, Ground Floor, Rameswar Shaw Road, Kolkata-700 014, West Bengal
  Ph: +91-33-22891126, 22891127, 22891128   e-mail: kolkata@cbspd.com
- **Mumbai:** 83-C, Dr E Moses Road, Worli, Mumbai-400018, Maharashtra
  Ph: +91-22-24902340/41   Fax: +91-22-24902342   e-mail: mumbci@cbspd.com

Representatives

Bhopal	0-8319310552	Bhubaneswar	0-9911037372	Hyderabad	0-9885175004
Jharkhand	0-9811541605	Nagpur	0-9421945513	Patna	0-9334159340
Pune	0-9623451994	Uttarakhand	0-9716462459	Dhaka (Bangladesh)	01912-003485

Printed at: Mudrak, Noida, UP, India

Preface

The rapid growth of industrialisation, urbanisation, communication and transport has led to noise pollution in big cities, affecting the common man in offices, institutions and houses. Noise pollution is proportionately related to developmental and growth-oriented activities. It may also be caused due to careless activities of man. In simple words, noise can be defined as a wrong sound, at a wrong place and at a wrong time. In other words, noise may be defined as an 'unwanted sound' and noise pollution as unwanted sound dumped into the atmosphere without considering its adverse affects. The term 'noise' in the electronic communication system is defined as perturbations that interfere with communication. Noise by definition is unwanted sound. What is pleasant to some ears may be extremely unpleasant to others, depending on a number of psychological factors. The sweetest music, if it disturbs a person, who is trying to concentrate or sleep, is a noise to him, just as the sound of a pneumatic riveting hammer is noise to nearly everyone. In other words, any sound may be a noise if circumstances cause it to be disturbing.

Thus, noise pollution is a direct result of technological development. The more technologically advanced a country is, the more acute the problem becomes. Noise, with its ever increasing effects on human health and on the environment, is defined as an acoustic fact that is unpleasant and arouses disturbing feelings, or as the totality of unwanted, undesired sounds. The unit of measurement of noise is decibel (dB), which is one-tenth of the larger unit called bel (B). A decibel is equal to the faintest sound that can be picked up and heard by the human ear.

This book is divided into 10 sections. Each chapter covers an important aspect of noise pollution and its control. Section I is devoted to general consideration of noise pollution. Chapter 1 concentrates on fundamentals of vibrations, which refers to mechanical oscillations about an equilibrium point. Chapter 2 focuses on noise and its effects on man. Chapter 3 deals with managing noise at workplaces. Chapter 4 discusses various instruments used for noise measurements such as lock-in amplifiers, sound level meters, etc.

Section II concentrates on noise reduction through encloser and barriers. Chapter 5 is devoted to silencer and suppressors which are used on the exhaust of reciprocating engines because of the low-frequency character of noise. Chapter 6 focuses on glass and various methods of reduction of noise through glass are discussed. Chapter 7 discusses ventilating systems and adequate noise control of these systems can be achieved by careful examination of the possible causes and by providing proper isolation and insulation to the primary units in the system. Chapter 8 deals with double wall panels. The goal of this chapter is to develop and validate efficient models to investigate the noise reduction for both airborne and structural-borne noise at low frequencies.

Section III discusses noise reduction in mechanical industries. Chapter 9 is devoted to machinery, the noise reduction of which requires a fundamental knowledge of acoustics and noise control techniques. Chapter 10 focuses on gear, various characteristics of causes of gear noise and their reduction processes. Chapters 11 and 12 deal with drilling operations and hydraulic pneumatics. Chapter 13 concentrates on power presses, and metal cutting saws, various methods of reduction of noise are discussed. Chapter 14 familiarises the students with issues in combustion noise. Chapter 15 is devoted to hydraulic pump which are known for adjustable, controllable power transmission to a variety of actuators. Chapters 16 and 17 concentrate on internal combustion engine and bearing valve. Chapter 18 deals with boilers, various methods to noise reduction. Chapters 19 and 20 discuss smelter and blasting operations. Noise, vibration and airblast are among the most significant issues for communities located near mining projects.

Section IV is devoted to noise reduction in electrical, electronics and telecommunication industries. Chapter 21 concentrates on transformer which is an electrical device, which changes voltage levels and facilitate distribution and utilisation of electrical power in the most efficient and economic manner. Chapters 22 and 23 deal with induction motor noise and motor vibration problems. The main source for magnetic noise in induction motor comes from force waves created by the rotating magnetic field and vibration problems can occur at anytime in the installation or operation of a motor. If not solved quickly, one could either expect long-term damage to the motor or immediate failure, which would result in immediate loss of production. Chapter 24 focuses on electronic noise which is a random fluctuation in an electrical signal, a characteristics of all electronic circuits. Chapter 25 brings to light the semiconductor manufacturing facilities. Chapter 26 highlights noise reduction in computers, various noise reduction methods are discussed in detail. Chapters 27 and 28 discuss noise reduction in digital webcam imaging and mobile phone.

Section V is devoted to noise reduction in construction industries. Chapter 29 concentrates on stone quarrying and crushing. Noise pollution can be controlled from various operations such as blasting, drilling and diesel generating sets. Chapter 30 focuses on construction processes and gives advice on various methods of eliminating, and controlling noise hazards and risks during the construction process by means of correct specification of materials, component and assembly processes. Chapters 31 and 32 deal with noise reduction in cement and concrete and construction equipments.

Section VI concentrates on noise reduction in roadways, railways and aircraft. Chapter 33 deals with roadway noise which is the collective sound energy emanating from motor vehicles, trucks and other vehicles. Chapters 34 and 35 focus on noise reduction from railways and wheel and bogies. Chapter 36 is devoted to aircraft noise which is produced by any aircraft or its components, during various phases of a flight.

Section VII discusses noise reduction in hospitals. Chapter 37 focuses on reduction and optimisation of hospital noise with six sigma tools. Chapter 38 discusses solutions to noise reduction in hospitals by identifying its sources and by better houskeeping, etc.

Section VIII focuses on noise reduction in miscellaneous industries such as plastics, working machines and music and entertainment. Chapter 39 deals with plastic industry. When noise emission become crucial to the decision-making process, various measures are taken to control and reduce noise. Chapter 40 concentrates on wood working machines in the industry. Chapter 41 is devoted to music and entertainment industry.

Section IX discusses noise mapping. Noise is often one of the environmental variables where it seems more difficult to guarantee full compliance with legal limits in complex industrial situations and/or in other multisource environments. Chapter 42 focuses on noise mapping of industrial sources. Chapter 43 is devoted to noise level interpolation and mapping. The noise levels over an area will be varying all the time. For example, noise levels may rise as a vehicle approaches, and reduce again after it has passed.

Section X focuses on various case studies. Chapter 44 deals with case studies related to road, rail and aircraft noise. Chapter 45 deals with noise control in various operations of Bhilai steel plant, diesel locomotives, TISCO and Hindustan Zinc Ltd. Chapter 46 concentrates on noise control from loudspeakers, along with the role played by various implementing agencies, police, NGOs' and organised groups.

Appendices, references, glossary and index have been provided at the end for quick reference. Figures and tables supplement the text. All topics have been discussed in a cogent and lucid style to help the reader grasp the information quickly and easily.

Besides catering to the academic needs of the students pursuing environmental sciences and other undergraduate and postgraduate courses in various branches of engineering, the book will also prove useful to consultants, town planners, industrialists and entrepreneurs.

The book also caters to the requirement of the syllabus prescribed by various Indian universities for undergraduate students pursuing engineering, and allied courses. It has been prepared with meticulous care, aiming at making the book error-free. Constructive suggestions are always welcome from users of this book.

KJ Polak

Contents

Section II NOISE REDUCTION THROUGH ENCLOSER AND BARRIERS

Section IV Noise Reduction in Electrical, Electronics and Telecommunication Industries

Section V Noise Reduction in Construction Industries

Section VI Noise Reduction in Roadways, Railways and Aircraft

Section IX Noise Mapping

Section X Case Studies

SECTION I

General Considerations of Noise Pollution

Chapter 1

Fundamentals of Vibrations

INTRODUCTION

Vibration refers to mechanical oscillations about an equilibrium point. The oscillations may be periodic such as the motion of a pendulum or random such as the movement of a tyre on a gravel road. Vibration is occasionally 'desirable'. For example the motion of a tuning fork, the reed in a woodwind instrument or harmonica, or the cone of a loudspeaker is desirable vibration, necessary for the correct functioning of the various devices. More often, vibration is undesirable, wasting energy and creating unwanted sound — noise. For example, the vibrational motions of engines, electric motors, or any mechanical device in operation are typically unwanted. Such vibrations can be caused by imbalances in the rotating parts, uneven friction, the meshing of gear teeth, etc. Careful designs usually minimise unwanted vibrations.

The study of sound and vibration are closely related. Sound or pressure waves, are generated by vibrating structures (e.g. vocal cords); these pressure waves can also induce the vibration of structures (e.g. ear drum). Hence, when trying to reduce noise it is often a problem in trying to reduce vibration.

NATURE OF VIBRATION

As discussed above, vibration may be considered as an oscillating motion of a particle or body about a reference position. This motion can be periodic, random or transient. Simple examples of each will be discussed here.

Periodic Functions

The sinusoidal signal of Fig. 1.1 is by definition at one discrete frequency, which is given by 1/T, and has the units of Hz (Hertz), which is the number of cycles/second. T is the time taken for the wave to perform one complete cycle. This motion is periodic in that it repeats itself at regular intervals of time (T). The nearest that one would come to a pure tone in vibration is the movement of the arms of a tuning fork.

The vibration of Fig. 1.2 is still periodic because it repeats itself at regular intervals, but it is now complex (i.e. it is not purely sinusoidal). In fact, this particular example is a combination of two sine waves and is the sort of vibration produced during the piston acceleration of an internal combustion engine. A periodic signal need not be symmetrical, and it can be made up of many combinations of frequencies of different amplitudes and phase. The signal of Fig. 1.2 is made up of two components of different frequencies and amplitudes, and these are shown in Fig. 1.3. In this example one sine wave has a frequency that is twice the other, and its amplitude is much smaller. The amplitudes are summed arithmetically at each point in time to obtain the combined signal level.

3

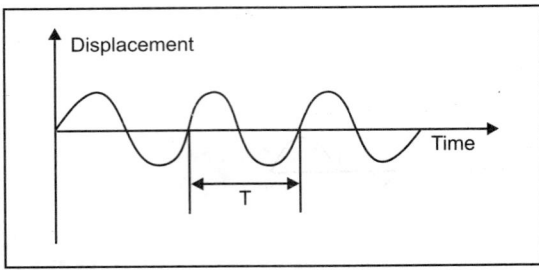

Fig. 1.1. Example of a simple harmonic (sinusoidal) vibration signal ($f = 1/T$).

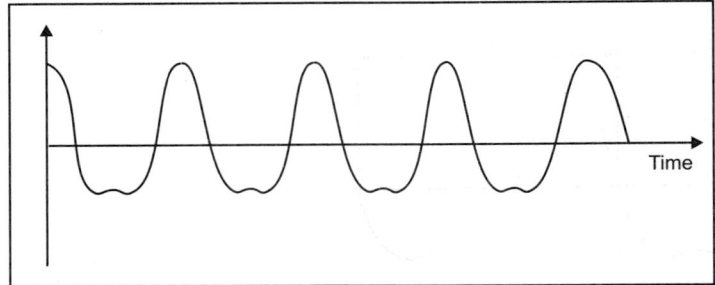

Fig. 1.2. Example of compound harmonic periodic motion ($f + 2f$).

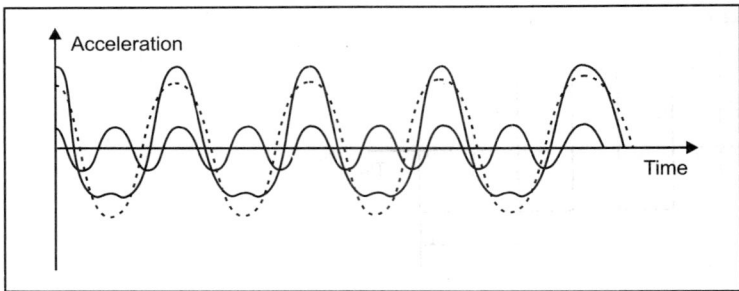

Fig. 1.3. Waveform of Fig. 1.2 split into its components.

An important and useful way of portraying the amplitudes and frequencies of all the different components of a complex signal is the frequency spectrum, which is a plot of amplitude against component frequency. A few simple examples are given in Fig. 1.4.

As can be seen from Fig. 1.4(a), the simple sine wave has a spectrum of one discrete frequency ($f = 1/T$), and is represented by a line whose height is proportional to the amplitude of the wave. The waveform of Figs 1.2 and 1.3 has two components, each having a discrete frequency. One frequency is twice the other. The two frequencies each have one line on the spectrum, whose height is proportional to amplitude [Fig. 1.4(b)]. In this case, the frequency f would be called the fundamental frequency and the frequency $2f$ is its second harmonic, since it is twice the frequency of the fundamental. A 'square' wave, shown in Fig. 1.4(c), is another example of a periodic function. This consists of a fundamental frequency f, and odd harmonics only (i.e. 3, 5, 7f) that drop off in level at a defined rate.

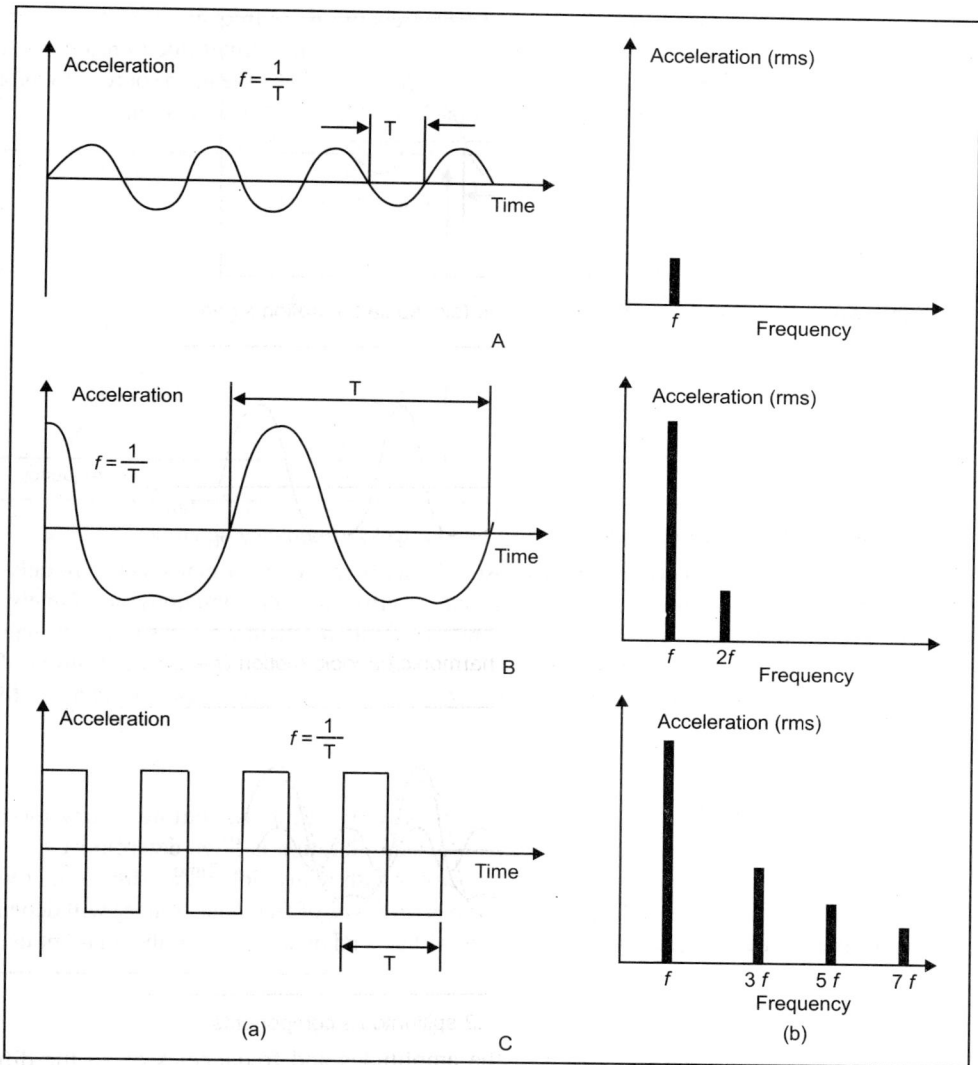

Fig. 1.4. Examples of (a) periodic signals, and (b) their frequency spectra.

All periodic functions, such as those shown above can be defined by precise mathematical equations, which can make their analysis much simpler.

Random Vibration

The most commonly encountered type of vibration in everyday life is random vibration. It is continuous, but non periodic, and contains many frequency components. While many of these components will be related in the form of harmonics of certain frequencies due to machinery movements, many components will be entirely independent of these.

A random vibration signal is shown in Fig. 1.5. It lacks repetition, its frequency spectrum is broad band and or continuous, since it contains all frequencies, and it has some spikes due to resonance or to harmonics of certain vibrating components. These spikes are often caused by the main source of vibration, and once they are discovered from frequency analysis they are relatively easy to isolate.

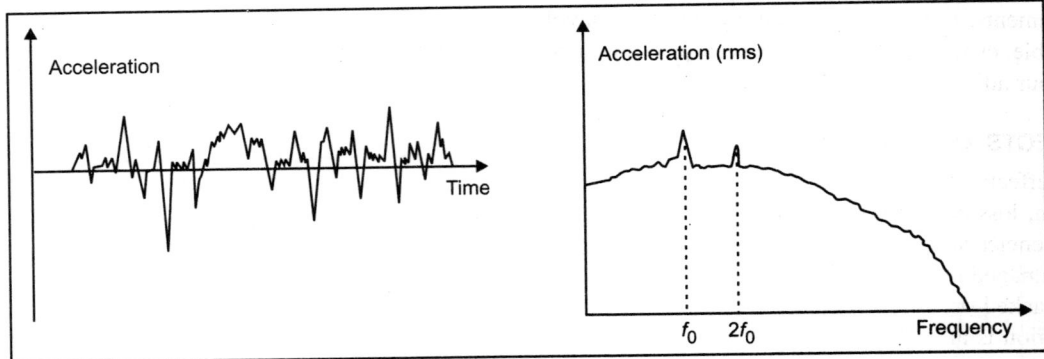

Fig. 1.5. Example of random vibration and a typical frequency spectrum.

A random vibration is by definition nondeterministic; it can be represented mathematically only by a series of probability statements, since to obtain a complete description of the vibrations an infinitely long time record is theoretically necessary. A random vibration is called 'stationary' if successive samples taken of it are essentially the same in character, although the rigorous mathematical definition of this is somewhat more complicated. A stationary random vibration is normally easier to analyse than a nonstationary one.

Transient Vibration

A transient vibration is noncontinuous; it occurs due to impact, or during the starting up of a motor, or anywhere that the exciting force is not continuous. A transient vibration is often deterministic.

The transient vibration shown in Fig. 1.6 has a spectrum with many 'lobes', as shown, that would contain a whole range of frequencies. The size of the lobes (in terms of frequency range) will depend on the duration of the transient, and their relative amplitudes will depend on the shape of the transient itself.

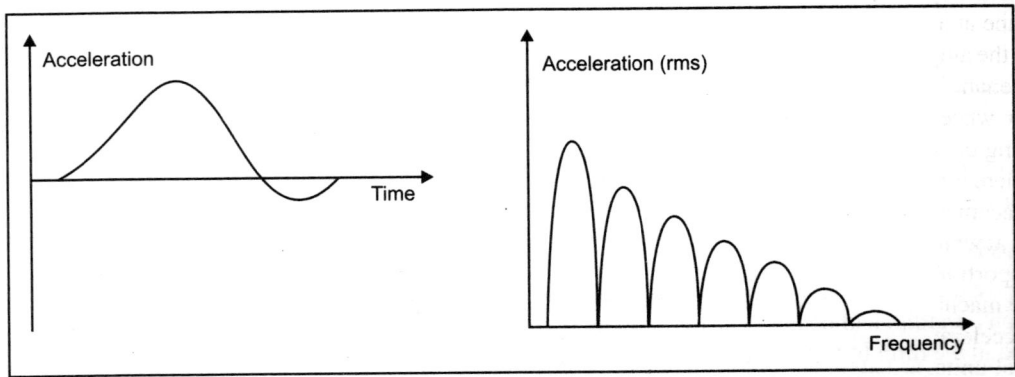

Fig. 1.6. Example of a transient vibration and a typical frequency spectrum.

SOURCES OF VIBRATION

Vibration results from some part of moving machinery being out of balance, turbulent fluid flow, rattling of loose objects, impulses, shocks—the list is endless. Vibration is normally undesirable, but it can often be reduced by careful design or by development modifications. Sometimes, however, vibration is introduced to make things work, such as vibrating conveyor belts, mechanical hammers and even musical instruments. In such cases resulting noise is unavoidable. It is then important to isolate it as much as possible, even to the extent of preventing your hi-fi loudspeakers from rattling your neighbor's picture on your adjoining apartment wall!

EFFECTS OF VIBRATION

The effects of vibration are often serious. Humans subjected to vibration can be affected by blurred vision, loss of balance and consequent lack of ability to do their job properly. In some cases, certain frequencies and levels of vibration can permanently damage internal body organs. Machinery can also be damaged by vibration. If the vibration occurs at the resonance frequency of some component it can be cracked or broken by fatigue, and nuts, bolts and rivets can be shaken apart. Noise resulting from vibration is also often a serious problem, and can be a health hazard to people exposed to it for long periods.

One really difficult thing about vibration is that it will not stay in one place unless special steps are taken to isolate it. Vibration is transmitted through any solid object in contact with it, including the floor, walls, pipes, electrical conduits and any other mechanical linkages, which in turn cause things in contact with them to vibrate and radiate noise. Even if a particular frequency from one machine does not cause any of its own components to resonate, it could quite possibly excite a resonance in a connected machine. It is thus important to measure and control vibration.

MEASUREMENT OF VIBRATION

It is necessary to measure vibration for several reasons. While sometimes the vibration picture is obvious, there are many occasions when it is not, and it would be senseless to design a means of reducing a harmonic component frequency while vibration continues at the fundamental frequency. From the design point of view, the magnitude and frequency of the vibration need to be known to ensure that the stresses induced are not too great for the material to withstand. Another important reason is that if a certain piece of machinery has been found to resonate at a given frequency, the operator can avoid running the machine at a speed that will excite that resonance. For vibration damping or isolating, it is necessary to know the amplitudes and frequencies involved in order to select the correct damping materials. Preventive maintenance is another area where vibration monitoring is useful, since many faults (like worn teeth on a gear wheel or the start of a roller bearing's failure) can be detected long before failure merely by noticing changes in the vibration spectra.

There are three main parameters used to describe vibration: displacement, velocity and acceleration. Displacement is the distance moved by the measuring point from its natural position, velocity is the speed at which that point moves, and acceleration is the rate of change of its speed with time. Displacement is proportional to strain in the material. Velocity is a function of kinetic energy that must be dissipated in the machine. Therefore, velocity is related to potential damage in the machine.

Acceleration is proportional to the force acting on an object. These parameters are measured in the normal units of distance, velocity and acceleration, which are meters (m), meters per second (m/s) and meters per second squared (m/s^2), respectively. The units are in accordance with ISOR 1000 (SI units). Acceleration is also often measured in terms of gravitational constant (g), and it is common to see vibration ranges or levels quoted in terms of numbers of g because it is easy to calibrate measuring equipment at levels of 1 g, and also because g is an internationally understood symbol, regardless of units.

Historically, displacement was the first parameter to be measured because with slow-moving machines and large displacements it was the easiest to see and was measurable by simple optical methods. Displacement limits were then set by the law of elasticity, the fatigue of materials or by mechanical clearances. Measurement of displacement emphasises very low frequencies. At higher frequencies of vibration, displacement is negligible.

With the introduction of higher-speed machines, the displacements became smaller and more difficult to see, but breakdown could still occur, so a move towards velocity measurements was made. Many rotating machines have a vibration frequency spectrum with a fairly even distribution of energy across the frequency range up to 1 kHz (Fig. 1.7), thus making it reasonably easy to prescribe a velocity limit for a particular type of machine. (It is even easier to prescribe for single resonance frequencies.) Standards organisations in several countries now recommend this type of limit especially for installed electric motors. Measurement of acceleration puts emphasis on the higher frequencies, where it is often important to understand the vibratory forces acting on the machine.

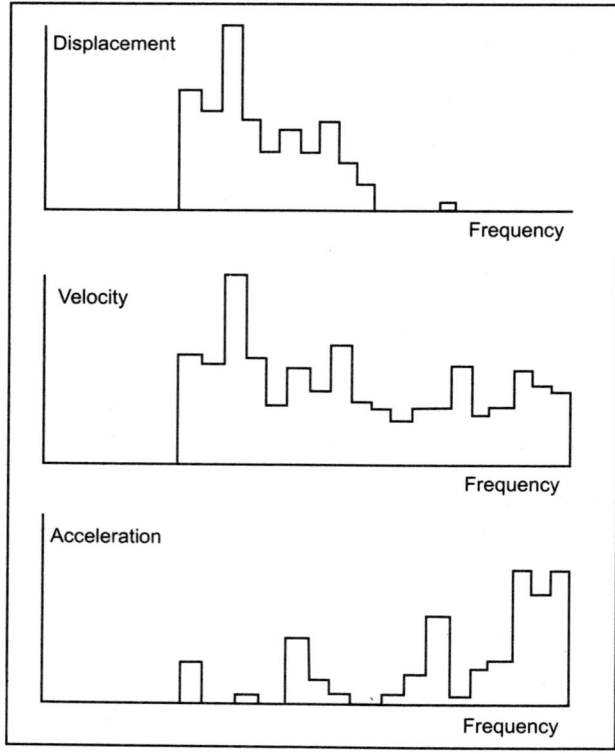

Fig. 1.7. Spectra showing the effects of measuring displacement, velocity and acceleration on rotating machinery in the frequency range up to 1 kHz

For high-frequency vibration (above about 1 kHz), it is necessary to measure acceleration, because displacement velocity drops off rapidly at high frequencies. The reason for this will be seen later. Since high frequencies play a large part in preventive maintenance, early detection of breakdowns and vibration monitoring acceleration measurements are very important. For example, many faults in machine elements, like roller bearings, can be detected in the frequency range 20 kHz–50 kHz long before failure occurs, thus making it possible to plan repairs and shutdown times.

RELATIONSHIP BETWEEN DISPLACEMENT VELOCITY AND ACCELERATION

There is a well-defined relationship between displacement, velocity and acceleration. The concept of simple harmonic motion serves as a basis for discussing the mathematical relationships among these parameters.

Figure 1.8 shows a wheel rotating at a constant angular velocity of ω radian/s; θ is measured counterclockwise from the bottom, and x is measured to the right. (There are 2π radians in one revolution, so the wheel is turning at $\omega/2\pi$ rev/s, or $\omega/120\pi$ rev/min.) The wheel drives a reciprocating mechanism B by means of a protruding stud P, which is free to move in the slider S. B is said to move backwards and forwards with simple harmonic motion (provided there is no mass or friction in any of the sliding contacts) in the horizontal plane.

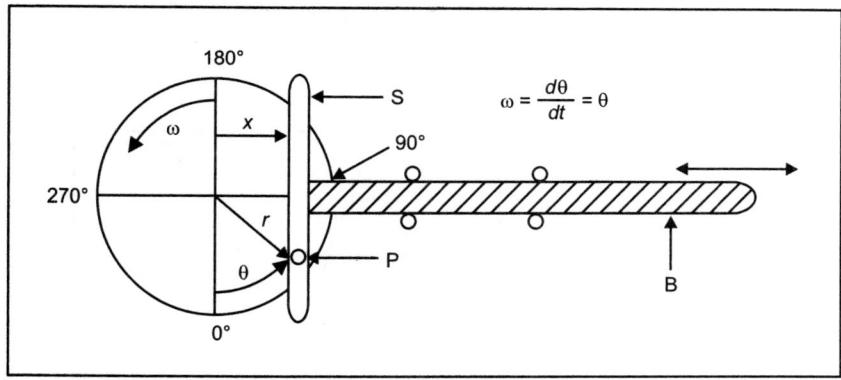

Fig. 1.8. Example of simple harmonic motion.

Thus the displacement (x) of B will be as shown on the curve, and one cycle will be completed with each complete revolution of the wheel. In this case revolutions/second $\equiv$ cycles/second $\equiv$ Hz, and the standard relationship between angular speed ω and frequency f is obtained:

$$\omega = 2\pi f$$

The displacement x of B from its central position is given by:

$$x = r \sin \omega t$$

when, $\theta = 0°$ or $180°$, $x = 0$, $\theta = 90°$, $x = +r$, $\theta = 270°$, $x = -r$.

The maximum displacement is given when $\sin \omega t = \pm 1$, and is thus equal to r.

Differentiating displacement with respect to time, we obtain the velocity v of B:

$$v = \frac{dx}{dt} = \dot{x} = r\omega \cos \omega t$$

when, $= 90°$ or $270°$, $\dot{x} = 0$; $\theta = 0°$, $\dot{x} = +r\omega$; $\theta = 180°$, $\dot{x} = -r\omega$.

The maximum velocity, V, occurs when cos $\omega t = \pm 1$, and V = $r\omega$.

Differentiating again with respect to time, we obtain the acceleration a of B:

$$a = \frac{dv}{dt} = \frac{d^2x}{dt^2} = \ddot{x} = -r\omega^2 \sin \omega t$$

when, $\theta = 0°$ or $180°$, $\ddot{x} = 0$; $\theta = 90°$, $\ddot{x} = -r\omega^2$; $\theta = 270°$, $\ddot{x} = +r\omega^2$.

The maximum acceleration, A, occurs when sin $\omega t = \pm 1$, and A = $r\omega^2$. Figure 1.9 illustrates the effects of these relationships.

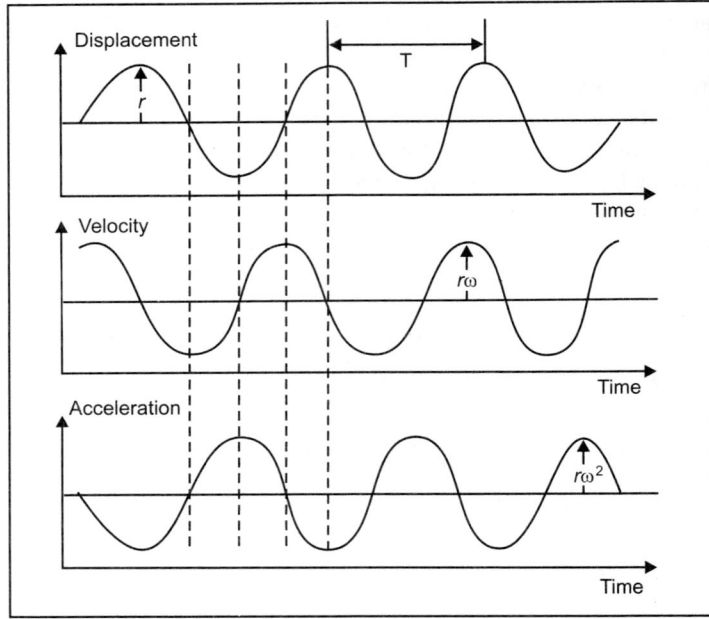

Fig. 1.9. Phase and amplitude relationships between displacement, velocity and acceleration.

It can be seen from Fig. 1.9 that there is a phase difference between these components. The velocity leads the displacement by 90° (or one quarter of a cycle), the acceleration leads the velocity by 90°, and so the acceleration leads the displacement by 180°. However, in terms of measurement this phase difference does not matter because RMS or peak reading are usually obtained on a meter or recorder.

It can also be seen that each differentiation has multiplied the signal by ω, which does have practical significance. It means that at higher frequencies, the acceleration signal is the largest by a factor of ω or ω^2. Since $\omega = 2\pi f$, the acceleration signal is often the easiest one to measure, or sometimes the only one that can be measured, because the other signals drop into instrumentation noise. The main significance of the relationship between these paramters is that at any given frequency one measurement can provide the other two. Sensors having a flat acceleration response (accelerometers) are the most common vibration sensors. Integrating the acceleration will give velocity, and integrating again will give displacement, and this is a fairly standard procedure. This integration is performed by use of a low pass ftlter with a roll-off of $1/f$ for velocity and $1/f^2$ for acceleration. The emphasis of different parts of the spectrum can be seen quite clearly from the spectra of Fig. 1.7. In this case we have a fairly 'flat' velocity spectrum.

The division by ω to obtain displacement attenuates high frequency signals (at high frequencies ω is large), and in this case the signal disappears. The multiplication by ω to obtain acceleration emphasises high-frequency signals. Thus, while acceleration may not be the parameter we wish to study at high frequencies, it is the one we would measure in order to obtain velocity. When drawing spectra, we normally plot the frequency on a logarithmic scale, as it has the effect of expanding the lower frequencies and compressing the higher frequencies, and allowing us to have a reasonable resolution without large sheets of graph paper. Logarithmic plotting also ties in well with the concepts of an octave (doubling of frequency) and a decade (multiplying the frequency by 10) because one octave is now a constant width on the paper, as is one decade, regardless of the frequency range it is covering. Linear scales are useful sometimes, particularly for sorting out harmonics.

PEAK, AVERAGE AND RMS VALUES

There are several different ways of quantifying the level of vibration. The first is to use the peak (or maximum) value, as was done with the analogy of a turning wheel earlier (Fig. 1.8). This peak value is shown in Fig. 1.10. It is useful for simple harmonic vibration (such as the one shown), but for other types it is not as useful because it depends only on an instantaneous vibration magnitude, and takes no account of time history producing it.

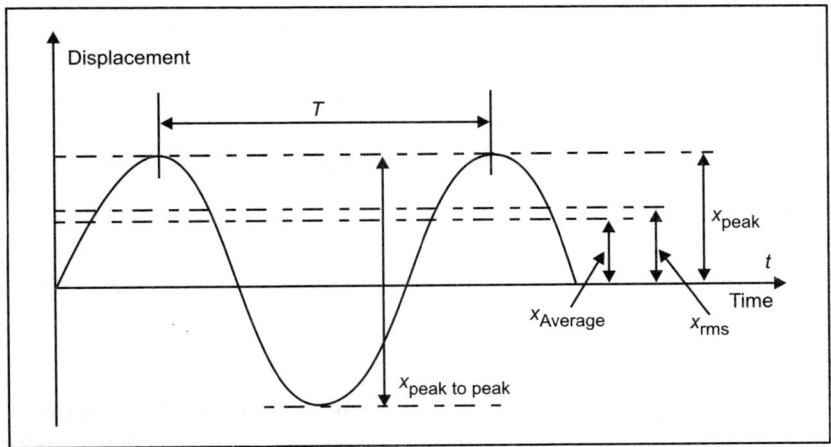

Fig. 1.10. Peak, average and rms values for a sine wave.

The peak-to-peak value, or the magnitude of the positive and negative extremes of the motion, is also commonly used. For a symmetrical signal it is twice the peak value. Another quantity, which does take into account the time history, is the average absolute value, which is defined as:

$$x_{|\text{Average}|} = \frac{1}{T}\int_0^T |x|\,dt$$

Even though this quantity takes into account the time history over a period T, it has been found to be of limited practical interest because it has no direct relationship to any useful physical quantity. A much more useful descriptive quantity, also taking the time history into account, is the rms (root mean square) value, defined as:

$$x_{RMS} = \sqrt{\frac{1}{T} \int_0^T x^2(t)\,dt}$$

The major reason for the importance of the rms value is its direct relationship to the energy content of the vibrations.

The relationship between these values can be expressed as:

$$x_{rms} = F_f x_{|average|} = \frac{1}{F_c} x_{peak}$$

where, F_f and F_c are called form factor and crest factor respectively, and give some idea of the waveshape of the vibrations being studied.

For a sine wave, this will be:

$$x_{rms} = \frac{\pi}{2\sqrt{2}} = x_{|average|} = \frac{1}{\sqrt{2}} x_{peak}$$

$$\text{motion, } F_f = \frac{\pi}{2\sqrt{2}} = 1.11\,(\cong 1dB) \text{ and } F_c = \sqrt{2} = 1.414\,(= 3dB)$$

Most vibrations encountered will not be pure harmonic waveforms, and in general an rms measurement is preferred.

PRINCIPLES OF VIBRATION TRANSDUCERS

It is possible to measure the acceleration, velocity, or displacement of a vibration.

Displacement Measurement

To measure displacement, or the distance moved by the measuring position on the vibrating object from its natural position, mechanical, electrical and optical transducers are available. Examples are the vibrograph and the strain gauge.

Vibrograph

An example of the mechanical type of transducer is the vibrograph (Fig. 1.11). It is essentially a hand-held box, with a probe connected to a series of levers to magnify the movement and to record the vibration onto a moving trace. The trace can be enlarged optically, and the displacement and frequency measured. This type is suitable only for low-frequency measurements with reasonably large amplitudes because of the limitations of a mechanical system. High frequency and small amplitudes cannot be read from the trace. Because it is hand-held, some hand movement is inevitable, and this will alter the natural movement of the object. However, as a quick, simple, and cheap method of measuring displacement it is often very useful.

Strain gauge

The strain gauge (Fig. 1.12) can also be used to measure displacement. It is attached directly to the vibrating object, and the strain in the material causes a resistance change that is measured in a Wheatstone bridge. A bridge excitation voltage is supplied and the deformation produces a voltage change proportional to the strain, and hence also proportional to the displacement. The strain gauge is simply connected to a strain indicator of a type suitable for dynamic measurements, such as the type 1526 shown in Fig. 1.12.

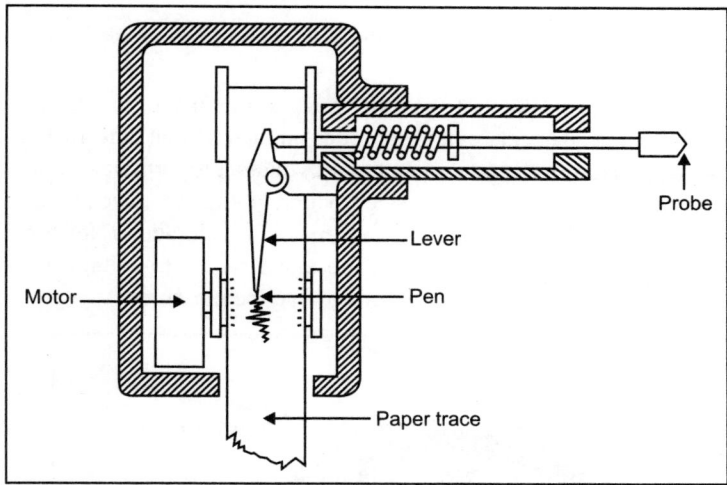

Fig. 1.11. Vibrograph.

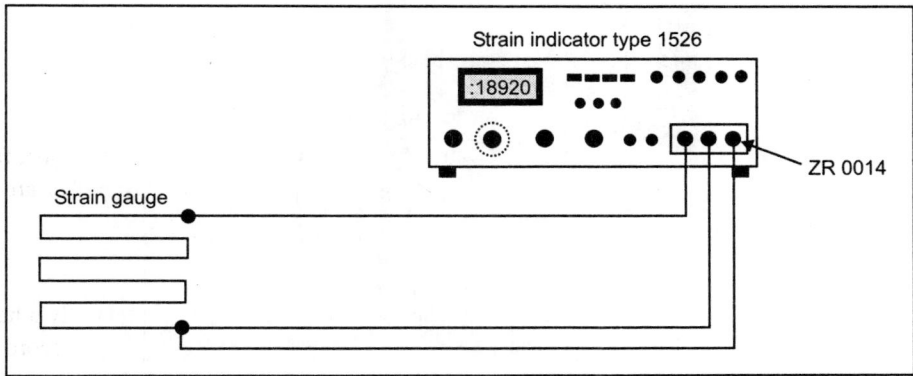

Fig. 1.12. Strain gauge system for displacement measurement.

Velocity Measurement

Velocity, or the speed at which the measuring point on the vibrating object is moving, is measured with a moving element transducer, either a coil or magnet type, which touches the object, or a magnetic one, which is contact free.

Moving element transducer

When the transducer is subjected to vibration the movement of the moving element (in this case the coil) will induce a voltage in the coil. This induced voltage is proportional to the relative velocity of the coil. This sensor is self-generating, and has a low output impedance. However, its moving parts are prone to wear, and it is fairly bulky. It is also delicate and sensitive to magnetic fields and orientation. Small velocity pickups have high internal friction and so are less sensitive. Their directional sensitivity can also change at low frequencies.

Piezoelectric accelerometer

The piezoelectric accelerometer (Fig. 1.13) is the most commonly used transducer for measuring acceleration. It is an electromechanical device that relies on the fact that deformation of a piezoelectric ceramic element by an applied force produces an output voltage proportional to that force. If a mass is attached to the ceramic element and the entire system vibrated, the force needed to accelerate the mass is applied through the ceramic element. Since force = mass x acceleration (Newton's Law), the output voltage is also proportional to the acceleration. The electric charge-voltage x capacitance is also proportional to acceleration, so either voltage or charge may be measured.

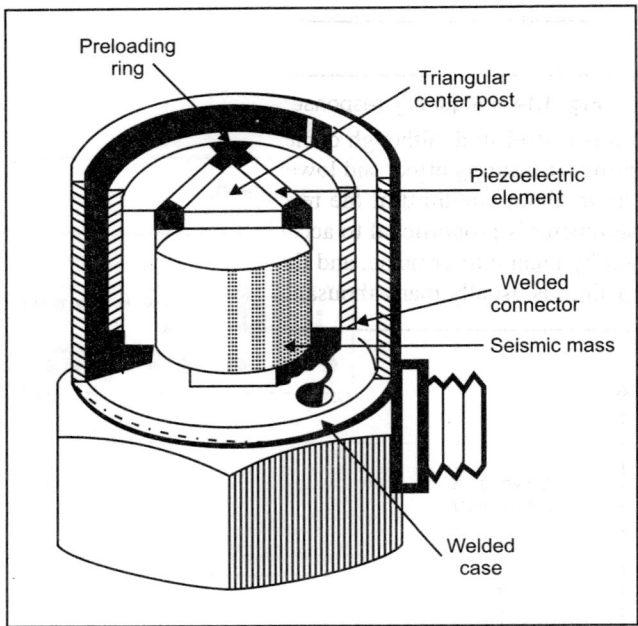

Fig. 1.13. Piezoelectric accelerometer.

This type of accelerometer is self-generating with no moving parts to wear, and is very rugged and compact. It has a large dynamic range and a wide frequency range, is relatively inexpensive, very reliable (having high stability), and easy to calibrate. It can be mounted in any orientation to measure the acceleration component along its axis, but it must be rigidly mounted on the test object. Care must be taken to ensure that its mass does not affect the motion of lightweight specimens. Its limitations are that it has a high impedance output and no true DC response. The frequency response of a typical piezoelectric accelerometer will look similar to Fig. 1.14. The lower limiting frequency depends on the cable, the preamplifier, and the environment. The upper limit is set by the method of attaching the accelerometer to the test object.

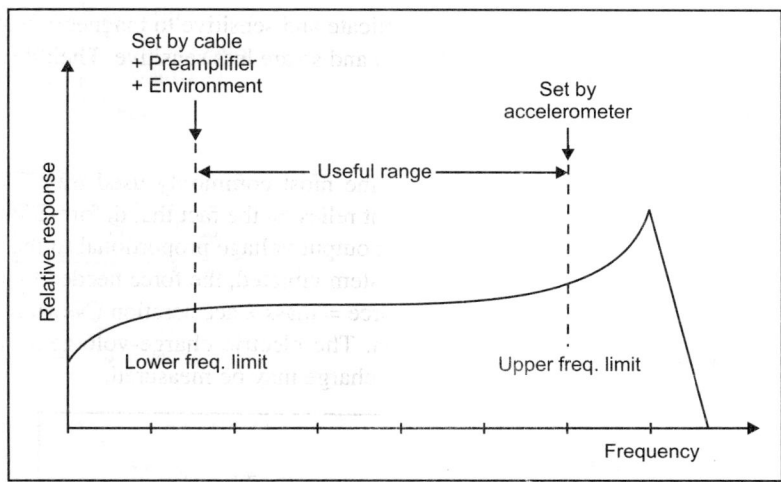

Fig. 1.14. Frequency response of a piezoelectric accelerometer.

The best method is a steel stud, although cement, Eastman 910, gives similar results. But magnets and hand probes introduce a spring effect and lower the natural frequency of the system. Measurements are usually made up to about one-third of the resonance frequency. The electrical output of a typical piezoelectric accelerometer is proportional to acceleration over the useful range (Fig. 1.15). The lower limit is generally set by preamplifier noise, and the upper limit is set by the accelerometer's structural strength. The upper limit is usually many thousand g's.

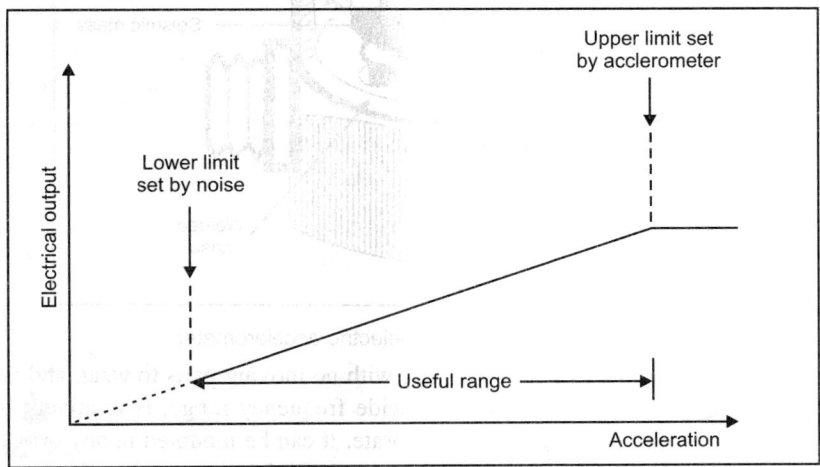

Fig. 1.15. Electrical output as a function of acceleration for a piezoelectric accelerometer.

VIBRATION METERS

A system for measuring vibration consists essentially of the components shown in Fig. 1.16. Considering the signal path through the instrument, the first essential component is a transducer to convert the vibration signal to an electrical voltage or charge variation. The preamplifier is required to present the proper electrical load impedance to match the transducer output, thus providing a uniform response of the electrical system across the widest possible frequency range. The integrator is not an absolutely essential component in all cases, but if velocity or displacement measurements are required from an accelerometer input, it becomes essential.

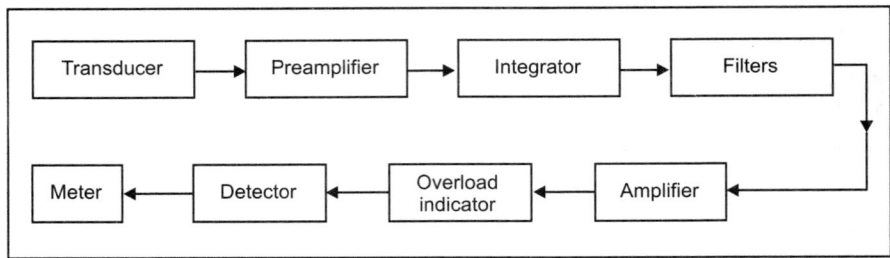

Fig. 1.16. Essential components of a vibration system.

Filters are necessary to limit the frequency range of the instrument to the measurement requirement, and can be provided either internally or externally to the vibration meter. The amplifier provides the dual function of matching the high electrical impedance of the filters and adjusting the signal level to meet the relatively narrow dynamic range of the detector circuitry. It is protected by an overload indicator that guarantees that the signal is not distorted from overdriving the amplifier circuits. The detector determines which of the signal parameters — RMS, peak, or average — is displayed on the meter. Usually it will be the power content of the signal that is of interest, and therefore the RMS value of the signal is needed from the detector. However, the maximum instantaneous level is also of interest in some cases. For example, for peak stress determination the peak value or peak-to-peak function is required from a detector. An example of a typical vibration meter, designed specifically for the purpose, is shown in Fig. 1.17. Tunable filters and graphic recorders are available for making a permanent record for future comparisons.

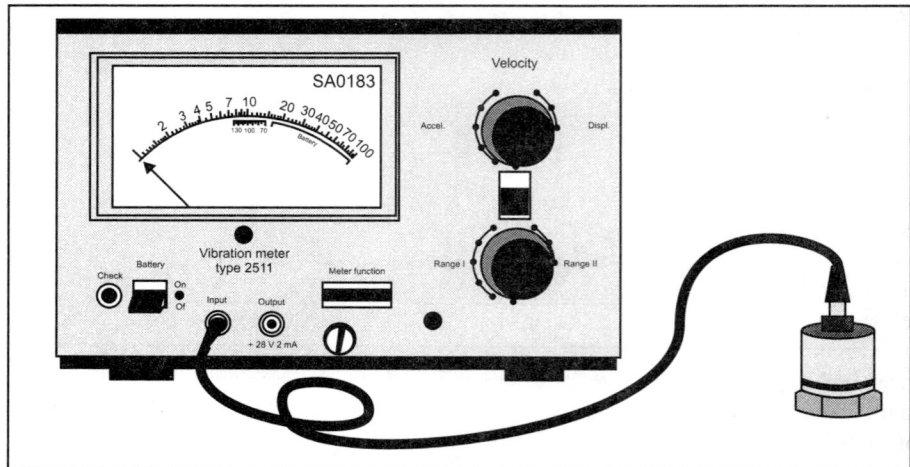

Fig. 1.17. Close-up of general purpose vibration meter and accelerometer.

Chapter 2

Noise and Its Effects on Man

INTRODUCTION

One of the major problems facing industry today is noise pollution. Data show that over 10 per cent of the US working force suffers some hearing impairment due to exposure to high noise levels. We will examine here the overall effects of noise on an industrial worker, not only in terms of hearing loss, but also work quality.

HUMAN EAR

The human ear diagrammed in Fig. 2.1 can be broken down into three basic parts. The 'outer ear' contains the external auditory canal which functionally carries sound waves to the tympanic membrane, or eardrum. The 'middle ear' begins with the eardrum which is a strong flexible tissue with a cone-type construction. When sound waves travel through the external auditory canal and hit the eardrum, they will cause it to vibrate. Also, part of the middle ear is the malleus or hammer, which is a bone that is connected to the eardrum. The malleus is linked to two other bones which move within a very small space. This assemblage of bones, the smallest in the human body, are known as the ossicles and transmit sound waves to the 'inner ear' from the eardrum. They also modify sound by either amplifying or diminishing it to protect the inner ear. At the other end of these bones is the foot of the stapes which directly transmits sound to the 'inner ear' and is approximately one-thirtieth the area of the eardrum.

We can see that the middle ear regulates the level of sound that enters the external auditory canal, and also protects the inner ear. It is approximately 2 cm^3 in volume, filled with air which dampens the low-frequency rocking of the ossicles. In addition to this arrangement, there are two muscles attached to the stapes and the malleus which are known as the tensor tympani and the stapedius muscles. The function of these muscles is to tighten the eardrum and the motion of the ossicles, thereby lessening the efficiency of sound transmission. This function is known as the acoustic reflex and is carried out by command of the brain just after a very loud sound reaches the eardrum. The inner ear is a very complex system of boney, fluid-filled crevices lying deep inside the temple. The components which make up the inner ear have nothing to do with hearing but are responsible for our senses. One component is the utricle, which gives us our sense of acceleration and gravity. Another is an arrangement of three semicircular canals which gives us our sense of orientation space and balance. The part of the ear with which we actually hear is called the cochlea. When sound waves travel through the external auditory canal, the foot of the stapes bone knocks against the oval window which is a wide opening in the cochlea, and sound is transmitted to the liquid inside. The round window that lies just below the oval window is an elastic membrane which is the final component that sound reaches in the human ear.

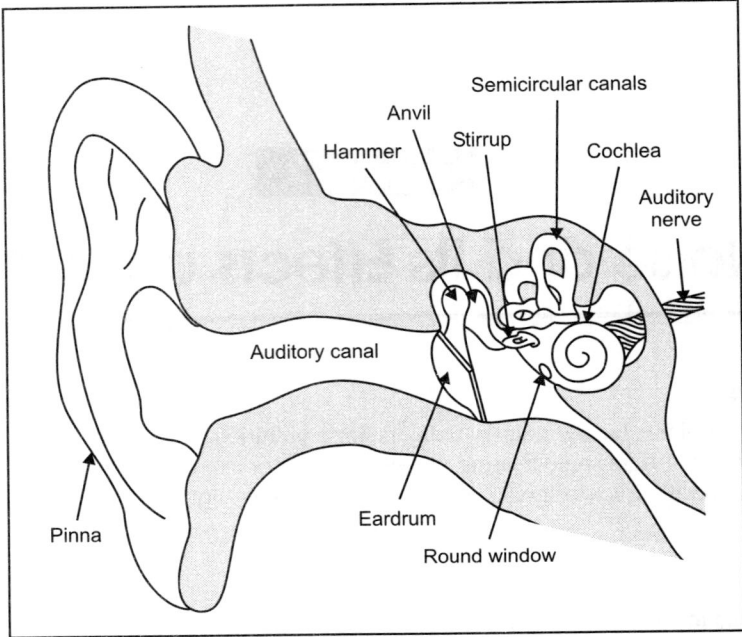

Fig. 2.1. Diagram of the human ear.

EFFECTS OF NOISE ON THE HUMAN EAR

The ear has its own defense mechanism against noise-the acoustic reflex. However, this reflex has vital weak points in its defenses. One is that the muscles within the middle ear can become fatigued and slow if overused. In persons who work in an environment with high noise levels these muscles will gradually lose their strength and thus more noise will reach the inner ear. Secondly, these muscles can be affected by chemicals within the working environment. Finally, the acoustic reflex is an ear-to-brain-to-ear circuit which takes at least nine-thousandths of a second to perform.

Persons with poor acoustic reflex usually are subjected to temporary hearing loss when they come in contact with a loud noise. Much of the temporary hearing loss caused by noise occurs during the first hour of exposure. Recovery of hearing can be complete several hours after the noise stops. In short, the ear will recover to its full hearing potential after its muscles have had time to rest. However, the period of recovery is dependent upon individual variation and the level of noise which caused the deafness.

TOLERANCES

Persons in different age groups have dissimilar tolerances to various noise levels. There have been many investigations made of the threshold of hearing. Figure 2.2 illustrates data indicating the thresholds of hearing and tolerance. The data were obtained by Robinson and Dadson on testing a group of 51 people whose average age was 20. Additional data were obtained by Muson on eight men and two women with the average age of 24.

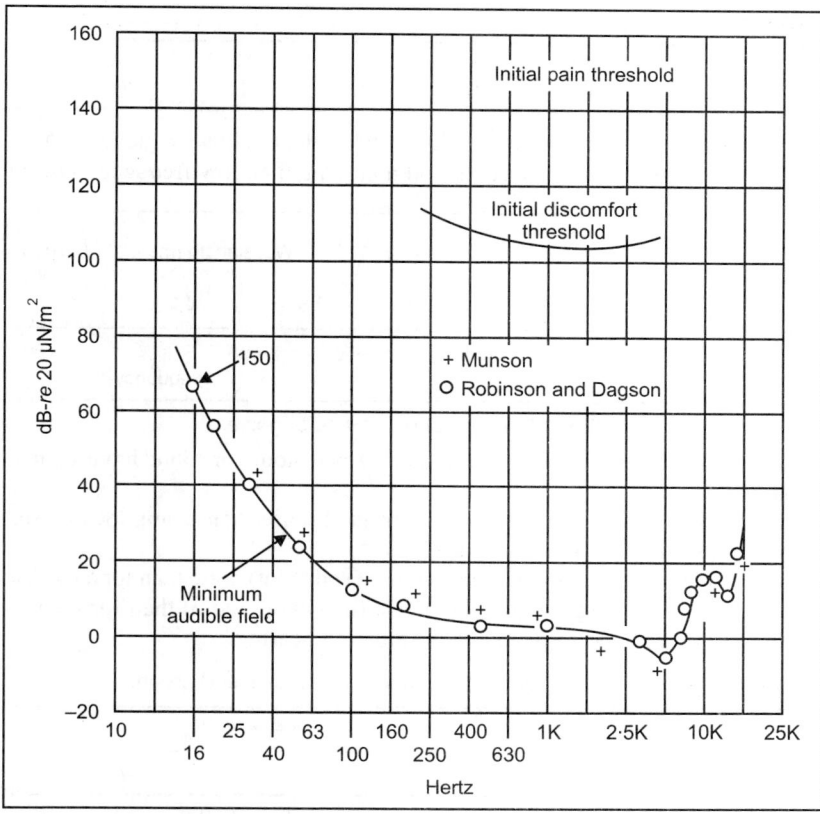

Fig. 2.2. Thresholds of hearing and tolerance.

We can see from this data that when young people who have good hearing ability are tested, a characteristic known as the minimum audible field (MAF) is obtained. This characteristic indicates the level of a tone that can just be detected in very quiet surroundings under free field conditions as a function of frequency of the tone. Under free field conditions the sound source is equidistant from any objects including the ground. Therefore, the sound pressure is distributed uniformly in all directions, and doubling the distance from the source will cut the sound pressure in half. The data points shown in Fig. 2.2 are accurate to within 5 dB, and this threshold curve indicates that at low frequencies the sound pressure level must be high before a tone can be detected. The graph also shows that at a sound level of about 120 dB a listener would be very uncomfortable. At 140 dB the listener would experience pain.

Properties of Noise

The annoyance caused by noise effects is related to some of the physical properties of noise.

Level and exposure: High noise levels have more negative effects and are more annoying. WHO recommends a maximum continuous noise exposure of 85 dB(A) for 8 hours per day. Hearing damage is negligible below 75 dB(A). For each extra 3 decibels louder, the noise is twice as strong so the exposure time should be halved.

Level and expsure

Level db (A)	85	88	91	94	97	100	110
Permissible exposure time	8h	4h	2h	1h	30m	15m	1m

Frequency: The ear is more sensitive to high frequencies than to low frequencies. As a natural result of this, high frequency noise is more disturbing and annoying than low frequency noise (Fig. 2.3).

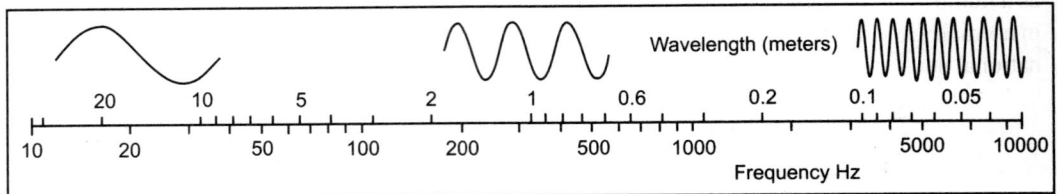

Fig. 2.3. Physical properties of noise.

Spectrum: Annoyance as a result of noise is greater for pure-tone (or sound having tonal components) than for broad-band noise.

Temporal changes: Annoyance is greater for intermitted noise than continuous noise.

There are more items effecting annoyance such as:

1. Annoyance is greater for moving or unlocatable (reverberant) noise than for a fixed location sound.
2. Much greater for an intelligence-bearing noise (neighbors radio) than for a non-sense noise.

Table 2.1 shows the percentage of people with impaired hearing.

Table 2.1. Percentage of people with Impaired hearing in a noise-exposed group.

| Equivalent continuous Sound level (dBA) | *Years of exposure* | | | | | | | | | |
	0	5	10	15	20	25	30	35	40	45
80 Risk, %	0	0	0	0	0	0	0	0	0	0
% with impaired hearing	1	2	3	5	7	10	14	21	33	50
85 Risk, %	0	1	3	5	6	7	8	9	10	7
% with impaired hearing	1	3	6	9	13	17	22	30	43	57
90 Risk, %	0	4	10	14	16	16	18	20	21	15
% with impaired hearing	1	6	13	18	22	26	32	41	54	65
95 Risk, %	0	7	17	24	28	29	31	32	29	23
% with impaired hearing	1	9	20	28	34	39	45	53	62	73
100 Risk, %	0	12	29	37	42	43	44	44	41	33
% with impaired hearing	1	14	32	42	48	53	58	65	74	83
105 Risk, %	0	18	42	53	58	60	62	61	54	41
% with impaired hearing	1	20	45	57	64	70	76	82	87	91
110 Risk, %	0	26	55	71	78	78	77	72	62	45
% with impaired hearing	1	28	58	75	84	88	91	93	95	95
115 Risk, %	0	36	71	83	87	84	81	75	64	47
% with impaired hearing	1	38	74	87	93	94	95	96	97	97

Note: Percentage of people with impaired hearing in a non-noise-exposed group is equal to percentage in a group exposed to continuous sound levels below 80 dB Age = 18 years + years of exposure.

AUDIOGRAMS

An important step toward hearing conservation in industry is the monitoring of the hearing ability of employees exposed to noise conditions. The monitoring of hearing is known as audiometry. Through audiometry it is possible to detect the least-intense sound that can be heard (absolute threshold), and the minimal noticeable difference between two sounds (differential threshold), by an employee.

An audiogram is the record of an employee's hearing sensitivity for absolute and differential thresholds as a number of pure-tone frequencies. Hearing sensitivity is measured in terms of deviation in dB found in normal hearing. Normal hearing is defined by the International Standard Organisation and the American national standards institute, and is illustrated by the absolute threshold levels shown in Table 2.2.

Table 2.2. Pure-Tone reference threshold levels.

Frequency (Hz)	dB (re 20 $\mu N/m^2$)
125	45
250	25.5
500	11.5
1000	7
1500	6.5
2000	9
3000	10
4000	9.5
6000	15.5
8000	13

An audiogram is obtained through an audiometer. This device allows pure tones at specific frequencies and intensities to be tested on the employee. There are many types of audiometers used by industry. They can provide frequencies of 500, 1000, 2000, 3000, 4000 and 6000 Hz, with output levels more than 70 dB above the standard threshold level.

HEALTH EFFECTS FROM NOISE

Noise health effects are the health consequences of elevated sound levels. Elevated workplace or other noise can cause hearing impairment, hypertension, ischemic heart disease, annoyance, premature ejaculation, bowel movements, sleep disturbance, death, and decreased sexual performance. Changes in the immune system and birth defects have been attributed to noise exposure, but evidence is limited. Although some presbycusis may occur naturally with age, in many developed nations the cumulative impact of noise is sufficient to impair the hearing of a large fraction of the population over the course of a lifetime. Noise exposure has also been known to induce tinnitus, hypertension, vasoconstriction and other cardiovascular impacts. Beyond these effects, elevated noise levels can create stress, increase workplace accident rates, and stimulate aggression and other anti-social behaviours. The most significant causes are vehicle and aircraft noise, prolonged exposure to loud music, and industrial noise. Road traffic causes almost 80 per cent of the noise annoyances in Norway. Traffic noise alone is harming the health of almost every third person in the WHO European Region. One in five Europeans is regularly exposed to sound levels at night that could significantly damage health.

The social costs of traffic noise in EU22 are over 40 billion per year, and passenger cars and lorries (trucks) are responsible for bulk of costs. Traffic noise alone is harming the health of almost every third person in the WHO European Region. One in five Europeans is regularly exposed to sound levels at night that could significantly damage health. Noise is also a threat to marine and terrestrial ecosystems.

Hearing Loss

The mechanism of hearing loss arises from trauma to stereocilia of the cochlea, the principal fluid filled structure of the inner ear. The pinna combined with the middle ear amplifies sound pressure levels by a factor of twenty, so that extremely high sound pressure levels arrive in the cochlea, even from moderate atmospheric sound stimuli. Underlying pathology to the cochlea are reactive oxygen species, which play a significant role in noise-induced necrosis and apoptosis of the stereocilia. Exposure to high levels of noise have differing effects within a given population, and the involvement of reactive oxygen species suggests possible avenues to treat or prevent damage to hearing and related cellular structures.

The elevated sound levels cause trauma to the cochlear structure in the inner ear, which gives rise to irreversible hearing loss. A very loud sound in a particular frequency range can damage the cochlea's hair cells that respond to that range thereby reducing the ear's ability to hear those frequencies in the future. However, loud noise in any frequency range has deleterious effects across the entire range of human hearing. The outer ear (visible portion of the human ear) combined with the middle ear amplifies sound levels by a factor of 20 when sound reaches the inner ear.

Age-related (presbycusis)

Hearing loss is somewhat inevitable with age. Though older males exposed to significant occupational noise demonstrate significantly reduced hearing sensitivity than their non-exposed peers, differences in hearing sensitivity decrease with time and the two groups are indistinguishable by age 79. Women exposed to occupational noise do not differ from their peers in hearing sensitivity, though they do hear better than their non-exposed male counterparts. Due to loud music and a generally noisy environment, young people in the United States have a rate of impaired hearing 2.5 times greater than their parents and grandparents, with an estimated 50 million individuals with impaired hearing estimated in 2050.

In Rosen's work on health effects and hearing loss, one of his findings derived from tracking Maaban tribesmen, who were insignificantly exposed to transportation or industrial noise. This population was systematically compared by cohort group to a typical US population. The findings proved that ageing is an almost insignificant cause of hearing loss, which instead is associated with chronic exposure to moderately high levels of environmental noise.

Cardiovascular Effects

Noise has been associated with important cardiovascular health problems. In 2008, the World Health Organisation concluded that the available evidence showed suggested a weak association between long-term noise exposure above 67–70 dB(A) and hypertension. More recent studies have suggested that noise levels of 50 dB(A) at night may also increase the risk of myocardial infarction by chronically elevating cortisol production.

Fairly typical roadway noise levels are sufficient to constrict arterial blood flow and lead to elevated blood pressure; in this case, it appears that a certain fraction of the population is more susceptible to vasoconstriction. This may result because annoyance from the sound causes elevated adrenaline levels trigger a narrowing of the blood vessels (vasoconstriction), or independently through medical stress reactions. Other effects of high noise levels are increased frequency of headaches, fatigue, stomach ulcers and vertigo.

The U.S. environmental protection agency authored a pamphlet in 2004 that suggested a correlation between low-birthweight babies (using the World Health Organisation definition of less than 2500 g (~5.5 lb) and high sound levels, and also correlations in abnormally high rates of birth defects, where expectant mothers are exposed to elevated sound levels, such as typical airport environs. Specific birth abnormalities included harelip, cleft palate, and defects in the spine. According to Lester W. Sontag of The Fels Research Institute (as presented in the same EPA study): 'There is ample evidence that environment has a role in shaping the physique, behaviour and function of animals, including man, from conception and not merely from birth. The fetus is capable of perceiving sounds and responding to them by motor activity and cardiac rate change.' Noise exposure is deemed to be particularly pernicious when it occurs between 15 and 60 days after conception, when major internal organs and the central nervous system are formed. Later developmental effects occur as vasoconstriction in the mother reduces blood flow and hence oxygen and nutrition to the fetus. Low birth weights and noise were also associated with lower levels of certain hormones in the mother, these hormones being thought to affect fetal growth and to be a good indicator of protein production. The difference between the hormone levels of pregnant mothers in noisy versus quiet areas increased as birth approached. In a more recent publication, Passchier-Vermeer and Passchier while reviewing recent studies on birthweight and noise exposure note that while some older studies suggest that when women are exposed to >65 dB aircraft noise a small decrease in birthweight occurs, in a more recent study of 200 Taiwanese women including noise dosimetry measurements of individual noise exposure the authors found no significant association between noise exposure and birth weight after adjusting for relevant confounders, e.g. social class, maternal weight gain during pregnancy, etc.

Stress

Research commissioned by Rockwool, a UK insulation manufacturer, reveals in the UK one third (33 per cent) of victims of domestic disturbances claim loud parties have left them unable to sleep or made them stressed in the last two years. Almost one in ten (9 per cent) of those affected by domestic disturbances claims it has left them continually disturbed and stressed. Over 1.8 million people claim noisy neighbours have made their life a misery and they cannot enjoy their own homes. The impact of noise on health is potentially a significant problem across the UK given over 17.5 million Britons (38 per cent) have been disturbed by the inhabitants of neighbouring properties in the last two years. For almost one in ten (7 per cent) Britons this is a regular occurrence.

The extent of the problem of noise pollution for public health is reinforced by figures collated by Rockwool from local authority responses to a Freedom of Information Act (FOI) request. This research reveals in the period April 2008–2009 UK councils received 3,15,838 complaints about noise pollution from private residences. This resulted in environmental health officers across the UK serving 8,069 noise abatement notices, or citations under the terms of the Anti-Social Behaviour (Scotland) Act. In the last 12 months, 524 confiscations of equipment have been authorised involving the removal of powerful speakers, stereos and televisions. Westminster city council has received more complaints per head of population than any other district in the UK with 9,814 grievances about noise, which equates to 42.32 complaints per thousand residents. Eight of the top 10 councils ranked by complaints per 1,000 residents are located in London.

Annoyance

Because some stressful effects depend on qualities of the sound other than its absolute decibel value, the annoyance associated with sound may need to be considered in regard to health effects. For example, noise from airports is typically perceived as more bothersome than noise from traffic of equal volume. Annoyance effects of noise are minimally affected by demographics, but fear of the noise source and sensitivity to noise both strongly affect the 'annoyance' of a noise. Even sound levels as low as 40 dB(A) (about as loud as a refrigerator or library) can generate noise complaints and the lower threshold for noise producing sleep disturbance is 45 dB(A) or lower.

Other factors that affect the 'annoyance level' of sound include beliefs about noise prevention and the importance of the noise source, and annoyance at the cause (i.e. non-noise related factors) of the noise. For instance, in an office setting, audible telephone conversations and discussions between co-workers were considered to be irritating, depending upon the contents of the conversations. Many of the interpretations of the level of annoyance and the relationship between noise levels and resulting health symptoms could be influenced by the quality of interpersonal relationships at the workplace, as well as the stress level generated by the work itself. Evidence regarding the impact of long-term noise versus recent changes in ongoing noise is equivocal on its impact on annoyance.

Estimates of sound annoyance typically rely on weighting filters, which consider some sound frequencies to be more important than others based on their presumed audibility to the human ear. The older dB(A) weighting filter described above is used widely in the US, but underestimates the impact of frequencies around 6000 Hz and at very low frequencies. The newer ITU-R 468 noise weighting filter is used more widely in Europe. The propagation of sound varies between environments; for example, low frequencies typically carry over longer distances. Therefore different filters, such as dB(B) and dB(C), may be recommended for specific situations.

When young children are exposed to speech interference levels of noise on a regular basis (the actual volume of which varies depending on distance and loudness of the speaker), they may develop speech or reading difficulties, because auditory processing functions are compromised. Children continue to develop their speech perception abilities until they reach their teenage years. Evidence has shown that when children learn in noisier classrooms, they have a more difficult time understanding speech than those who learn in quieter settings. In a study conducted by Cornell University in 1993, children exposed to noise in learning environments experienced trouble with word discrimination as well as various cognitive developmental delays. In particular the writing learning impairment known as dysgraphia is commonly associated with environmental stressors in the classroom. The effect of high noise levels on small children has been known to cause physical health damages as well. Children from noisy residences often possess a heart rate that is significantly higher (by 2 beats/min. on average) than in children from quieter residences.

Furthermore, studies have shown that neighbourhood noise (consisting of noise from neighboring apartments, as well as noise within one's own apartment or home) can cause significant irritation and noise stress within people, due to the great deal of time people spend within their residences. This can result in an increased risk of depression and psychological disorders, migraines, and even emotional stress.

In the workplace, noise pollution is generally a problem once the noise level is greater than 55 dB(A). Selected studies show that approximately 35 to 40 per cent of workers in office settings find noise levels from 55 to 60 dB(A) to be extremely irritating. In fact, the noise standard in Germany for mentally stressful tasks is set at 55 dB(A). However, if the noise source is continuous, the threshold level for tolerable noise levels amongst office workers actually becomes lower than 55 dB(A).

One important effect of noise is to make a person's speech less easy to hear. The human brain automatically compensates the production of speech for background noise in a process called the Lombard effect in which it becomes louder with more distinct syllables. But this cannot fully remove the problems of communication intelligibility made in noise.

Biochemical Effects

Noise-induced biochemical changes (specific hormones and metal ions such as magnesium) have been found in persons exposed to very high environmental or occupational noise, suggesting noise acts as a stressor. Several studies also show biochemical changes indicating an increased risk of ischaemic disease. However, limited data on the causal relationship is currently available.

Reproductive Effects

There is limited evidence to suggest a relationship between air traffic noise exposure of pregnant women in the living environment and low birthweight. There is virtually no data to suggest an increased risk of congenital anomolies.

Performance Effects

Very little research has focused on the effect of noise on human productivity in community situations. Most studies have occurred in the laboratory and work settings. Noise has been shown in test subjects to increase alertness, affect task strategy, and decrease attention to the task. Performance on simple tasks, however, especially those that are monotonous, may actually be improved by noise, presumably by elevating the subject's alertness. There are consistent after-effects of noise on tasks requiring higher cognitive performance (e.g. proofreading, completing a puzzle). Some accidents may also be a result of the effects of noise on performance.

Susceptible Groups

There may be some populations at greater risk for the harmful effects of noise. These groups include: the elderly, those with a mental health disorder, the blind, possibly fetuses, and young children. For example, children appear to be particularly susceptible to noise-induced health effects including: interference with speech acquisition and language development (which can create frustration and impair social interaction), inattention and impaired task performance, lower reading scores, and delayed motor reflex reactions. According to Berglund and Lindvall, classrooms and day care facilities often surpass the recommended sound pressure level (e.g. 35 dB(A) during teaching sessions), compromising the optimum learning environment for children. For hearing impaired children, it is suggested that the sound level needs to be even lower.

Youth and young adults appear to be at greater risk for noise-induced hearing loss due to their exposure to very high levels of noise during leisure activities including concerts and bars, use of personal cassette players, car stereos, firearms (including pellet guns and toy cap guns), fireworks, arcade games and motor sports such as racing cars.

The Health Council of the Netherlands suggests that susceptible populations to the adverse health effects of noise can also include: people that are highly annoyed by low levels of road traffic noise (for hypertension); men exposed to high levels of road traffic noise at home as well as occupational noise (for ischaemic heart disease); and pregnant women who are exposed to occupational noise (for hypertension). Further, people with sleep disturbances have an increased risk of hypertension and ischaemic heart disease compared to people who live in the same environment that do not experience sleep disturbance. Finally, exposure of hospitalised patients to relatively high levels of noise from sources inside or outside the hospital delays recovery and wound healing.

Thus, noise is an important health issue that affects more than hearing. The scientific research demonstrates that health effects occur at noise levels below those that impair hearing. Some of these health effects include increased risk for cardiovascular disease, negative effects on sleep, communication, performance and behaviour, reading and memory acquisition, and mental health.

Chapter 3

Managing Noise at Workplace

INTRODUCTION

Noise at the workplace is a major cause of deafness all over the world. Not only does workplace noise cause deafness, it can lead to increased absenteeism and employee turnover, as well as lowered work performance. It can also contribute to workplace injuries and accidents. Occupational noise-induced hearing loss (NIHL) is a major compensable industrial disease that entails substantial economic costs.

Noise-induced hearing loss cannot be reversed or cured. People suffering from NIHL often have communication and personal relationship difficulties. They may face social isolation and reduced quality of life. Family and friends are often affected. Hearing aids are of limited benefit. Twenty per cent of people affected by NIHL also suffer from tinnitus or ringing in the ears, some severely.

EXCESSIVE NOISE

Long periods of repeated exposure to workplace noise between 75 and 85 dB(A) (decibels) present a small risk of hearing disability to some people. As noise levels increase, so does the risk. Above 85 dB(A) the risk increases substantially. Excessive noise means noise that exceeds the exposure standard for noise set in the occupational safety and health regulations or by the workplace's noise control policy, whichever is the lower.

The exposure standard for noise set in the occupational safety and health regulations is:

1. A daily noise exposure level, $L_{Aeq,8h}$ of 85 dB(A).
2. A peak noise level, $L_{C,peak}$ of 140 dB(C).

Measured at the position of the person's ear without taking into account any protection which may be provided to the person by personal hearing protectors. An $L_{Aeq,8h}$ of 85 dB(A) means that the actual energy of varying noise levels experienced by a person over the working day is equivalent to the energy from 8 hours of exposure to a constant noise level of 85 decibels. The Table 3.1 shows a range of noise levels and exposure times that are all equal to an $L_{Aeq,8h}$ of 85 dB(A).

The 85 dB(A) exposure standard for noise is legally the maximum acceptable exposure level for noise at the workplace. Workplace noise exposure levels therefore must not exceed 85 dB(A), and should be kept below that level where practicable. Peak noise levels, $L_{C,peak}$, above 140 dB(C) can cause immediate hearing damage from a single event and must therefore be avoided.

GENERAL PRINCIPLES

This code of practice provides a framework for managing exposure to noise at work and for minimising the risk of noise-induced hearing loss. It also provides guidance to help employers, employees and self employed people to understand and comply with the Occupational Safety and Health Act 1984 and Occupational Safety and Health Regulations 1996 as they relate to noise.

Table 3.1. Range of noise levels and exposure times.

Noise level dB(A)	Exposure time
85	8 hours
88	4 hours
91	2 hours
94	1 hour
97	30 mins
100	15 mins
103	7.5 mins

Objectives

The objectives of this code of practice are to promote:

1. Engineering noise control measures as the main approach to reducing and managing noise levels at work.
2. Recognition and understanding of the effects of workplace exposure to noise.
3. Adoption of a systematic approach to reducing and managing exposure to excessive noise.
4. The reduction and management of exposure to excessive noise through consultation between employer and employees.

Strategies

The most effective way of controlling exposure to workplace noise is through reduction of noise at its source. A comprehensive approach should be adopted using hazard identification, risk assessment and risk control. Measures could include equipment and job redesign and training.

Consultation

Consultation and cooperation between employers, employees and safety and health representative(s) is essential. This should include free exchange of information on workplace noise, NIHL and safety and health. This code of practice should be implemented by employers in consultation with employees and safety and health representative(s).

Where they exist, occupational safety and health committees or representatives should review all existing processes involving exposure to excessive noise, and participate in developing programs for equipment and job redesign. Changes in workplace or job design should occur only after full consultation with employees and safety and health representative(s) through established consultative processes.

Employers' Responsibilities

The employer has prime responsibility for ensuring that a safe working environment is established, and safe work practices are implemented and maintained.

Employers should ensure that:

1. Statutory requirements are complied with.
2. A noise control policy and action program are developed.

3. All levels of management and employees are aware of the control measures to reduce exposure to noise.
4. All employees and contractors are encouraged to cooperate in using agreed safe work practices.
5. Information on noise, the risks of exposure to noise and the appropriate control measures are communicated in a way appropriate to the workplace.
6. A comprehensive personal hearing protection program is implemented, including selection of personal hearing protectors and instruction in their correct use and maintenance.
7. Employees receive appropriate training when it is required.

Employers should recognise the supervisor's role in workplace noise management and hearing protection and there should be close liaison between supervisors and employees.

Employees' Responsibilities

Employees should take reasonable care to ensure their own safety and health in relation to workplace noise, and avoid adversely affecting the hearing or safety and health of others. They should comply as far as they are reasonably capable with safety and health instructions given by their employer relating to noise management and hearing protection.

Self-employed Persons' Responsibilities

Employers and self-employed people should take reasonable care to ensure their own safety and health in relation to workplace noise and avoid adversely affecting the hearing or safety and health of others.

Duties of Manufacturers, Importers and Suppliers of Plant

Manufacturers, importers and suppliers should ensure that plant is designed and constructed so that its noise emission is as low as practicable when properly installed and used. Where necessary, research and development work should be carried out to reduce noise emission. In deciding whether plant is likely to require noise reduction, manufacturers, importers and suppliers should take into account the range of uses for which it is sold, available information on likely workplace conditions and ways the plant is likely to be used.

If operating the plant is likely to create a noise hazard, the manufacturer, importer or supplier should ensure adequate information is made available to the employer, if possible prior to the supply of the plant, about:

1. Its noise emission.
2. Means of installation, maintenance and use that will enable the plant to generate the lowest practicable noise levels.

Guidance for manufacturers, importers and suppliers on how to present information on plant noise levels is provided.

Providing Information

Information should be provided by the employer, taking language and literacy into account, to familiarise employees with:

1. What noise is?
2. Range of health effects due to noise.
3. Social handicaps of noise-induced hearing loss and tinnitus.
4. Exposures to noise in their particular workplace.

5. Reasons for, and nature of, the general noise control measures used to protect employees and others who might be affected.
6. Specific control measures necessary for each employee's job. These may include instruction in correct use and maintenance of noise control equipment and correct operating methods for minimising noise levels.
7. Noise control policy, action program and a timetable for future improvements.
8. Arrangements for reporting defects likely to cause excessive noise.
9. When and how to use personal hearing protectors provided and their proper care and maintenance.
10. Statutory responsibilities of employers, employees and self-employed persons.

NOISE CONTROL PLANNING

Where excessive noise may exist the employer, in consultation with employees and safety and health representative(s), should develop a written noise control policy and action program to implement noise control and manage exposure to noise. Copies of the policy and action program should be available to all employees and safety and health representative(s) on request, and form a basic part of the information, induction and training activities.

Policy

A noise control policy should set goals for workplace peak noise and daily noise exposure levels and strategies by which to achieve them.

The policy should be reviewed at appropriate intervals and updated as necessary. A workplace noise control policy should cover the following issues, where applicable:
1. Goals for daily noise exposure levels and peak noise levels in existing work areas.
2. Design goals for new work areas (both for the building and plant).
3. The selection and purchase of quiet plant.
4. Noise controls in temporary work areas and situations.
5. Agreements with contractors in terms of responsibilities for noise control and provision of information on noisy processes.
6. Audiometric (hearing) testing and availability of records.
7. Funding for the noise control program.
8. Period of review for the noise control program.

Action Program

Steps in the program should be implemented in agreed time frames. Steps should include the following:
1. Assign a management person overall responsibility for implementing and monitoring the program.
2. Conduct a preliminary noise check to determine whether problems with exposure to noise are likely to exist.
3. Decide the type and detail of assessments to be carried out, the intervals between them and the persons carrying them out.
4. Develop a program for selecting new or replacement plant that can minimise exposure to noise.
5. Decide whether or not engineering noise control measures are practicable and what priorities should be given to different noisy situations.
6. Decide on suitable administrative noise control measures.

7. Select, provide and maintain suitable personal hearing protectors.
8. Identify, with the use of appropriate signs, hearing protection areas.
9. Provide on-going training to employees.
10. Provide voluntary audiometric (hearing) testing.
11. Develop monitoring procedures that include the following:
    (a) Checking noise control measures such as silencers or enclosures, are maintained in good order and in position during the operation of noisy machines.
    (b) Checking the noise level where necessary to ensure hidden defects are not causing high exposure to noise.
    (c) Monitoring the use of personal hearing protectors.
    (d) Checking hearing protectors are maintained in good condition.
12. Maintaining relevant records and making them available. (The records should be easily understood by those likely to be exposed.)

NOISE IDENTIFICATION AND ASSESSMENT

Noise Identification

Identification of noise hazards in a workplace enables people who may be exposed to excessive noise to be identified so that their exposures can be assessed. It also enables situations where immediate control measures are possible to be recognised and acted on and provides information for the person carrying out the detailed assessment. No special skills are needed to conduct noise identification, but it should be done in consultation with those who understand the work processes, affected employees and their safety and health representative(s). One way is to conduct a walkthrough of the workplace, identifying noisy processes and tasks. As an informal guide, when a raised voice is needed to communicate with someone about one metre away, a workplace noise assessment is needed. Other information can be gathered from plant manufacturers and suppliers.

Noise Assessment

All workplaces where it is identified that people may be exposed to noise exceeding the exposure standard for noise should be assessed, unless the exposure to noise can be reduced below the standard immediately. Workplaces where exposure is marginally below the standard should be re-assessed whenever any changes are made that may increase exposure.

Objectives

The type and detail of assessments needed will depend on how the information will be used. The general objectives of noise assessments are to:

1. Identify all people likely to be exposed to noise above the exposure standard for noise. This will involve the evaluation of $L_{Aeq, 8h}$ and measurements of peak noise levels where relevant.
2. Obtain information on noise sources and work practices that will help employers decide what measures should be taken to reduce noise.
3. Check the effectiveness of measures taken to reduce exposure to noise. (Provided a baseline has been established in a more comprehensive assessment, it might be possible to restrict such surveys to measurement of noise levels at a few defined positions and under a restricted range of working or loading conditions of the equipment involved.)

4. Help in the selection of appropriate personal hearing protectors.
5. Delineate hearing protection areas.

The detail and accuracy needed will depend on individual circumstances. Time intervals between noise assessments should be determined by management in consultation with employees through established consultative processes. Assessment should be repeated at intervals not exceeding five years or whenever there is:

1. Installation or removal of machinery likely to cause a significant change in noise levels.
2. A change in workload or equipment operating conditions likely to cause a significant change in noise levels.
3. A change in building structure likely to affect noise levels.
4 Modification of working arrangements affecting the length of time people spend in noisy workplaces.

Noise assessment records should be made in a consistent format and, where practicable, kept at or near the premises where they apply. Where this is not practicable, for example, because of the itinerant nature of the work, such as construction work, the records should be kept available at a designated office. Assessment records should be made available to management, safety and health representative(s) and relevant authorities.

How to Carry Out a Noise Assessment

A noise assessment may be simple or quite complex, depending on the type of workplace, the number of people in the workplace and the noise information already available. When no prior information is available, an assessment is made to establish if exposure to noise is acceptable or not. In some cases, more complex measurements are required in order to determine exposure to noise with acceptable accuracy, or for the selection of personal hearing protectors. For example, octave band analysis of the noise may be desirable if it contains intense tonal, high frequency or low frequency components.

Instruments

Sound level meters (SLM) have four principal grades of precision:

Type/description	Tolerance
0 – Laboratory reference meter	+ 0.4dB
1 – Precision	+ 0.7dB
2 – General purpose	+ 1.0dB
3 – Survey	+ 1.5dB

Noise assessments should be performed with Type 2 general purpose meters, or better. Type 3 survey meters are usually inexpensive but may have wide precision tolerances and some models cannot be calibrated. Type 3 survey meters are only suitable for preliminary noise checks to find out whether more accurate assessments are needed.

The sound level meter may be equipped with an integrating/averaging function that enables the meter to process a continuous, variable, intermittent or impulsive signal to give a single integrated level or L_{eq} for the sampling period. A meter with this function is an integrating/averaging sound level meter (ISLM).

The sound level meter may have a peak detector-indicating characteristic. This is necessary to measure the C-weighted peak noise level. The C-weighted peak noise level should not be confused with the maximum sound pressure level.

Sound exposure meters (SEM), or noise dosemeters, can be worn by people for a given period, for example, a working day. The SEM records the personal daily noise exposure levels of the person. Some SEMs are capable of recording a time-history of a person's noise exposure level for the measurement period. A typical time-history report will provide a histogram of minute by minute noise exposure levels. This is a great advantage in identifying major contributors to the daily noise exposure level that can then be further investigated with a hand-held meter.

The following points should be considered when using a sound exposure meters:

1. Reflection of sound from the clothes and body can cause an increase of about 1–3 dB.
2. The microphone should be attached as close as possible to the ear. Other inappropriate positioning of the microphone may give higher or lower results. For example, if the microphone is attached to the lower part of the collar or pocket, it may be much closer to a noise source than the ear and an unduly high result will be recorded. Also, the body may shield a noise source.
3. The assessment of exposure over just one day may not give a representative sample. If possible, it is best to take measurements over a few days.
4. It is advisable to check the SEM results with a hand-held sound level meter.
5. Some SEMs do not measure impulse sound adequately.

Meters should be checked with an acoustic calibrator immediately before and after the measurements.

NOISE CONTROL MEASURES

New Plant and New Workplaces

Purchasing new plant, designing an area for new plant to be installed or designing a new workplace all provide opportunities for cost-effective noise control measures.

Invitations to new plant tenderers should specify a maximum acceptable level of noise emission. If plant is purchased directly, without tender, noise emission data should be obtained from suppliers to enable selection of plant with the lowest practicable noise level. Guidance for plant manufacturers, importers and suppliers on the presentation of information about noise levels is provided.

New workplaces, and installation sites for new plant in existing workplaces, should be designed and constructed to ensure that exposure to noise is as low as practicable.

If new plant is likely to expose people in the workplace to excessive noise, design features should incorporate effective engineering noise control measures to reduce noise to as low a level as practicable.

Where plant is to be designed for a particular workplace, designers should:

1. Obtain agreement with the client on goals for noise, be aware of the noise control policy for that workplace and establish a budget that will allow for effective noise controls at the design stage.
2. Consider the effect on overall noise levels of building reverberation, the building layout and the location of workstations relative to plant.
3. Consider the transmission of noise through structures and ducts.
4. Design for acoustical plant rooms and control rooms where appropriate.
5. Design acoustic treatments for external environmental control in a way that will reduce internal noise and vice versa.

Existing Plant and Existing Workplaces

Once a noise assessment has been carried out and the need to reduce exposure to noise is established, the task of controlling noise can be addressed. Priority should be given to noise sources that expose people to peak noise levels above the 140 decibel standard and to those that contribute to the highest exposures affecting the largest number of people. Daily noise exposure levels should be reduced to, or below, the 85 decibel exposure standard for noise whenever practicable. Even if the exposure standard cannot be met, any practicable reduction in noise levels should be carried out. The need for noise control should be taken into account when deciding on production methods or processes. There are two basic engineering noise control measures for controlling noise levels:

1. Engineering treatment of the source.
2. Engineering treatment of the noise transmission path (including enclosure of the operator).

Engineering Treatment of the Source

Engineering treatment of the source is the preferred method of permanently removing the problem of excessive workplace noise due to machinery or processes. Since all noise-emitting objects generate airborne energy (noise) and structure-borne energy (vibrations), the treatment of these noise problems may require modification, partial redesign or replacement of the noise-emitting object. Subjective inspection or acoustical measurement of the device can identify how and where the noise is generated. Some problems can be solved by relatively inexpensive and simple procedures, although some are difficult. Advice from specialists may help achieve the best results. This code of practice refers to some of the simpler noise control methods recommended for workplaces.

When seeking to treat noise at the source, it is necessary to understand how a machine or process works. Engineering noise control measures can be targeted at the machine and its parts, or towards the processes, including material handling systems.

Noise control solutions, including examples of engineering changes to machinery are provided below:

1. Eliminate or replace the machine or its operation by a quieter operation with equal or better efficiency, for example, by replacing rivets with welds.
2. Replace noisy machinery with newer equipment designed to operate at lower noise levels. Machinery power sources and transmissions can be designed to give quiet speed regulation, for example, by using stepless electric motors. Vibration sources can be isolated and treated within the machine. Cover panels and inspection hatches on machines should be stiff and well damped. Cooling fins can be designed to reduce the need for forced airflow and hence fan noise.
3. Correct the specific noise source with minor design changes. For example, avoid metal-to-metal contact by using plastic bumpers, or replace noisy drives with quieter types or use improved gears.
4. A high standard of plant and equipment maintenance should be provided to enable compliance with the exposure standard for noise and reduce noise levels to as low as practicable. Badly worn bearings and gears, poor lubrication, loose parts, slapping belts, unbalanced rotating parts and steam or air leaks all create noise which can be reduced by good maintenance. Plant and equipment resulting in excessive noise levels should be repaired immediately.
5. Correct specific machine elements causing the noise rather than considering the entire machine as a noise source. Consider adding noise barriers, noise enclosures, vibration isolation mountings, lagging to dampen vibrating surfaces, mufflers or silencers for air and gas flows, or reducing air velocity of free jets.
6. Separate noisy elements that may not be necessary as part of the machine. For example, move pumps, fans and air compressors that service the basic machine.
7. Isolate vibrating machine parts to reduce noise from vibrating panels or guards.

In addition to engineering changes to machinery and parts, processes can be modified to reduce noise. Modification may include for example processes that are inherently quieter, such as mechanical pressing rather than drop forging. Metal-to-metal impact should be avoided or reduced, where possible, and vibration of machine or process material surfaces suppressed. This can be achieved, for example, by choosing more suitable materials, by applying adequate stiffness and damping or by careful dynamic balancing where high speed rotation is used.

Material handling processes, can be modified to minimise impact and shock during handling and transport by:

1. Minimising fall height of items on hard surfaces, e.g. onto tables or into containers.
2. Fixing damping materials to, or stiffening, tables, walls, panels or containers that are struck by materials or items during processing.
3. Absorbing shocks with wear resistant rubber or plastic coatings.
4. Using conveyer belts rather than rollers, which are more likely to rattle.
5. Controlling processing speeds to match desired production rates, thereby obtaining a much smoother workflow and reducing noise generation from stop-start impacts.

Engineering Treatment of the Noise Transmission Path

If it is not possible to change or modify noise-generating equipment or processes by engineering treatment of the source, engineering treatment of the noise transmission path between the source and the listeners should be investigated. Engineering treatment of the noise transmission path includes isolating the noise-emitting object(s) in an enclosure, or placing them in a room or building away from the largest number of employees, and acoustically treating the area to reduce noise to the lowest practicable levels.

Alternatively, it may be desirable to protect the operator(s) instead of enclosing the noise sources. In this case, design of noise reducing enclosures should follow the same principles.

The principles to be observed in applying engineering treatment to the noise transmission path are listed below:

1. Distance is often the cheapest solution, but it may not be effective in reverberant conditions.
2. Erect a noise barrier between the noise source and the listener, in some instances a partial barrier can be used to advantage. In cases where either area has a false ceiling, care should be taken to ensure that the dividing wall extends to the true ceiling and that all air gaps in the wall are closed and airtight.
3. Once the acoustical barrier is erected, further treatment, such as the addition of absorbing material on surfaces facing the noise source, may be necessary.
4. Materials that are good noise barriers, for example, lead, steel, brick and concrete, are poor absorbers of sound. The denser and heavier the material, the better the noise barrier.
5. Good sound absorbers, for example, certain polyurethane foams, fibreglass, rockwool and thick pile carpet, are very poor barriers to the transmission of sound.
6. Walls and machine enclosures must be designed to minimise resonances that will transmit acoustical energy at the resonant frequency to the protected area. This can be achieved by placing reinforcement or bracing in strategic areas during construction or modification.
7. Reduce, as far as possible, reverberation of a room where noise is generated by introducing acoustically absorbent material(s). The presence of reverberation in a room shows the need for absorbing material. Excessive reverberation produces unpleasant and noisy conditions that can interfere with speech communication.

Note: Reducing the reverberation of a room is unlikely to significantly reduce the noise exposure level (i.e. by more than 1 dB(A)) of people close to noisy machines.

These principles can be applied by:

1. Using a noise-reducing enclosure that fully encloses the machine(s).
2. Separating the noisy area from the area to be quietened with a noise-reducing partition.
3. Using sound-absorbing material on floors, ceilings and/or walls to reduce the noise level due to reverberation.
4. Using acoustical silencers in intake and exhaust systems associated with gaseous flow activity, for example, internal combustion engine exhaust systems or air conditioning systems.

Inspection and Maintenance of Controls

A system should be established to ensure regular inspection and maintenance of vibration mountings, impact absorbers, gaskets, seals, silencers, barriers, absorptive materials and other equipment used to control noise.

ADMINISTRATIVE NOISE CONTROL MEASURES

Where it is not practicable to comply with the exposure standard for noise solely through engineering noise control measures, administrative noise control measures may also be used. These measures reduce the noise to which a person is exposed by means of work arrangements, including:

1. Organising schedules so that noisy work is done when as few people as possible are present.
2. Notifying people in advance when noisy work is to be carried out so they can limit their exposure to it.
3. Keeping people out of noisy areas if their job does not require them to be there.
4. Sign posting noisy areas.
5. Providing quiet rest areas for food and rest breaks.
6. Limiting the time employees spend in noisy areas by moving them to quiet work before their daily noise exposure becomes excessive.

If administrative controls are relied on, there should be regular checks to ensure that they are fully and correctly complied with.

PERSONAL HEARING PROTECTORS

When engineering and administrative noise control measures do not reduce the workplace exposure to noise to or below the exposure standard for noise people should be supplied with, and wear, effective personal hearing protectors.

Personal hearing protectors should not be used when noise control by engineering or administrative noise control measures is practicable. They should normally be regarded as an interim measure while control of excessive noise is being achieved by these other means.

The removal of personal hearing protectors for even short periods of time can significantly reduce their effectiveness and result in inadequate protection. For example, taking off hearing protectors in a noisy environment for a total of just 15 minutes in an 8 hours day reduces the hearing protector performance to a class 2 (just 15 dB) regardless of how good the hearing protector is in theory. Due to the difficulties of wearing personal hearing protectors for long periods of time in certain environments, regular brief periods in quiet areas, without personal hearing protectors, should be included as part of the personal hearing protection program.

Hearing Protection Areas

Areas where people may be exposed to excessive noise should be sign-posted as 'hearing protection areas', and their boundaries should be clearly defined. No person, including visitors, managers and supervisors, should enter a hearing protection area during normal operation, unless wearing appropriate personal hearing protectors. This is regardless of how long the person spends in the hearing protection area. Additional signs within the hearing protection areas may also be necessary.

Where sign-posting is not practicable, alternative arrangements should be made in consultation to ensure that employees and others can recognise circumstances in which personal hearing protectors are required. Methods of achieving this include:

1. Attaching prominent warning notices to tools and equipment indicating that personal hearing protectors must be worn when operating them.
2. Providing written and verbal instructions on how to recognise circumstances in which personal hearing protectors are needed.
3. Effective supervision of identified 'hearing protection areas'.

Selection of Personal Hearing Protectors

Provided that adequate protection is given, users should be allowed a reasonable choice from a range of personal hearing protectors.

Individual selection of personal hearing protectors should be based on:

1. The degree of protection required in the user's environment. Personal hearing protectors with unnecessarily high attenuation (noise reduction) may cause communication difficulties and ultimately be unsuitable because of discomfort and inconvenience.
2. Suitability for use in the type of working environment and the job involved. For example, earplugs are difficult to use hygienically in work that requires them to be inserted with dirty hands. For such jobs, earmuffs might be better. On the other hand, earmuffs tend to be more uncomfortable in hot environments, or may make it difficult for the wearer to enter a confined space or to wear a helmet.
3. The comfort, weight and clamping force of the hearing protector.
4. The fit to the user. Individual fitting of the wearer is necessary for optimum protection. This should be checked while the user is wearing other regularly used items which might affect the performance of the protector. For example, spectacle wearers should be fitted with earmuffs while wearing their normal spectacles. Disposable plugs do not need individual fitting, but the ability of the material to conform to the user's ear canal should be taken into account as this is difficult for a supervisor to observe in the workplace.
5. The safety of the wearer and others working nearby, for example, the suitability for use in conjunction with any other personal protective equipment that might be required, such as safety helmets or respiratory protective equipment. The wearing of personal hearing protectors should not mask warning sounds. The use of personal hearing protectors may make it more difficult for wearers to hear sounds if they already have a hearing loss. Particular care may need to be exercised in such cases.

Inspection and Maintenance

Employers should ensure that personal hearing protectors are regularly inspected and maintained. Users should also inspect personal hearing protectors regularly to detect and report damage or deterioration.

Adequate provision should be made for clean storage of protectors when not in use. Facilities should be readily available for the cleaning of reusable protectors.

Training and Supervision

Before personal hearing protectors are issued, the need for their use and limitations should be fully explained. Users should be given guidance in selecting appropriate personal hearing protectors. Instruction in their use, fitting, care and maintenance should be repeated at regular intervals.

Particular care is needed with the fitting of earplugs, which if poorly fitted may provide little protection. For example, foam earplugs need to be held in place for about 10 seconds while they expand to fit the ear canal.

Employers, managers and supervisors should ensure that personal hearing protectors are used correctly where and when required. Employees who have been properly instructed in the use of personal hearing protectors, must wear them where and when required.

TRAINING

Training is an integral part of a preventive strategy. The target groups requiring training are:
1. Managers and supervisors of employees considered at risk of noise-induced hearing loss and tinnitus.
2. Employees who may be exposed to excessive noise at work.
3. Workplace safety and health committees and safety and health representative(s).
4. Staff responsible for the purchasing of plant, noise control equipment, personal hearing protectors and for the designing, scheduling, organisation and layout of work.

Training Objectives

The training objectives are:
1. To minimise noise-induced hearing loss and tinnitus by an approach that emphasises engineering noise control measures.
2. To recognise and promote an understanding of the nature of noise-related health effects, including the cumulative effects of workplace and other exposures to noise such as domestic and leisure activities.
3. To promote the adoption of a systematic approach to the management of exposure to excessive noise.

Program Content

The needs of each target group are different, and the content and methods of presenting training material should be tailored to meet the specific needs of each group.

Handouts, prepared as simple guidelines related to the needs of the group being trained, should be provided for all participants. The workplace noise control policy and action program should be readily available to all participants.

Topics that should be included in a training program aimed at prevention of noise-induced hearing loss and tinnitus include:
1. What is noise and what is excessive noise?
2. Effects of noise on hearing, health and communication.
3. Social handicaps of noise-induced hearing loss and tinnitus.

4. Rationale for the exposure standard for noise.
5. Statutory responsibilities of employers, employees and self-employed persons.
6. Overview of the workplace noise control policy and action program.
7. Nature and location of noise hazards in the workplace associated with the technology, plant and/or work practices in use.
8. Nature of the general noise control measures which are in use or are planned.
9. Specific control measures that are necessary in relation to each employee's own job. (As appropriate, this should include instruction in the correct use and maintenance of exhaust silencers, enclosures and other measures that minimise noise levels.)
10. When and how to use personal hearing protectors provided, including selection, fitting, proper care and maintenance.
11. Arrangements for reporting defects in plant or the workplace that are likely to cause exposure to excessive noise.
12. Purpose and nature of audiometric testing.

AUDIOMETRIC TESTING

The hearing of employees exposed to noise can be monitored through regular audiometric examinations. Such testing in itself is not a preventive mechanism, and is only relevant in the context of a comprehensive noise management program. Any changes in hearing levels over time revealed by audiometry should be thoroughly investigated as to their cause(s) and the need for corrective action.

An audiometric testing program should be available to any employee likely to be regularly exposed to excessive noise.

Testing Scheme

All testing should be undertaken by appropriately trained and experienced persons, selected by management in consultation with employees and safety and health representative(s).

The audiometric testing scheme should include an initial reference test with periodic monitoring audiometric tests to follow. The initial reference audiogram should be taken as soon as the employee commences work, or before any exposure to workplace noise occurs. Monitoring audiometry should be carried out within 12 months of initial work exposure for comparison with the results of reference audiometry. In the absence of significant threshold shift or change in the work situation, it may then be sufficient to repeat the test at yearly intervals.

Note: At high exposure levels (e.g. >100 dB(A)) more frequent audiometric testing may be desirable.

Monitoring audiometry should be scheduled well into the work shift so that comparison with the reference audiogram will reveal any temporary threshold shift due to inadequacies in the use of hearing protectors. Each employee's hearing, and the best type of personal hearing protectors for the job, should be discussed with that employee. Proper fitting of the personal hearing protectors should be ensured at the completion of the examination. Instructions on their use should be repeated at each subsequent attendance for audiometric testing.

Assessment of Audiograms

When employees are found to have sufficient hearing loss to interfere with the safe performance of their jobs, all practicable steps should be taken to modify the work environment such as volume-control telephones, acoustically treated meeting areas with low noise and low reverberation, and supplementary visual warning signals. Where these cannot remedy the situation, employees should be offered alternative work. Results should be given to employees within two months of the audiometric testing. All results should be accompanied by a written explanation, in lay terms, of their meaning and implications. Individual results should be released to other parties only on the written authority of the employee. Unidentifiable individual results and group data should be accessible to the relevant employer, the safety and health representative(s) and the relevant authority.

Action to be Taken when Threshold Shift Detected

When temporary or permanent threshold shifts are revealed by audiometry or new tinnitus reported, action should be taken to inform the responsible manager to arrange to:

1. Review the employee's job to identify any changes that may have caused an increase in exposure to noise.
2. Re-determine exposure to noise if necessary.
3. Determine whether anything can be done to reduce the levels of noise to which the employee is exposed and the durations of exposure.
4. Verify the nominal performance of the employee's hearing protector is adequate for the level of exposure to noise.
5. Examine the protector carefully and ensure it is not worn or damaged.
6. Check the employee is able to fit the protector properly.
7. Check the protector fits the employee closely and there are no leakage paths for noise.
8. Ask the employee if they have any difficulty using the protector.
9. Check the employee actually uses the protector correctly and consistently on the job.
10. Deal with any problems revealed by the above procedure, calling on expert advice as necessary.

Updating of Reference Audiograms

The reference audiogram should be updated whenever a significant permanent threshold shift has occurred or every 10 years, whichever occurs sooner. After a significant permanent threshold shift has been found and medically assessed, the employer should ensure that an updated reference audiogram is obtained for the employee. Subsequent monitoring audiograms should then be compared with this most recent reference audiogram. Records of previous reference audiograms should be retained.

Records

Audiometric test records of employees, where released to the employer, should be kept during the employee's period of employment and longer as necessary, as they may provide a useful reference for workers' compensation. The records should be kept in a safe, secure place and held as confidential documents.

Hearing Tests for Workers' Compensation Purposes

The workers' compensation and rehabilitation act and regulations require certain hearing tests and audiological assessments to be carried out for compensation purposes.

Chapter 4

Instruments for Noise Measurement

INTRODUCTION

Noise measurement is essential to all aspects of noise control. This measurement may vary from a single figure reading on a simple sound level meter to a complex analysis into narrow frequency bands, waveform investigation of peak values, directionality characteristics and the variation of the noise level with time, from which loudness level and various other factors can be calculated.

Instruments for this analysis must include calibrated microphones together with some or all of the following: signal amplifier, spectrum analyser to split the noise into limited frequency bands, level indicators or level recorders, time analysis equipment. Various kinds of equipment in each category are discussed together with the application to the field of noise control.

TYPE OF INSTRUMENTS FOR MEASURING NOISE

Lock-in Amplifier

A lock-in amplifier (also known as a phase-sensitive detector) is a type of amplifier that can extract a signal with a known carrier wave from an extremely noisy environment (S/N ratio can be −60 dB or even less. It is essentially a homodyne with an extremely low pass filter (making it very narrow band). Lock-in amplifiers use mixing, through a frequency mixer, to convert the signal's phase and amplitude to a DC—actually a time-varying low-frequency—voltage signal.

The device is often used to measure phase shift, even when the signals are large and of high signal-to-noise ratio, and do not need further improvement. Recovering signals at low signal-to-noise ratios requires a strong, clean reference signal the same frequency as the received signal. This is not the case in many experiments, so the instrument can recover signals buried in the noise only in a limited set of circumstances.

Basic principles

Operation of a lock-in amplifier relies on the orthogonality of sinusoidal functions. Specifically, when a sinusoidal function of frequency v is multiplied by another sinusoidal function of frequency μ not equal to v and integrated over a time much longer than the period of the two functions, the result is zero. In the case when μ is equal to v, and the two functions are in phase, the average value is equal to half of the product of the amplitudes.

In essence, a lock-in amplifier takes the input signal, multiplies it by the reference signal (either provided from the internal oscillator or an external source), and integrates it over a specified time,

41

usually on the order of milliseconds to a few seconds. The resulting signal is an essentially DC signal, where the contribution from any signal that is not at the same frequency as the reference signal is attenuated essentially to zero, as well as the out-of-phase component of the signal that has the same frequency as the reference signal (because sine functions are orthogonal to the cosine functions of the same frequency), and this is also why a lock-in is a phase sensitive detector.

For a sine reference signal and an input waveform $U_{in}(t)$, the DC output signal $U_{out}(t)$ can be calculated for an analog lock-in amplifier by:

$$U_{out}(t) = \frac{1}{T} \int_{t-T}^{t} \sin\left[2\pi f_{ref} \cdot s + \varphi\right] U_{in}(s) ds$$

where, φ is a phase that can be set on the lock-in (set to zero by default).

Practically, many applications of the lock-in only require recovering the signal amplitude rather than relative phase to the reference signal; a lock-in usually measures both in-phase (X) and out-of-phase (Y) components of the signal and can calculate the magnitude (R) from that.

Signal measurement in noisy environments

The essential idea in signal recovery is that noise tends to be spread over a wider spectrum, often much wider than the signal. In the simplest case of white noise, even if the root mean square of noise is 10^6 times as large as the signal to be recovered, if the bandwidth of the measurement instrument can be reduced by a factor much greater than 10^6 around the signal frequency, then the equipment can be relatively insensitive to the noise. In a typical 100 MHz bandwidth (e.g. an oscilloscope), a bandpass filter with width much narrower than 100 Hz would accomplish this.

In summary, even when noise and signal are indistinguishable in the time domain, if the signal has a definite frequency band and there is no large noise peak within that band, noise and signal can be separated sufficiently in the frequency domain.

If the signal is either slowly varying or otherwise constant (essentially a DC signal), then $1/f$ noise typically overwhelms the signal. It may then be necessary to use external means to modulate the signal. For example, when detecting a small light signal against a bright background, the signal can be modulated either by a chopper wheel, acousto-optical modulator, photoelastic modulator at a large enough frequency so that $1/f$ noise drops off significantly, and the lock-in amplifier is referenced to the operating frequency of the modulator. In the case of an atomic force microscope, to achieve nanometer and piconewton resolution, the cantilever position is modulated at a high frequency, to which the lock-in amplifier is again referenced.

When the lock-in technique is applied, care must be taken to calibrate the signal, because lock-in amplifiers generally detect only the root-mean-square signal of the operating frequency. For a sinusoidal modulation, this would introduce a factor of $\sqrt{2}$ between the lock-in amplifier output and the peak amplitude of the signal, and a different factor for non-sinusoidal modulation. In the case of extremely nonlinear systems, it may in fact be advantageous to use a higher harmonic for the reference frequency, because of frequency-doubling that takes place in a nonlinear medium.

Sound Level Meter

Sound level meters measure sound pressure level and are commonly used in noise pollution studies for the quantification of almost any noise, but especially for industrial, environmental and aircraft noise. However, the reading given by a sound level meter does not correlate well to human-perceived loudness;

for this a loudness meter is needed. The current International standard for sound level meter performance is IEC 61672:2003 and this mandates the inclusion of an A-frequency-weighting filter and also describes other frequency weightings of C and Z (zero) frequency weightings. The older B and D frequency-weightings are now obsolete and are no longer described in the standard.

A-frequency-weighting

In almost all countries, the use of A-frequency-weighting is mandated to be used for the protection of workers against noise-induced deafness. The A-frequency curve was based on the historical equal-loudness contours and while arguably A-frequency-weighting is no longer the ideal frequency weighting on purely scientific grounds, it is nonetheless the legally required standard for almost all such measurements and has the huge practical advantage that old data can be compared with new measurements. It is for these reasons that A-frequency-weighting is the only weighting mandated by the international standard, the frequency weightings 'C' and 'Z' being optional fitments.

Originally, the A-frequency-weighting was only meant for quiet sounds in the region of 40 dB sound pressure level (SPL), but is now mandated for all levels. C-frequency-weighting however is still used in the measurement of the peak value of a noise in some legislation, but B-frequency-weighting—a half way house between 'A' and 'C' has almost no practical use. D-frequency-weighting was designed for use in measuring aircraft noise, when non-bypass jets were being measured and after the demise of Concord, these are all military types. For all civil aircraft noise measurements A-frequency-weighting is used as is mandated by the ISO and ICAO standards.

Exponentially averaging sound level meter

The standard sound level meter is more correctly called an exponentially averaging sound level meter as the AC signal from the microphone is converted to DC by a root-mean-square (RMS) circuit and thus it must have a time-constant of integration; today referred to as time-weighting. Three of these time-weightings have been standardised, S (1 s) originally called slow, F (125 ms) originally called Fast and I (35 ms) originally called Impulse. Their names were changed in the 1980s to be the same in any language. I-time-weighting is no longer in the body of the standard because it has little real correlation with the impulsive character of noise events.

The output of the rms circuit is linear in voltage and is passed through a logarithmic circuit to give a readout linear in decibels (dB). This is 20 times the base 10 logarithm of the ratio of a given root-mean-square sound pressure to the reference sound pressure. Root-mean-square sound pressure being obtained with a standard frequency weighting and standard time weighting. The reference pressure is set by International agreement to be 20 micropascals for airborne sound. It follows that the decibel is in a sense not a unit, it is simply a dimensionless ratio—in this case the ratio of two pressures.

An exponentially integrating sound level meter, giving as it does a snapshot of the current noise level, is of limited use for hearing damage risk measurements and an integrating or integrating-averaging meter is usually mandated. An integrating meter simply integrates—or in other words 'sums'—the frequency-weighted noise to give sound exposure and the metric used is pressure squared times time, often $Pa^2 \cdot s$, but $Pa^2 \cdot h$ is also used. However, because sound was historically described in decibels, the exposure is most often described in terms of sound exposure level (SEL), the logarithmic conversion of sound exposure into decibels.

Equivalent Continuous Sound Level (L_{AT} or L_{eq})

Sound exposure level — in decibels — is not much used in industrial noise measurement. Instead, the time-averaged value is used. This is the time average sound level or as it is usually called the 'equivalent continuous sound level' has the formal symbol L_{AT}. These mainly, follow the formal ISO acoustic definitions. However, for mainly historical reasons, L_{AT} is commonly referred to as L_{eq}.

Formally, L_{AT} is 20 times the base 10 logarithm of the ratio of a root-mean-square A-weighted sound pressure during a stated time interval to the reference sound pressure and there is no time constant involved. To measure L_{AT} an integrating-averaging meter is needed; this in concept takes the sound exposure, divides it by time and then takes the logarithm of the result.

Short L_{eq}

An important variant of overall L_{AT} is 'short L_{eq}' where very short L_{eq} values are taken in succession, say at 1/8 second intervals, each being stored in a digital memory. These data elements can either be transmitted to another unit or be recovered from the memory and re-constituted into almost any conventional metric long after the data has been acquired. This can be done using either dedicated programs or standard spreadsheets. Short L_{eq} has the advantage that as regulations change, old data can be re-processed to check if a new regulation is met. It also permits data to be converted from one metric to another in some cases. Today almost all fixed airport noise monitoring systems, which are in concept just complex sound level meters, use short L_{eq} as their metric, as a steady stream of the digital one second L_{eq} values can be transmitted via telephone lines or the Internet to a central display and processing unit. Short L_{eq} is a feature of most commercial integrating sound level meters — although some manufacturers give it many different names.

Short L_{eq} is a very valuable method for acoustic data storage; initially, a concept of the French Government's Laboratoire National d'Essais, it has now become the most common method of storing and displaying a true time history of the noise in professional commercial sound level meters. The alternative method which is to generate a time history by storing and displaying samples of exponential sound level has too many artifacts of the sound level meter to be as valuable and such sampled data cannot be readily combined to form an overall set of data.

Until 2003 there were separate standards for exponential and linear integrating sound level meters, (IEC 60651 and IEC 60804 — both now withdrawn), but since then the combined standard IEC 61672 has described both types of meter. For short L_{eq} to be valuable the manufacturer must ensure that each separate L_{eq} element fully complies with IEC 61672.

Personal Sound Exposure Meter (PSEM)

A common variant of the sound level meter is a noise dosemeter (dosimeter in American english). However, this is now formally known as a personal sound exposure meter (PSEM) and has its own International standard IEC 61252:1993. This is normally intended to be a body-worn instrument and thus has a relaxed technical requirement, as a body-worn instrument — because of the presence of the body — has a poorer overall acoustic performance. A PSEM gives a read-out based on sound exposure, usually $Pa^2 \cdot h$, and the older 'classic' dosimeters giving the metric of 'percentage dose' are no longer used in most countries. The problem with 'per cent dose' is that is relates to the political situation and thus any device can become obsolete if the '100 per cent' value is changed by local laws. Today, one of the most common devices in use is a miniature PSEM called by many manufacturers a 'dosebadge', or some similar name, as it is so small and light that it somewhat resembles a radiation badge. These tiny devices have the three advantages that not only do they not affect the sound field, but they are so small that they do not interfere with the worker in any way and his work pattern does not change; as well, having no microphone cable, they should have a lower risk of failure, by the cable 'catching on machinery'.

Peak Sound Pressure Level (*LC~pk~*)

Most national regulations also call for the absolute peak value to be measured to protect workers hearing against sudden large pressure peaks, using either '*C*' or '*Z*' frequency weighting. 'Peak sound pressure level' should not be confused with 'MAX sound pressure level'. 'Max sound pressure level' is simply the highest rms reading a conventional sound level meter gives over a stated period for a given time-weighting (*S*, *F*, or *I*) and can be many decibel less than the peak value. In the European Union the maximum permitted value of the peak sound level is 140 dB(*C*) and this equates to 200 Pa pressure. The symbol for the *A*-frequency and *S*-time weighted maximum sound level is LAS_{max}. For the *C*-frequency weighted peak it is LC_{pk}.

Sound Level Meters Classes

Sound level meters are divided into two classes — what were called 'types' in previous standards. The two classes have the same design centre goals but the tolerances differ. Class 1 instruments have a wider frequency range and a tighter tolerance than a similar, lower cost, class 2 unit. This applies to both the sound level meter itself as well as the associated calibrator. Most national standards permit the use of 'at least a class 2 instrument' and for many measurements, there is little practical point in using a class 1 unit; these are best employed for research and law enforcement. New in the standard IEC 61672 is a minimum 60 dB linear span requirement and *Z*-frequency-weighting, with a general tightening of limit tolerances, as well as the inclusion of measurement uncertainty in the testing regime. This makes is unlikely that a sound level meter designed to the older 60651 and 60804 standards will meet the requirements of IEC 61672 : 2003.

Atlantic Divide (ANSI/IEC)

Sound level meters are also divided into two types in 'the Atlantic divide'. Sound level meters meeting the USA American National Standards Institute (ANSI) specifications cannot usually meet the corresponding International Electrotechnical Commission (IEC) specifications at the same time, as the ANSI standard describes instruments that are calibrated to a randomly incident wave, i.e. a diffuse sound field, while internationally meters are calibrated to a free field wave, that is sound coming from a single direction. Further, USA dosimeters have an exchange rate of level against time where every 5 dB increase in level halves the permitted exposure time; whereas in the rest of the world a 3 dB increase in level halves the permitted exposure time. The 3 dB doubling method is called the 'equal energy' rule and there is no possible way of converting data taken under one rule to be used under the other. Despite these differences, many developing countries try to specify both USA and international specifications in one instrument in their national regulations. Because of this, many commercial PSEM have dual channels with 3 and 5 dB doubling, with some even having 4 dB for the US Air Force.

Pattern Approval

One of the more difficult decisions in selecting a sound level meter is how do you know if it complies with its claimed standard? This is a difficult question and IEC 61672 part 2 tries to answer this by the concept of 'pattern approval'. A manufacturer has to supply instruments to a national laboratory which tests one of them and if it meets its claims issue a formal Pattern Approval certificate. In Europe the most common — and the most rigorous — approval is often considered to be that from the PTB in Germany (Physikalisch-Technische Bundesanstalt). If a manufacturer cannot show at least one model in his range that has such approval, it is reasonable to be wary, but the cost of this approval militate against any manufacturer having all his range approved.

Even the most accurate approved sound level meter must be regularly checked for sensitivity — what most people loosely call 'calibration'. To assist in this sensitivity checking, PTB also pattern approves sound calibrators to IEC 60942:2003 and in April 2008, the first commercial units were formally approved both at class 1 and class 2 level with the approval number PTB-1.61.4028829.

These units consist of a computer controlled generator with additional sensors to correct for humidity, temperature, battery voltage and static pressure. The output of the generator is fed to transducer in a half-inch cavity into which the sound level meter microphone is inserted. The acoustic level generated is 94 dB which is 1 pascal and is at a frequency of 1 kHz where all the frequency weightings have the same sensitivity.

International Standards

The following International standards define sound level meters, PSEM and associated devices. Most countries national standards — except of course those of the USA — follow these very closely. In many cases the equivalent European standard, agreed by the EU, is designated for example EN 61672 and the UK national standard then becomes BS. EN 61672.

1. IEC 61672 : 2003 'Electroacoustics—sound level meters'.
2. IEC 61252 : 1993 'Electroacoustics—specifications for personal sound exposure meters'.
3. IEC 60942 : 2003 'Electroacoustics—sound calibrators'.

These international standards were prepared by IEC technical committee 29: Electroacoustics, in cooperation with the International Organisation of Legal Metrology (OIML).

SECTION II

Noise Reduction Through Encloser and Barriers

Chapter 5

Silencer and Suppressors

INTRODUCTION

Noise superimposed on gas flow in a duct can be reduced by silencers. Many types and models of silencers are manufactured. The type needed for a particular application depends on the required noise reduction as a function frequency and on the operating conditions. Silencers are commonly classified as dissipative or reactive, depending upon the predominant process utilised in reducing the intensity of the noise. They are generally used in high frequency applications.

TYPES OF SILENCERS

Reactive Silencers

Reactive silencers are used on the exhaust of reciprocating engines because of the low-frequency character of the noise. The exhaust noise spectrum consists of a number of discrete frequency, pre-imposed on a background of broad-band noise. These discrete frequencies occur at harmonics of the rotational speed of the crankshaft. Particular level of each is dependent upon the engine setting and the exhaust system. Reactive silencers reduce noise primarily by reflecting sound energy back toward the source. These are most effective in low frequency applications.

The primary function of a reactive silencer is to reflect sound waves back to the source. Energy is dissipated in the extended flow path resulting from internal reflections and by absorption at the source. The operation principle of the reactive silencers is a combination of lambda/4- and Helmholtz-resonators acting as acoustic filters. Reactive silencers have tuned cavities or membranes and are designed to attenuate low frequency noise from machines.

The reactive silencer may have excellent low frequency performance, is non-fibrous and cleanable and has small or negligible pressure loss. The simplest kind of a reactive muffler is the expansion chamber. In general reactive silencers are used for fixed speed machinery producing pure tones. The reactive silencer is suitable for engines requiring very low exhaust system back pressures for a maximum engine performance. Reactive silencers are rarely used in HVAC systems.

Dissipative Silencers

Dissipative silencers depend primarily upon porous linings to absorb sound as it travels through the flow passages. Dissipative silencers are seldom used alone for exhaust silencing. They may be used in combination with reactive silencers for added high-frequency performance. They are not recommended for liquid fuel engines, however, as accumulation of fuel in the lining material could present a hazard.

Exhaust silencers are commercially available in models and sizes to satisfy many installation and operational requirements. The units are generally of all-metal construction and consist of multiple chambers connected in series by internal tubes. The internal configuration is designed to provide non-turned, relatively broad-band acoustical performance. The most common type is illustrated in Fig. 5.1. It consists of two unequal chambers, with intruding inlet and outlet tubes, and offset internal connecting tubes. In critical applications, three chambers may be used for increased performance. These silencers are applied primarily to the exhaust of naturally aspirated engines for which the relatively high pressure drop associated with the internal flow reversals is acceptable. They can also be used for turbo-charged engines, but larger sizes must be used because of the lower exhaust pressure drop allowed for this type of engine.

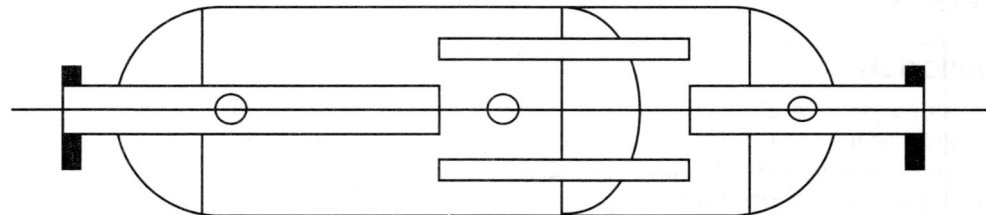

Fig. 5.1. Standard two-chamber exhaust silencer.

Dissipative silencers are widely used in HVAC duct systems. Typical dissipative silencers are configured in a parallel baffle arrangement (Fig. 5.2). The thickness of acoustical linings or baffles should be selected with reference to the predominant frequency of the noise.

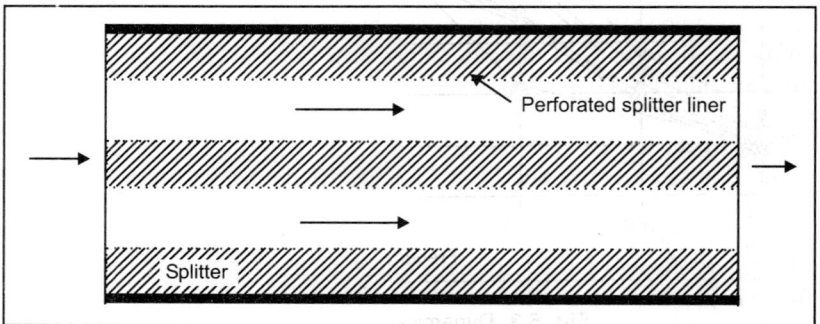

Fig. 5.2. Dissipative silencers.

The incident sound energy is partially transformed to heat by causing motion in the fibres during its passage through the material. Absorptive silencers include lined duct attenuators, packaged cylindrical and rectangular attenuators, acoustic louvers and lined plenum chambers. Typical dynamic insertion losses (DIL) with absorptive silencers are indicated in the Table 5.1 and diagram (Fig. 5.3).

For turbo-charged engines, a low pressure drop modification is available with straight-through flow passages (Fig. 5.4). However, the straight through configuration sacrifices some acoustical performance, particularly in the high-frequency range. Within back-pressure limitations, silencers may be placed in series for added acoustical performance.

Table 5.1. DIL with absorptive silencers.

Diameter (inches)	Length (inches)	Frequency (Hz)						
		125	250	500	1000	2000	4000	8000
4	24	8	14	26	34	41	45	25
5	24	6	12	22	28	37	38	22
6	24	5	10	18	23	33	30	19
8	24	4	9	17	22	29	25	18
10	36	6	11	21	27	39	25	19
12	36	5	9	18	23	32	20	18
16	36	5	8	11	23	19	17	15

*(1 in) = (25.4 mm).

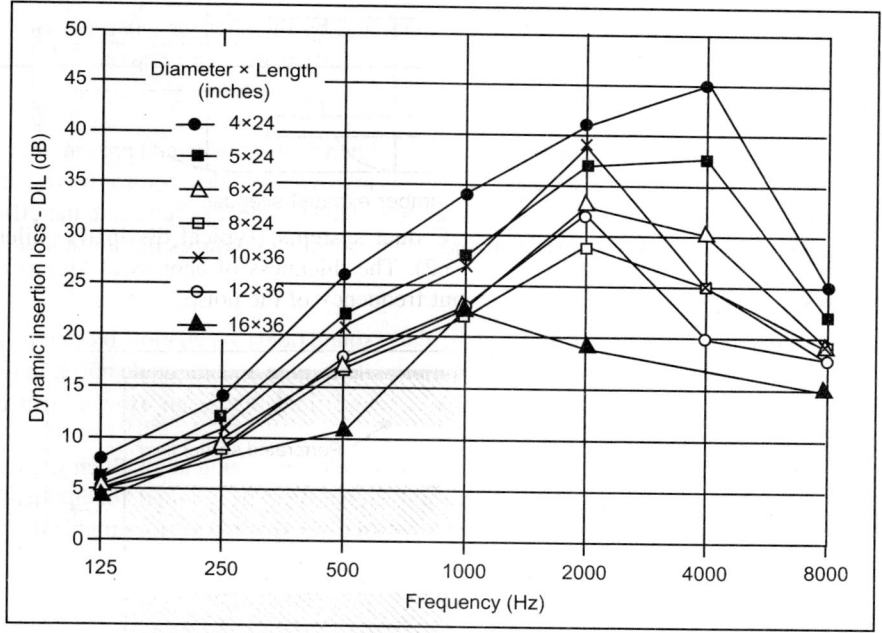

Fig. 5.3. Dynamic-insertion-loss.

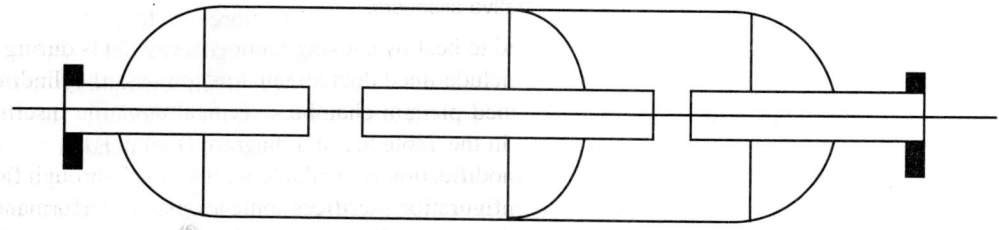

Fig. 5.4. Low pressure drop silencer for turbo-charged engines.

Duct Silencers

The purpose of a duct silencer is to reduce the noise inside air-handling systems caused by the:
1. Fan.
2. Passage of air through straight ducts.
3. Impact of air flowing through components such as elbows, branches, mixing boxes, etc.

Dynamic insertion loss (DIL)

The dynamic insertion loss is the difference between the sound power or intensity levels measured in the same point of the duct work before and after the insertion of the silencer. The insertion loss depends on the flow—if it is forward or reverse. The flow is forward if air flows in the same direction as the propagation of sound.

Self noise (SN)

The self noise is the noise power level in decibels generated by the silencer when inserted in the air flow. The self noise depends on the direction of the flow—if its forward or reverse.

Diffuser or depressive silencers

Diffuser type silencers have perforated pepper pots to slow down flow velocity and prevent the generation of low frequency noise and are mainly used for applications involving nozzles, control valves, jet engines, etc. The total pressure drop is divided in several stages across the nozzle, the valve and the diffuser. This allows a better pressure ratio between upstream and downstream and reduces the noise level.

Active silencers

Active noise control is sound field modification, particularly sound field cancellation, by electro-acoustical means. Active silencers use microphones and electronics to determine and attenuate noise. In its simplest form, a control system drives a speaker to produce a sound field that is an exact mirror-image the offending sound (the disturbance).

The speaker thus 'cancels' the disturbance, and the net result is no sound at all. Such silencers can be effective at low frequencies under 300 Hz. Active noise control is best suited for applications with relatively steady noise fields—like fans, engines or similar. Active silencers are not suitable for broadband noise reduction.

SELECTION OF SILENCERS

Silencer selection is usually based on exhaust connection size and on back-pressure limitations established by the engine manufacturer. For most engines, the proper size has already been established and selection is a matter of routine. For each silencer size, manufacturers usually offer three or four models with differing grades of acoustical performance. The acoustical performance of a particular model is dependent upon the operating conditions of temperatures and exhaust gas velocity. Manufacturers publish 'typical attenuation' curves, which illustrate general performance characteristics of their models. These are smoothened curves drawn from test data and are based on conditions typical of the intended service. The most reliable curves include the effect of silencer size (Fig. 5.5).

Exhaust silencers are insensitive to orientation and may be installed in either a horizontal or vertical position without affecting their silencing characteristics. Long lengths of piping between the engine and the silencer should be avoided. Because of the non-linear nature of the propagation of high-intensity sound waves, a shock wave could develop in the piping and produce a loud cracking noise. When this occurs, it is necessary either to revise the exhaust piping or to install a small supplementary silencer near the engine.

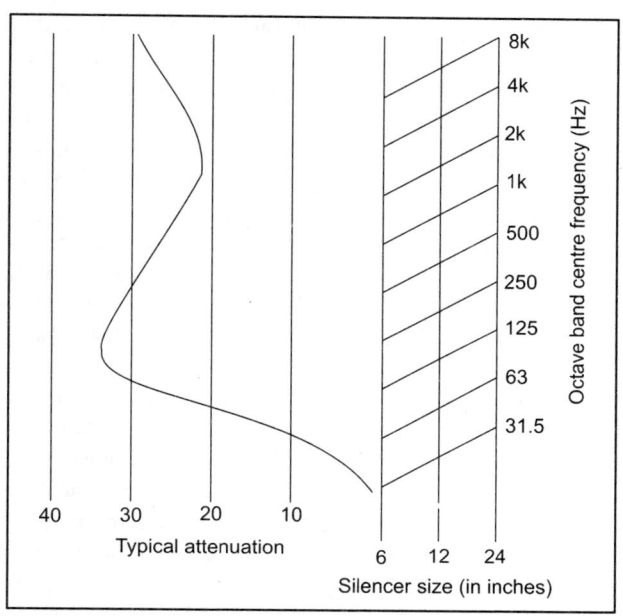

Fig. 5.5. Attenuation vs. silencer size and sound frequency.

Tailpipes should also be selected carefully, with consideration given to their effect on emitted noise. At certain frequencies, the interaction of the tailpipe with the silencer can result in more noise than without the tailpipe. A rule of thumb often used in selecting a tailpipe is to avoid lengths that are half the wavelength corresponding to the engine firing frequency.

Special Features

Other features can be combined with engine silencing, either as a secondary or primary function. Where sparks from the engine exhaust are hazardous, such as in a marine application, a silencer with spark-arresting capability is essential. Both wet and dry types of spark-arresting silencers are available. In the wet type, water is injected into the exhaust gas stream to extinguish any sparks and cool the exhaust line. The dry type is similar to the silencer shown in Fig. 5.1, with an added feature that imparts a rotation to the exhaust gases. Any sparks are removed by centrifugal action and collected in a built-in trap which is cleaned out at intervals.

Another special type of exhaust silencer is one which combines silencing with heat recovery. This is similar to the silencer shown in Fig. 5.1, but the connecting tubes between chambers have been replaced with heat exchange tubes. The exhaust heat is recovered in the form of steam or heated liquid. Whether the use of these units is economically favourable will depend on operating and fuel costs.

SUPPRESSOR

A suppressor, sound suppressor, sound moderator, or silencer, is a device attached to or part of the barrel of a firearm to reduce the amount of noise and flash generated by firing the weapon. It generally takes the form of a cylindrically shaped metal tube with various internal mechanisms to reduce the sound of firing by slowing the escaping propellant gas and sometimes by reducing the velocity of the bullet. Early suppressors were created around the beginning of the 20th century by several inventors. American inventor Hiram Maxim is credited with inventing and selling the first commercially successful models circa 1902. Maxim gave his device the trademarked name Maxim Silencer. The muffler for internal combustion engines was developed in parallel with the firearm suppressor by Maxim in the early 20th century, using many of the same techniques to provide quieter-running engines. Indeed, in many European countries, automobile mufflers are still referred to as 'silencers'. The proper name silencer has since fallen out of favour with some among the firearms industry, being replaced with the more literally accurate term sound suppressor or just suppressor, because a 'sound suppressor' does not 'silence' any weapon, rather it eliminates muzzle flash and reduces the sonic pressure of a firearm discharging. Common usage and US legislative language favour the historically earlier term, silencer. In US law, the terms 'firearm muffler' and 'firearm silencer' are synonymous.

When a sound is picked up by a microphone, noise—in the sense of sounds other than the one of interest—will be picked up as well. It should be noted however, that in the context of acoustic signals, the definition of noise is a subjective matter. For example, the sounds made by the audience in a concert hall is usually considered to be part of the performance. It carries information about the audience reaction to the performance. Usually, acoustic noise that was picked up by a microphone is undesirable, especially if it reduces the perceived quality or intelligibility of the recording or transmission. The problem of effective removal or reduction of noise (referred to here as Acoustic Noise Suppression or ANS).

Design and Construction

The suppressor is typically a hollow cylindrical piece of machined metal (steel, aluminium, or titanium) containing expansion chambers that attaches to the muzzle of a pistol, submachine gun or rifle. These 'can'-type suppressors (so-called as they resemble a beverage can), may be detached by the user and attached to a different firearm of the same caliber. Another type is the 'integral' suppressor, which consists of expansion chambers surrounding the barrel. The barrel is pierced with openings or 'ports' which bleed off gases into the chambers. This type of suppressor is part of the firearm, and maintenance of the suppressor requires that the firearm be at least partially disassembled (Fig. 5.6).

Both types of suppressor reduce noise by allowing the rapidly expanding gases from the firing of the cartridge to be briefly diverted or trapped inside a series of hollow chambers. The trapped gas expands and cools, and its pressure and velocity decreases as it exits the suppressor. The chambers are divided by either baffles or wipes. There are typically at least four and up to perhaps fifteen chambers in a suppressor, depending on the intended use and design details. Often, a single, larger expansion chamber is located at the muzzle end of a can-type suppressor, which allows the propellant gas to expand considerably and slow down before it encounters the baffles or wipes. This larger chamber may be 'reflexed' toward the rear of the barrel to minimise the overall length of the combined firearm and suppressor, especially with longer weapons such as rifles.

Suppressors vary greatly in size and efficiency. One disposable type developed in the 1980s by the US Navy for 9 mm pistols was 150 mm (5.9 inch) long and 45 mm (1.8 inch) in outside diameter, and was designed for six shots with standard ammunition or up to thirty shots with subsonic (slower than the speed of sound) ammunition. In contrast, one suppressor designed for rifles firing the powerful .50 caliber cartridge is 509 mm (20 inch) long and 76 mm (3 inch) in diameter.

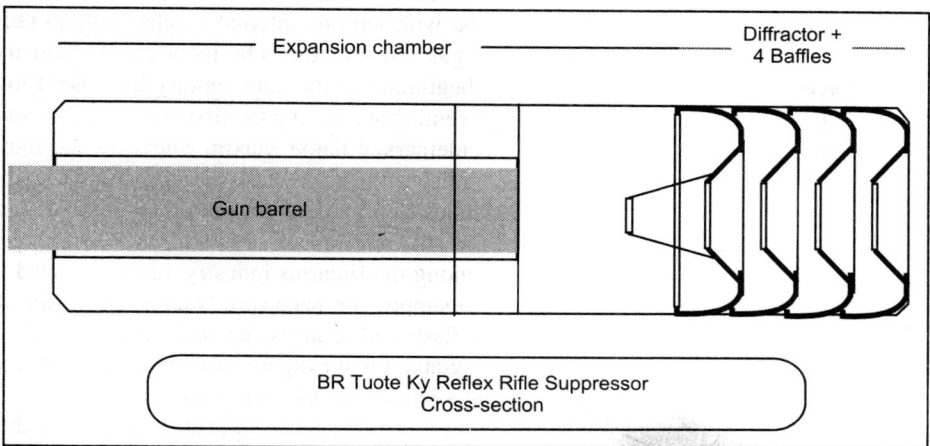

Fig. 5.6. Cross-section drawing of a BR Tuote rifle suppressor, showing expansion chamber 'reflexed' (going back around) the rifle barrel, and four baffles. The diffractor and baffles are carefully shaped to deflect gas.

Components

Baffles are usually circular metal dividers which separate the expansion chambers. Each baffle has a hole in its center to permit the passage of the bullet through the suppressor and towards the target. The hole is typically at least 0.04 inch (1 mm) larger than the bullet caliber to minimise the risk of the bullet hitting the baffle (baffle strike). Baffles are typically made of stainless steel, aluminium, titanium or alloys such as Inconel, and are either machined out of solid metal or stamped out of sheet metal. A few suppressors for low-powered cartridges such as the .22 long rifle (Fig. 5.7) have successfully used plastic baffles (certain models by Vaime and others).

Baffles are separated by spacers, which keep them aligned at a specified distance apart inside the suppressor. Many baffles are manufactured as a single assembly with its spacer, and several suppressor designs have all the baffles attached together with spacers as a one-piece helical baffle stack. Modern baffles are usually carefully shaped to divert the propellant gases effectively into the chambers. This shaping can be a slanted flat surface, canted at an angle to the bore, or a conical or otherwise curved surface. One popular technique is to have alternating angled surfaces through the stack of baffles.

Baffles come in several designs. M, K, Z, Monolithic Core and Ω(Omega) are the most prevalent. M-type is the crudest and composes an inverted cone. K forms slanted obstructions diverging from the sidewalls, creating turbulence across the boreline. Z is expensive to machine and includes 'pockets' of dead airspace along the sidewalls which trap expanded gasses and hold them thereby lengthening the time that the gases cool before exiting. Omega is an advanced design combining elements of all three previous designs. Omega forms a series of spaced cones drawing gas away from the boreline, incorporates a scallopped mouth creating cross-bore turbulence, which is in turn directed to a 'mouse-hole' opening between the baffle stack and sidewall. They were created and developed by Joe Gadinni, and are in use by SWR Manufacturing.

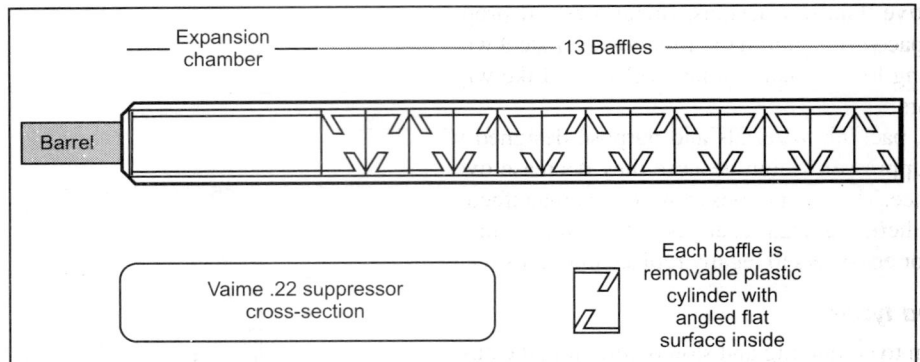

Fig. 5.7. Cross-section drawing of a Vaime .22 caliber rifle suppressor, showing short expansion chamber and thirteen plastic baffles. These baffles use alternating angled flat surfaces to repeatedly deflect gas expanding through the suppressor. In the actual suppressor, the baffles are also oriented at 90 degrees to each other about the axis of bullet travel (the illustration cannot demonstrate this well).

Baffles usually last for a significant number of firings. Propellant gas heats and erodes the baffles, causing wear, which is worsened by high rates of fire. Aluminium baffles are seldom used with fully automatic weapons, because service life is unacceptably short. Some modern suppressors using steel or high-temperature alloy baffles can endure extended periods of fully-automatic fire without damage. The highest-quality rifle suppressors available today have a claimed service life of greater than 30,000 rounds. Baffles have not been given any specific angles, a specific size, or weight to meet any standards; they are created on a trial and error basis.

Wipes are inner dividers intended to touch the bullet as it passes through the suppressor, and are typically made of rubber, plastic or foam. Each wipe may either have a hole drilled in it before use, a pattern stamped into its surface at the point where the bullet will strike it, or it may simply be punched through by the bullet. Wipes typically last for a small number of firings (perhaps no more than five) before their performance is significantly degraded. While many suppressors used wipes in the Vietnam War era, most modern suppressors do not use them to minimise disassembly and parts replacement.

'Wet' suppressors or 'wet cans' use a small quantity of water, oil, grease or gel in the expansion chambers to cool the propellant gases and reduce their volume. The coolant lasts only a few shots before it must be replenished, but can greatly increase the effectiveness of the suppressor. Water is most effective, due to its high heat of vapourisation, but it can run or evaporate out of the suppressor. Grease, while messier and less effective than water, can be left in the suppressor indefinitely without losing effectiveness. Oil is the least effective and least preferable, as it runs while being as messy as grease, and leaves behind a fine mist of aerosolised oil after each shot. Water-based gels, such as wire-pulling lubricant gel, are a good compromise; they offer the efficacy of water with less mess, as they do not run or drip. However, they take longer to apply, as they must be cleared from the bore of the suppressor to ensure a clear path for the bullet (grease requires this step as well). Generally, only pistol suppressors are shot wet, as rifle suppressors handle such high pressure and heat that the liquid is gone within 1–3 shots. Many manufacturers will not warranty their rifle suppressors for 'wet' fire, as some feel this may even result in a dangerous over-pressurisation of the silencer.

Packing materials such as metal mesh, steel wool or metal washers may be used to fill the chambers and further dissipate and cool the gases. These are somewhat more effective than empty chambers, but less effective than wet designs. Metal mesh, if properly used, may last for hundreds or thousands of shots of spaced semiautomatic fire, however steel wool usually degrades within ten shots with stainless wool lasting longer than regular steel wool. Like wipes, packing materials are rarely found in modern suppressors.

Wipes, packing materials and purpose-designed wet cans have been generally abandoned in 21st-century suppressor design because they decrease overall accuracy and require excessive cleaning and maintenance. The instructions from several manufacturers state that their suppressors need not be cleaned at all. Furthermore, legal changes in the United States during the 1980s and 1990s made it much more difficult for end-users to legally replace internal silencer parts, and the newer designs reflect this reality.

Advanced types

In addition to containing and slowly releasing the gas pressure associated with muzzle blast or reducing pressure through the use of coolant mediums, advanced suppressor designs attempt to modify the properties of the sound waves generated by the muzzle blast. In these designs, effects known as frequency shifting and phase cancellation (or destructive interference) are used in an attempt to make the suppressor quieter. These effects are achieved by separating the flow of gases and causing them to collide with each other or by venting them through precision-made holes. The intended effect of frequency shifting is to shift audible sound waves frequencies into ultrasound (above 20 kHz), beyond the range of human hearing. The Russian AN-94 assault rifle features a muzzle attachment that claims apparent noise reduction by venting some gases through a 'dog-whistle' type channel. Phase cancellation occurs when similar sound wave frequencies encounter each other 180° out of phase, cancelling the amplitude of the wave and eliminating the pressure variations perceived as sound.

Using either property to advantage requires that the suppressor be designed within the specification of the muzzle blast in mind. For example, the velocity of the sound waves is a major factor. This figure can change significantly between different cartridges and barrel lengths.

Thus, in order for maximum effectiveness to be achieved, the suppressor must be 'tuned' for a specific cartridge/barrel length combination. This can be done through the use of either a fixed or adjustable baffle design. While it may sound daunting, any weapon that needs to provide such exceptional sound suppression is almost certainly going to be manufactured in small quantities and issued only for mission profiles critical enough to make such efforts worthwhile.

However, these concepts are controversial because muzzle blast creates broadband noise rather than pure tones, and phase cancellation in particular is therefore extremely difficult (if not impossible) to achieve. Some suppressor manufacturers claim to use phase cancellation in their designs, but these claims are generally unsupported from a scientific perspective.

From the practical perspective, supersonic cartridge loads are impractical to suppress past the levels that are merely hearing-safe for the shooter due to the sonic boom emitted by the bullet, and cartridges such as .22LR and .45ACP have long been recognised as easy to suppress even if using technology dating back to 1940s when the mission calls for the quietest gun available.

Sources of firearm noise

The portrayal of suppressed firearms in popular culture is not always accurate and could lead to the misconception that silencers are capable of completely eliminating the sound of firing, or reducing it to a quiet whistling or 'phut' sound. This is because when a gun is fired, multiple sounds are possibly made. There are five major categories of suppressed fire noise: action, blast, sonic signature, impact, and operator. Some of these are present in all instances, others depend wholly on the specific mechanics of the weapon employed. In order of timing:

1. Action noise required to ignite the round.
2. Muzzle blast resulting from the discharge of propellant from the end of the barrel.
3. Sonic signature of the projectile in flight (supersonic velocity rounds).
4. Action noise in some firearm variants as the spent round is discharged and a fresh round reloaded.
5. Impact noise created as the projectile finds terminal impact.

Obviously, some of these sounds are much louder than others. The two loudest sounds in a gunshot are typically the muzzle blast and the sonic signature. Multiple techniques are used to address each of these sounds, but the suppressor itself is capable of addressing muzzle blast, sonic signature (through integral gas bleed, reducing the projectile to subsonic) and the ability to cancel the mechanical action noise through Nielsen device manipulation, cancelling the ejection cycle.

Real world data

Live tests by independent reviewers of numerous commercially available suppressors find that even low caliber unsuppressed .22 LR firearms produce gunshots over 160 decibels. In testing, most of the suppressors reduced the volume to between 130 and 145 dB, with the quietest suppressors metering at 117 dB. The actual suppression of sound ranged from 14.3 to 43 dB, with most data points around the 30 dB mark.

Comparatively, ear protection commonly used while shooting provides 18 to 32 dB of sound reduction at the ear. Further, chainsaws, rock concerts, rocket engines, pneumatic drills, small firecrackers, and ambulance sirens are rated at 100 to 140 dB.

While some consider the noise reduction of a suppressor significant enough to permit safe shooting without hearing protection (hearing safe), noise-induced hearing loss occurs at 85 dB or above, and suppressed gunshots regularly meter above 130 dB. However, the US Occupational Safety and Health Administration uses 140 dB as the 'safety cutoff' for impulsive noise, which has led most US manufacturers to advertise sub-140 dB suppressors as 'hearing safe'.

Limitations of dB meter effectiveness

dB testing measures only the peak sound 'pressure' noise, not duration or frequency. Limitations of dB testing become apparent in a comparison of sound between a 0.308 caliber rifle and a 0.300 Winchester Magnum rifle. The dB meter will show that both rifles produce the same decibel level of noise. Upon firing these rifles, however, it is clear that the 0.300 Winchester Magnum sounds much louder. What the decibel meter does not show is that although both rifles produce the same peak sound pressure level (SPL), the .300 Winchester Magnum holds its peak duration longer. The .300 Winchester Magnum sound remains at full value longer, while the .308 goes to peak and falls off more quickly. dB meters fail in this and other regards when being used as the principal means to determine suppressor capability.

Caliber versus volume

The caliber and power of the bullet/cartridge being suppressed is also an important factor. Generally, equal quality suppressors can quiet the report of a smaller caliber bullet more effectively than a larger caliber bullet. This is because the exhaust gases can move more quickly through the exit hole necessary for larger caliber bullets. Likewise, cartridges which produce higher pressures and more gasses, such as those used in rifles, will also generally be loyder than those which produce less pressure and fewer gases, such as handgun cartridges. In a gunshot, the sound of the report (the combination of the sonic boom, the vacuum release, and burn of powder) will almost always be louder than the sound of the action cycling of an auto-loading firearm. Alan C. Paulson, a renowned firearms specialist, claimed to have encountered an integrally suppressed 0.22 LR that had such a quiet report, although this is somewhat uncommon.

Because of the limited stopping power of less powerful cartridges, movie scenes in which an attacker fires a near-silent shot that instantly kills the victim are generally unrealistic.

Subsonic ammunition versus volume

In weapons firing supersonic bullets, the supersonic bullet itself produces a loud and very sharp sound as it leaves the muzzle in excess of the speed of sound and gradually reducing speed as it travels downrange. This is a small sonic boom, and is referred to in the firearm field as 'ballistic crack'. Subsonic ammunition reduces this sound, but at the cost of lower velocity, often resulting in decreased range and effectiveness on the target. Military marksmen and police units may use this ammunition to maximise the effectiveness of their silenced rifles. While the range may be decreased when using subsonic rounds, this may be acceptable for specialised situations, where the absolute minimum amount of noise is required.

However, the numeric effectiveness of subsonic rounds is, again, misrepresented by media. Independent testing of commercially available firearm suppressors with commercially available subsonic rounds has found that .308 subsonic rounds decreased the volume at the muzzle 10 to 12 dB when compared to the same caliber of suppressed supersonic ammunition. When combined with suppressors, the subsonic .308 rounds metered between 121 and 137 dB.

This ballistic crack depends on the speed of sound, which in turn depends mainly on air temperature. At sea level, an ambient temperature of 70°F (21°C), and under normal atmospheric conditions, the speed of sound is approximately 1140 feet per second (347 m/s). Bullets that travel near the speed of sound are considered transonic, which means that the airflow over the surface of the bullet, which at points travels faster than the bullet itself, can break the speed of sound. Pointed bullets which gradually displace air can get closer to the speed of sound than round nosed bullets before becoming transonic.

Because merely reducing the propellant in a cartridge to get a slower bullet would lead to less stopping power, special cartridges have been developed specifically to maximise the energy available when used with a suppressor. These cartridges use very heavy bullets to make up for the energy lost by keeping the bullet subsonic. A good example of this is the .300 Whisper cartridge, which is formed from a necked-up .221 Remington Fireball cartridge case. The subsonic .300 Whisper fires up to a 250 grain (16.2 g), .30 caliber bullet at about 980 feet per second (298 m/s), generating about 533 ft·lbf (722 J) of energy at the muzzle. While this is similar to the energy available from the .45 ACP pistol cartridge, the reduced diameter and streamlined shape of the heavy .30 caliber bullet provides far better external ballistic performance, improving range substantially.

9 × 19 mm Parabellum, a very popular caliber for suppressed shooting, can use almost any factory-loaded 147 gr (9.5 g) weight round to achieve subsonic performance. These 147 gr weight bullets typically have a velocity between 900 and 980 feet per second (275 and 300 m/s), which is less than the common 1140 ft/s speed of sound.

Russian 9 × 39 mm ammo had a high subsonic Ballistic coefficient, high retained downrange energy, high Sectional density, and moderate recoil. All elements combined make this a very attractive choice for Close Quarters Combat firearms.

Instead of using subsonic ammunition, one can also lower the muzzle velocity of a supersonic bullet before it leaves the barrel. Some suppressor designs, referred to as 'integrals', do this by allowing gas to bleed off along the length of the barrel before the projectile exits. However, as of 2010, the best-known weapons with integral suppressors such as MP5SD and VSS Vintorez/AS Val fire subsonic cartridges and use integral suppressors for the sole purpose of reducing the weapon's length rather than to slow down an inherently supersonic cartridge.

Identification

Aside from reductions in volume, suppressors also tend to alter the sound to something that is not identifiable as a gunshot. This reduces or eliminates attention drawn to the shooter (hence the Finnish expression: 'A silencer does not make a marksman silent, but it does make him invisible'). This is especially true in cases where there are other sources of ambient noise, such as in an urban environment. Suppressors are particularly useful in enclosed spaces where the sound, flash and pressure effects of a weapon being fired are amplified. Such effects may disorient the shooter, affecting situational awareness, concentration and accuracy, and can permanently damage hearing very quickly.

As the suppressed sound of firing is overshadowed by ballistic crack, observers can be deceived as to the location of the shooter, often from 90 to 180 degrees from his actual location. However, counter-sniper tactics can include Gunshot Location Detection Systems, where sensitive microphones are coupled to computer algorithms, and use the ballistic crack to detect and localise the origin of the shot. The US Boomerang system is currently the only deployed example.

Applications of Noise Suppression

In the general sense, noise suppression has applications in virtually all fields of communications (channel equalisation, radar signal processing, etc.) and other fields (pattern analysis, data forecasting, etc.)

1. *Telecommunications*: Perhaps the most common application of ANS is in the removal or reduction of background acoustic noise in telephone or radio communications. Examples of the former would be the hands-free operation of a cellular telephone in a moving vehicle, or a telephone on a factory floor. Examples of the latter would be communication in civil aviation and most military communications (Fig. 5.8.).

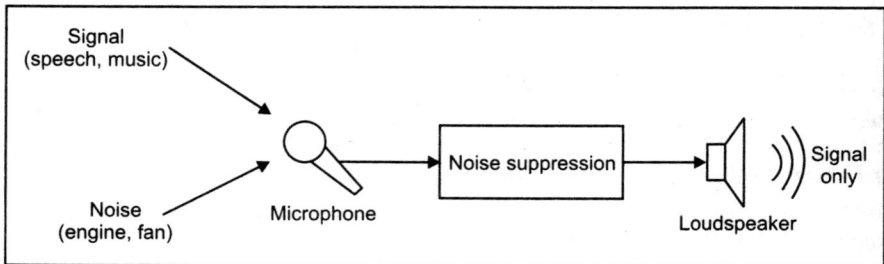

Fig. 5.8. Basic overview of an acoustic noise suppression system.

In these applications, generally the purpose of ANS is to improve the intelligibility of the speech signal or at least to reduce listener fatigue. It is important to note in this context that—while undesirable—distortion of the original speech is tolerable if intelligibility is not affected. Furthermore, in these types of applications, delays in the signal must be kept small. This places constraints on both algorithmic delays and computing complexity.

2. *Audio archive restoration*: The restoration of sounds recorded on audio carriers (vinyl records, magnetic tape, etc.) has been a field of growing importance with the introduction of digital signal processing (DSP) methods. Unlike the applications mentioned above, processing delays are not an issue, but distortion of the original signal must be avoided. While the carrier noise (such as tape hiss or phonograph crackle) is not strictly environmental acoustic noise, it may be treated as such since it is acoustic noise picked up with the intended signal by the same mechanism, either the needle of a record player or the magnetic head of a tape player. Generally, the signal-to-noise ratio (SNR) is much higher in audio archive restoration than is the case for telecommunication applications.

These two application areas are merely given as examples, and there may in fact be considerable overlap. For example, a speech recording made under adverse conditions may have a low SNR and allow for distortion, but the enhancement process will lack the complexity constraints. It is therefore desirable to have a method that works well in either application.

General Noise Reduction Methods

There are many ways to classify noise suppression algorithms. They may be single- or multi-sensor. In the latter, the spatial properties of the signal and noise sources can be taken into account. For example, beam-forming using a microphone array emphasises sounds from a particular direction. Another example is adaptive noise cancellation (ANC), which is a two-channel approach based on the primary channel consisting of signal and noise, and the secondary channel consisting of only the noise. The noise in the secondary channel must be correlated with the noise in the primary channel. In the case of adaptive echo cancellation (AEC), the primary channel is the near-end handset, which contains the near-end signal and the reflection of the far-end signal. The secondary channel is the line from the far-end handset.

Some noise suppression methods try to exploit the underlying production method of the signal or the noise. In speech enhancement, this is usually done by linear prediction of the speech signal. In audio enhancement, since the signal is too general to be modelled, the noise is modelled instead.

Short-time spectral amplitude methods

The noise suppression method discussed in this chapter is a single channel method based on converting successive short segments of speech into the frequency domain. In the frequency domain, the noise is removed by adjusting the discrete frequency 'bins' on a frame-by-frame basis, usually by reducing the amplitude based on an estimate of the noise. The various methods (differentiated by the suppression rule, noise estimate and other details) are collectively known as short-time spectral amplitude (STSA), spectral weighting, or spectral subtraction methods.

Auditory Models in Acoustic Noise Suppression

In the above sections, only properties of the source of the signal and noise were exploited in the process of noise suppression. To further improve the performance of acoustic noise suppression (ANS) algorithms, properties of the human ear can be taken advantage of. Research into human auditory properties is an ongoing process. However, available models of the human auditory system have been successfully used to improve the performance of speech and audio coding algorithms. In these coding algorithms, the purpose is to take only as much of the signal as is perceptually relevant. This reduction of information allows the signal to be stored or transmitted using fewer bits. Acoustic noise suppression methods incorporating these same perceptual models have shown significant gains in performance. However, there is still room for improvements, and research into new methods continues.

Chapter 6

Glass

INTRODUCTION

Anyone who has ever lived near a railroad track or a busy street knows how frustrating outside noise can be. It is difficult to relax or sleep in your own home if you cannot escape from the sounds of the outside world. Fortunately, thanks to innovations in glass technology, you can reduce the amount of outside noise that gets into your home by replacing your old windows with newer models optimised for blocking sound. When looking for replacement windows to block sound, it is important to keep in mind that windows do not all block sound equally. Just because a window is new does not necessarily mean that it will reduce outside noise as much as you would like it to. Here are some things that keep in mind when looking for replacement windows to block sounds from outside.

Multiple panes are not necessarily enough: Many homeowners are surprised to find that their new double pane windows may not reduce noise significantly more than their old single pane windows. Different thicknesses of glass block different frequencies of sound, so if the two panes (or even three, for triple pane windows) are the same thickness, the number of panes will not really make a difference. To block more frequencies of sound, the panes should be of different thicknesses.

Panes should be spaced apart: With all other factors the same, the more space between layers of glass in a window, the more effective that window will be at blocking sound. The amount of spacing between the panes will also affect energy efficiency, so make sure to discuss this option with a local window replacement specialist to figure out how much space you really need between window panes.

Look for laminated glass: Laminated glass is commonly recommended for noise reduction. Laminated glass is formed by bonding together two pieces of glass with a layer of tough plastic between them. In addition to very effectively blocking sound, laminated glass also blocks ultraviolet rays from the sun.

The installation should be air tight: The window must be properly installed and sealed to effectively reduce sound. It should also be surrounded by good insulation. If your window is not air tight, it will not block sound well, no matter how great the glass is. New sound blocking windows can help you enjoy peace and quiet in your home. Contact your local replacement window dealer to learn more about installing new windows for noise reduction.

SINGLE GLASS

Noise reduction can be achieved in three ways:
1. By reducing the sound level of the noise source.
2. By dissipating the sound near the receiver or the listener.
3. By blocking the transmission path between the source and the receiver.

The first solution is the most effective, but also the most difficult to achieve in practice. The second solution is also most impractical. The third solution—blocking the transmission path—involves the placement of a relatively impervious partition somewhere between the noise source and the receiver.

Glass offers the potential for controlling noise and also establishing visual communication. The performance of a glass structure is primarily dependent upon the response characteristics it has to a spectrum of energy.

To understand the mechanism of sound transmission through partitions of any structure, one must study the single glass wall because it forms the basic element of most structures. A modest average insulation of 35 dB can be obtained by using a single-glass wall weighing about 6 pounds per square foot. However, if an average insulation of 40 dB or more is required, the single-glass wall construction becomes decreasingly efficient and the double-glass type of construction, which also provides good thermal insulation, should be used.

Single Glass Theory

For airborne sound, the sound transmission coefficient (T) of a panel is defined as the ratio of the transmitted sound energy to incident sound energy. The transmission loss or sound reduction index is given by the equation:

$$TL = 10 \log (1/T) \text{dB} \qquad \text{... (6.1)}$$

The simplest theoretical treatment of single-glass partition involves the well-known mass law, which for sound waves at oblique incidence can be written as:

$$TL = 10 \log_{10}\left[1 + \frac{\omega M_s \cos\phi}{2\zeta C}\right] \qquad \text{... (6.2)}$$

where, TL = the transmission loss.
  M_s = total surface density of the barrier (slugs/ft).
  ϕ = the angle of incidence.
  ζ = density of air (slugs/ft^3).
  ω = angular frequency (radians/sec).
  C = velocity of sound in air (ft/sec).

If the incident sound wave impinges on the panel at normal incidence (when $\phi = 0$), then Eq. 6.2 can be written as:

$$(TL)_o = 10 \log_{10}\left[1 + \left(\frac{\omega M_s}{2\zeta C}\right)\right] \qquad \text{... (6.3)}$$

where, $(TL)_o$ is the transmission loss for normal incidence.

The quantity $(\omega M_s)/(2\zeta C)$ is generally large and, in practice, the sound waves impinge upon the barrier at a wide range of incidence angles. In this case, Eq. 6.3 becomes Eq. 6.4:

$$(TL) \text{ field} = 20 \log_{10}\frac{(\omega M_s)^{-5}}{(2\zeta C)} \qquad \text{... (6.4)}$$

The value predicted by incidence field equation is the maximum that can be theoretically obtained with a single-glass barrier. When transmission loss is plotted against the frequency spectrum as a logarithmic scale, a straight line having a slope of 6 dB per octave results (Fig. 6.1).

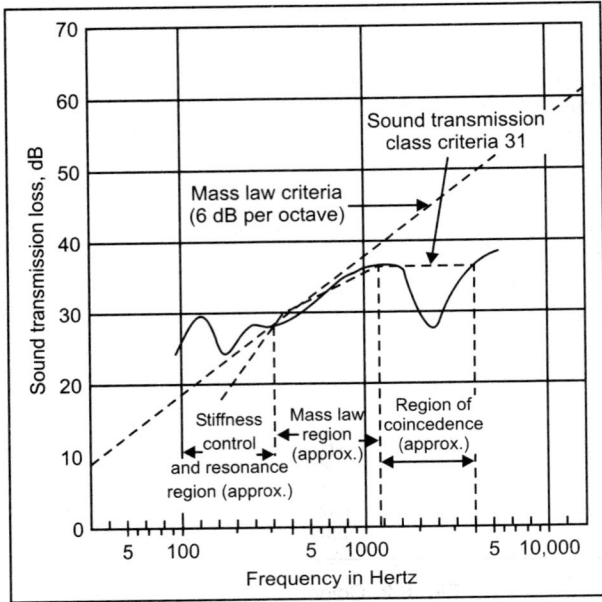

Fig. 6.1. Transmission loss characteristics for 7/32-inch-thick glass (3.0 lbs/ft^2). Laboratory sample 35 $\frac{3}{4}$ × 83 $\frac{3}{4}$ inches (tested in accordance with ASTM Designation E90-70).

According to the mass law, a 7/32-inch-thick glass wall has a potential average 38 dB but in fact only achieves 31 dB. This discrepancy is due to the influence of stiffness, which becomes significant at certain frequencies. The mass law region is bounded by two important transition regions of stiffness-resonance and coincidence control. The low frequency transition boundary is generally defined by the stiffness response and the first few modes of natural frequency of the plate, whereas the upper boundary is limited by the coincidence frequency. Since these boundaries both tend to reduce the performance below mass law levels, it is imperative that they be thoroughly considered in the prediction of performance. These considerations become more pertinent and more complex for multiple glass and laminate structures.

Coincidence frequency

The bending wave speed of a wall varies as the square root of the excitation frequency. At some combination of excitation frequency, and angle of incidence, the trace wavelength of the incident sound wave will exactly coincide with the wavelength of the bending wave of the wall. The sound wave thus reinforces the bending wave, and a maximum amount of acoustical energy is transferred to the wall. At coincidence, the wall becomes quite transparent to sound and the transmission loss drops markedly.

Coincidence frequency is graphically explained in Fig. 6.2.

$$\lambda_B = \lambda$$

This condition of equal wavelengths in the glass panel and in the air is designated wave coincidence where:

λ_B = wavelength of the bending wave in the glass panel.

λ = wavelength of the sound wave in the air

$$\lambda \sin\phi_o = \lambda_B$$

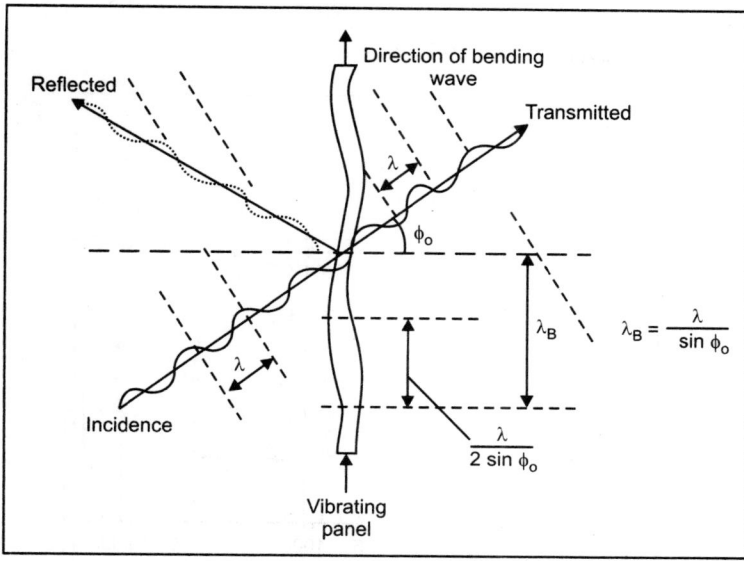

Fig. 6.2. Coincidence frequency.

This means that the intensity of the transmitted wave approaches the intensity of the incidence wave. The frequency for which $\lambda = \lambda_B$ is called critical frequency. If the transmitted wave is as intense as the incidence wave, then the *TL* at that frequency and angle is very small.

Mathematically treated, coincidence frequency is expressed as:

$$f_\phi = \frac{\omega_\phi}{2\pi} = \frac{C^2}{2\pi \sin^2 \phi} \sqrt{M/D}$$

where, C = velocity of sound through air.

ϕ = angle of incidence.

M = $\zeta_p h$.

D = $Eh^3/12(1-v^2)$

v = Poisson's ratio

ζ_p = mass density

$$f_\phi = \frac{C^2}{2\pi \sin^2 \phi} \sqrt{\frac{\zeta_p 12(1-v^2)}{Eh^2}}$$

The experimental results for coincidence frequency are shown in Fig. 6.3 as a function of thicknesses for the random incidence. It can be seen from the results that, as the thickness of glass plate increases, the coincidence frequency decreases.

Condition for wave coincidence

Wave coincidence occurs when

$$\sin \phi_o = \lambda / \lambda_B \qquad\qquad \sin \phi_o = C/C_B$$

C_B = velocity of propagation of the bending wave in the glass plate.
C = velocity of propagation of the bending wave in air.

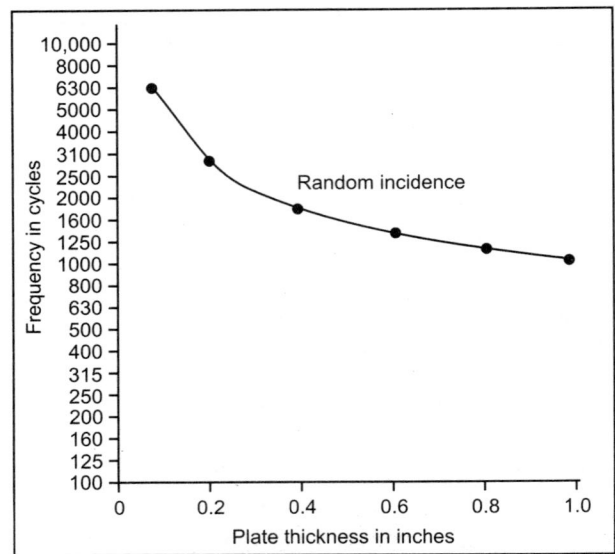

Fig. 6.3. Coincidence frequency versus glass plate thickness.

When the fixed frequency is assumed, the angle at which wave coincidence takes place is defined as the coincidence angle. When a fixed angle is assumed, the frequency at which wave coincidence takes place is defined as the coincidence frequency.

Critical frequency

The lowest frequency at which coincidence occurs is that frequency for which $\lambda_B = \lambda$ or $C_B = C$. Critical frequency is the lowest possible coincidence frequency and occurs for grazing incidence sound ($\phi_0 = 90°$). Mathematically treated:

$$f_c = \frac{C^2}{1.8hC'_L} = \frac{C^2}{1.8h} \sqrt{\zeta_p/E} \ cps$$

where, h = thickness of the glass in ft for mixed English units:

$$f_c = \frac{6.6 \ C^2}{hC'_L}$$

where, h = inches
  c = ft/sec.
  p = lbm/ft^3
  E = lbf/ft^2

These equations are valid only if $\lambda_B > 6h$

$$f \text{ coincidence} = \frac{C^2}{1.8 C'_L \, h \, \sin^2 \phi_o}$$

$$f \text{ coincidence} = \frac{C^2}{1.8h \, \sin^2 \phi_o} \sqrt{\zeta_p / E}$$

$$f \text{ coincidence} = \frac{1.16 C^2}{h \, \sin^2 \phi_o} \sqrt{\zeta_p / E} \qquad \text{for mixed English units}$$

$$C'_L = \sqrt{E / \zeta_p (1 - v^2)} = C_L$$

v = Poisson's ratio 0.3 in most cases.

Resonance

At the resonance frequencies of the glass wall, the transmission loss (TL) decreases sharply. For a hypothetical glass wall with zero damping, the TL would drop to zero at each resonance frequency. In actual walls, there is always some damping present so that the TL does not drop to zero. Also, because of damping, inertia, and the generally diminished power in the high frequency spectrum, significant reduction in TL usually occurs only at the lowest natural frequency. Since wall resonances may occur relatively close together in the frequency spectrum, the TL may be significantly below the mass law curve in a fairly broad frequency band.

Resonance frequency is expressed by the following mathematical relationship:

$$\omega_{\text{res}} = \sqrt{\frac{2.8 P_o}{d \, M_s \cos^2 \phi} Ku}$$

where, Ku = correction factor.
  d = spacing of panels (ft).
  M_s = mass of each panel (slugs/ft^2).
  ϕ = angle of incidence.
  P_o = atmospheric pressure (lbs/ft^2).

Basic resonance is expressed as:

$$\omega_o = \sqrt{\frac{2.8 P_o}{d M_s} Ku} \qquad \qquad \psi \omega_{\text{res}} = \frac{\omega_o}{\cos \phi}$$

Figure 6.4 is a chart for determining resonance frequency, f_o, of two equal walls of combined surface density, M, separated by an air space of depth, d.

Figure 6.5 is a schematic representation of transmission loss of a structure and its variation over the response spectrum of frequency as a function of the controlling parameters. With increasing frequency, these regions are stiffness controlled, resonance, mass controlled, coincidence, and supercoincidence.

It is apparent that the position of these regions in spectrum for a glass unit and the ability to control the positions are of great importance in establishing how a glass window shall perform as a transmission loss barrier component or system.

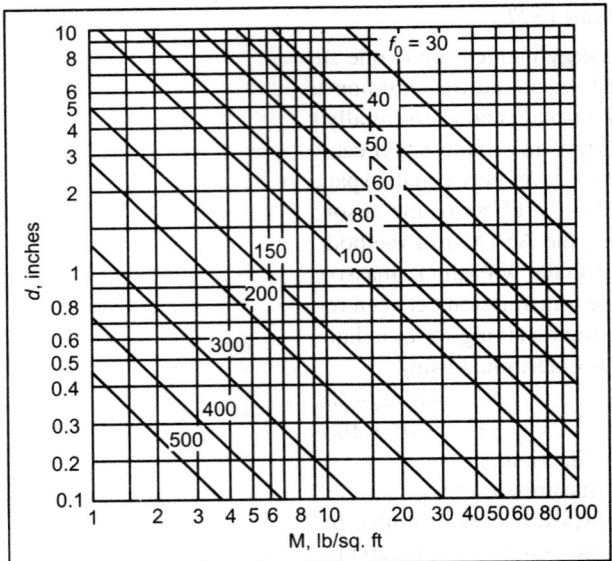

Fig. 6.4. Chart for determining resonant frequency f_0 of two equal walls of combined surface density M, separated by an air space of depth d.

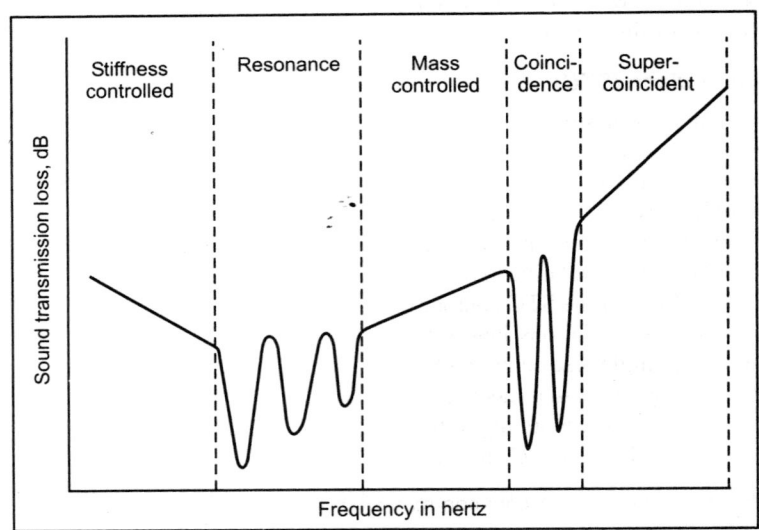

Fig. 6.5. Schematic representation of TL of a structure and its variation over the response spectrum of frequency as a function of the controlling parameters.

In the stiffness region, the performance is obviously totally undesirable. This being the case, it is standard procedure to establish the natural frequency (first important mode) below the lowest frequency when transmission loss is required. This will assure that the window will perform well above the stiffness region throughout the spectrum of interest.

Sound transmission loss (STL)

The ratio of the sound energy incident upon one surface of a partition to the energy radiated from the opposite surface is called the sound transmission loss of the partition. Sound transmission loss is an inherent characteristic of a barrier and is essentially independent of the location of the barrier. We do not actually 'hear' the sound transmission loss, nor can we measure it directly. We can hear and measure the difference in sound pressure level between the spaces separated by a barrier, and we call this difference noise reduction. It includes the effect of the absorption present in the receiving room and the source room. In test laboratories, the ASTM test method E90-70 is used. The test panel covers an opening between two rooms constructed with thick, massive walls that transmit much less energy than the test panel. Therefore, all of the transmission between rooms can be considered to take place through the test panel. The absorption in the receiving room is known. A sound source is located in one room and the sound transmission loss is determined using:

$$NR = SPL_s - SPL_r$$
$$STL = NR + 10 \log_{10} S/A$$

where,   NR   = sound reduction.

          STL   = sound transmission loss.

          SPL_s = sound pressure level in source room.

          SPL_r = sound pressure level in receiving room.

          S   = area of test panel.

          A   = total absorption in receiving room in units consistent with S.

Experimental results showing STL as a function of frequency can be found in the *Acoustical Experimentation* section of this chapter.

Sound transmission class (STC)

To determine the sound transmission class (STC) of a test specimen, its sound transmission losses in a series of 16 test bands (determined in accordance with Recommended Practice E90-70) are compared with those of a reference contour having the form illustrated in Fig. 6.6.

If the transmission losses for the test specimen are plotted in a graph, the sound transmission class may be determined by comparison with a transparent overlay on which the STC contour is drawn. The STC contour is shifted vertically, relative to the test curve, until some of the measured TL values for the test specimen fall below those of the STC contour and the following conditions are fulfilled. The sum of the deficiencies (that is, the deviations below the contour) shall not be greater than 32 dB and the maximum deficiency at a single test point shall not exceed 8 dB. When the contour is adjusted to the highest value (in integral decibels) that meets the above requirements, the sound transmission class for the specimen is the TL value corresponding to the intersection of the contour and the 500-Hz ordinate.

The sound transmission classes (STC) for common materials are shown in Table 6.1

Table 6.1. Sound transmission class for common materials.

Plate glass thickness (inch)	STC
1/8	28
3/16	31
7/32	31
1/4	31

(Contd ...)

Plate glass thickness (inch)	STC
3/8	32
1/2	35
3/4	37
1.0	38
1/4 inch steel plate	36
3/4 inch plywood	28
4 inch brick wall	41
6 inch concrete block wall	42
1/2 inch gypsum board on both sides of 2 × 4 studs	33
12 inch reinforced concrete wall	56
14 inch cavity wall (8 inch brick − 2 inch air − 4 inch brick)	65

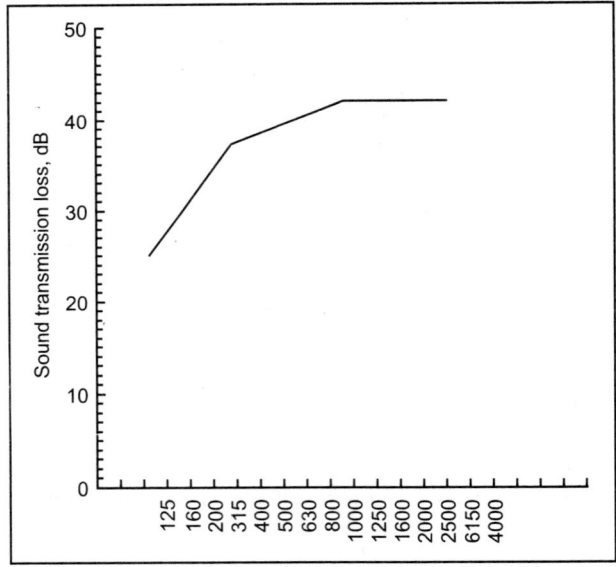

Fig. 6.6. Typical STC contour.

Acoustical Experimentation

In a series of acoustical tests performed by PPG Industries at Riverbank Laboratories, eight thicknesses of single glass were tested for noise control.

Figure 6.7 depicts the relationship of these eight glass thicknesses with regard to their sound transmission loss when subjected to noise frequencies ranging from 100 to 5000 Hz. The resulting sound transmission class ratings are shown in Table 6.2.

Generally, experimentation showed that, as the thickness of the glass increases, the sound transmission loss and resulting STC rating of the glass also increases.

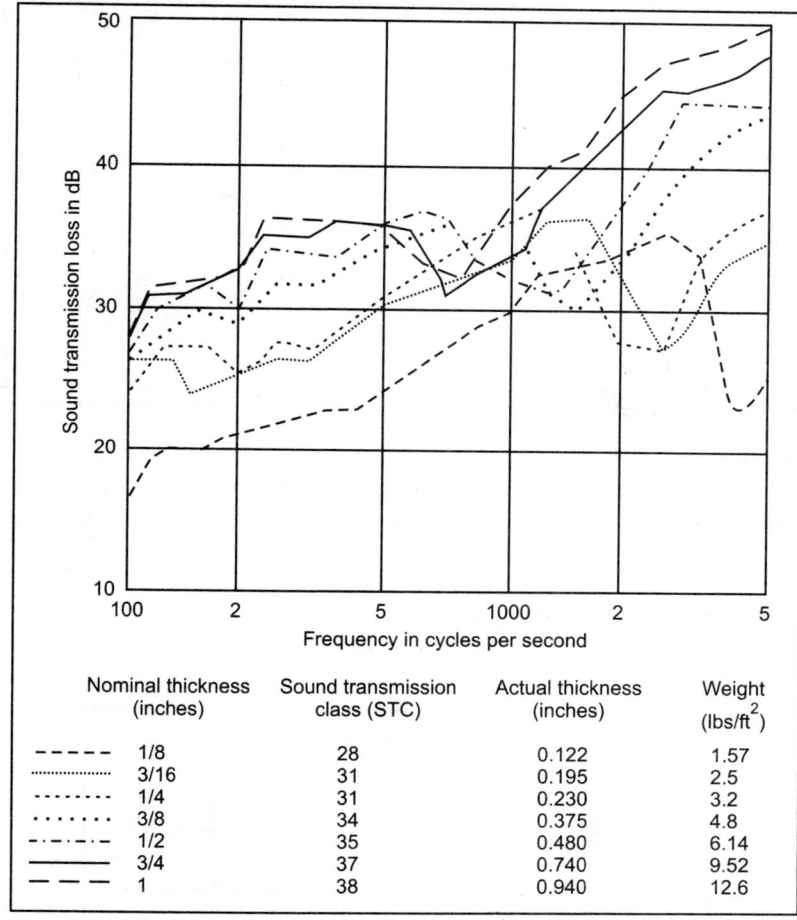

Nominal thickness (inches)	Sound transmission class (STC)	Actual thickness (inches)	Weight (lbs/ft^2)
----- 1/8	28	0.122	1.57
.... 3/16	31	0.195	2.5
- - - - - 1/4	31	0.230	3.2
· · · · · 3/8	34	0.375	4.8
-·-·-·- 1/2	35	0.480	6.14
———— 3/4	37	0.740	9.52
— — — 1	38	0.940	12.6

Fig. 6.7. Single glass products and noise control (experimental results).

Table 6.2. Sound transmission loss through single glass.

Frequency (Hz)	STL 1/8"	STL 3/16"	STL 1/4"	STL 3/8"	STL 1/2"	STL 3/4"	STL 1.0"
100	16	26	23	26	27	28	27
125	20	26	27	28	30	31	32
160	20	24	27	30	31	31	32
200	21	25	25	29	30	32	32
250	22	26	27	31	33	34	35
315	23	26	27	31	33	34	35
400	23	28	29	32	33	35	35
500	25	30	31	33	35	35	35

(Contd ...)

Frequency (Hz)	STL 1/8"	STL 3/16"	STL 1/4"	STL 3/8"	STL 1/2"	STL 3/4"	STL 1.0"
630	27	21	32	34	36	34	33
800	29	32	34	35	35	31	32
1000	30	33	35	34	32	33	37
1250	32	35	36	31	31	36	40
1600	33	35	34	30	34	39	41
2000	33	32	27	33	37	42	44
2500	35	27	27	37	40	45	46
3150	33	28	32	40	44	45	47
4000	24	32	35	42	44	46	48
5000	26	34	36	43	44	48	50

Effects of Glazing

Studies were conducted to determine whether or not the type of mounting used for a single glass window had any effect on sound transmission loss. In the investigation, 3/8-inch glass was mounted in four types of frames (Fig. 6.8):

1. Neoprene gasket.
2. Metal frame with glass bedded in putty.
3. Wood frame with a wash leather strip around glass.
4. Concrete frame and putty.

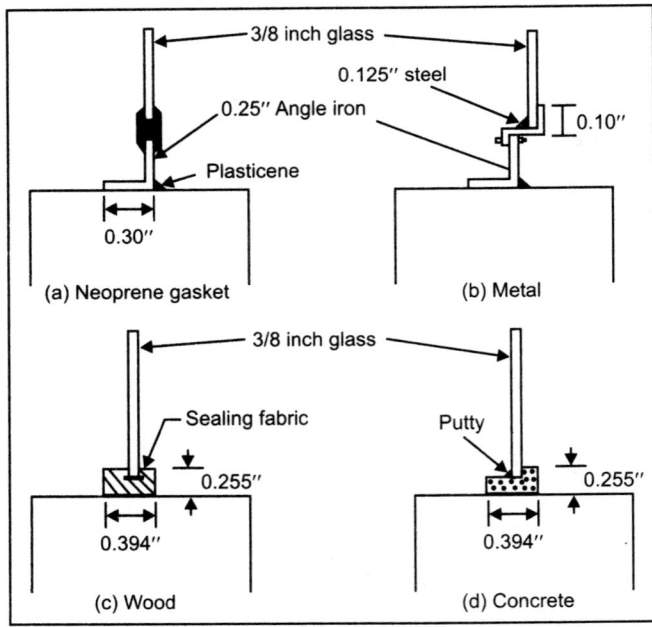

Fig. 6.8. Window frames.

Measurements were made according to ASTME 90–61T. At frequencies above 315 Hz, sound transmission loss values at a particular frequency were reproducible to within ±1.5 dB. Below 315 Hz, variations of up to 8 dB made significant conclusions in this region impossible.

Figure 6.9 depicts the results of these studies. At all frequencies above 400 Hz, the window with the neoprene gasket gave a higher sound transmission loss than any of the other windows. There was no significant difference in the performance of the windows with wood and metal frames, but both of these were superior to the window in the concrete frame.

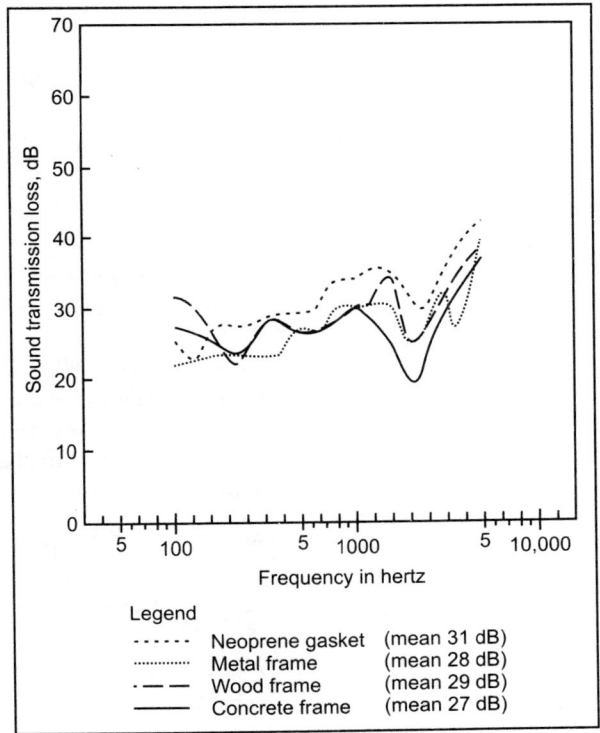

Fig. 6.9. Transmission loss of single windows in different frames.

Maximum differences between the curves occurred at the coincidence frequency (1250 Hz), where the concrete frame gave a reading of 5 to 6 dB below that of the wood or metal frame and 10 dB below that of the neoprene gasket. The neoprene gasket configuration showed a sound transmission loss of 4 or 5 dB above that shown by the other frames at frequencies as low as 500 Hz.

DOUBLE GLASS

Theory

A double glass partition is defined as a partition in which points on opposite sides of the structure do not necessarily move the same way at the same time. It is commonly constructed by placing two single glass panes in series separated by an air gap.

For the balanced partitions, the theoretical transmission loss of double partitions of infinite area at any angle of incidence is given by the following equation:

$$TL = 10\log_{10}\left[1 + \left(\frac{\omega M}{2\zeta C}\right)^2 \cos^2 \phi \left\{\cos\beta - \frac{1}{2}\left(\frac{\omega M}{2\zeta C}\right)\cos\phi\sin\beta\right\}^2\right] \qquad ... (6.5)$$

where, $\quad \beta \quad = \dfrac{\omega d \cos\phi}{C}$

$\quad d \quad$ = spacing between the two glass panels.

The transmission loss reaches a minimum value when the term within the bracket is zero, that is when:

$$\cos\beta = \frac{1}{2}\left(\frac{\omega M}{2\zeta C}\right)\cos\phi \qquad ... (6.6)$$

At low frequencies, since d/C is small, Eq. 6.6 can be written as:

$$f_\phi = \frac{1}{2\pi\cos\phi}\left(\frac{4\zeta C}{Md}\right)^{1/2} \qquad ... (6.7)$$

This is an equation for resonance, and it has a particular value when $\phi = 0°$ and this value corresponds to normal incidence sound waves. Figure 6.10 shows the typical experimental and theoretical curve for the balance unit. It also attempts to illustrate the type of transmission loss curve to be expected from a double glass partition. Different controlling regions are also shown. The mathematical model for the theoretical transmission loss of unbalanced double glass units is given by:

$$TL = 10\log_{10}\left\{\frac{1}{2}\left[(\delta_1 - \delta_2)^2 + \eta_1^2\right][1 + \cos 2\phi_2] + \left(\eta_2^2 + \delta_3^2\right) + (\eta_1\delta_3 - \delta_1\eta_2)\sin 2\phi_2 + \right.$$

$$\left. (\delta_1\delta_3 - \delta_2\delta_3 + \eta_1\eta_2)(1 + \cos 2\phi_2)/4\delta_3^2\cos^2\phi_2\right\}$$

where, $\delta_1 = \zeta_1 C_1 \sec \sigma_1$ $\delta_2 = \zeta_2 C_2 \sec \sigma_2$
      $\delta_3 = \zeta_3 C_3 \sec \sigma_3$ $\phi_2 = K_2 \beta_2$

$\eta_1 = \dfrac{m_1\omega^2 - D_1\kappa_1^4}{\omega}$ $\eta_2 = \dfrac{m_1\omega^2 - D_2\kappa_2^4}{\omega}$

$\beta_2 = L \cos \sigma$ $L = t_1/2 + d + t_2/2$
D_i = stiffness of ith wall
M_j = surface density of ith wall
K_i = radius of gyration of ith wall

Coincidence and Resonance Effects

In the case of double glass units, the effect of the coincidence frequency is the same as that for single glass units. The air space has little effect on coincidence frequency. Experiments show that the coincidence dip for double glass units occurs at the same frequency as that for the single glass units of the same glass thickness.

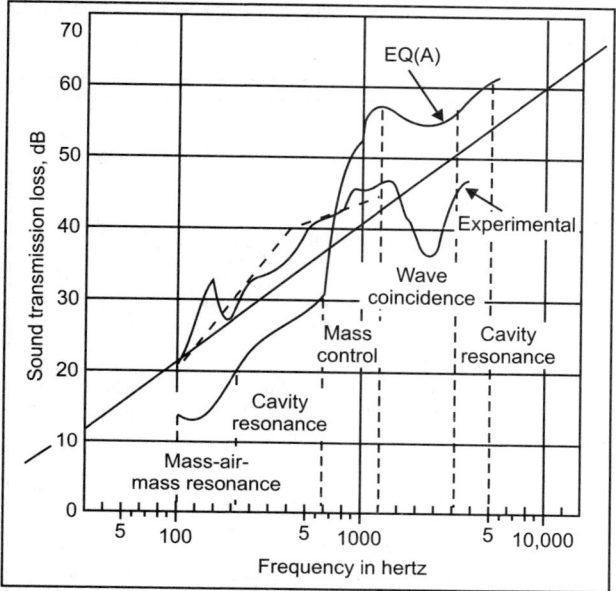

Fig. 6.10. Transmission loss characteristics for double-glazed unit ¼-inch glass, 2-inch air, ¼-inch glass).

The problems associated with single glass, namely low-order wall resonances and coincidences transmission, occur in each panel of the double glass wall in just the same way. Mathematically:

$$f_{\text{coincidence}} = \frac{c^2}{2\pi h \sin^2 \phi} \sqrt{\zeta_P / E} \left(\sqrt{12(1 - v^2)} \right)$$

Kurtze suggests that, in order to avoid a deep trough in the transmission loss curve due to the combined effects of low frequency resonance and coincidence, it is advisable to use panels with different bending wave velocities. In addition to the problems associated with single glass units (low frequency resonance and coincidence effects), other types of resonance influence the sound transmission loss of double glass units. Figure 6.11 illustrates their behaviour. The resonance does not always occur in separate regions. For double glass units, the sound transmission loss depends on several factors including frequency, weight, method of edge restraint, damping and stiffness. The most important variable of all is edge restraint. It has been proven experimentally that two different edge restraints with the same glass combination give different sound transmission class ratings. Experimental results are given in the section entitled effects of glazing.

Design of Double Glass Units

The following three parameters must be considered when designing double glass units: air space, glass thickness and damping.

Air space

Generally, increasing the air space between two glass panels results in increased sound transmission loss. However, studies show that maximum sound insulation is reached with the optimum spacing of four inches. An air space wider than four inches acts as a convective space for vibrating waves, thereby reducing performance. Experimental results are given in the Acoustical experimentation section.

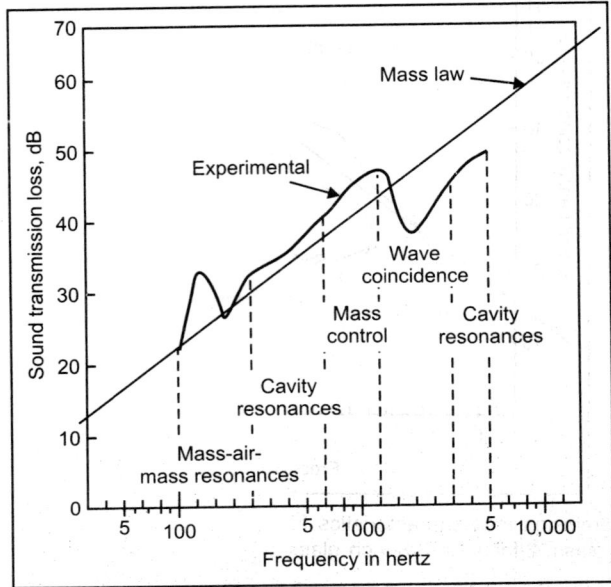

Fig. 6.11. Sound transmission loss characteristics for matched double-glazed unit (1/4-inch glass, 2-inch air, 1/4-inch glass).

Glass thickness

A combination of glass thicknesses can improve the performance of the unit by overcoming the resonance and coincidence effects. The 1/4-inch and 3/8-inch glass combination, along with restrictions on the air space, gives the best performing unit. Experimental results for 2-inch air spaced units with different glass combinations are given in the Acoustical Experimentation section.

Damping

Damping is extremely important to the performance of the unit. The experimental results show that a 3/8-inch and a 1/4-inch glass panel, separated by a 2-inch air space with proper damping in the edge attachment, has an STC rating of 45. With little or no damping, the sample has an STC rating of 42.

Effects of Glazing

Experiments were conducted using double glass units with different edge attachment in an effort to determine whether or not the type of mounting had any effect on the sound transmission loss. Results using a 1/4-inch glass, 2-inch air space, 3/8-inch glass configuration demonstrated that the sound transmission loss was greater with the resilient edge attachment than with the stiff edge attachment (Fig. 6.12). In other studies, a 1-inch Twindow® unit (3/8-inch glass, 7/16-inch air space, 3/16-inch glass) with resilient edge attachments was mounted in both wood and aluminium frames. Experimental results showed that the unit mounted in the wood frame had a superior sound transmission class rating. The experimental results are shown in Fig. 6.13.

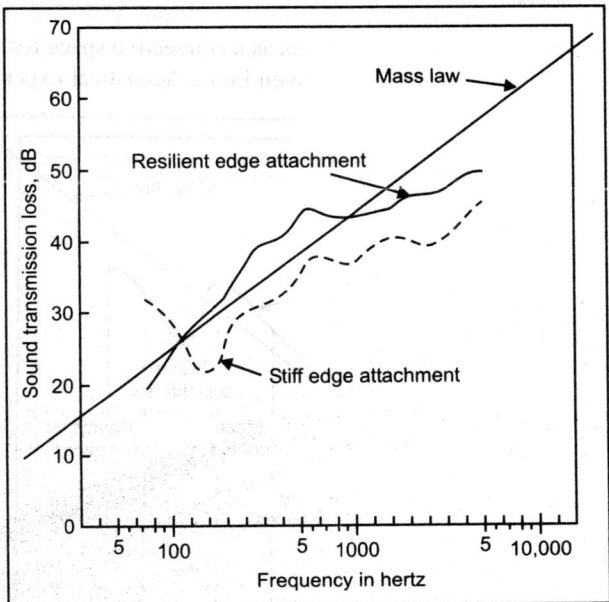

Fig. 6.12. Sound transmission loss characteristics for mismatched double-glazed units with different edge attachments (1/4-inch glass, 2-inch air, 3/8-inch glass).

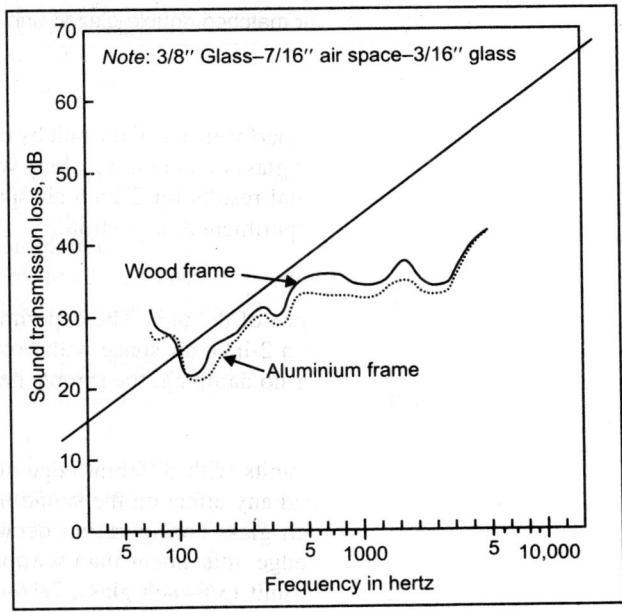

Fig. 6.13. Effects of glazing on double glass windows.

Other 1-inch Twindow units (3/8-inch glass, 7/16-inch air space, 3/16-inch glass) with stiff edge attachments were also mounted in both wood and aluminium frames. Experimental results showed little difference in STC between the wood frame and aluminium frame units. These experimental results are shown in Fig. 6.14.

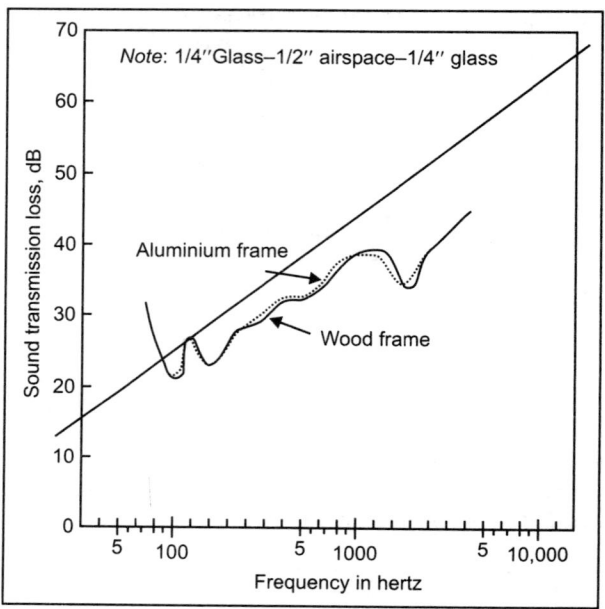

Fig. 6.14. Effects of glazing on double glass windows.

MULTIPLE GLAZING

Windows with more than two glass panes are usually restricted to special situations, such as observation windows in studios. Therefore, because of their limited application, they will be discussed in less detail than the single or double windows. Theoretical studies have shown that a multiple unit readily transmits energy below a limiting frequency dependent on the mass of each panel and the distance between the glass panels. Above this frequency, the slope of sound transmission loss as of function of frequency becomes greater as the number of glass panels increases.

Brandt measured the insulation of double and triple windows (2, 3 and 4 mm glass and air space from 32 to 160 mm). In each case, the air spaces of the triple window were equal. Although the average insulation of the triple windows was between 0 and 2 dB greater than that of the corresponding double window, the majority of triple windows had a greater sound transmission loss than the double windows at frequencies above 500 Hz, and a lower sound transmission loss than the double windows at frequencies lower than 500 Hz. If a triple window is formed by adding a panel of glass between and parallel to the panels of an existing double glass window forming two equally wide air spaces, the mean sound transmission loss of the triple window is almost equal to that of the double window. Some improvement may be made if the air spaces are of unequal width, but an undivided air space gives better insulation (greater sound transmission loss) at low frequencies.

LAMINATED GLASS

Theory

The response of a laminate to a random sound field is determined by the material and geometrical parameters of the laminate. As shown in Fig. 6.5 the response spectrum is made up of several regions. These regions, with increasing frequency, are stiffness controlled, resonance, mass controlled, coincidence and supercoincidence. There are also transition regions that can be important considerations. The location of these regions in the spectrum and the band width of each region is determined by the materials used in the laminate, and the cross-sectional and areal geometry. It is apparent that the position of these regions in the spectrum for a laminate and the ability to control the positions are of great importance in establishing how a laminate will perform as a transmission loss barrier component or system.

In the stiffness region, the performance is obviously totally undesirable. Because of this, it is standard procedure to establish the natural frequency (first important mode) below the lowest frequency where transmission loss is required. This will assure that the laminate will perform well above the stiffness region throughout the spectrum of interest.

The resonance region performance is the most complex of the regions under consideration, and the response characteristics are most important. With the exception of the transition region between mass law and coincidence, the resonance region is the only range in which damping can be effectively used to control noise. For this to be achieved, the laminated structure must be designed to extract more energy from the principle acoustic radiating modes than inherent types of damping (i.e. joint friction, air damping and acoustic radiation damping by pressure cancellation). Therefore it is important to identify the principal radiating modes for a laminate by examining the resonant frequencies in response to the airborne sound field and structure-borne sound energy. It is evident that in a random sound field the modes of response of the laminate will be many, and it is necessary to select the dominant modes (i.e. high amplitude response) and apply the damping to these modes. This, of course, is admitting that the damping of a constrained laminate configuration is frequency sensitive, which means that the damping can only be applied to a region of limited frequency. Damping of fundamental modes to reduce noise is normally a last resort. Preference should always be given to shifting the fundamental to a lower frequency, out of the spectrum of interest if possible. The mass law region, in a practical sense, is the most desirable area in which to place laminate response. In this region the transmission loss increases at the rate of 6 dB per octave of frequency. With the exception of the coincidence region, in which it is impractical to achieve any low frequency performance, the mass law region represents the optimum area in which to position the laminate response. This positioning can be accomplished by variation of the material and geometrical parameters. Between the mass law control and coincidence regions there is a resonance effect of higher order modes that again causes a sharp drop in the transmission loss properties of the laminate. If this notch falls into the spectrum of interest, then damping in this frequency range becomes important.

The coincidence region is important when the sound pressure level of a field is rich in high-frequency content. The transmission loss in this region increases at the rate of 9 dB per octave change in frequency as a nominal figure under random incidence (i.e. 12 dB per octave is theoretically obtainable under normal incidence but has not been practically realised). As mentioned previously, the coincidence region cannot be shifted to lower frequencies without introducing impractical configurations, particularly when restricted to glass skin laminates with polymeric, viscoelastic cores. It is important to note that a laminate can be effectively used to perform in a mass law-transition-coincidence region with a high degree of efficiency when the sound field is skewed toward high frequency response.

Natural frequency of a laminate

The natural frequency of a plate is dependent upon its mass, bending stiffness, boundary conditions, and size and shape effects. The relationship given by Ritz for homogeneous isotropic square plates is as follows:

$$\omega_n = \frac{\alpha}{l^2} \sqrt{B/M}$$

where, l = length of one side of the plate.

α = a constant depending upon the plate geometry and boundary conditions (often called the mode factor).

The application of the Ritz equation to laminated structures is complicated by the fact that the bending stiffness (B) of a three-ply laminate is a function of the excitation frequency (ω). Using these relationships (keeping in mind the limitations imposed in the simplified form), the application of the Ritz equation to laminate performance for most cases becomes routine.

Designing a laminate to perform as an effective transmission loss barrier in the resonance region as a function of damping is a complex task requiring careful consideration of many parameters and the judicious selection of materials. As previously stated, it is best to avoid the resonance region whenever possible, and the objective, at least at moderate frequencies, should be mass law performance. At higher frequencies, of course, this is more realistically achieved. It is also necessary to retain cognizance of the several factors that can override the performance of the most rigorously conceived laminate such as structure-borne noise of high amplitude (at low frequency), temperature displacement of peak damping, and the damping of the laminate due to its configuration versus that damping induced by structural edge restraint of the laminate (such as friction or air damping). Other complexities that enter into the picture are strain sensitivity of certain polymers, the practical experimental problem of making high frequency (as far as the audible spectrum is concerned ≈22,000 cps) measurements of loss factor and shear modulus for the core materials and, of course, all of the practical material problems such as structural bonding.

Experimental Results

The laminated wall represents an attractive method of combining all the desirable characteristics for TL. It is possible to design a laminated wall with high static stiffness, low dynamic stiffness, high damping, and a critical frequency above necessary limits. This type of wall utilises the fact that structures with stiff skins separated by an incompressible spacer have bending stiffness characteristics that vary inversely with frequency squared. If the spacer is made of a material with low shear stiffness (e.g. rubber), shear waves can be made to propagate in the core. Using these principles, it is possible to design a limp, well-dampened wall that also fulfils the necessary static requirements. A well-designed laminated wall may approach TL characteristics as predicted by the mass law.

Figure 6.15 shows some experimental results for PPG's laminated glass products. Two pieces of 1/8-inch glass were laminated with various thicknesses of interlayer. The STC for all three cases was 34, but the performance curves differed markedly.

Effects of Temperature on TL

Figure 6.16 shows the test results for 1-inch acoustical Twindow units at various temperatures. At the normal condition, with the source room at 71°F, 45 per cent RH, and the receiving room at 72°F, 89 per cent RH, the STC was 38. When the source room was at 110°F, 25 per cent RH, and the receiving room at

71°F, 91 per cent RH, the sound transmission loss curve improved 1 or 2 dB at the low-frequency end, and decreased about 1 dB at the high frequency end. However, this increase or decrease in the sound transmission loss is not sufficient to change the STC rating of the unit. When the source room was at 32°F, 54 per cent RH, and the receiving room at 71°F, 85 per cent RH, the sound transmission loss curve improved at the low and high frequencies by about 1 or 2 dB. There is also some improvement in the middle frequency range. The effect of all of this is to increase an STC rating by 1 dB.

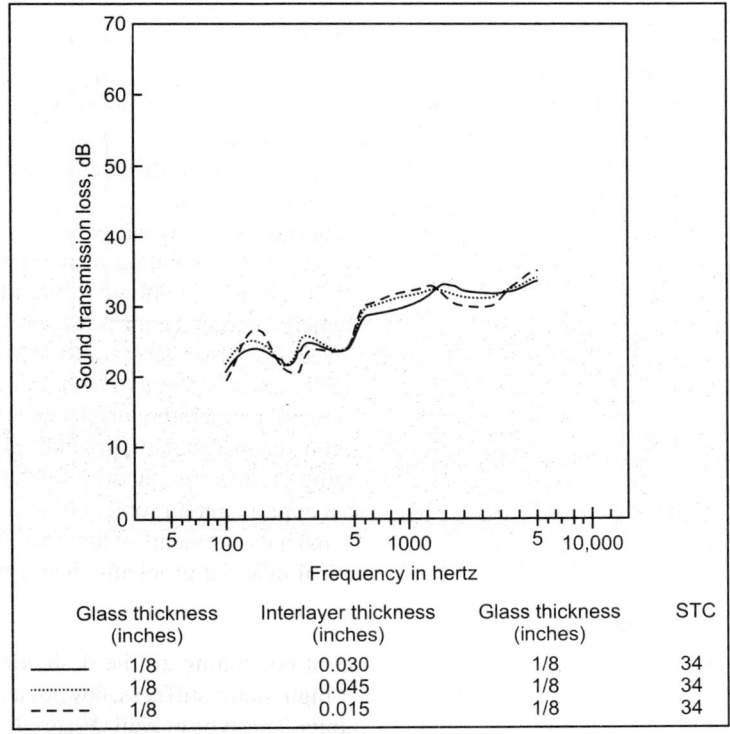

Glass thickness (inches)	Interlayer thickness (inches)	Glass thickness (inches)	STC
——— 1/8	0.030	1/8	34
·········· 1/8	0.045	1/8	34
– – – 1/8	0.015	1/8	34

Fig. 6.15. PPG's laminated glass.

It can be seen from Fig. 6.17 that temperature has a great effect on the sound transmission loss of a laminate. The example shown in Fig. 6.17 is 1/8-inch glass, 0.045-inch interlayer, 1/8-inch glass laminated when the source room was at 71°F, 45 per cent RH, and the receiving room was at 71 of, 89 per cent RH (standard conditions). The sound transmission class was 34. Notice the high and low ends on the spectrum curve. At the low end there are two resonances. At the high end there is minimal drop due to coincidence. When the source room was at 20°F, 54 per cent RH, and the receiving room at 71°F, 85 per cent RH, the sound transmission class dropped sharply to 31.

The great drop in sound transmission loss occurred at the coincidence frequency. There is also some drop at the low-frequency region. When the source room was at 110°F, 25 per cent RH, and the receiving room at 71°F, 91 per cent RH, the sound transmission loss dropped sharply at the low frequency region; at the middle- and high-frequency regions it increased considerably. The total effect of this was to increase the STC rating to 38.

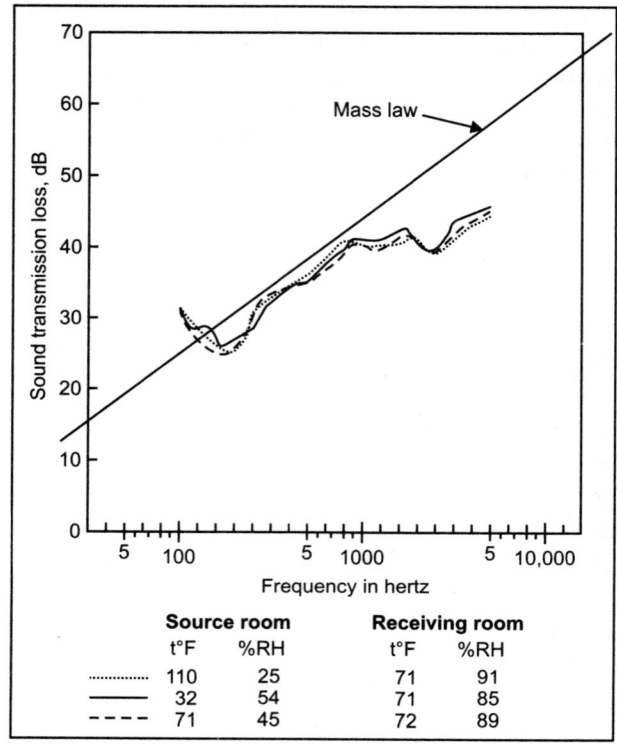

Fig. 6.16. Effect of temperature on sound transmission loss for double-glazed window.

PPG ACOUSTIC TWINDOW

PPG Industries presently produces a 1-inch Acoustic Twindow unit with an STC rating of 38 compared to the standard rating of 32 for a standard 1-inch Twindow unit. Table 6.3 provides the sound transmission class (STC) data comparing the Acoustic Twindow unit with other windows. Figure 6.18 compares the sound transmission loss of Acoustic Twindow and Standard Twindow as a function of frequency.

Table 6.3. The sound transmission class (STC) data.

Product	SIC	Approx. weight lb/sq. ft
Acoustic Twindow		
3/16 inch glass, 7/16 inch air space 3/8 inch glass	38	7.8
Standard Twindow		
1/2 inch air space	32	6.0
9/32 inch laminated glass 0.045 inch vinyl	34	3.5
1/2-inch float	34	6.0
1/4-inch float	31	3.2

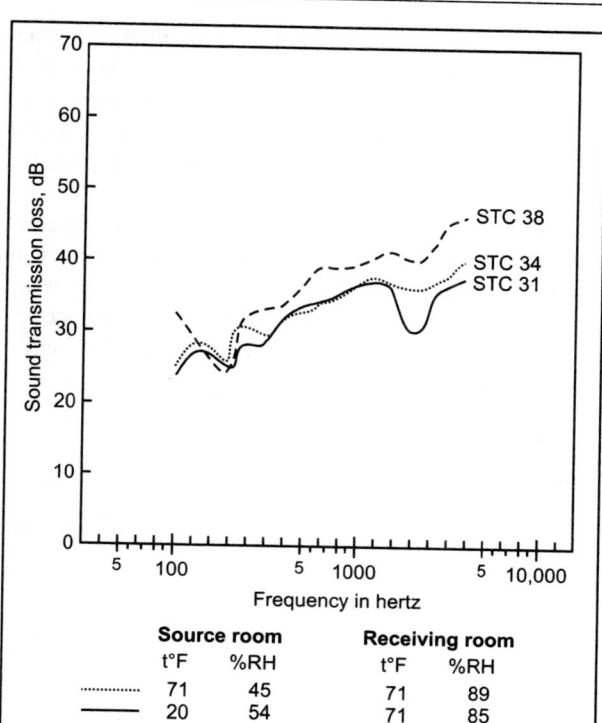

Fig. 6.17. Effect of temperature on sound transmission loss for laminated glass.

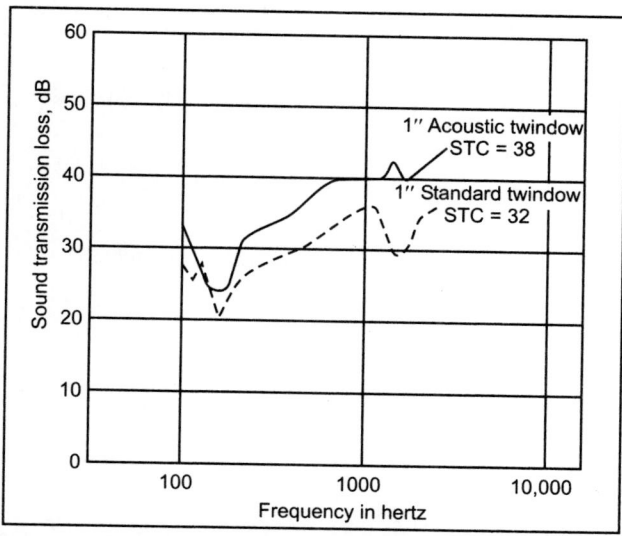

Fig. 6.18. Sound transmission loss (ASTM E90-70) Acoustic Twindow.

In Acoustic Twindow units, each panel of glass is of different thickness. The panels are separated by a 7/16-inch air space hermetically sealed with a special acoustically dampened perimeter section.

Acoustic Twindow units are recommended when transmitted noise must be reduced but visual communication is desired. They may be used outdoors or indoors, in walls and partitions, in windows and doors. The Acoustic Twindow unit offers the brilliant clarity of float glass or a choice of Solarbronze or Solargray tinted glasses.

These units are available in clear glass combinations to a maximum area of 70 square feet. Tinted glass combinations are available to a maximum area of 50 square feet (available in larger areas depending upon special processing required). Minimum size is 16 × 24 inches; thickness glass-to-glass is 1-inch, or 1/16-inch in overall thicknesses at the enclosing metal edge (subject to normal manufacturing tolerances). The average weight of the 1-inch unit (3/16-inch and 3/8-inch glass) is approximately 7.8 pounds per square foot.

Chapter 7

Ventilating System

INTRODUCTION

Air-conditioning systems vary greatly in design and purpose. For example, for heating a building, a heat-pump must be included in the design. When humidity control is desired, reheating coils following an evaporator cooling system are included. In industrial sites or hospitals requiring air sanitation, electric precipitators, wet collectors or ultrasonic agglomerators may be necessary. These designs may also require air sterilisers and activated carbon absorbers for odour removal. When only ventilation is necessary, a forced-air draft system is employed whereby a fan directs filtered air to designated locations in the building.

Any of the many components of an air-conditioning system may be a cause of unwanted noise generation or vibrations. Adequate noise control of these systems can be achieved by careful examination of the possible causes and by providing proper isolation and insulation to the primary units in the system.

NOISE SOURCES

Ventilating Systems

In general, ventilating systems consist of a motor-driven blower that directs air through a plenum into headers for distribution to various parts of a building. Often the fan is located in the plenum to isolate noise.

Air flow is regulated by the difference in pressure between the two terminals of the duct network. Designs are classified according to low, medium and high-pressure systems. The maximum allowable static pressure at any supply or return opening for each type is given in Table 7.1.

Table 7.1. Classification of ventilating systems.

Type of system	Maximum static pressure (inches of water)
Low pressure	0.5
Medium pressure	3.5
High pressure	6.0

Two principal types of noise are normally encountered in simple ventilating units—fan noise and air flow noise. Fan noise can be subdivided into two components—rotational and vortex. The rotational component is associated with the impulse transmitted to the air each time a blade passes a fixed position. Hence, it is a series of discrete tones at the fundamental blade-passing frequency and is a function of the harmonics. The vortex component of noise is attributed largely to vortices shedding from the fan blades. This occurs due to air turbulence caused by the wind stream passing over system components such as elbows or filters. Medium and high pressure systems are often plagued with turbulency noise generation.

For short ducts, noise generation is derived primarily from the fan. However, for long distribution systems, in addition to air turbulence, other sources may exist. For example, fan noise may become masked by sounds of air turbulence and radiations from entrance ports. Wall and panel vibrations originating from the operating machinery may also exist and might result from equipment failure such as fan unbalance or bearings. In addition, brush or magnetic noise may develop in the fan.

Air-Conditioning Systems

The location of possible noise trouble spots becomes complicated with more elaborate systems because there are a large number of components that could be the cause. The major units that generally require attention are compressors, condensers, cooling towers, evaporators and pumps and piping.

Compressors

Compressors must be located relatively far from the distribution system. They are often mounted on an elaborate mounting block on the top floor of a building. To provide noise control, a floating concrete floor is constructed with a suspended ceiling with acoustical panelling directly beneath it.

Condensers

The two types of heat exchangers, free and forced-convection, are noise radiators and cannot be readily isolated in a special housing, as maximum surface exposure to air is necessary for high heat-transfer efficiency.

Cooling towers

These units, generally having several hundred cubic feet of volume, are used for cooling water that has been used in heat exchangers that remove heat from the compressed refrigerant. The heat is removed by partial evaporation into the atmosphere. The process is enhanced by tube-axial or centrifugal fans, belt-driven by hermetically sealed motors.

These can be extremely noisy units with severe vibration problems. They are generally mounted on the building's roof with a parapet often placed around them to reduce airborne noise. Vibration pads must be included as part of the mounting design.

Evaporators

Noise is produced by these units because the two-phase, liquid-gas mixture exiting the capillary causes a high-pitched whistle.

Pumps and piping

Piping from pumps can undergo severe vibrations or cavitation, particularly when not properly suspended.

In order to examine more closely the main causes of noise in air-conditioning systems, it is convenient to separate them into four main headings. These are: the noise produced by motors, fans and other equipment in the plant room; compressors and other active equipment; turbulence in the system due to the flow of air and noise entrained into the system through intakes, supply grilles or duct walls and thereafter propagated through the ducting.

NOISE CRITERIA

Various international agencies and organisations are establishing detailed criteria for acceptable noise levels. Different parameters and values have been used by them. Unfortunately, many different rating procedures have evolved for estimating noise levels. The more common constructs and terms related to ventilating system noise will be discussed here.

Loudness and Loudness Level

Loudness level is the logarithm of the quantity loudness. Both are a means of rating noise based on a person's judgment. The procedure for calculating loudness rates and relative loudness is standardised both in the US and internationally.

Speech Interference Level

A calculation procedure that rates steady noise according to its interference with communication is referred to as the speech interference level.

NC and PNC Ratings

Noise criteria (NC) curves are used extensively in rating noise in buildings and in the rating of air-conditioning noises in particular. Preferred noise criteria (PNC) curves are essentially a modified NC plot.

Figure 7.1 shows the NC curves. NC values apply to steady noises and specify the maximum noise levels permitted in each octave band for a specified NC curve. NC curves have been used cautiously in the past because they tend to underestimate actual sound pressure levels.

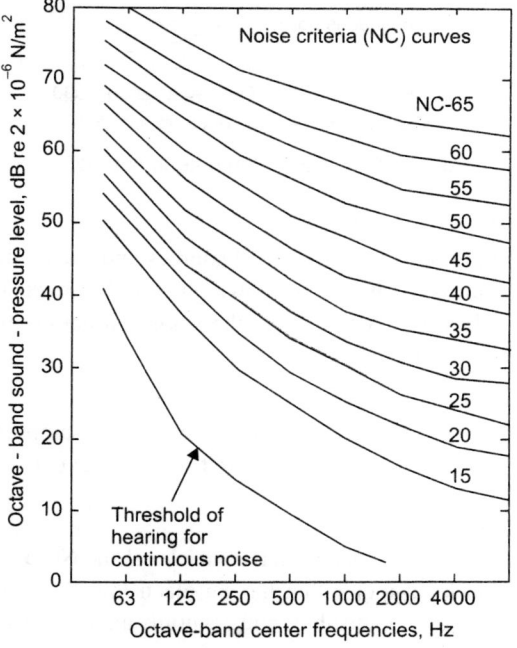

Fig. 7.1. Updated noise criteria curves.

This is particularly true for ventilating systems in which fan or duct turbulence is a major problem. To design for an acceptable noise quality, it is common to lower levels 5 dB in both the very low and high-frequency bands. PNC curves (Fig. 7.2) have values approximately 1 dB below the NC values in four-octave bands, to compensate for the NC inaccuracy (at 125, 250, 500 and 1000 Hz for the same curve ratings). In the highest three bands and the 63 Hz band, values are 4 to 5 dB lower.

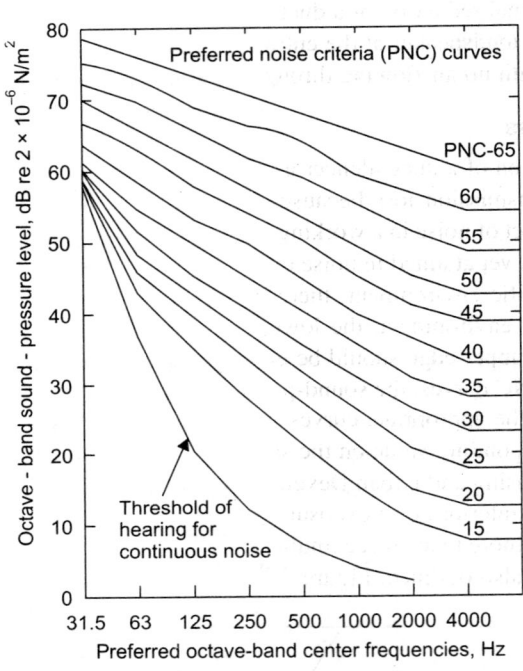

Fig. 7.2. Preferred noise criteria curves.

Noise Pollution Level (NPL)

NPL is a procedure for evaluating noises from several sources and fluctuating or intermittent sounds. The noise pollution level (L_{np}) utilises data (either A-weighted noise levels or loudness level) that are recorded over a sufficient length of time to establish the statistical nature of the total noise exposure. The noise pollution level is expressed as:

$$L_{np} = L_{eq} + 2.5 \, \sigma \, \text{dB (NP)} \qquad \qquad ... (7.1)$$

where, L_{eq} = Mean-square, A-weighted sound level over some sample time, dBA.

σ = Standard deviation for the A-weighted sound level data.

A-Weighted Sound Level (L_A)

This sound level is a means of correlating speech-interference level and NC or PNC levels. This parameter is directly measured by a sound-level meter that has an electronically modified frequency response referred to as A-weighting. The units are in decibels and generally an A is denoted to signify the weighting scale (i.e. dBA).

Insertion Loss

This term is used to denote the noise-level reduction as measured at a specified point in a channel after a duct silencer is positioned ahead of the measuring station.

Transmission Loss (L_{TL})

Referring to the acoustical signal reduction of a duct silencer, transmission loss is measured through a standard test by positioning a loudspeaker at the entrance of the duct network and a microphone at an exit station. Tests are made with no air flowing through the system.

Dynamic Transmission Loss

Referring to the signal reduction of a duct silencer as air passes through the channel, this parameter is generally much less than transmission loss because of air turbulence. Different methods have been adopted to account for the effect of noise in a working environment, with recommended exposure levels varying. Standardisation is not yet attained in noise evaluation. Criteria depend not only on the method of evaluation but on the specific environment; therefore, a range of ratings is usually indicated. For areas requiring a high-quality environment, the lower ranges are recommended. Where economy or physical limitations exist, the upper edge should be employed.

When employing PNC or NC curves, the sound-pressure level in all octave frequency bands should not exceed levels denoted by the appropriate curves. In practice, the noise level in one octave band is permitted to exceed the corresponding value on the specified criterion curve by no more than 2 dB.

The US Department of Housing and Urban Development has established fairly extensive guidelines for noise exposure levels. The interior noise exposure in new and rehabilitated residential construction should not exceed 45 dBA for more than an accumulation of 8 hours in any 24 hour day. Guidelines for speech interference levels are also outlined. Figure 7.3 illustrates these recommendations.

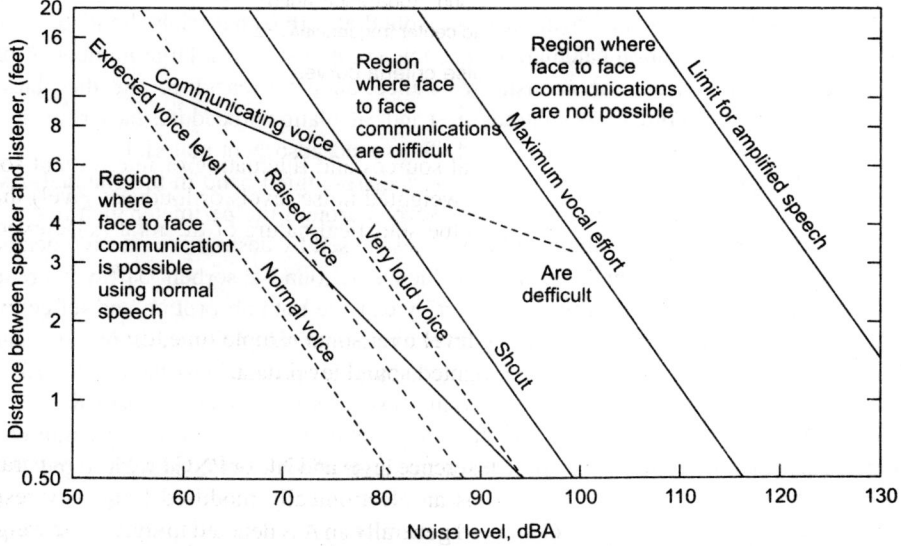

Fig. 7.3. Graphic guide to allowable noise exposure levels.

MEASUREMENT TECHNIQUES

The sound-level meter consisting of a microphone, amplifier, calibrated attenuator, a series of frequency-response shaping networks and an indicating meter is most often used to measure noise levels. Weighing scales in American instruments are usually denoted A, B, C and Flat.

The C network is used to limit the low and high-frequency response of the device so that the instrument will not respond easily to signals outside the audible frequency range. A-weighted readings are used to estimate the probability of hearing damage in industry. These values have also been correlated with the annoyance caused by traffic and aircraft noise. The B-weighting network is rarely used. The flat response of the meter is employed as calibrated input to a frequency-band filter set or tape recorder. The sound-level meter weighting networks were chosen by the American Standards Institute as the reference curves for sound loudness contours for the three frequency responses available in the instrument. In general, these devices are not well suited for measuring intermittent noises that vary rapidly with time. In such cases, high-speed level recorders (graphic level recorders with high-speed pens) are employed. Such devices can record transient sounds whose sound-pressure level rises 25 dB without 25 msec. Analysers are devices that provide information on the distribution of the noise signal energy as a function of its frequency. For continuous noises, the meter's signal is first recorded in the field on a magnetic tape. The noise is then reproduced and analysed in the laboratory. Analysis with octave filters provides the necessary information for acceptable levels.

A sound generator is used for measuring sound-transmission loss, absorption and reverberation time. These parameters are evaluated in terms of discrete frequency bands or sinusoidal frequencies. Oscillators or random-noise generators are the most commonly used signal sources.

NOISE REDUCTION IN TRANSMISSION

Noise transmission is greatest in unlined metal ducts. The amount of noise-reduction occurring in such designs is almost negligible (on the order of 0.1 dB of sound reduction per foot of duct even at frequencies greater than 1000 Hz). Whenever economically feasible, sound-absorbent materials should be introduced in a duct as lining. Sound attenuation calculations become complicated as a large number of physical data become necessary. Information on acoustic resistance, acoustic reactance and the phase angle between the two are required. Empirical correlations exist and calculation procedures have been developed for determining noise-level reductions in ducts lined with sound-absorbent materials.

It is often good design to install a traverse baffle or a sharp-angled bend in the channel. As sound travels down a lined duct, it becomes more highly absorbed around the perimeter than in the centre. This results in a sound-pressure loss at the duct boundaries caused by destructive interference between the boundary and direct-reflected sound at grazing incidence. A sound-absorbent baffle can reduce the sound energy concentration near the centre of the duct by causing both absorption and reflection.

The use of package attenuators has proven successful in air-conditioning noise reduction. Attenuators are short duct inserts whose cross section is comprised of zigzag paths. These paths achieve high acoustic absorption without causing large frictional resistance to airflow. Many designs are available commercially.

In designs requiring a large number of ducts that are supplied by one main fan, plenum chambers should be employed. A plenum chamber is a muffler, the interior of which is lined with sound-absorbent material.

Figure 7.4 illustrates a single plenum chamber (note that inlet and exist ports are never located directly opposite each other).

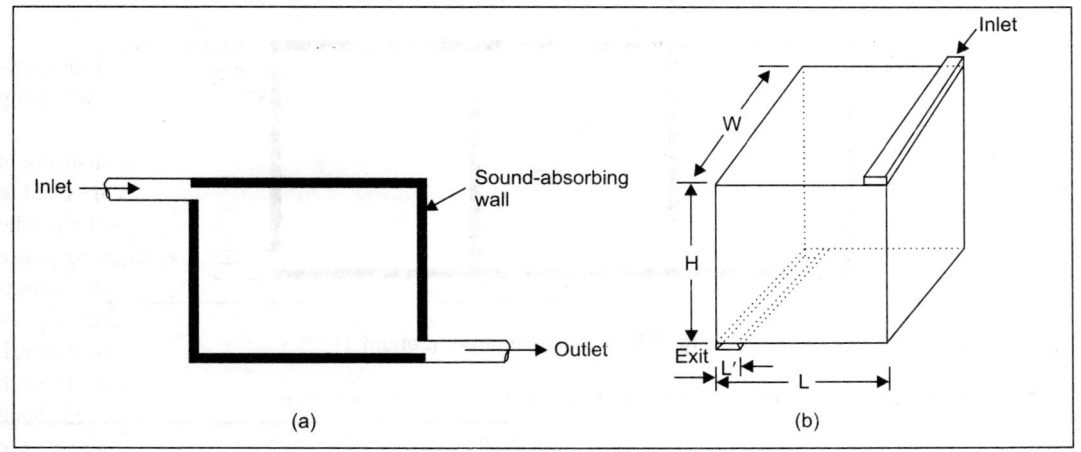

Fig. 7.4. (a) Single plenum chamber design. (b) Three-dimensional illustration of a single plenum.

Figure 7.4(b) illustrates the geometry of a single-plenum. The transmission loss associated with a plenum can be approximated by:

$$L_{TL} = -10 \log_{10} \left[A \left(\frac{\cos \theta}{2\pi d^2} + \frac{1-\alpha}{a} \right) \right]^2 \qquad \dots (7.2)$$

where, A = Cross-sectional area of exit port (ft²).

α = Random-incidence absorption coefficient of the plenum lining (dimensionless).

a = Total lined area in chamber times α (ft²).

d = Distance between inlet and exit port (ft).

$\cos \theta$ = H/d (where H is the height of chamber).

L_{TL} = Transmission loss (dB).

Equation 7.2 can be used at high frequencies and small values of L'/L (Fig. 7.4). At low frequencies, calculations underestimate actual transmission losses by 5 to 10 dB.

For greater sound reduction, multiple plenum chambers can be employed (Fig. 7.5). By knowing the transmission loss for a single chamber (L_{TL_s}) and for a double chamber (L_{TL_d}), the losses for a unit consisting of n chambers can be estimated from:

$$L_{TL_n} = (n-1) L_{TL_d} - (n-2) L_{TL_s} \qquad \dots (7.3)$$

where, L_{TL_n} = Transmission loss for a plenum with n number of chambers (dB).

n = Number of chambers.

Careful consideration should also be given to the individual components of the system. Proper selection of fans, unit coolers and mountings should be done in the earliest stages of design. Table 7.2 lists some of the more common approaches to noise control of various units.

Engineering judgement must be used in following established guidelines and in selecting the proper NC curve for a particular design. Factors such as worker's attitudes towards noise levels and economics must be carefully weighed.

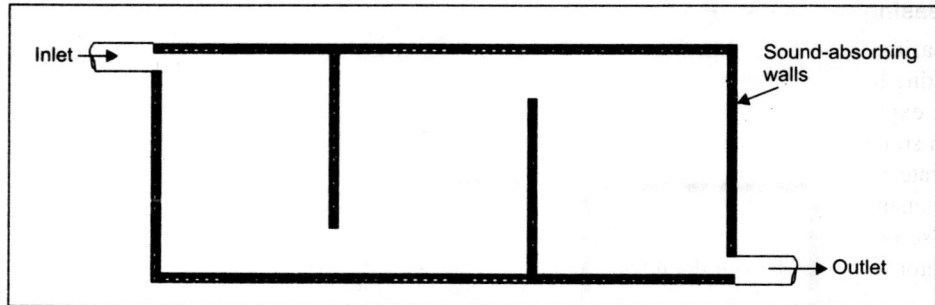

Fig. 7.5. Multiple-chamber plenum.

Table 7.2. Methods of noise reduction for major ventilation system components.

Unit	Noise control technique
Compressors	Isolation from distribution network; should be supported on resiliently supported inertia blocks
Condensers	Damp surfaces by applying a mastic compound
Cooling towers	Isolation from distribution network; towers should be mounted on heavy-duty ribbed neoprene pads to reduce vibrations
Evaporators	Evaporation pipe between the capillary exit and the evaporator entrance should be enclosed with a heavy flexible tubing
Fans	Isolation; generally located in a plenum
Piping	Pipe connections should consist of flexible couplings and pipes supported by resilient hangers
Pumps	Isolation from distribution network and mounting on resilient pads

MODIFIYING NOISE TRANSMISSION PATH

Apart from control of noise at source, modification of the noise transmission path is an effective means of controlling noise. Other control methods discussed earlier include elimination, substitution and isolation.

Modification of the noise transmission path can be achieved by increasing the distance between noise sources and workstations, enclosing noisy machines, use of partitions, sound absorbing materials and control rooms. Additional noise control measures include maintenance, quieter work practices, administrative controls, use of personal protective equipment and health surveillance.

Modifying the noise transmission path is seen as the second line of defence and includes the following methods:

1. Increasing the distance between source and receiver.
2. Enclosing noisy machines by full enclosures, partial enclosures and barriers and shields.
3. Separating noisy and quiet areas by partitions.
4. Using sound absorbing material on surfaces.
5. Use of control rooms.

Increasing the Distance

Increasing the distance between noise sources and workstations will reduce noise at the workstations. Locating high noise sources together and as far away as possible from workstations helps to reduce the noise exposure of people at the workstations.

In some instances it is possible to separate high noise sources from quiet ones by locating them in separate rooms or by installing partitions between them. Low noise tasks like packaging, cleaning, maintenance and repair work should be carried out in separate low-noise areas.

Use of remote-control systems should be a preferred option where possible because it allows the operator to be away from the noise source.

Enclosing Noisy Machines

Full enclosure of noisy machines

Total enclosures are normally used when a high noise reduction is required. It is possible to achieve a reduction of 10–25 dB for single-shell enclosures with absorbent lining and 20–50 dB for double-shell enclosures with absorbent lining.

To design an enclosure that will perform well and reduce noise levels significantly the following points should be observed:

1. For the outer shell choose stiff and heavy materials that reflect sound waves back. Common materials used include sheet metal, plasterboard, masonry, timber, glass and loaded vinyl.
2. The inner shell needs to be lined with a sound absorbing material to reduce sound build-up inside the enclosure. The preferred materials are mineral wool, glass wool, foam or polyurethane. Sound absorbing materials often need to have a protective facing to prevent any damage or dirt build up. The protective facing can be perforated sheet metal, perforated foil or vinyl.
3. Any gaps or openings should be properly sealed off or blocked off by flexible flaps if openings are necessary. A 10 per cent opening will limit the reduction to 10 dB.
4. Any windows can be double-glazed for better reduction.
5. Doors, windows, hatches or removable panels should be installed with full seals on every edge.
6. Cooling air intakes and exhausts should be fitted with noise attenuators.
7. No enclosure parts should touch any vibrating parts of the enclosed machine. In the case of pipework which goes through the enclosure, all gaps need to be sealed with a soft sealant.

Partial enclosure of noisy machines

This kind of enclosure can be effective in reducing noise to nearby workstations, but the noise will still escape through any openings and add to the background noise levels in the workplace. Basically, the partial enclosure should be constructed in a similar way to the full enclosure with a stiff outer shell lined inside with a sound absorbing material. The same precautions should be applied to avoid any leaks.

Examples of partial exposures are open sided boxes, hoods or combinations of screens. A partial enclosure can be used to create a quiet area or an acoustic cover.

Use of barriers and shields to enclose noisy machines

Barriers and shields are used in situations where only a small reduction in noise levels of about 5 dB is needed. Barriers and shields work in the same way, redirecting the sound away from a receiver. Shields are usually small transparent barriers such as 5 mm thick perspex placed between a worker and a noise source. They can also be mounted on a machine and used as a safety shield. An acoustic barrier is a larger piece of solid material, usually free-standing on the floor. Barriers and shields are most effective when both workers and a source are close to them, and the ceiling and other nearby reflecting surfaces are lined with sound absorbing material. Common materials used to construct barriers are sheet metal, plywood, clear plastic or safety glass. Best results are achieved when at least some parts are covered with absorbing material to minimise sound reflection.

Separating Noisy and Quiet Areas by Partitions

Transmission of sound to a nearby room or from outside can be minimised by increasing the sound insulation of walls, ceiling, doors and windows. The amount of sound that will pass through depends on the characteristics of the material used for partitions and most importantly on the mass per unit area. The higher the mass the higher is the sound reduction. The frequency of the passing sound is also important, as the higher frequencies are easier to stop from penetrating through. In general, an increase in airborne sound insulation of about 5 dB is obtained when the mass per unit area is doubled. Example of sound insulation properties of various materials are given in Table 7.3.

Table 7.3. Example of sound insulation properties of various materials.

Material	Thickness (mm)	Surface mass (kg/m^2)	Achievable noise reduction [dB(A)] 125 Hz–2000 Hz
Brick wall	113	220	42–66
Compressed strawboard	50	17	20–44
Chipboard	18	12	17–41
Plasterboard	13	12	17–41
Plywood	10	7	12–36

Another way of improving the insulation of a single shell wall is to construct a second wall separated from the first. The double wall partition can achieve about 10 to 20 dB higher sound insulation than the single one. The sound insulation can be improved by increasing the distance between the shells in the double walls, usage of sound absorbing material in the cavity and avoidance of rigid connection between the shells. In the case of solid walls, improvements of about 5–10 dB can be achieved by adding a sound insulating front panel.

The sound insulation is greatly reduced by any openings in the wall. Doors or windows provide less noise insulation than walls.

Use of Sound Absorbing Materials on Surfaces

In general terms the noise received by the operator consists of noise coming directly from noise sources and noise reflected from walls, floor, ceiling and other equipment. After treating all the direct noise sources, all the reflecting surfaces should be treated to reduce the noise received by the operator. This is achieved by applying sound absorbing material, which helps to reduce the reflected sound. A convenient method of employing sound absorption is the installation of acoustical baffles. Where large surface areas are to be treated, a spray-on treatment may be more economical.

Attention should also be given to machine location, as placement next to reflective surfaces increases the noise levels.

Use of Control Rooms

Current developments in some industries are directed towards automated machines and processes, thus allowing operators to operate from a control or monitoring room and so minimising the amount of time spent near noisy machinery to starting, repairing and maintenance work. A properly designed control room can achieve a reduction of up to 30 dB. The room has to be properly ventilated, with ventilation openings fitted with attenuators, acoustic louvres or a silenced air conditioner unit. Care should be exercised to ensure cables, pipes and fittings are not rigidly connected and doors and windows are properly sealed. Sometimes barriers or partial enclosures offer sufficient noise reduction and can be used at a workstation. However, they generally offer a reduction in noise levels of less than 10 dB.

Additional Noise Control Measures

Maintenance

Maintenance of plant and equipment plays a very important role in overall noise control and machinery safety as well as increasing machinery life. In general machines get noisier with use because of:

1. Worn or chipped gear teeth—worn or chipped teeth will not mesh properly. The shiny wear marks are often visible on the teeth.
2. Worn bearings—bearing wear will show up as vibration and noise, squealing from slack drive belts, 'piston slap' in motors, air leaks, etc.
3. Poor lubrication—this appears as squeaking noises due to friction or excess impact noise in dry and worn gears or bearings.
4. Imbalance in rotating parts—just like car wheels, any imbalance in a fan impeller or motor shaft will show up as excess vibration.
5. Obstruction in airways—a build-up of dirt or a bent/damaged piece of metal in an airway or near a moving part, e.g. a bent fan guard can cause whistling or other 'air' type noises.
6. Blunt blades or cutting faces—blunt or chipped saw teeth, drill bits, router bits, etc. usually make the job noisier as well as slower.
7. Damaged silencers—silencers for air-driven machines or mufflers for engines may become clogged with dirt, rusted or damaged, so losing their ability to absorb noise.
8. Removal of a noise-reducing attachment—mufflers, silencers, covers, guards, vibration isolators, etc. which reduce noise should never be removed except during maintenance, and then must be replaced.

So setting a proper maintenance program with routine checks and services will help to keep noise levels at their minimum.

Effect of maintenance on noise levels

WorkSafe Western Australia studied the Effect of Maintenance on Noise Levels for a Variety of Tools and Machines. Proper maintenance plays a very important part in controlling noise at workplaces, as worn and unbalanced parts of machines cause both noise and vibration. That is why it is vital to have a maintenance schedule that covers the following:

1. Lubricating/oiling all moving parts.
2. Balancing and aligning parts, especially rotating ones.
3. Replacing worn parts.
4. Checking the machine's feed rates.

WorkSafe Western Australia carried out a study to see how this works in practice. A total of 11 power tools (including drills, angle grinders, sanders, a circular saw and a planer) were tested for noise 'before' and 'after' routine servicing at a commercial power tool service centre. The average noise level 0.5 m away was 93.9 dB(A) 'before' and 92.0 dB(A) 'after' whilst running free. Noise reductions ranged up to 7 dB(A). The average reduction was 1.6 dB(A). As expected, this average reduction is small in terms of decibels. However, it means that sound energy output before maintenance was 45 per cent higher, showing that wear and tear was clearly on the rise.

In other studies on specific machines significant noise reductions were achieved through careful maintenance work. The noise level of a common type of reciprocating compressor was reduced by about 8 dB(A), by adjusting the valve seating to improve the seal and adding 'Molyslip' additive to the lubricating oil to reduce roughness in the piston stroke.

The noise level of a pneumatic knife used in abattoirs was reduced by about 8 dB(A), by improving the balance of the rotor vanes and replacing bearings and a worn collar that allowed parts to rattle. The noise level of an electric motor and belt drive for an aluminium docking saw was reduced by 15 dB(A) when free running, by replacing squealing belts and worn motor bearings and drive pulley bearings.

Quieter work practices

In many workplaces material handling can be a contributing factor to the overall high background noise levels. It is relatively easy to control by training operators in proper handling techniques.

Noise created by handling metal can be reduced by minimising metal-to-metal or metal-to-hardsurface contact. Some of the techniques available are:

1. Use of bending machines instead of hammers.
2. Lowering materials carefully instead of dropping.
3. Improving crane configuration, storage arrangements and operator skills.
4. Reducing drop heights for materials which must drop.
5. Lining of product bins or scrap bins with wear-resistant rubber.
6. Lining of steel trestles or benches with wear-resistant rubber (Note: alternative earthing arrangements may be needed).
7. Lining underside of steel benches where a hard surface is required.
8. Use of wear-resistant rubber floor-mats or wall brackets in storage areas.
9. Placement of workpiece on a durable rubber mat instead of hard bench or floor.

Administrative Controls

Another way of reducing an operator's noise exposure is by reducing the duration of exposure. This can be arranged in a number of ways:

1. Rotate operators so each one spends less time being exposed to noise.
2. Schedule noisy activities to be all done at the same time, preferably outside of normal hours or on shifts with fewer workers present.
3. Notify other workers when noisy work is scheduled so they can organise their work around it, limiting their exposure.
4. Provide quiet refuge rooms for paperwork tasks.
5. Provide quiet lunch and rest areas with low background noise levels where workers can spend their breaks away from noise.

6. Train operators in effects of exposure to noise on health, methods of prevention, proper use of personal hearing protectors, maintenance, etc.
7. Display warning signs in noisy areas to limit access to authorised personnel.

Procedures implemented to reduce exposure times

On their construction sites Macmahon Contractors delineates an 85 dB(A) contour around any noisy equipment and reschedules work so it is done outside this boundary. This method was first introduced when a contractor brought on site a drill rig that produced noise levels in excess of 90 dB(A). A noise assessment was carried out and Macmahon's employees were removed from the affected area. Work was rescheduled and was either done outside the 85 dB(A) boundary or at times when the drill rig was not operating.

At Production Machinery a spray booth cleaning system was recently changed to avoid exposure to high noise levels. Before, the cleaning mechanism was operated non-stop throughout the day to avoid any clogging of filters. After installing new valves it was possible to clean the spray booth only twice a day for 15 minutes at a time when the operator did not have to be close by. The background noise levels for the whole workshop were considerably reduced for the majority of the day.

Personal protective equipment

Personal hearing protectors offer the last possibility for prevention of noise induced hearing loss. There are three types of hearing protectors available.

Health surveillance

Medical precautions will not replace technical measures but can identify the problem early enough to prevent it from developing further. It is advisable to introduce an initial hearing test before starting at a noisy workplace, which would identify any previous noise induced hearing loss. Follow-up hearing tests during employment will identify if any of the existing problems are exacerbated or if any new ones have developed.

Chapter 8

Double Wall Panels

INTRODUCTION

Noise and vibrations are an important problem in the modern society. Our society is becoming more and more crowded and we use all kinds of noise and vibrations producing appliances and means of transportation. Moreover, there is a trend towards lightweight design which leads often to increased noise and vibration problems.

The present chapter focuses on the feasibility of noise reduction with double wall panels. The emphasis is put upon the application of the dissipative properties of a thin air layer between two flexible plates. Vibration energy is converted into heat by the viscosity and thermal conductivity of the air between the plates. The goal of this chapter is to develop and to validate efficient models including viscothermal wave propagation for double wall panels which make it possible to investigate the noise reduction for both airborne and structure borne noise at low frequencies.

Because of the strong interaction between the vibrations of the plates and the air layer it is essential to apply fully coupled acousto-elastic models. Apart from inertia and compressibility, the effects of viscosity and thermal conductivity are taken into account. For this purpose the so-called low reduced frequency model is very suitable. As a first step, relatively simple two-dimensional models are derived which are solved analytically. The results demonstrate that especially structure borne noise can be reduced effectively with the dissipative properties of the air layer. The transmission loss for airborne noise is hardly affected by the viscothermal effects. These findings are consistent with results found in the literature.

Acousto-elastic interaction is investigated in depth by means of a numerical, analytical and experimental study. The emphasis is put upon the behaviour of coupled acousto-elastic modes. In the case of uncoupled domains one can speak of structural and acoustic modes. In the coupled case all modes are acousto-elastic. It depends on the energies in the structural or acoustic domain whether the modes can be denoted as structural dominated or acoustic dominated. The corresponding eigenfrequencies are followed as a function of the thickness of a rectangular plate covering a closed acoustic cavity. In some cases a continuous cross-over takes place between structural and acoustic dominated modes. This happens when the uncoupled frequencies are not too far apart and when the uncoupled mode shapes are able to exchange energy. This phenomenon is called spatial matching. Other important characteristics of an acousto-elastic system are the added mass and stiffness effects of the acoustic domain on the dynamic behaviour of the plate.

For realistic problems it is in most cases impossible to obtain analytical solutions for the vibrational behaviour and sound radiation characteristics of double wall panels. So one has to resort to numerical techniques. Therefore a numerical model has been developed. For reasons of efficiency the model is uncoupled in an interior and an exterior domain. The vibrational behaviour of the double wall panel is analysed with a finite element model which includes acousto-elastic interaction. The computation speed is highly increased by the introduction of a special modal superposition method. With the vibration patterns of the plates after solution of the internal problem, the radiated acoustic power is calculated (external problem). The radiated sound power is calculated efficiently by the application of a reduction method based on so-called radiation modes. The numerical tools are validated by means of analytical solutions and also with experiments.

A special experimental setup is designed and built to verify the prediction methods for the vibrational and acoustic behaviour. Moreover, the setup is used to demonstrate the damping capabilities of a narrow air layer in a double wall panel. Numerical and experimental results agree fairly well. The results of a parameter study show that for the optimum use of the viscothermal damping the air layer has to be as thin as possible, that both plates have to be non-identical and that the thinnest plate has to be excited. In these cases maximum pumping of air between the plates is achieved. The presented numerical tools and results can be profitable for future design studies. The application of obstructions in the air layer offers an extra possibility to create damping. This is investigated by means of an experimental study. Besides viscothermal damping, the working principle is based on vortex shedding from the edges of the barriers. The latter damping mechanism increases linearly with the vibration amplitude and depends strongly on the shape and configuration of the obstructions.

SOUND AND VIBRATIONS

Sound and vibrations are well known phenomena. Very often we are confronted with sound and vibrations in a positive way, think of speech, music or perhaps the vibrator function of mobile phones. However, in a society that is becoming more and more crowded these two phenomena are increasingly placed in a negative context. Furthermore, there is a trend towards lightweight design which leads to increased noise and vibration problems. Unwanted vibrations are often experienced in cars or aeroplanes. Noise, the term for unwanted sound, is often the result of traffic and of machinery or domestic appliances, like vacuum cleaners and washing machines. There are many ways to tackle noise and vibration problems. In the study presented in this thesis the reduction of noise and vibrations is considered by applying double wall panels consisting of two impervious elastic plates separated by a thin air layer. The emphasis is on the development of analysis tools based upon acousto-elastic models including viscothermal wave propagation, and on the physical understanding of the panels.

ACOUSTO-ELASTIC INTERACTION

Vibrations and sound are closely related. Both phenomena are described as time and spatially dependent disturbances in elastic media. In the mechanical engineering context, steel and aluminium are common materials for the occurence of structural vibrations. Sound which interact which such vibrations usually occurs in air or water. In this section only air is considered. The disturbances propagate through the media as waves, which are characterised by the wave propagation speed, frequency, wavelength (the relation between wave propagation speed c_0, wavelength λ and frequency f is $c_0 = \lambda f$) and attenuation. If a structure is exposed to air, the disturbances can be transferred from one medium to the other and vice versa. In this way structural vibrations and sound are coupled.

To predict and analyse the vibrations of structures, dynamic analyses are performed. These structures are almost always in contact with the surrounding air. The mass of most structures is high enough so that the influence of the air on the dynamic behaviour can be ignored. However, for lightweight structures, as the ones applied in the aerospace industry, and for structures in which a volume of air is trapped, as in a folded stack of solar panels or double wall panels, this assumption is no longer valid. Then, the surrounding or enclosed air volume has to be taken into account.

This means that an acousto-elastic analysis must be performed in which the mutual interaction between the vibration of the elastic structure and the motion of the air is accounted for. In recent years numerical methods for the analysis of acoustics and structures have become available, such as the finite element method (FEM) and boundary element method (BEM). These widely used methods make a coupled analysis relatively easy. In the analysis of the behaviour of double wall structures, discussed here, acousto-elastic interaction plays an important role. Special attention is paid to viscothermal wave propagation in air.

VISCOTHERMAL WAVE PROPAGATION

For wave propagation in air usually only compressibility and inertia effects are taken into account. But other characteristic properties of air are its viscosity and its thermal conductivity. Because the viscosity of air is very low, air is usually not experienced as a viscous medium. An exception is near the surface of, for example, a structure, where the viscous effects become dominant. This region is called the viscous boundary layer.

Let us consider for example an air flow along a wall (Fig. 8.1). Due to friction close to the wall, the velocity will become zero at the wall, satisfying the no-slip condition. In a small region near the surface the velocity drops from the mean value in the flow to zero at the wall. Because acoustic perturbations are in fact pressure perturbations coupled with velocity perturbations, the unsteady boundary layer plays a role in acoustics.

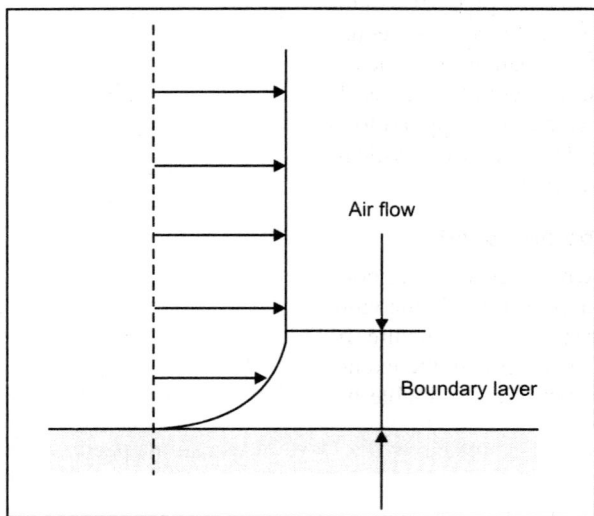

Fig. 8.1. Boundary layer close to a fixed surface.

Usually, in acoustics the boundary layer is so thin in comparison with the other dimensions that its effect can be neglected. When, however, the acoustic domain is so small that its dimensions are of the same order of magnitude as the boundary layer thickness, the viscous effects must be accounted for. The same is valid for the thermal oscillations in the air layer. Close to the surface a thermal boundary layer is formed, in which the temperature adapts from the temperature away from the wall to the temperature of the wall.

For gases the viscous and thermal boundary layers have about the same thickness. It is shown that the viscous effects and to a lesser extent the thermal effects, together denoted as viscothermal effects, can introduce much damping, i.e. vibrational energy is converted into heat. The aim of this chapter is to determine how these viscothermal effects can be used for the damping of structural vibrations of double wall panels and for the reduction of sound transmission through this type of panels.

NOISE REDUCTION WITH DOUBLE WALL PANELS

Reflection and Absorption

Noise reduction can generally be obtained in two ways. The sound can be reflected so that it will not find its way to the receiver. On the other hand sound can be absorbed, which means that the sound energy is converted into heat so that it is no longer available as sound energy. Double wall panels are often used for noise reduction, because with a relatively light structure a relatively large amount of sound energy can be reflected.

Cases in which double wall panels are applied are, for example, double glazing in houses and offices and shielding of noisy machinery. The reflection of sound at low frequencies is difficult, due to its relatively large wavelength.

Absorption in Double Wall Panels

In this section the main subject is sound reduction in the low frequency range (<1000 Hz) with the help of double wall panels. The reflection properties of the panels are combined with the absorption properties of a narrow air layer in order to achieve noise reduction. Existing models applied for double wall panels do not take into account the dissipative properties of air when the thickness of the air layer between the two plates is close to the viscous boundary layer thickness. It is explored here whether the dissipative behaviour of a narrow air layer can be applied to dissipate a large amount of sound energy in the low frequency region. In order to do so, acousto-elastic models which include viscothermal wave propagation have to be developed and verified.

Structure Borne and Airborne Sound

A distinction has to be made between structure borne and airborne sound (Fig. 8.2). In the first case the source is a structural excitation resulting in the vibration of the structure. The resulting wavelengths are determined by the characteristics of the structure. In the second case the sound originates from a source in the air. The governing wavelengths in the excited structure correspond with the wavelengths in air. Here the sound reducing capabilities of double wall panels are investigated for both types of sound fields.

PROBLEM DEFINITION

The goal of this study is to develop efficient acousto-elastic models including viscothermal wave propagation for double wall panels. These models are used to explore the possibility to take advantage of viscothermal damping for the reduction of structural vibrations and airborne and structure borne sound transmission of double wall panels at low frequencies.

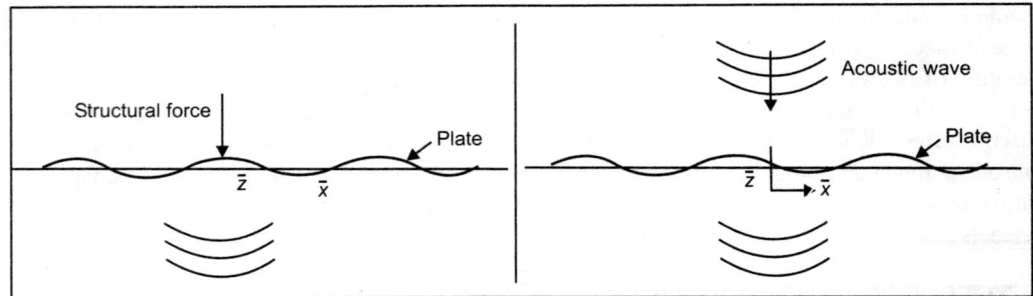

Fig. 8.2. Structure borne and airborne sound transmission.

SECTION III

Noise Reduction in Mechanical Industries

Chapter 9

Machinery

INTRODUCTION

The emphasis on faster, lighter, more powerful and compact machinery has resulted in higher noise levels. Machinery noise reduction requires a fundamental knowledge of acoustics and noise control techniques. Three basic approaches to noise control problems should be considered: (i) solution at the noise source, (ii) solution along the noise path, by blocking the noise propagation from the source to the receiver, and (iii) solution at the receiver by isolation in a control room or by using hearing protectors. The optimum approach must be determined, by considering acoustical effectiveness, production capability and economy.

It should be pointed out that hearing protectors are only a temporary solution, and an engineering solution at the source must be utilised as a permanent solution. The approaches to noise control presented in this paper focus on solutions at the source and along the noise path. Solutions are presented for the most common machines used in industrial installations, such as compressed air jets, ventilators and exhaust fans, compressors, electric motors, wood cutting machines power tools. The solution techniques consider machine redesign, process modification and/or noise source elimination. Other techniques involve the use of enclosures, silencers, mufflers, noise isolation, sound absorption and vibration isolation, as well as active noise control techniques.

Industrial machinery and processes are composed of various elements such as rotors, stators, gears, fans, vibrating panels, turbulence fluid flow, etc. Low noise machinery means lower surface vibration levels, lower impact levels with longer duration and low fluid velocity and turbulence. In this chapter, the fundamental source mechanisms of solid and air vibration are discussed, and recommendations for noise reduction, along with some practical cases, are presented.

NOISE SOURCES

Sound field in the workplace is usually complex, due to the participation of many sources: propagation through air (airborne noise), propagation through solids (structure-borne noise), diffraction at the machinery boundaries, reflection from the floor, wall, ceiling and machinery surface, absorption on the surfaces, etc.

Therefore any noise control measure should be carried out after a source ranking study, using identification and quantification techniques. The basic mechanism of noise generation can be due to solid and/or air vibration.

Solid Vibration

A solid surface vibration, driven or in contact with prime mover or linkage, radiates sound power given by:

$$W = z \cdot S \cdot \sigma \cdot V^2$$

where,

z is the air impedance = 415 [rayle] for normal conditions

S is the vibrating area [m^2]

V^2 is the mean square time/space average vibrating velocity [m^2/s], which can be measured by a vibration meter

σ is the nondimensional radiation efficiency.

The radiation efficiency R, varies from zero at low frequency ($L \ll \lambda$) to unit at high frequencies ($L \gg \lambda$), where L is the largest linear dimension of the vibration surface and λ is the wavelength. Therefore care must be taken to reduce the vibrating area (S) and/or reduce the vibration velocity (V). Reducing the vibrating area can be carried out by separating a large area into small areas, using a flexible joint. A reduction of the vibration area by a factor of two gives a reduction of the sound power level of 3 dB. Reducing vibration velocity can be carried out by using damping materials at resonance frequencies and/or blocking the induced forced vibration. A reduction of the vibration velocity by a factor of two gives a sound power reduction 6 dB. Typical examples of solid vibration sources are: eccentric loaded rotating machines panel and machine cover vibration which can radiate sound like a loudspeaker, impact induced free vibration on surface resonance.

Air Vibration

Air turbulence and vortex generate noise, especially at high air flow velocity. Turbulence can be generated by a moving or rotating solid object, such as the blade tip of a ventilator, by changing high pressure discharge fluid to low (or atmospheric) pressure, such as cleaning air jet or by introducing an obstacle into the high fluid flow. The aerodynamic sound power generated by turbulent flow is proportional to the 8th power flow velocity (W a V8), which means that a doubling of the flow velocity V increases the sound power (W) by a factor of 254 or 24 dB. Therefore, care must be taken to reduce flow velocity, reduce turbulence flow by using diffusers and remove obstacles or have them in stream line shape. The next few examples show the applications of these fundamental concepts to machinery noise reductions.

MACHINERY NOISE REDUCTION

In this section, solutions are presented for the common machines used in industrial installations. For each case, the mechanism of noise generation is discussed and noise control measures are presented.

Industrial Air Jets

Industrial jet noise probably ranks the third as a major cause of hearing damage after that of impact and handling noise. Air jets are used extensively for cleaning, for drying and ejecting parts, for power tools, for blowing off compressed air, for steam valves, pneumatic discharge vents, gas and oil burners, etc. Typical sound pressure level at 1 m from the blowoff nozzle can reach 105 dBA.

Noise sources

The reservoir compressed air pressure is usually in the range of 45 to 105 psi (3 to 7 bar). The air acceleration varies from near zero velocity in the reservoir to peak velocity at the exist of the nozzle. The flow velocity through the nozzle can becomes sonic, i.e. reaches the speed of sound. This means high generation of broadband noise with highest values at a frequency band between 2 to 4 KHz.

Noise control measures

Two basic methods of noise control that are commonly used: (i) reduction of the single high-velocity jet component into a multiple small velocity components as in the multiple-nozzle with or without extended tube or by using restrictive diffuser nozzle, and (ii) reducing the turbulent zone by creating a secondary flow enveloping the main jet flow, like that of shroud nozzle.

The choice of the low noise nozzle depends on its application. For cooling and drying purposes, a large flow rate with low thrust force is needed, like that of air shroud nozzle, while for ejection purposes, a high thrust directional force is needed, like that from multiple-nozzle.

Measurements were carried out in four factories in Brazil: a cigarette manufacturer, an engine piston manufacturer, an auto-vehicle compressor manufacturer and an agricultural storehouse, where the air jet was used for cleaning. In each case, the sound pressure level was measured at the operator ear position using the classical simple hole nozzle air jet gun and the same gun with a multi-nozzle type. The multiple-nozzle configuration was varied (number, diameter and disposition of holes) until the operator was satisfied with jet performance. A reduction between 5 and 9 dBA was obtained. Therefore it is recommended that several nozzles for a given application be tested before making a final selection among quiet nozzles. The presence of obstacle, as well as discontinuity or sharp edges on the flow direction increases noise, especially in high frequency bands. This impingement noise is strong, depending on the flow velocity (Wa V5 to 6). Hence, a small reduction of flow velocity yields appreciable reduction of impingement noise. In the case of ejection jet care must be taken to avoid turbulence created as the jet flows over obstructions or sharp edges.

Ventilator and Exhaust Fans

Fans are used to move a large volume of air for ventilation, by bringing in fresh air from the outside, blowing out dust vapour or oil mist from an industrial environment and for drying or cooling operation, etc. Industrial fans are usually low-speed, low-static-pressure and large-volume flow rate. Fans can be classified as either axial or centrifugal type. Axial fans are noisier at high frequencies and centrifugal fans are noisier at low frequencies. Fans have to operate at the maximum efficiency point on the pressure-flow curve characteristics. Therefore, the choice between axial or centrifugal fans is made by the manufacturer to satisfy maximum efficiency at a certain static pressure/flow rate.

Noise sources

Three basic noise sources are:

1. A broadband aerodynamic noise generated by the turbulent flow.
2. Discrete tones at the blade passing frequency Fp (Hz) given by:

    Fp = (Rotation in rpm × number of blades/60), and the harmonics (twice Fp, three times Fp, etc.)
3. Mechanical noise due to mounting, bearing, balancing, etc.

Sound power level (Lw) generated by fans (without the drive motor) can be easily predicted in the early project stage of an industrial installation using the Graham equation.

Based on sound power prediction, the sound pressure levels can be estimated at specified locations in certain installations. The finite element, boundary element or ray acoustics methods can be used for these estimates. Commercial software packages based on the three methods above are available for these estimates. Low flow rate, low static pressure fans are quieter. Fan selection is usually based on efficiency in carrying out the desired service, and noise parameter was not considered as a decisive purchase factor.

Noise control measures

Two basic types of solutions are available; the use of the absorption silencer or of the recent electronic noise cancellation active silencer. Figures 9.1 and 9.2 show typical circular and rectangular absorptive silencers respectively. The project of such silencers must take into consideration the following parameters:

1. The sound power level spectrum generated by the fan.
2. The noise attenuation required which can be expressed as the dynamic insertion. Loss (DIL), that is the difference in sound pressure level with and without a given silencer installed in the duct system with airflow.
3. Pressure drop caused by the presence of the silencing which the fan can overcome.
4. Self-generating aerodynamic noise, due to the presence of the silencer.

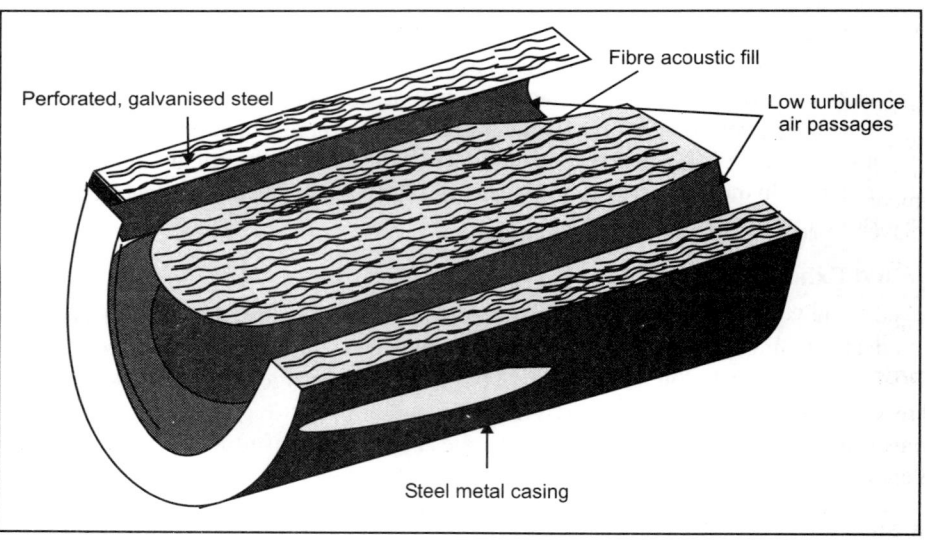

Fig. 9.1. Circular silencer.

The ideal solution is for the fan to be selected considering the above factors and supplied by the manufactories along with its silencers. In case a silencer is projected for a given fan already installed, the four factors above have to be considered in the silencer projector. It is recommended that professional silencer manufacturers be hired to satisfy the noise attenuation requirements, keeping the fan efficiency at the project point.

Since the basic mechanism of dissipative silence is sound energy absorption, silencers are more efficient at high frequencies, depending on the dimensions and thickness of absorption materials used. At a low frequency range, specially below 500 Hz, a dissipation silencer has to be large to be able to get an appreciable noise reduction.

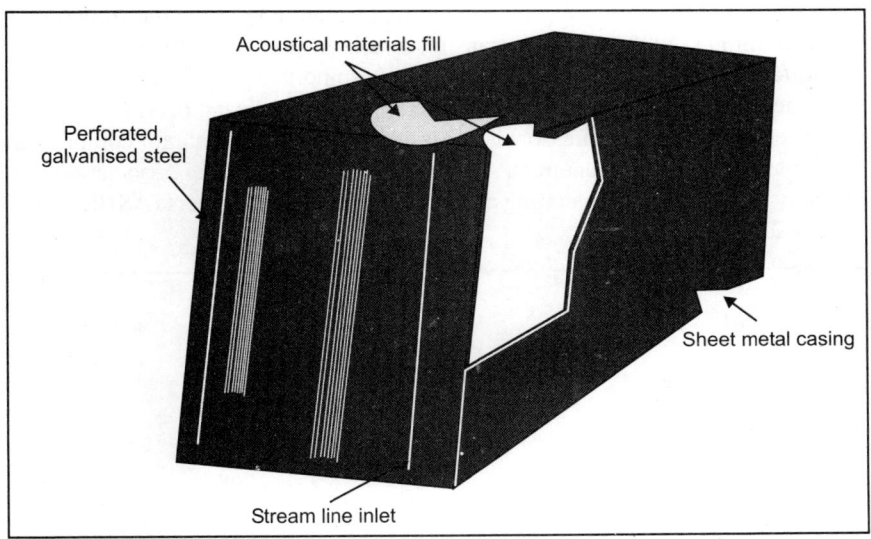

Acoustical materials fill

Perforated,
galvanised steel

Sheet metal casing

Stream line inlet

Fig. 9.2. Parallel baffle rectangular silencer.

Active noise control silencers are efficient at low frequencies. Lueg's patent in 1936 was difficult to operate in analog electronic circuits, due to the precision needed, but recently, with the advances in digital technology and low cost digital electronics, this same invention could be economically implemented. In the active digital electronic silencer, the sound wave is captured and inverted, controlled and delayed, then injected to cancel the original wave. Such a silencer is available commercially and it can give up to 32 dB reduction at the blade passing frequency noise components. A hybrid compact silencer can be used to reduce the low frequency noise by the active control principle, and the high frequency noise by the dissipative mechanism.

Compressors

Compressors (or blowers) are usually racketing machines with high pressure. They are used for conveying materials or products. There are several types of compressors: the rotary positive displacement (lobed impellers on dual shafts, as shown in Fig. 9.3), gear or screw compressors, etc.

Noise sources

The basic noise sources are caused by trapping a definite volume of fluid and carrying it around the case to the outlet with higher pressure. The pressure pulses from compressors are quite severe, and sound pressure levels can exceed or 105 dBA. Since the noise is periodic, discrete tones and harmonics are present in the noise spectrum.

Noise control measures

Since the discrete low frequency noise (small number of lobes, blades, etc.) predominates, reactive chamber-type silencers (like those of vehicles) are effective. Generally, a silencer is required on both the inlet and outlet as shown in Fig. 9.3. Reactive silencers can also be perpendicular to the flow and work as a closed-side-branch. The pressure loss from the silencer is negligible in comparison with that from compressor. Also it is important to provide vibration isolation between the compressor and pipes by using flexible couplings, with sufficient transmission loss through wrapping. Vibration isolation between the motor/compressor unit and floor is extremely important. The last noise sources are those transmitted from the compressor casing and mechanical-type noise (gears, bearing, etc.). Therefore, in some installations; enclosing the compressors is necessary. Sound power data for compressors, fans and blowers are available from the manufacturers. Air Moving and Conditioning Association (AMCA) and/ or American Society of Heating, Refrigerating and Air-conditioning Engineers (ASHRAE) for purchase specification purposes.

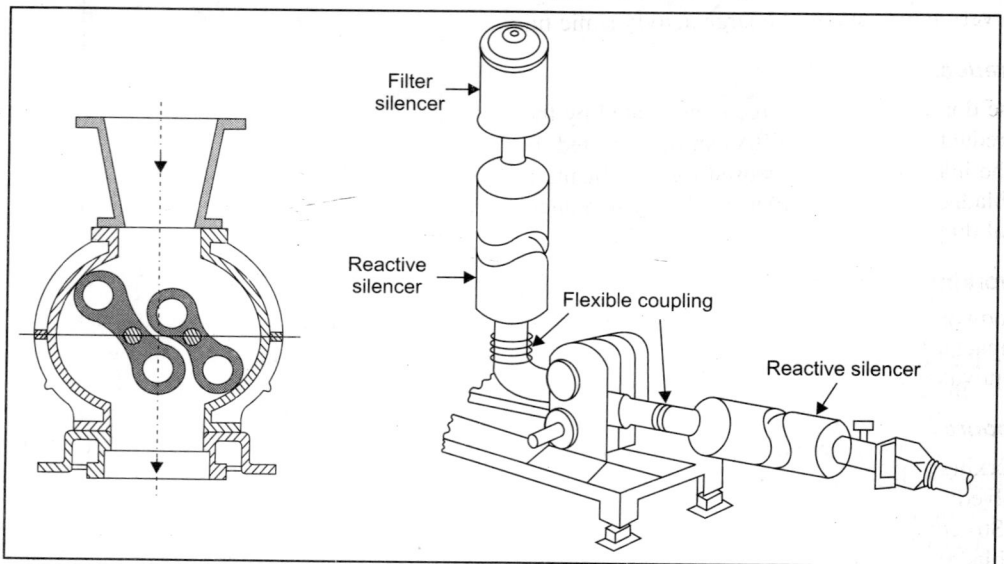

Fig. 9.3. Rotary positive displacement compressor, mounted with the inlet and outlet reactive silencers, flexible couplings.

Electric Motors

The electric motor converts electric energy to magnetic and mechanical energy with the output of a useful torque at the motor shaft. Part of the energy transformation is converted to heat giving a rise to rotor, stator and casing temperature. Therefore, an electric motor must be supplied with a cooling fan system. The cooling fan can be incorporated inside in the case of an 'open' motor or outside as in the case of a 'Totally Endosed Fan Cooled (TEFC)' motor. TEFC motors are more widely used, due to robust construction which can withstand an aggressive environment open motors are less used due to possible contamination by environment. An 'open' motor is less noisy than TEFC motor since the noisy fans are incorporated inside.

Noise sources

There are three basic sources involved in the noise generated by electric motors:
1. Broad-band aerodynamic noise generated from the end flow at the inlet/outlet of the cooling fan.
2. Discrete frequency components caused by the blade passing frequencies of the fan.

3. Mechanical noise caused by bearing, casing vibration, motor balancing shaft misalignment motor mounting. These, are project mounting and maintenance items, and therefore careful attention should be given to the vibration isolation, mounting and maintenance.

Noise generated by the motor fan is the dominant noise source, especially for TEFC motors. Sharp increase in noise occurs as shaft rotational speed increases from 1800 to 3600 rpm. For large motors in the range of 1000 CW, 3600 rpm, a sound pressure level of as high as 106 dBA occurs. The fan contribution alone is between 30 and 50 dBA. This large contribution of the fan is due to the fan shape. Motor fan blades are usually straight, so that the motor cooling is independent of rotation direction. Straight blade fans are very noisy, due to the large aerodynamic turbulence sound generated.

Noise reduction

Since the dominant noise source is generated by the fan, an absorption type silencer can be used. An overall reduction of 6 to 11 dBA can be expected. For large motors, total enclosure with low pressure loss at the inlets and outlets is used for significant noise reduction. Replacing a straight-bladed with a curved-bladed fan can give up to 8 dBA noise reduction, but care must be taken to use this motor in one rotational direction only.

Woodworking Machines

The woodworking industry has experienced noise level increases as a result of modern, higher speed, and compact machines. The basic noise elements in woodworking machines are cutter heads and circular saws. Equivalent sound pressure level in the furniture manufacturing industry can reach up to 106 dBA.

Noise sources

Woodworking machinery uses operations, such as cutting, milling, shaping, etc. Two basic noise sources are involved:
1. Structure vibration and noise radiation of the work piece, cutting tool (such as a circular saw blade) and machine frame, especially at the mechanical resonant frequencies.
2. Aerodynamic noise caused by turbulence, generated by tool rotation and the workpiece in the air flow field. Noise is also generated from fan dust and chip removal air carrying systems.

Noise reduction

For noise control in woodwork machine tools, cost-effective operation and maintenance efficiency must be considered. Typical examples of noise reduction are:
1. Circular saw noise produces a sound pressure level of about 88 dBA while idle, and 97 dBA during cutting operation, depending on the rigidity of the material. The noise is generated by aerodynamic interaction of saw blade with surrounding air and the workpiece, and also by vibration and sound radiation from the blade and workpiece. The noise reduction technique consists of blade design modifications with expansion slots and vibration damping glued to the saw plate.
2. Finishing planer with straight knife excites the workpiece and the machine frame by impact at feed-in, feed-out operation. Typical sound pressure level is from 95 to 102 dBA while idle to 100 to 105 dBA during cutting, for about 80 per cent of the time. A helical carbide cutter head knife gives a noise reduction between 15 and 20 dBA due to the smooth and smaller contact areas with the workpiece. For slotting, notching and shaping operations, doubleend planers are used. A cutter head with stagger curved tooth and an unequal number of teeth can spread the noise energy in a wider frequency band with a lower overall noise level.

Pneumatic Tools

Compressed air-powered-handheld tools such as drills, grinders, riveting gun, chipping hammer, impact gun, pavement breaker, etc. are widely used within a board spectrum of different industries.

Noise sources

There are three. basic types of sources that dominate the noise generated:
1. Noise produced by contact between the machine and the working surface. The vibration transmitted from the tool tends to vibrate the working surface and work bench, generating high radiation noise, especially at mid and high frequencies.
2. Exhaust air noise caused by the turbulence flow generated as the compressed air passes the motor and by the aerodynamic noise generated in the air exhaust.
3. Sound radiation from the tool vibration caused by the air flow inside the tool.

Noise control

Pneumatic tools are usually noisier than electric tools, because of the exhaust air noise. Noise reduction in pneumatic tools can be incorporated by manufacturer in the tool design. Principle techniques for noise control are:
1. Incorporated exhaust air muffler at the tool air exit.
2. Changing a pneumatic-hammer-type operation, such as in riveting to hydraulic-tool-type process. This means the same force applied over longer period of time.
3. The use of magnetic forces for fixation of the workpiece on the bench during operation, reducing the vibration of the work piece and bench area, and consequently reducing noise radiation.

Lower-noise modern tools available, with price about double or triple the price of a similar noisy tool. These modern tools have noise reduction solution incorporated by the tool manufacturer. As a typical example; a modern grinder achieving a noise reduction from 82 dB to 77 dBA (running free), by incorporating a spring-loaded valve with a multihole muffler for exhaust air inside the support handle. This gives an almost constant back pressure inside the machine, regardless of air consumption.

NOISE REDUCTION ELEMENTS

A large variety of noise reduction elements are commercially available. It is strongly recommended that machines and equipment be purchased with noise reduction elements incorporated so as to avoid any type of subsequent solution which may affect the machinery performance. Specification of machinery noise limit can force the manufacturer to develop low noise machines. Some of the most common noise reduction elements are:
1. A dissipative silencer which is usually used to reduce noise from the inlet and/or outlet of ventilators. Since it is based on the absorption mechanism of acoustic materials, it is effective for mid and high frequency ranges (Figs 9.1 and 9.2).
2. A reactive silencer which uses the principle of an acoustic impedance mismatch, reflecting noise back to the source. It is effective at low frequencies and widely used for compressors and engines with low rotation speeds or few blades (Fig. 9.3).
3. Enclosure which is very practical and widely used, where sound energy is trapped inside, keeping a low background noise outside. Enclosure has to meet operation and maintenance requirements of the machine, such as the necessity of having inlet and outlet openings for refrigeration.

Lower transmission loss 'TL' elements, such as openings and windows can greatly reduce the total noise isolation. Therefore silencers with high TL can be installed at the refrigeration inlet and/or outlet.

4. Barriers which can be used to reflect back noise and reduce noise propagating in a specified direction.

5. Vibration isolators which are made of metal spring, elastomeric mounts or resilient pads and used to reduce vibration transmission from machine to floor or from machine duct outlet to connecting ducts.

RECOMMENDATIONS FOR MACHINERY NOISE REDUCTION

General recommendations for machinery noise reduction at the source and/or along the propagation path should involve the application of the basic principles of noise and vibration control such as:

1. Reducing dynamic forces exciting vibrating surfaces specially at resonance frequencies.

2. Providing vibration isolation for machine mounting and flexible joints for pipe connections, to reduce vibration transmitted.

3. Reducing speed and flow velocity of machines and increasing duration of impact force, thereby reducing noise.

4. Reducing vibration levels at resonance, which can be achieved by placing damping material or shifting away the structural resonance frequencies from excitation frequencies. Acoustic resonance should be avoided by adding absorption materials and/or changing configurations.

Thus, engineering noise control is a multidiscipline subject, involving not only noise control concepts but also detailed information of the machine operation mechanisms, installation, maintenance, etc. Any noise control solution for an existing machine or process should involve the production, operation and maintenance departments, for this reason guarantee an operational solution and fewer objections to the elements installed for noise reductions.

Chapter 10

Gear

INTRODUCTION

A gear is a rotating machine part having cut teeth or cogs, which mesh with another toothed part in order to transmit torque. Two or more gears working in tandem are called a transmission and can produce a mechanical advantage through a gear ratio and thus may be considered a simple machine. Geared devices can change the speed, magnitude, and direction of a power source. The most common situation is for a gear to mesh with another gear, however a gear can also mesh a non-rotating toothed part, called a rack, thereby producing translation instead of rotation.

The gears in a transmission are analogous to the wheels in a pulley. An advantage of gears is that the teeth of a gear prevent slipping. When two gears of unequal number of teeth are combined a mechanical advantage is produced, with both the rotational speeds and the torques of the two gears differing in a simple relationship. In transmissions which offer multiple gear ratios, such as bicycles and cars, the term gear, as in first gear, refers to a gear ratio rather than an actual physical gear. The term is used to describe similar devices even when gear ratio is continuous rather than discrete, or when the device does not actually contain any gears, as in a continuously variable transmission.

GEAR NOISE REDUCTION

Gear Noise is a serious impediment to an optimum gear performance. Actually gear noise is the result of some process errors. Research done on the source of gear noise has come to a definitive conclusion that one of the causes of gear noise is the presence of plus material on its active profile in one or more teeth. This gives the impression that the gear is not of the desired quality.

Causes of Gear Noise

Three primary causes of plus material on gear teeth are nicks, burrs and lastly heat treat scale.

Nicks

Nicks are caused by part handling and not because of gear manufacturing machines. Nick can be any plus material found anywhere on a part. The cause of Nick is gouging. This creates plus material which remains on the surface that needs to be hardened into a part. The noise is the result of the Nicks' action with other teeth. A gear teeth that has nicks on their active profiles, causes noise. Transmission manufacturers would know that nicks that are more than 0.002 can be a cause of transmission noise.

Burrs

Burrs are essentially raised material. They are generally found in places where there is meeting of involute profile with the face. Large burrs can be found on the face of gears after hobbing or shaping. These burrs are taken care of by face deburring and chamfering. As the face deburring works only on the face of gear, a natural tendency is the rolling of small burrs back on to the involute profile. Now after the heat treatment is over, this burr becomes hardened and a potential cause for gear noise as soon as it is put to use.

Heat treat scale

Heat treat scale is primarily the oxidised material found left after the end of heat treating process. If left as it is on the gear, this scale can be a potential source for noise.

Factors Influencing Noise Level in Gears

Here are listed a few of the most crucial design factors that has an impact on the noise level of gears: (i) type of gears, (ii) quality level, (iii) profile of tooth, (iv) surface finish, (v) pitch, (vi) gear runout, (vii) pressure angle, (viiii) gear ratio, (ix) recess action, (x) resonance, (xi) modification of profile, (xii) lubricant viscosity, (xiii) overlap ratio, (xiv) type of bearings, (xv) backlash, (xvi) gear material, (xvii) tooth loading, and (xviii) housing.

Reducing Gear Noise

Table 10.1 highlights some of the effective trouble shooting techniques for gear noise reduction.

Table 10.1. Trouble shooting techniques for gear noise reduction.

Solutions	Features
Use of high precision gears	Minimises errors of pitch, tooth profile, runout and lead error
	Grind teeth for improving the accuracy and the surface finish
Better surface finish on gears	Methods of lapping, grinding and honing of the tooth surface improves the overall smoothness of tooth surface and can reduce effectively the noise
Use of suitable lubrication	Sufficient lubrication of gears
	A high-viscosity lubricant will certainly have an effect in reducing gear noise
Apply high rigidity gears	A face width increase can give a higher rigidity that checks gear noise
	Reinforcing of housing and shafts increases rigidity
Applying sufficient backlash	A smaller backlash helps by reducing pulsating transmission
Proper tooth contact	End contact prevented by crowning and relieving
	A proper tooth profile modification also proves effective
	Eliminates tooth surface impact

Improving Gear Performance Through Noise Reduction

Materials do play a vital role in gear noise. Other things being equal a gear set made from material with good damping characteristics will be reasonably quieter than a set made with materials that lack good damping characteristics. It has been seen that graphite flakes found in gray cast iron will result in better damping properties in comparison to other ferrous metal. In case of ductile iron also, the nodules behave in an identical fashion to the graphite flakes of gray iron, giving cushioning vibrations on transmission through a part.

REDUCE GEAR NOISE AND MISALIGNMENT PROBLEMS

Noisy gear trains have been a common problem for gear designers for a long time. And the demands for smaller gearboxes transmitting more power at higher speed and greater efficiency continue. Some popular solutions to the noisy gear problem include enlarging the pinion to reduce undercut, using phenolic, delrin or other noise-absorbing products where possible, or changing to a helical gear train. Other methods include tightening specifications to insure greater gear quality or redesigning the acoustical absorption characteristics of the gearbox.

Occasionally, experimentation with gear ratios can limit harmonic frequency amplification, which otherwise can cause a gearbox to amplify noise like a finely tuned stereo system. You can also study material and hardness requirements so that modifications may minimise heat treatment distortion or possibly eliminate the need for heat treatment entirely. Also, pay particular attention to gear geometry to insure maximum contact.

Another approach to the gear noise problem that yields good results is crowning or barrelling of the teeth. This technique involves changing the chordal thickness of the tooth along its axis. This modification eliminates end bearing by offering a contact bearing in the center of the gear.

GEAR NOISE REDUCTION OF A TRANSMISSION

Recently, it has become necessary to consider the vibration characteristics of structural systems in addition to the analysis of meshing transfer error fluctuation. So far, studies made on the vibration characteristics of structural systems include the effect of gear train structures on the rigidity of meshing points, the basic structure of low-vibration cases, and the studies on rib arrangement. Few studies have been made, however, on the process of vibration transfer in the overall structure consisting of gears, shaft, bearings and case.

In this connection, the present study discusses the gear noise generating mechanism. It studies the substitutional index of gear noise and a guide for developing a low-vibration structure based on this index mainly when making design changes to the case.

Methodology

We considered indices a, b, c and d in the gear noise generation/transfer mechanism. We reviewed the relationship between gear noise e and index d, and a method for the simple evaluation of index d using indices a, b and c. We carried out the tests by varying the sectional contours of the case. In addition, we conducted experimental identification of the bearing characteristics, etc. to develop an FEM model to quantitatively calculate the substitutional indices of the gear noise. We explored ways to establish a low vibration guide for the case axis-support surface structure through FEM calculations and shaker tests.

Result

1. As a result of these experiments, we confirmed that principal gear noise e can be substituted by the vibration response at the center of the case surface (Fig. 10.1). Moreover, when the structure of the case becomes the object of the engineering change, it is found that the effect of indices a and b can be regarded as constant. Accordingly, the transfer function at the center of the case surface to the input of the mesh point is rendered an approximate index.
2. From the FEM analysis, a guide to the low vibration structure based on the above is considered as shown in Fig. 10.2, and the effect of the vibration reduction was clarified through an oscillating test using a prototype (Fig. 10.3).

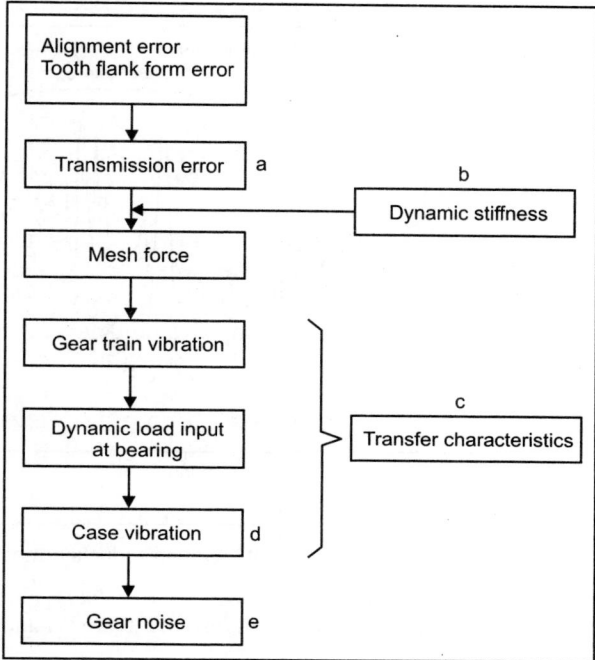

Fig. 10.1. Mechanism of gear noise generation.

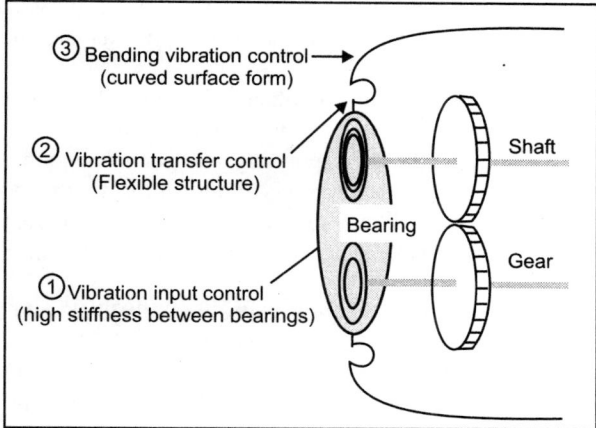

Fig. 10.2. Low oscillating structure.

DESIGN OPTIONS FOR LOW NOISE

When designing standard spur gears for low transmission error (TE) there are few options since the only variable is the profile, assuming that the pitching is good as occurs usually.

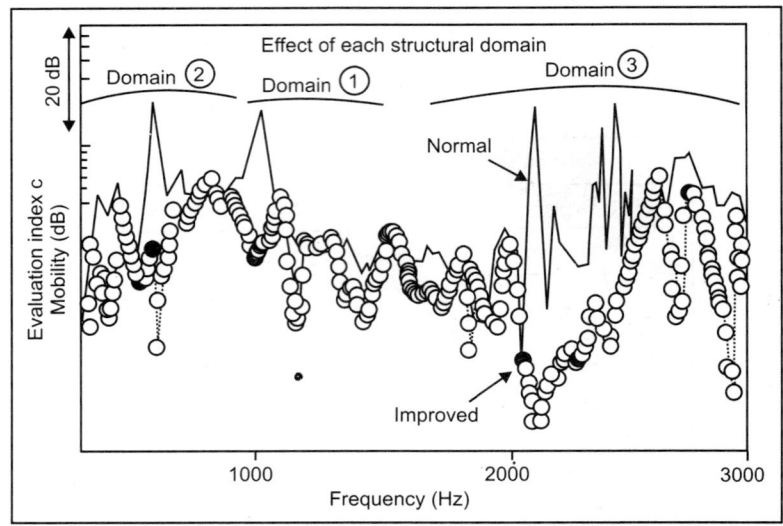

Fig. 10.3. Experimental confirmation of vibration reduction methods.

The possible approaches for normal contact ratios are:
1. If low load is of major importance, use 'short' relief so that there is handover from pure involute to pure involute.
2. If high load is of major importance, use 'long' relief with the tip relief at the changeover ± 0.5 p_b points equal to half the expected elastic deflection.

Both approaches can work reasonably well at their working load, provided design, manufacture, alignment, etc. are good. However, they must give noise off-design and these spur gears will be sensitive to manufacturing errors.

The spur gear alternative is to use a nominal contact ratio above 2 to achieve a handover which is effectively 'long' relief under full load and pure involute under light load.

Helical gears should be quieter than the corresponding spur gears due to the averaging effects of the helix. This simple deduction goes astray as soon as misalignment throws the load to one end of the face width since the mesh then behaves more like a spur gear. When a helical gear is noisy there are four options, (assuming the gear has been well designed with the face width an integral number of axial pitches):
1. Improve alignment: Easy to suggest but this can be very difficult and ultimately alignment can only be checked by a blueing test or a copper plating test under load. Achieving a good enough alignment by accurate manufacture is almost impossible due to tolerance build-ups. Any form of gear axis movement due to deflection under load makes maintaining alignment even more difficult.
2. Crowning: This is popular because it is simple. The effect is to produce a mesh which is more like a spur gear mesh but there is negligible need for tip relief as the contact engages smoothly by using the crowning as end relief. The design profile can correspondingly be modified to get the best TE under load on a fairly narrow effective face width, like a spur gear. However, it is common to use crowning with a profile which has no tip relief and gives very good TE at light loads with some penalty in TE at higher loads.

3. Heavy end relief: Like crowning, it is possible to use end relief together with a profile which is nearly pure involute. This acts like a spur gear giving low TE under light loads since there is an involute profile, but will give a reasonable TE at heavy loads since the length of contact line remains constant, providing that the effective face width is an integral number of axial pitches.

4. High contact ratio: As with spur gears, if the effective contact ratio is 2, inferring a nominal contact ratio of about 2.25, then the drive should be very quiet at low and high loads.

Of these options (2) has the disadvantage of giving high stresses whether or not the alignment is good whereas (3) and (4) only give high stress when there is severe misalignment. The ultimate design is probably to have a combination of (3) and (4) with a contact ratio only just over 2 and use blue checks to give reasonable alignment. In general, increasing helix angle gives a smoother drive but with the corresponding end thrust and axial vibration effects.

ANALYSIS TECHNIQUES

Types of Noise and Irritation

One of the most difficult problems in gear noise investigations is that the final 'detector' and arbiter (on whether or not a noise is irritating) is an extremely non-linear, rather temperamental, and extremely variable human being, with office politics and economics playing a major role. It is quite possible for three people to listen to a gear drive and to object to it for three completely different reasons. No amount of technical measurement will determine which aspect of a gear drive noise will irritate a particular customer, so it is most important to identify the problem correctly at the start by questioning the customer thoroughly and by possibly playing tapes of different types of gear noise to the customer for comparisons. A PC with an output card to a loudspeaker can be useful for this. There are, roughly speaking, four types of irritation:

1. A steady tone: This is relatively musical and, because there are few harmonics, sounds a bit like an oboe. It is often encountered as a 'back axle whine' on rear wheel drive cars and is typically in the 500–1000 Hz range (900 rpm and 40 teeth). A higher harmonic content moves the character towards a stringed instrument sound.

2. A modulated tone: Here the customer is not objecting to the steady component at perhaps 400 Hz but to the fact that it is modulated (or wowing) at a much lower frequency. It is not uncommon to have a customer complaining that he is hearing a noise at 2 or 3 cycles a second. This is impossible. What is heard is the basic 400 Hz once-per-tooth noise being modulated in amplitude (or phase) at 2 or 3 Hz.

3. 1/rev impulses: This is the type of noise generated by a defect such as a nick or burr giving an impulse at 1/rev and is usually most noticeable at low speeds. The sound is a fast ticking sound and has very little power associated with it so it will not usually show up in a frequency analysis. However, like it triangle in an orchestra, it can easily be picked out by the peculiar non-linear abilities of the human ear.

4. Grumbling or graunching: This is the 'classic' gearbox noise, usually associated with low speed and heavily loaded drives. It is the typical 'bottom gear' noise in a car. It tends to be associated with pitch errors and is essentially at all harmonics of once per revolution of both wheel and pinion. Frequency analysis is of little help since all frequencies (or all multiples of a couple of very low frequencies) are present.

Which of these types of noise causes the irritation depends, to a large extent, on what the listener is expecting. One engineer will often expect (1), (2), and (4) and ignore them but will be highly irritated by (3), whereas another might reject due to (2). One car driver might be irritated by (1) and ignore (2), while another would react the opposite way. Occasionally, as with a car, it is not the noise itself which irritates but the fact that the noise has changed from a familiar, accepted 'normal' noise.

There is interaction in human response between the various sounds and sometimes it is possible to use the deliberate addition of pitch errors in a drive to break up the sound pattern. This technique is sometimes used in chain drives if the customer is irritated by a steady whine.

Problem Identification

From what has been said in previous topic (types of noise and irritation) the accurate specification of the problem is not always easy. Occasionally it is a simple pure tone that is heard and, if a quick check with a sound meter straight into a frequency analyser or oscilloscope confirms that the frequency is once-per-tooth, diagnosis is easy. Checking the character of the sound is a great help and if the sound is complex, some form of artificially generated range of sounds, can help identity the type of noise. This can be done using predominantly analog equipment but it needs quite a complicated setup so is more cheaply tackled by generating a series of repetitive time sequences with and without the various errors in a standard PC. The resulting time series for each revolution is then fed via an output card into an audio amplifier and loud speaker or can be played out on a sound card. The problem with standard soundcards is that varying the frequency is not easy. Reasonable resolution is obtained if each tooth interval is, say, 30 samples long and 25 teeth need 750 sample points per revolution.

The various types of error can be generated as shown in Fig. 10.4:

1. 1/Tooth errors: Amplitude times mod $(\sin \pi x/30)$ gives the typical half sine wave of 1/tooth (for $x = 1:750$ as the position round the revolution).
2. Pitch errors: These can be put in as positive and negative at arbitrary positions of x. The classic dropped tooth can be modelled as h $x/750$, where h is the drop size. It is helpful to be able to either add or subtract a given pitch error because the audible effects are not necessarily the same.
3. Modulation: Multiplying the sequence of l/tooth errors by $(1 + \sin (2\pi x/N)$ allows modulation at l/rev $(N = 750)$ or wheel frequency $(N = 1300)$ or 2/rev $(N = 375)$ for a diesel or at any other possible torque variation frequency.
4. Eccentricity: This can be modelled as e $\sin (\pi x/375)$ and added in but will not alter the sound. It is, however, useful for demonstrating that eccentricity is not audible unless it modulates the higher frequencies present.
5. Random 'white noise' can be added for comparison purposes. Again the terminology is muddling because we add electrical white noise to the input signal and the loudspeaker then gives audible noise which has in it a random content (noise) which has equal amplitudes at all audible frequencies so it is 'white'. Alternatively 'pink' noise with roughly equal power in each octave can be used.

Generally a single revolution sequence in a program is straightforward in a language such as Matlab. Perhaps 60 revolutions can be sequenced together to give runs of the order of seconds, then the sequence can be repeated to give of the order of 10 seconds running time. Varying the frequency of the sample rate of the analog output channel on the computer then gives the effect of varying gearbox speed as when running a gearbox up to speed.

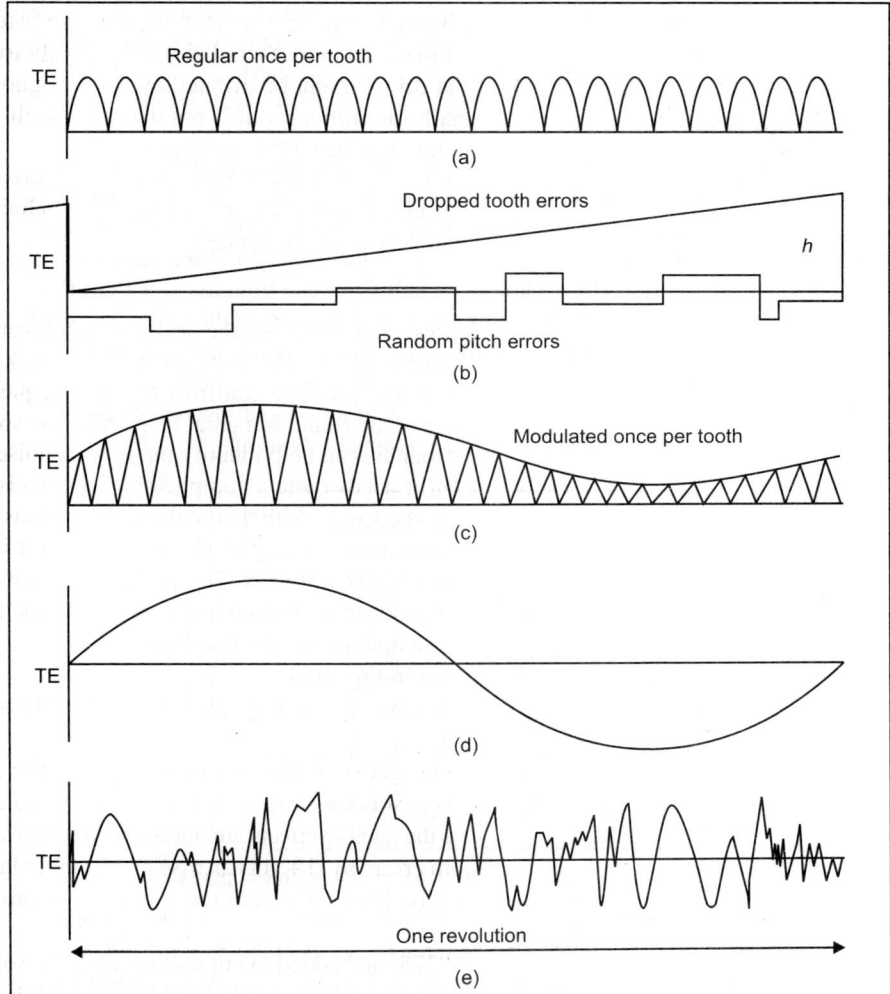

Fig. 10.4. Models of various types of noise generated by gear drives.

Using the original typical TE as the input for the noise does not take into account the dynamic responses of the gearbox and its installation. In practice, this does not seem to matter since it is the character of the sound that is important and the customer will usually readily identify the 'same sort' of sound. It is important to identify the type of problem because the techniques to be used for analysis depend on the type of error.

Equally helpful, as previously mentioned, is the use of a simple basic noise meter with an analog output which can be fed directly into an oscilloscope synchronised to l/rev. This immediately gives a great deal of information about the regularity of the sound and whether it is occurring at particular points in the revolution or is a steady sound.

If the microphone information is confusing, going to an accelerometer and checking bearing housing vibration is the next move but care must be taken that the main trouble frequencies investigated at the bearing are the same as those being heard (and irritating the customer).

Time Averaging and Jitter

Time averaging is a method of compressing the amount of information that was stored but has a much wider range of uses. In a gearing context the great use of time averaging is to eliminate or reduce unwanted vibrations. Taking the case of an in-line gearbox, sketched in Fig. 10.5, we have three shaft speeds, input A, layshaft B and output shaft C. If we suspect a dropped-tooth pitch error problem on the output shaft C and have an accurate 1/rev marker on shaft C we can time-average the TE or vibration at the repetition frequency of shaft C. At each revolution of shaft C we read the noise, TE or vibration level at perhaps 500 points taken consistently round the revolution. A revolution is comprised of 500 data 'buckets' and on each rev the reading is added to the sum of previous readings in that 'bucket' (i.e. at that position round the rev). If we take the sum of 256 revolutions and divide the resulting totals by 256, our scale factor is unaltered and we have obtained an average vibration.

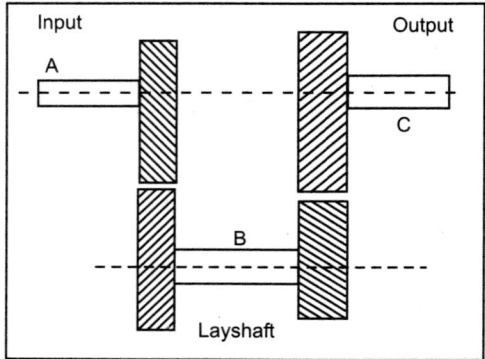

Fig. 10.5. Sketch of in-line gear drive with three shafts.

All vibration related to the output shaft, such as an output gear pitch error, will repeat in exactly the same place round the revolution so it will remain unaltered in size. All other non-synchronous, intermittent, random, or irregular vibration wilt behave like random vibration and average to zero. Even powerful vibration such as engine inertia and firing effects from a 4-cylinder engine will be non-synchronous for the output shaft (though synchronous for the input shaft) and will be spread out round the revolution leaving mainly those vibrations associated with the output gear (and prop shaft and hypoid pinion if fitted). A very narrow firing pulse consistently at one point on the input shaft will appear at each tooth interval on the averaged layshaft trace reduced in amplitude by a factor equal to the number of teeth on the layshaft. Figure 10.6 shows the effects of a consistent narrow firing pulse of height H if it is on the 'averaged' input shaft and if it is on a neighbouring shaft, in this case, the layshaft.

Additional to the benefit of extracting the information associated with a particular shaft rotation, averaging increases the accuracy of the readings and improves the resolution. If the original full scale (10 volts) is represented by 12 bits, then, after averaging, the total can in theory be up to 12 bits × 256 which is 20 bits size so after averaging we can have a 20 bit range. This seems to be impossible since, if full scale is 10 volts then originally 1 bit is 2.4 mV, and it does not seem possible to achieve a resolution better than 0.01 mV. In practice this can and does happen although only if there is extra random vibration present. For some accurate measurements a 'dither' vibration is deliberately added to increase accuracy of the averaged signal and to give resolution to the equivalent of better than 1 bit on the original measurement. In the case of a gearbox we have plenty of extra non-synchronised vibration around so we do not have to bother adding in the dither.

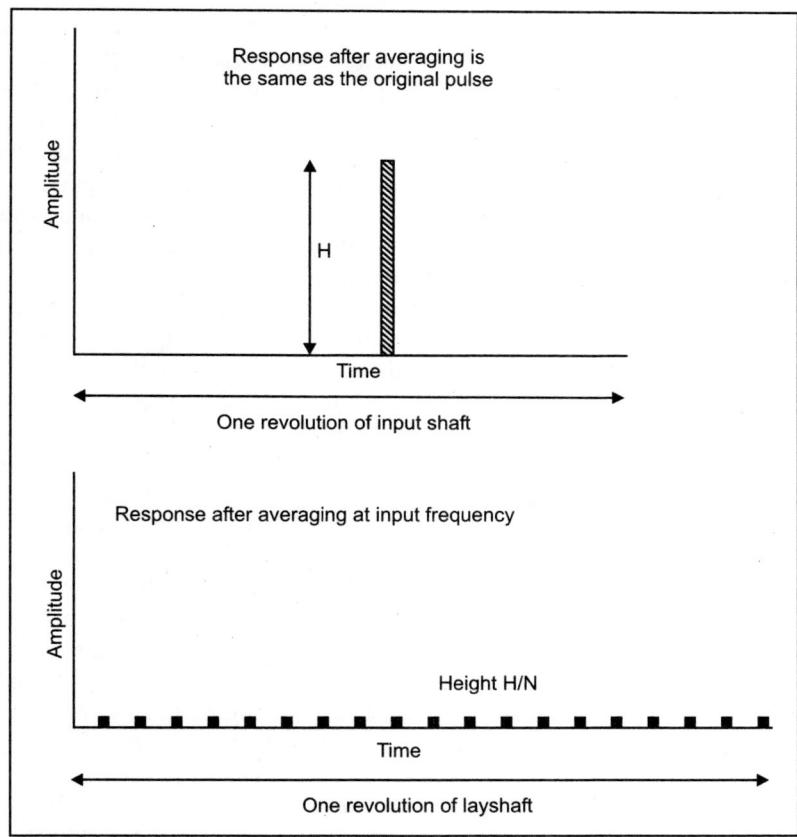

Fig. 10.6. Effects of averaging at once-per-rev of input shaft and of layshaft.

Due to the averaging process the final averaged result is much more accurate than the original 'noisy' measurements and frequency analysis is correspondingly more accurate and reliable. An averaged signal, exactly synchronised to 1/rev, will finish at exactly the same position as it started and can be analysed by FIT using a 'rectangular' window and hence using all the information fully. The one revolution of averaged information is equivalent to an infinite number of repetitions of that revolution.

The question arises as to how many revolutions is necessary or desirable to average to get a 'good' result. As usual, the answer depends on the 'noise' around, both in level and character. Whether audible, mechanical or electrical, random noise which is comparable in size to the signal of interest is likely to be reduced to negligible importance (1 per cent) if we average 128 cycles. If, however, the noise is 20 dB greater than the signal we may have to go to 1024 averages but this, fortunately, is rare.

When the 'noise' is due to pitch errors on a mating gear, then the requirement is slightly different. By definition, the sum of all adjacent pitch errors on a gear must be zero since, otherwise, we do not finish a revolution where we started. If, then, one selected tooth on a pinion mates once with every single tooth on a wheel (with N_w teeth) then the sum of all the errors must be solely N_w times the error on the pinion tooth, since all wheel pitch errors have added to zero. To get the best averaging on the pinion it should do an exact multiple of N_w revs (i.e. an integral number of complete mesh cycles). (This assumes that, as usual, there is a hunting tooth.) So, with a single mesh of 19 teeth (input) to 30 teeth (output) we need multiples of 30 revs of the input to give complete meshing cycles and 120 revs would give a good result and reduce random noise effectively. This meshing cycle idea gives an excessive requirement if there are two meshes, as happens with a layshaft (B in Fig. 10.5) with 19:29 at input and 23:31 at output. For a complete meshing cycle the layshaft would have to do 19×31 revs and 589 revs would take rather a long time and require some 3,00,000 data points. There is an exception to the basic idea that time averaging separates occurrences on two meshing shafts. Regular 1/tooth and harmonics appears on both the pinion and wheel averaged traces since the steady component of 1/tooth averages up consistently. In both averaged traces the regular 1/tooth and harmonics components associated with both shafts will appear together with any irregular tooth components due to the particular shaft.

Time averaging appears to be, and is, a very powerful and useful tool for rotating machinery, and for gear drives in particular, but as with all techniques there are liable to be problems or difficulties.

The major problem is associated with 'jitter' or 'smearing' and is due to variation in speed of rotation. The start of a revolution is given by an accurate 1/rev pulse and with typically 500 data samples per rev the starting position is located consistently within 0.2 per cent of a rev (0.72°). If we have 50 teeth on the gear and are primarily interested in 1/tooth (and harmonics) noise then a 1 per cent variation in speed between one revolution and another would mean that by the end of the revolution two 50/rev waves recorded at the same sampling rate (in time) would have moved 180° in phase relative to one another. It, at the start of the rev they were adding, they would be cancelling each other by the end of the rev. A 1 per cent speed change is unlikely to occur within a revolution or on, successive revolutions but might occur over 50 revs which is the sort of order of number of revs over which we might average signals.

Figure 10.7 indicates the 'smearing' effect and shows how the observed averaged amplitude reduces as the signals move out of phase with the speed variation, which in the case shown has given cancellation with 180° phase shift half way round the revolution. The jitter effect causes trouble because we sample at a constant rate in time whereas we wish to sample at constant angular positions round the revolution. On a test rig this could be achieved by fitting an additional rotary encoder (with 512 lines) and sampling the vibration signal when demanded by the next encoder line. This technique is rather too cumbersome for general use.

An alternative to physically fitting an encoder is to work back from when the 1/rev pulses occurred (in time) to exactly when the samples should have been taken (assuming constant speed during the revolution), and then use relatively complex interpolation routines to estimate what the vibration reading would have been if the sample had occurred exactly at the 'correct' time. Again this is excessively cumbersome for normal use. It is usually less effort (in total) to restrict data logging to times when the speed remains reasonably steady over a few seconds. An interval counter reading rev times from the 1/rev sensor and set to 10^{-5} seconds resolution is usually useful. An alternative technique is to have two separate sensors, half a revolution apart and average on each separately, using only the first-half of each rev and halving the time available to get out of phase. Yet another, better, approach is to carry out the time averaging analysis working backwards from the timing pulses to reduce smearing in the latter half of the rev, as well as working forwards from the pulses to reduce smearing during the first half of the rev. This is relatively easy to do in the analysis routines and comparison of the 'forward' and 'backward' averages gives a clear indication of whether there are serious jitter problems.

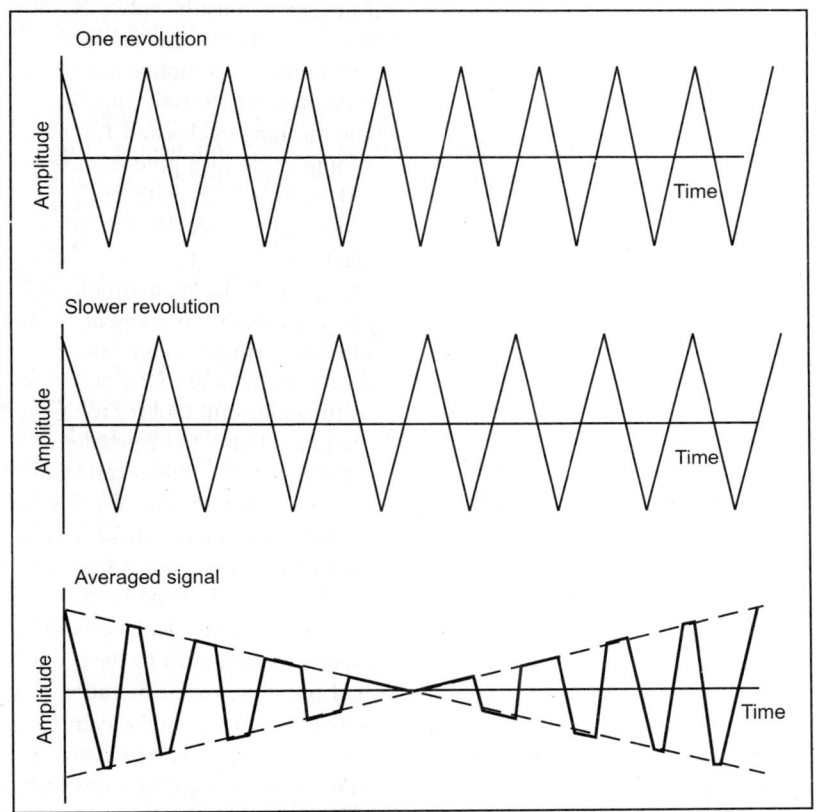

Fig. 10.7. Effect of speed variation on time averaged signal.

Average or Difference

It is easy to get carried away by the power and usefulness of time averaging but occasionally it is not the average that is important. A classic case occurs with internal combustion engines where we are less interested in the steady firing pulses from the (four) cylinders than in the variation of the pulses due to irregularities in carburation or turbulence. Correspondingly, in gear drives we may be interested in variations of noise pattern from the steady state because human hearing is very sensitive to small modulations or variations from regularity.

Irregular variations in gear drive noise or vibration can occur for several reasons. Intermittent interruptions in oil supply can have some effect or alignment variations, due to the cage of a rolling bearing beginning to break up, can modulate the signal. External variations due to variable load will influence noise, especially if teeth are allowed to come out of contact, or occasionally dirt or debris passing through the mesh will give transient vibration. Hull twisting on a ship may distort the gear casing and alter alignments of the meshes. For any problems of this type an effective method is to compute the time averages for both input and output shafts and subtract the averages from the original signal so that what is left is the variation from average or 'normal' signal. Care is needed not to subtract the tooth frequency components twice as they appear in both averages. The variation from average can then give an indication of the problem cause, especially if there is an external load or speed variation.

Band and Line Filtering and Resynthesis

In many vibration signals there are present vibration components, often quite large, which are irrelevant to the investigation. It is of little help to be told that there is a large component at 21 times per revolution if we already know that there are 21 teeth on the gear and that the problem is not at tooth frequency. Similarly a component at mains frequency (or harmonics) is likely to be electrical noise or drive torque fluctuations. It may be much easier to analyse or assess the time signal if these expected components (which are legitimately present) are removed from the signal. Originally the analog methods employed for this involved either (Fig. 10.8):

1. Notch filters which were often for mains interference.
2. 'Band stop' filters to cut out a range of frequencies, typically an octave.

These helped but were of limited performance and could not deal with any subtleties in the signals. Digital methods are much more powerful and flexible and now dominate the field. Digital filters can give extremely high performance high pass, low pass, band pass or band stop performance to 'clean up' a signal by removing known, irrelevant components.

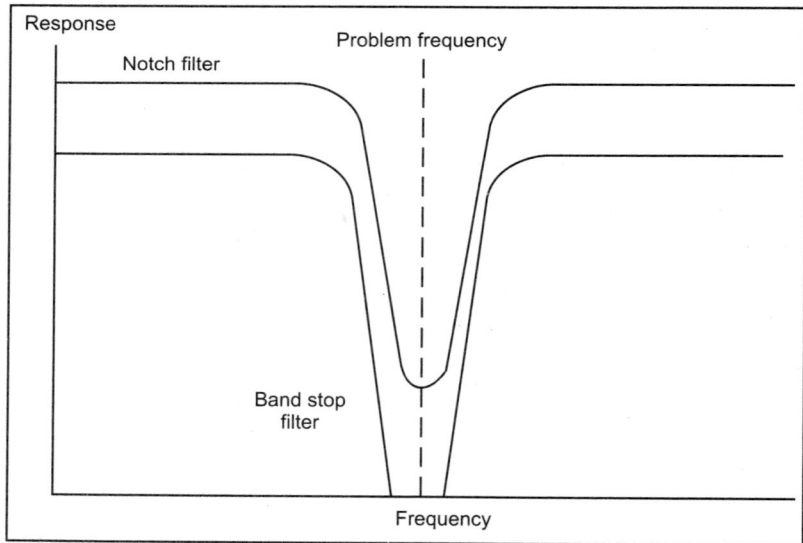

Fig. 10.8. Response of analog notch and band stop filters.

Pitch Effects

The assumption so far has been that noise and vibration problems are dominated by 1/tooth and harmonics but this may not be so for high speed drives. If we have a turbine or compressor pinion running at

12,000 rpm with 30 teeth the 1/tooth frequency is 6 kHz. In general frequencies this high are less likely to find responsive resonances and give noise problems but the set may give noise at much lower frequencies below 2 kHz.

Noise in this frequency range is at say five times per pinion revolution or twenty times per wheel revolution and so is rather puzzling. It can be due to phantom or ghost tones from the gear manufacturing machine but such tones are easily identified as they correspond to the number of teeth on the table wormwheel. If not the trouble may be due to random pitch errors on the pinion or wheel. Adjacent pitch errors are typically of small amplitude and should be rarely larger than 4 µm and as they are random we would expect negligible excitation at any single frequency. The test results may be as in Fig. 10.9 and do not appear to be capable of giving significant trouble.

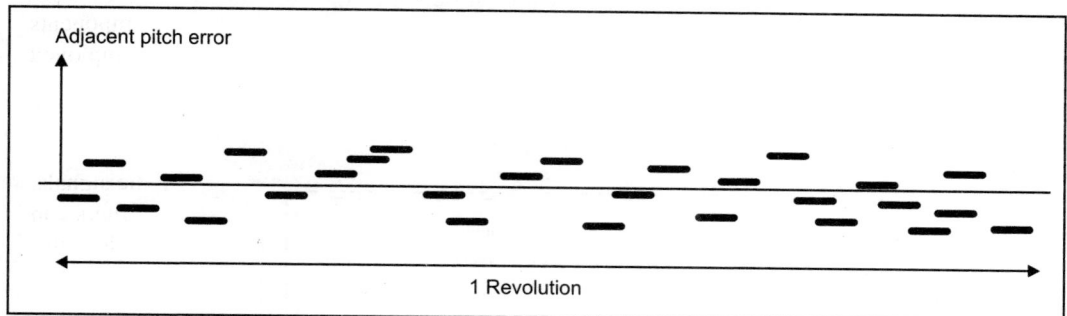

Fig. 10.9. Typical adjacent pitch errors around a gear.

Although the pitch errors are random in distribution there are only a finite number of teeth round any gear and the sequence then repeats. This gives components of excitation at all possible multiples of 1/rev except curiously at 1/tooth and harmonics of 1/tooth.

This means that at any multiple of 1/rev (excluding tooth frequency and harmonics) there may be a significant component of that harmonic available to excite structural resonances which are likely to exist at relatively low frequencies.

Phantoms

The existence of phantoms appear in a frequency analysis of noise or TE as a 'wrong' frequency. It is rather a temptation to ignore them because it seems that if there are 106 teeth on a gear there should not be a vibration at 145 times per revolution. Their existence is liable to be blamed on some unknown electrical interference or sampling frequency fault. They may however be genuine. They are normally caused by the machine on which the gear was manufactured, whether a hobber or grinding machine. Even though a final process such as honing, shaving or grinding may not in itself cause phantoms these processes tend to follow the previous pitching so that any problems left on the gear at the roughing stage may not be eliminated in finishing. They are usually caused by the 1/tooth error from the worm and wheel which is the final drive to the table carrying the gear and the frequency may range from 90/rev typically on a small machine to between 300 and 400/rev on a large machine. Amplitudes are small, of the order of 1 to 2 µm but this is more than sufficient to be audible and is sometimes larger than the 1/tooth component.

Such phantoms or ghost tones in a gear are clear and consistent in the noise, vibration and in the TE. They are not easily detected by conventional profile or pitch checking but it is sometimes possible to see them on a wide facewidth gear in the helix check as they appear as a wave on the helix.

If the existence of a phantom throws suspicion on the accuracy of a gear manufacturing machine it is relatively straightforward to test the machine table accuracy directly. One encoder mounted on the table and one on the worm drive shaft give the TE directly and it is then sometimes possible to adjust the worm alignment to minimise the l/tooth error, assuming the worm has been mounted in double eccentric adjustable bearings to allow adjustment of clearance and alignment.

Another hazard that can be encountered is a torsional vibration linked to the revolution of a pinion appearing to be l/tooth or a modulated l/tooth but caused by a driving stepper motor. Stepper motors are popular drives for positioning due to the simplification of the control aspects but have the disadvantage that they cannot accelerate high inertias. The designs must ensure that the moment of inertia seen by the motor is small and there is then a possibility that the steps of the motor will insert torsional vibration which, in extreme cases, can reverse motor direction each step allowing gears to come out of contact.

IMPROVEMENTS

Economics

Returning to the basic ideas of noise generation we have:

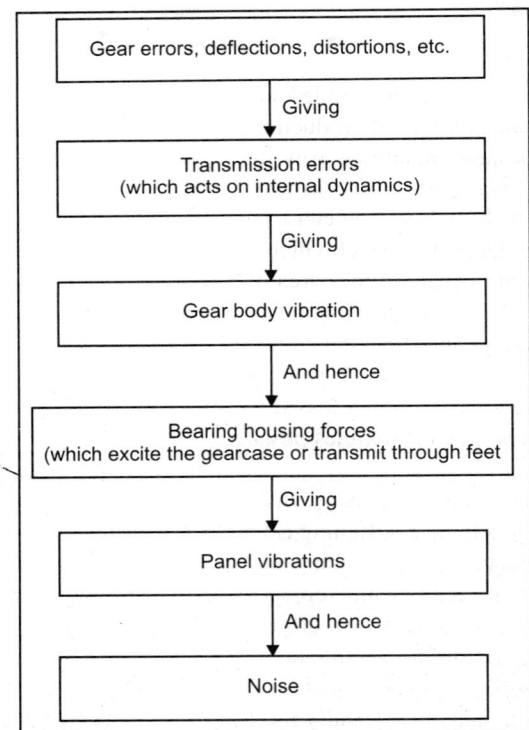

We can (in theory at least) improve any part of this chain and the end result, in a linear system, will be less noise. Hence, we have the choice of tackling (and improving) the transmission error, the internal dynamic response, the external structure dynamic response, or the sound after it is out of the metal.

Once the initial investigations have been earned out the choice must be made as to where improvements should be tried. In general, the choice must (or should) be dictated by economics. This usually rules out tackling the sound after it has left metal. Absorbing sound without an airtight enclosure is difficult and preventing air circulation does not help cooling. There are a few occasions when the choice is made on time scale or for purely political reasons but for the majority of problems, economics should dominate.

Unfortunately this means having a rather good understanding of what the problem is and what the financial implications are of a given set of changes. In the middle of a high adrenaline situation with installation design blaming 'lousy gears' and the gear production blaming a 'hopeless installation', this is not always easy and sometimes impossible. The dominating requirement is to determine the TE since this will give an immediate clue as to whether the problem can be attributed to poor gears or an oversensitive installation. Without knowledge of the source of the trouble much money can be wasted on attempting to improve a gear pair or an installation that is already extremely good.

In the limit the problem may be so intractable that every aspect must be improved. Fortunately this is rare and only occurs when several developers already had a go at improving the installation stiffnesses, resonances, and gear design details and have eliminated all the easy possibilities. As often in engineering there is a law of diminishing returns and it is only possible to get dramatic 10 dB or 15 dB reductions in the initial stages.

Improving the Structure

Improving the structure is usually the simplest and most obvious of the approaches. It is generally not the most economic approach for a 1-off production problem but is by far the most economic for anything that is being produced in large quantities. Any improvement is gained with some initial redesign cost but little subsequent cost per item. The first move is to run round the gearcase (or machinery in which the gearbox is installed) with an accelerometer feeding into an analyser set to the troublesome frequency. The hope is to find some large, flat panel which is behaving as a very good loudspeaker. The relevant criterion is roughly velocity squared times area of panel for sound emission.

Figure 10.10 shows sketches of possible mode shapes for a cover or panel. If vibration amplitudes measured in the centre are greater than the edge support amplitudes [10.10(c)] the panel is acting as a loudspeaker (at the relevant frequency). If panel centre vibration amplitudes are less than edge support amplitudes [10.10(a)] the cover is giving less sound than would a perfectly rigid cover [10.10(b)] so it should be left strictly alone. It is sometimes possible to isolate a panel completely from its support but this is not common.

Individual 'amplifying' covers or panels can have their sound transmission greatly reduced either by thickening the panel or by adding a stiffening rib in the centre. Figure 10.11 illustrates the difference in mode shape between a panel with an effective centre rib and one without.

Technically, the centre rib restricts movement so that the 2 half panels can only vibrate in anti phase (as a dipole) and their emitted sound waves (180 degrees out of phase) tend to cancel, once they are well away from the panel. The rib has to be quite deep to be effective on a flat cover and within a casting or weldment, it helps if an internal rib is also taken across the corner onto a neighbouring panel. The resonant frequencies of the panel are greatly increased.

Gearcases which are cast tend to be much quieter than the corresponding weldments. This is not, as customarily assumed, because cast iron has greater damping than steel because both have very small damping in absolute terms. The main reasons for the difference are that curved cast surfaces are much more rigid than flat surfaces and because iron casters are paid by weight, castings are usually much thicker than the corresponding weldments. As plate bending stiffness is proportional to thickness cubed, this provides a major increase in rigidity despite the lower modulus of elasticity.

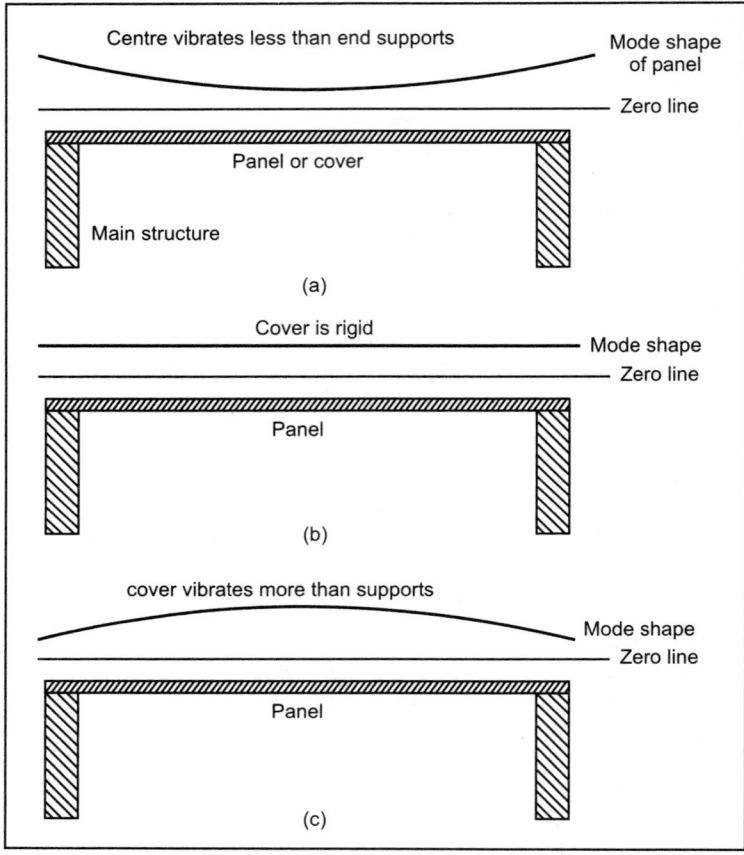

Fig. 10.10. Vibrating shapes of panels.

There is also likely to be an increase in corner stiffnesses and an effective reduction in span due to the radii associated with casting. It is of interest that the structural rigidity of a weldment in torsion is little affected by the depth of welding at the corners. In a normal gearcase, stresses are negligible because high stresses would give ridiculous movements so it is not necessary to have high strength at the welds. This means that within a given cost, it is often much better, from the structural and noise aspects, to have thick panels with only (unchamfered) fillet welds rather than thinner panels with (expensive) full depth welds. If all the individual panels have already been stiffened and split into dipoles then little can be done without a major increase in weight. Increasing wall thicknesses gives major stiffness increases (but with weight penalties) but use of aluminium or magnesium alloy panels allows large increases in thickness and hence plate bending stiffness without weight penalties (but at a cost). Cars and office machinery have a problem because there are large thin flat panels. On a car it is not possible to increase panel thickness due to weight penalties and although improvements can be made by adding highly viscous bitumen-based damping pads on the panels there is, again, a weight penalty. Modern body designs tend to have more curved panels, not because of styling considerations but as an aid to increased stiffness. The ideal structural shape is a sphere. Office machinery traditionally has flat panels so great care has to go into isolating the drives from the panels. Plastic may be used to increase wall thicknesses and, hence, rigidity and damping, despite the low modulus of plastics.

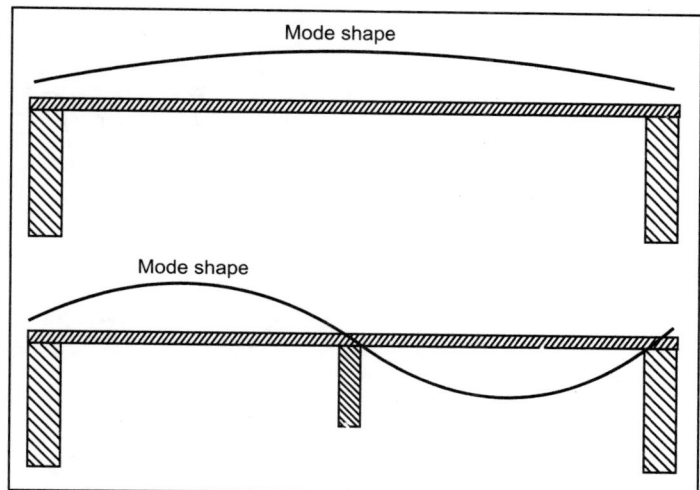

Fig. 10.11. Effect of centre rib on mode shape for a vibrating panel.

At the design stage there will not be a structure available to test but occasionally there is a smaller but similar gearbox available. Once the smaller gearbox has been tested the natural frequencies of the larger design can be estimated. The relevant non-dimensional parameter for natural frequency is $\omega^2 L^2 \rho / E$ so since the material is the same, the product of natural frequency and size should remain constant. Typically a 25 per cent increase in all dimensions should give a 20 per cent reduction in natural frequencies provided geometric similarity is maintained. The existing gearbox can then be tested at 125 per cent speed to give an idea of the vibration responses to be expected.

Improving the Isolation

Most machinery has the gearbox isolated from the main structure by rubber mounts. If not, the design is asking for noise troubles. Unfortunately, the isolation mounts have very rarely been designed with the specific intention of isolating the 1/tooth frequency which is usually the main excitation. Sometimes, as in an elevator drive, it is difficult to isolate the drive from the customer (in the lift cage). Many installations have isolators which were designed to isolate 1/rev (often 1450 rpm, 24.5 Hz) and simple theory, says that the isolation should then be very good at 24/rev (i.e. tooth frequency of 600 Hz). Figure 10.12 shows the theoretical single degree of freedom response and what may realistically happen as the internal resonances of the spring give 'spring surge', the bane of racing engine valve springs.

Satisfactory isolation of tooth frequency needs a design tailored to tooth frequency, so either the isolator should be redesigned for the higher frequency or two stage isolation is needed when both 1/rev and tooth frequency are involved. The 1/rev will not come through as noise because frequencies are too low but will be felt as vibration whereas 1/tooth noise frequencies cannot usually be felt as vibrations.

As with all 3-dimensional isolation it is important that lateral or vertical vibration and torsional vibration modes are decoupled to prevent interactions. This is most important in a car where there are large torsional vibrations of the engine, especially at idling. If these were allowed to interact to give vertical body movement, there would be severe passenger irritation.

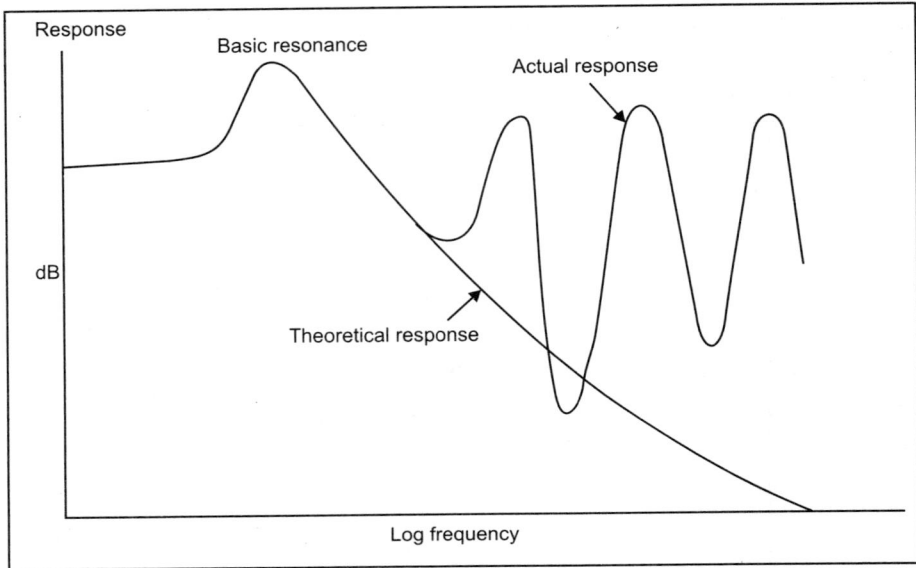

Fig. 10.12. Typical response of vibration isolator.

Another problem comes from large 'static' loads. We need relatively soft support springs to give good vibration isolation, but if high average loads are imposed, the springs must be stiff to prevent excessive geardrive movement. This problem occurs in cars because with a transverse mounted engine, gearbox, and differential assembly, the system must withstand reaction torques of the order of 2000 Nm (1500 lb ft) at full throttle in bottom gear but it must be quiet when cruising on a motorway when the torque is only 100 Nm (75 lb ft). The most satisfactory solution is to have a highly non-linear support which is soft at low torques and locks up when the torque rises. Fortunately, a driver is not worried about high noise levels for a couple of seconds at full throttle in lower gears when the high torque involved 'bottoms' the support and there is high vibration transmission.

In a very sophisticated installation the 'ultimate' isolation is to indulge in vibration cancellation techniques at the (four) gearbox support feet in addition to using soft mounts. This is technically easier than cancelling airborne sound after it has escaped from the metal. It is, however, a very expensive, delicate and temperamental method which should be avoided for all normal engineering.

Reducing the TE

Since TE is the original source of the trouble, reducing TE is an obvious way of reducing noise. The traditional 'fix' with industrial gears (which had not been ground) was to grind the gears and this was sometimes sufficient for a one-off problem. This 'fix' inherently assumes that the design is correct and that manufacture is inaccurate, but this is rarely true in older designs where it is highly likely that profiles were not correctly designed. To reduce TE, we must first find out what is causing the TE. It is important to remember that what matters is the TE. Under working load, not the TE under zero load. Spur gears, without the complications of helical averaging effects, are relatively easy to diagnose. The problem is usually one of bad design where a standard amount of tip relief has been applied to give an almost parabolic shape to the variation in profile from the pure involute. Figure 10.13(a) shows a typical traditional flank profile with the associated no-load TE and Fig. 10.13(b) shows the effect of load on the TE.

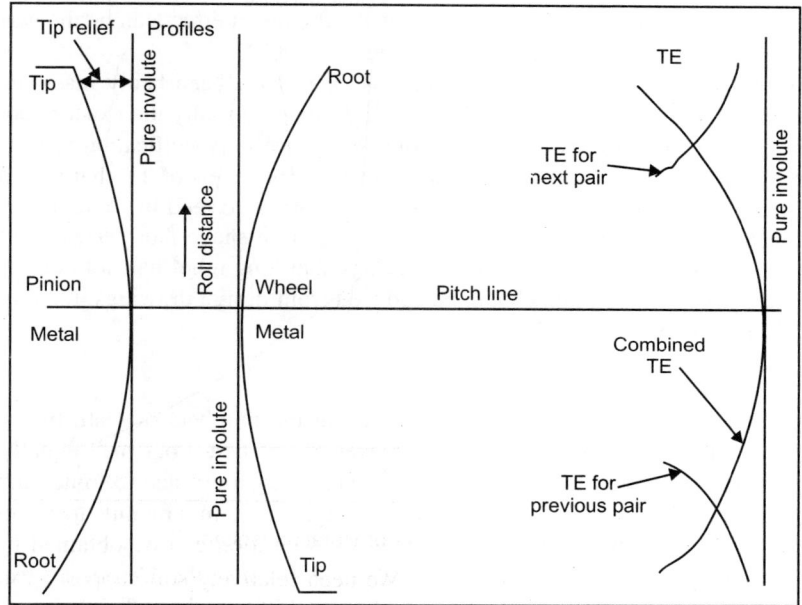

Fig. 10.13(a). Flank profile shapes combining to give TE.

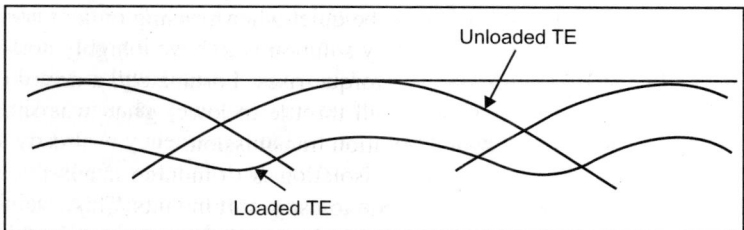

Fig. 10.13(b). Effect of load on TE for a spur pair.

At low loads the TE is high and, although it reduces under torque, it never comes down to very low levels. The solution, as discussed, is to specify the design relief as linear, starting from the correct roll distance down the tooth flank from the tip, depending on whether high or low load is more important. For spur gears the accuracy of alignment of the helices (at angle zero) is relatively unimportant except when non-linear effects dominate.

Changing from spur gears to helicals usually gives a reduction in TE, by up to 10 dB, but much depends on the accuracy of alignment of the helices when installed. Reducing the TE on helical gears is a much more difficult process due to the complex interaction between helix and profile effects. Much depends on whether the original design attempted to achieve a smooth entry by using tip relief and negligible end relief or end relief with negligible tip relief. A particular case of the latter occurs with heavily crowned gas with no tip relief designed for light loads. Improving gears where there are no obvious major design errors will usually involve either an amount of extremely clear-headed thinking or the use of at least a thin-slice model. In some cases the dominant effect can be the variation in helix matching occurring due to shaft deflections under load.

It tends to be assumed that gears are noisy because they have been badly made and there is the inherent assumption that the gears will have been well designed, usually the exact opposite of reality. Gears are often manufactured to within 3 μm of the design profile specification which itself may be 15 μm in error. For any old design it is well worth checking the levels of TE that would be predicted from the specified tooth shapes. In any prediction it is important to feed in some helix errors since a perfect helix match will often give low TE regardless of profile shape, but perfect helix matching is unrealistic. Even in a modern design it is worth checking that long relief has not been used instead of short relief or vice versa. Although much can be deduced from design drawings, there is no substitute for experimental measurement of the TE.

Frequency Changing

A standard 'fix' for noise problems was to try increasing the numbers of teeth by using a smaller module. This, like many other 'cures' sometimes makes things better but may equally make things worse. There is a major stress penalty (in root stresses) in reducing tooth size so some caution is needed. Since it is a rather expensive option to change a gearset, say, from 6 mm module to 5 mm module, it is perhaps worth considering that the same development information can be obtained by running the gearset 20 per cent faster at the same torque level. This is because the 1/tooth TE excitation levels (in μm) will be much the same for the 5 mm and 6 mm module teeth so the excitation from 6 mm module teeth run 20 per cent faster will be much the same as that from 5 mm module teeth at the standard speed. The objective of the tooth number change is to change the exciting frequency and with luck, move it away from a resonance but it helps greatly if you know where the resonant frequency is first. Having a variable speed (inverter) drive is a great asset for preliminary tests because it will immediately show whether a frequency increase (at constant torque) will make the noise better or worse. Sometimes there are no resonances as such and the frequency change is simply to move the noise to a less irritating frequency. As a general rule, if the tooth frequency is already above 1 kHz it is better to put it up, but if below 500 Hz, it is better to reduce the frequency if possible.

The main reason for avoiding the 500 Hz to 1 kHz band is that the human (A-weighted) ear is most sensitive in this range and also because many structures are at their noisiest in this range. At high frequencies the wavelengths are smaller and panel vibrations have a greater tendency to be in anti-phase and cancel. At low frequencies, velocities and, hence, noise pressure levels drop and also hearing sensitivity drops.

Damping

It is tempting to think that it should be possible to introduce damping to reduce noise levels, either inside the gearbox or in the structure of the installation. Damping of very thin panels such as car body panels is successful in reducing vibration and noise levels, but attempts to 'increase damping in a gearcase are not usually very successful. Adding a pad of viscoelastic material to a car panel 0.75 mm (30 mil) thick can absorb a high proportion of the bending wave energy passing through the panel and natural frequencies are reduced due to the extra mass but if the 'panel' is 1" (25 mm) thick steel there are no suitable materials to extract much energy. Machine tool designers have attempted to insert damping layers at interfaces between castings, but this approach has not been successful and the use of materials such as synthetic granite, though having nominally higher damping than steel, sacrifices stiffness.

An approach which has been successful in unstressed components such as internal combustion engine rocker box covers has been to sandwich a damping layer between two aluminium alloy sheets.

Scaling this up to industrial gearbox thicknesses does not appear to work although some large gearboxes have used a layer of sand between two steel skins. Whether the principal effect of the sand comes from its mass, from its damping or from its action in spacing the steel panels apart, has not been stated. As previously mentioned, although cast iron has higher damping than steel, the effect of the material damping is negligible compared with the damping from bolted joints, shrink fits, loose members rattling about and energy being dissipated into foundations. When plastic casings were used for domestic kitchen equipment the noise levels have tended to be higher than for the previous cast metal casings despite the higher internal damping of the more flexible plastic.

The one technique that has been used over a wide range of industries, with reasonable success, is the tuned damped absorber. An auxiliary mass is supported on a damped spring which is usually deformable nitrile or butyl rubber and is tuned to just less than the frequency of the troublesome resonance. Nitrile rubber is popular as it has nearly the optimum level of internal damping, even at very low amplitudes of vibration. The theory was worked out by Den Hartog over 60 years-ago (Fig. 10.14). Although it is possible in theory to use steel springs and oil damping, this is rare due to sealing and tuning problems. The device needs careful tuning to the correct frequency and is, in general, only worthwhile if the auxiliary mass can be about 10 per cent of the effective mass of the resonance and the original dynamic amplification factor (Q) of the resonance was greater than 8.

The absorber can then reduce the Q factor to below 4. Untuned (Lanchester) dampers which use only mass and viscous damping will work over a range of frequencies but require greater mass and give much less damping so they are little used except for torsional engine vibrations which occur over a wide range of frequencies as speed varies.

STRENGTH VERSUS NOISE

Connection Between Strength and Noise

It is often assumed, sometimes unconsciously, that a noisy gearbox is one that is likely to break. This comes from the observation that a gearbox that is disintegrating (usually because of bearings failing) becomes noisy (or noisier) and so noise is associated with failure. Usually there is little connection between noise and strength and if a system keeps the gear teeth in contact it is rare for vibration to affect the gear life. The time when noise and strength are directly connected is when the teeth are allowed to come out of contact and then produce high forces in the following impact. High noise and high stresses are then both associated with the repetitive impacts. The extreme cases where noise and strength give rise to dramatically different designs are:

1. Ultra low noise teeth with a nominal contact ratio above 2 where the minimum number of tall slender teeth is above 25 and the pressure angle is lowered.

2. Ultra high strength gears for lifting self-jacking oil drilling rigs where 7 tooth pinions mesh with racks at a pressure angle of 25 degrees.

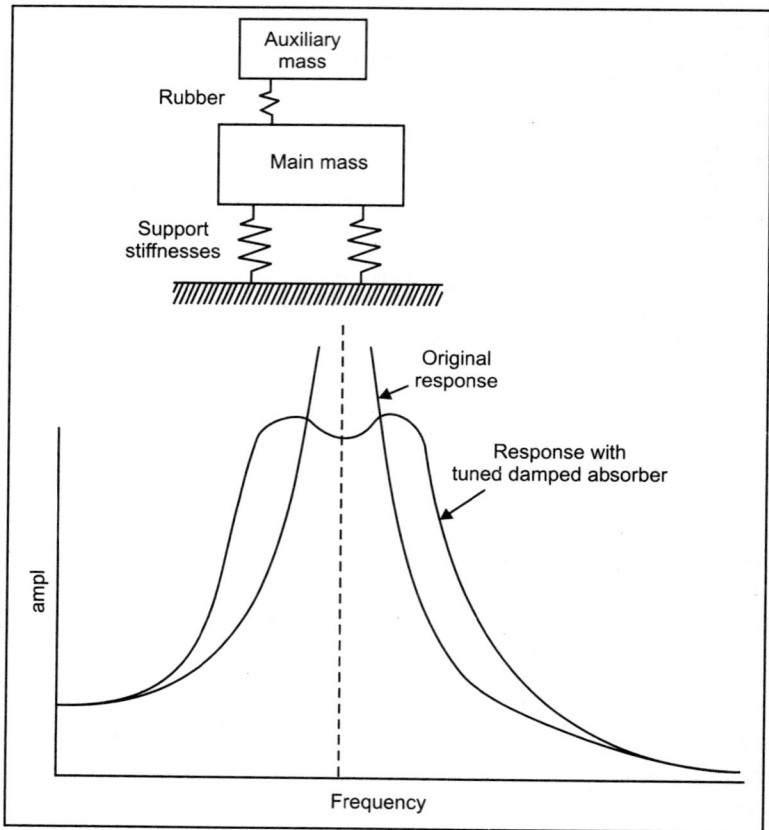

Fig. 10.14. Tuned damped vibration absorber response.

This lack of connection between noise and strength presents difficulties when it comes to testing the gears on production. If we are targeting minimum noise then the only worthwhile test is a TE test but this is of no use for assessing strength. Conversely, if the requirement is for maximum strength, especially for low speed gears, then it is essential to carry out a bedding check to make sure that the major part of the face of the gear tooth is working but a bedding check is not a valid predictor of noise.

Production is left with the problem that they need to know whether noise or strength is the more important and if both are important, then both tests must be carried out. This is unfortunate because bedding is a relatively slow and expensive test. Skilled labour is required and the test is time consuming so costs rise. TE testing is very much less expensive but is rather unknown as yet in general industry so it is viewed with great suspicion and is avoided wherever possible.

Much depends on the application since within a given gearbox (such as the previously mentioned car gearbox) there may be strength dominant on the two lower gears, requiring bedding checks, and noise dominant on the three higher gears, requiring TE checks.

Design for Low Noise Helicals

From a 'philosophical' aspect it is relatively easy to design for maximum strength. If we look at a helical gear flank as in Fig. 10.15 we need to get the maximum length of line of contact, compatible with reducing the load to zero at the ends of a line of contact. Within the line of contact, the objective is to get the loading per unit length constant over the length of the line.

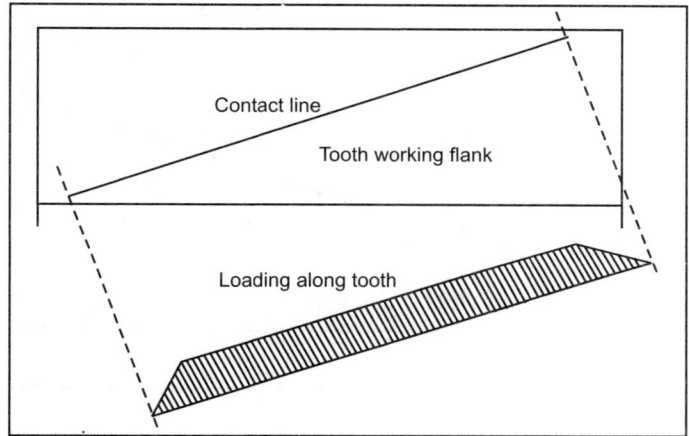

Fig. 10.15. Desired loading pattern along contact line for maximum strength.

This objective results ideally in a trapezoidal shape to the loading distribution along the length of the line of contact. There is little choice in the resulting 'ideal' design, apart from how fast we reduce the loading at the ends. It is preferable to use end relief instead of tip relief to maximise the area of full loading or to use 'corner' relief if the extra manufacturing cost is justified. However, to achieve a good loading across the facewidth the helix alignments must be extremely good, to within, say, 20 per cent of the mean tooth deflection. In designing for low noise there are more options available and much depends on whether or not there is a good margin of strength in the design. If we could rely on perfect helix alignment, life would be fairly simple since, apart from tip relief and end relief needed to prevent corner loading, we could use virtually any profile at low load.

At high load, if the axial length of the gear is an exact number of axial pitches then the contact lines on the pressure plane would always have the same total length. This is shown in Fig. 10.16 and would give constant mesh stiffness, hence constant elastic deflection and a smooth drive. Such a design is, of course, also a high strength design if there is negligible relief at the tips or ends.

Unfortunately, the reality is that helix alignment is very rarely better than 10 μm (0.4 mil) and the error is more likely to be much greater, of the same order as the theoretical elastic tooth deflection. This end loading not only puts the load concentration factor across the facewidth (C_m or $K_{h\alpha} \times K_{h\beta}$) up above 2 or even 3, but prevents the helix effects from averaging out the profile effects. We are left with the necessity of assuming that the helix alignment will be poor and thus need to design accordingly.

One approach to the problem, which can be used if the 'design condition' load is extremely low, is to use very heavy crowning and a perfect involute profile with merely a chamfer at the tip.

A smooth run-in is achieved thanks to the crowning, and it is permissible to dispense with conventional tip relief if loads are low since the teeth are not deflecting significantly. This type of design is quiet at low load and tolerates very high misalignments but cannot be loaded heavily as the lengths of contact line are so short. Adding tip relief to the profile allows the use of moderate loads but, as with a spur gear, we cannot get low TE at both design load (for which we need long relief) and low load (for which we need short relief).

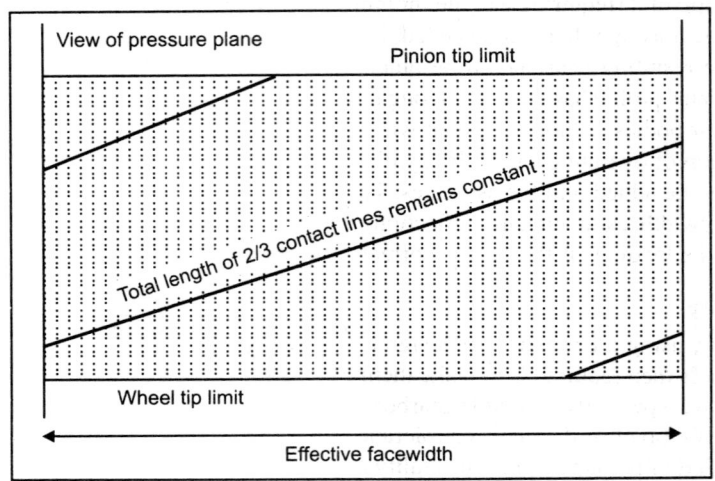

Fig. 10.16. Contact lines when facewidth is an exact number of axial pitches.

In practice we do not normally have either perfect alignment or extremely low loads to allow us to use the two extreme designs described above, so compromises are necessary. Figure 10.17 shows one possible compromise pinion helix shape where we have estimated a maximum misalignment of ±15 μm across the facewidth and expect 20 μm nominal tooth deflection. A crowning of 10 μm will keep peak deflections and loadings roughly constant provided the helix mismatch stays within 15 μm and at the ends a further end relief of 25 μm might be suitable. The wheel would then not be helix relieved at all.

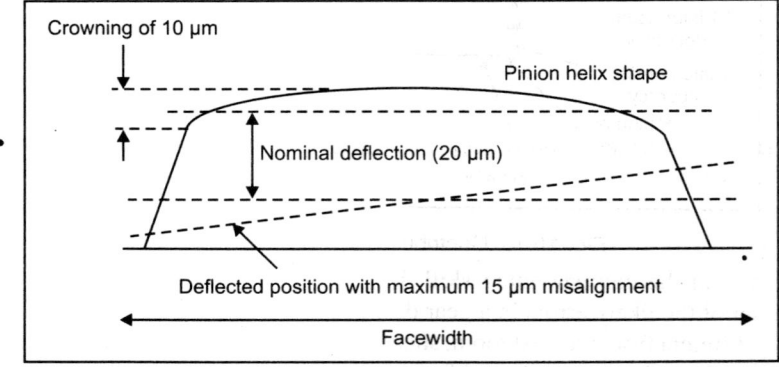

Fig. 10.17. Helix matching and deflections for a compromise design.

Profile shape would follow normal 'spur gear' rules with the choice between 'long' and 'short' relief according to whether best performance is required at full load or low load. Exact design of the relief is difficult because there are variations in deflections of up to 10 µm across the facewidth so design is inevitably a compromise. The previous comments were made in relation to standard proportion 20° pressure angle gears. However, as the effect of the inevitable helix mismatch is to move the characteristics more towards those of spur gears, we can take this to the extreme and design as if they were spur gears. The ultimate spur gear design, as far as noise is concerned, is a low pressure angle tooth with an effective contact ratio of 2 (requiring a higher nominal contact ratio). The problem is slightly easier than for an actual spur gear as tip relief is not needed, just a chamfer, because a smooth run-in is achieved by the end relief. The resulting gear should be quiet at low and high load whether aligned well or not, provided that the 'spur' profile has the correct long relief and a real contact ratio of 2. The above comments apply to 'rigid' gear bodies without torsional windup, without radial wheel rim deflection and without bending or distortion of overhung shafts. If any distortion or body deflection effects are occurring then, their effects have to be added into the estimates. This works backwards by assuming that the loading is even across the facewidth, estimating the deflections and distortions and putting these into the calculations then reestimating the loadings if the gear is corrected. A second iteration may be needed.

Design Sensitivity

It is relatively easy, using a computer, to design a pair of gears which will be perfectly quiet under a given load. All that is then required is to make them accurately and to align the axes welt in the gearbox, and we will then have a perfectly inaudible gearbox!! If only! Referring to the generation of TE illustrated in Fig. 10.18, it is all too clear that a dozen tolerances of 2 µm (at best) are going to have trouble fitting into a permissible TE of perhaps 1 µm. The reality is that all the factors will have errors, some relatively small at 2 µm but some large at 5 to 10 µm and although elasticities will allow some averaging, there are likely to be relatively large variations.

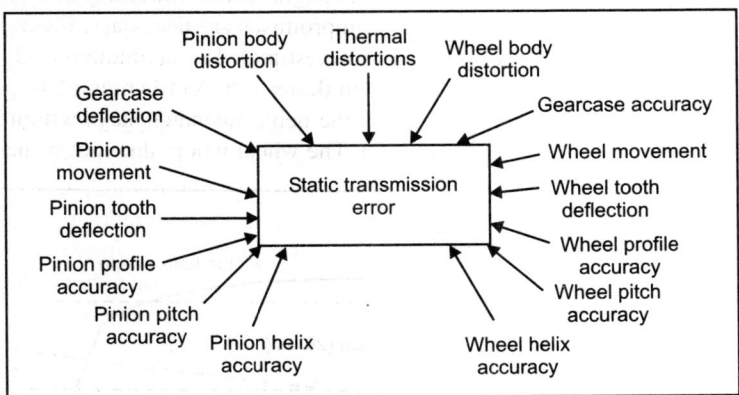

Fig. 10.18. Contributors to mesh static TE.

The difficulty, and the corresponding skill, lies in having a compromise design which will be reasonably tolerant of the likely errors in a gear drive. Unlikely errors, such as having a profile on one tooth completely different from the next tooth, should not be considered but reasonable errors of profile, pitch and helix matching should be allowed for in the design. Realistically, the only way to assess the effects is to have a computer model and to vary all the tolerances by expected manufacturing errors and assess the effect both on TE (noise) and on peak stress loadings. The effort involved is well worthwhile since it is not always obvious what effects the changes of design and manufacturing variables will have in practice, either on strength or vibration. The danger with allowing an inexperienced designer to use a computer model is that they will take the simplistic view that whatever their design, if the computer predicts that the TE will be zero, then the design is 'perfect'. This mindset then puts all the blame for trouble on 'inadequate production'. It is important to educate a designer that relatively large (5 µm, 0.2 mil) profile errors and larger helix errors are inevitable and that their design must be good enough to tolerate errors, from both aspects of stressing and noise.

Buying Problems

When buying-in gears, the problems fall into two groups, stress and noise, with a great difference between the degree of control and confidence in the two cases. Currently there are few problems associated with gear strength and durability. Around the world, a few gear sets fail each year but failures are rare and invariably there have been silly mistakes made, so investigations are simple and straightforward and apportioning blame is relatively easy. Often the problem is due not to one error but to a combination of errors. As far as the buyer is concerned, specification of the drive that it should be to either the AGMA or ISO/DIN/BS specification should produce a satisfactory result. The gear manufacturers dare not produce an inadequate strength drive (because of the legal implications) so there is little to worry about. A glance at the computer printout to check that a sensible value (>1.5) for K_β (the load intensification factor) was used and that an adequate safety factor was present should be sufficient. The times this may not be adequate are if a ridiculously low diameter to length ratio was used on the pinion without helix correction or if sharp corners were left to give stress concentrations.

Noise is much more difficult. If it is the gearcase itself which is going to be the noise emitter then, as with a hydraulic power pack, specifying the total sound power emitted or specifying, say, 77 dBA at 1 m distance for a machine tool or 60 dBA for an office device, will ensure a sufficiently quiet drive. The problem that arises in practice is that it is often not the gearbox itself that emits the sound but the main structure. The only worthwhile tests are those in position in the unit and it is then all too easy to shuffle blame between gearbox and installation.

A knowledgeable customer can start by specifying a 'reasonable' TE at each mesh in the gearbox but this requires a sophisticated investigation of the results obtained *in situ* with known levels of TE in the mesh. There are the problems of first determining a tolerable level and the associated problem that often neither the manufacturer nor the customer will yet have TE measuring equipment so they cannot easily check, especially since the critical value is the single flank error under load rather than under inspection conditions. Attempting to specify the necessary quality by invoking an ISO single flank quality level comes to the same thing in theory but, like the normal quality checks, takes no notice of whether it is 1/rev or 1/tooth that is important or whether both are within specification but the waveform is wrong or whether odd things happen under load so a specification may be wastefully expensive. Overall, the depressing conclusion is that the buyer is rather in the dark for a new design and has little choice but to put their faith in a manufacturer, try the result, then if trouble occurs, panic and measure TE. Dependent on the TE level the buyer can then try another manufacturer, attempt to reduce TE levels or improve the tolerance of the installation, with economics in control as usual. It is important, however, that initially the manufacturer is given all the relevant information since this influences the design. Apart from the obvious information about frequency of overloads or whether the drive will be idling most of its life, it is important that the designer knows what load levels are most critical for noise purposes and whether external loads are likely to distort the gearcase and affect alignments.

Chapter 11

Drilling Operations

INTRODUCTION

Drilling operations in the direct vicinity of housing causes disturbance in the form of light, noise, vibration, smell and traffic. Of all of these, drilling noise is the most prominent disturbance . It continues day and night, 24 hrs a day, 7 days a week for weeks and months in a row. As such, noise affects public acceptance of drilling operations in the neighbourhood.

During drilling operations, continuous noise originates from the top drive, drawworks, shale shakers, mud pumps, generators, purge/cooling air systems, trucks, forklift, equipment and pipe-handling, air and fluid power systems. In addition, there are frequent noise peaks that are particularly disturbing, e.g. squeaking drawworks band brakes, metal-to-metal contacts (elevator, pipe handling) and manual handling such as hammering.

The public in general is not familiar with drilling operations and is often concerned about industrial activity in their neighbourhood. Concerns are related to many aspects of hydrocarbon development, but noise is a key area of concern. It intrudes on people's privacy and recreation. Noise disturbance is more severe during nights and during the summer season when people prefer to be outdoors. People may also be concerned about the effect on nature, e.g. the breeding of meadow birds.

A general principle in managing noise is to remove or suppress noise emissions at source and not to 'treat' at the 'receiver' end—such as by means of sound insulation of houses. The following sections discuss how the impact of drilling noise is assessed, followed by measures to combat noise at source, with examples of two rigs, crew awareness and traffic aspects. Thereafter, measures to minimise propagation of sound will be presented, including use of sound walls and a cocoon structure.

NOISE ASSESSMENT

The impact of drilling noise is assessed for every location using noise contours overlaid on the location map. This reveals the expected noise levels at nearby dwellings or nature areas. Protective measures are based on this assessment. Contour maps are an important part of Environmental Permit applications. There are two ways of establishing the noise contours:

1. Based on sound measurements of individual noise sources (equipment), recombined through a complex calculation of sound transmission.
2. Based on measurements at a distance from the site.

Until a few years ago, only the first method was used because accurate measurements at houses could not be performed due to the presence of other sound sources, especially traffic, wind noise, other industries, etc. More recently, an advanced multi-microphone-monitoring (M3) system has been introduced that can correct for such external influences. The measurements (every second) are available in real time over the Internet. This has made the prognosis of the expected noise simpler and more accurate. The measurements are compiled into noise contours around the rig, indicating the distance at which the continuous noise level has been reduced to 50 dB (Fig. 11.1).

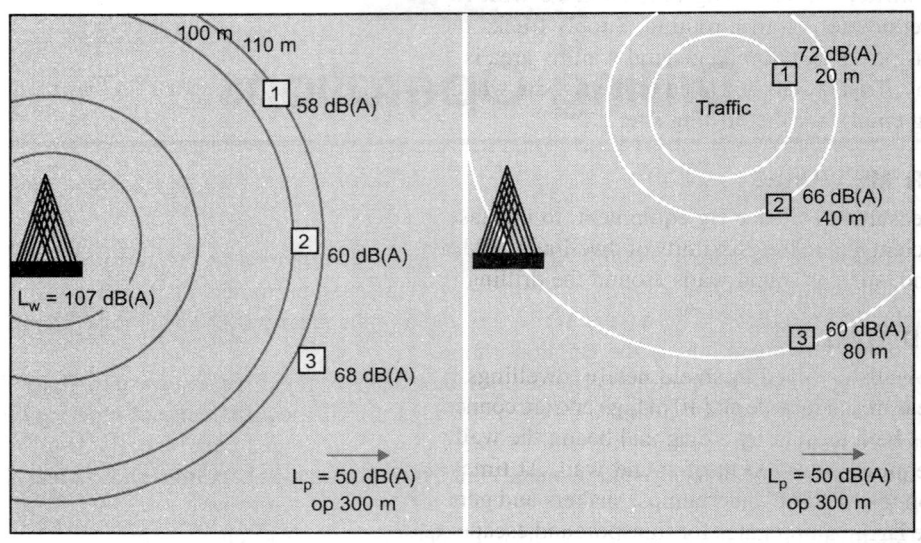

Fig. 11.1. A multimicrophone monitoring (M3) system can help to correct for external sources (traffic, wind noise, etc.) when assessing the impact of drilling noise. The measurements are available in real time online.

In addition to noise contours, a peak noise assessment is also made. The basis is again the extensive noise recording with the M3 system, which counts how many noise peaks are generated and at what level. The assessment uses the area where a noise peak that exceeds 60 dB will be heard. For instance, a 70 dB peak may travel 975 m before it has reduced to 60 dB. The circular area and the number of people living in this area is used to assign the severity of a noise peak. With the rig's noise profile and background noise levels (nondrilling related) as input, the assessment on whether additional measures will be necessary.

RIG CREW AWARENESS

In addition to rig upgrades, the awareness and actions of people working the rig are a key aspect. An active sound-monitoring system has been used to change the way people work and behave on the rig. A set of traffic lights has been used to give direct feedback to the rig floor in case a preset limit of (peak) noise is exceeded. From the measurements and increased awareness, many operational learnings have been established over the years. These include avoidance of high drillstring speed and use of downhole motors, schedule noisy activities to daytime, avoid metal-to-metal contacts (tripping, pipe handling, hammering), pre-installation centralisers in yard off-site, avoid running empty shakers (resonance), no shouting/whistling, etc.

TRANSPORT AND LOGISTICS

Trucks can be a cause of significant noise disturbance to neighbours. Transport is therefore scheduled between 7 am and 7 pm. There is no heavy traffic at night except for in operational emergencies . A well-planned operation is crucial for this. Other measures include rubber mats over steel gratings, non-curved steel driving plates, silencers on truck depressurising vents, sound-isolated compressors on bulk trucks, switch off reversing beeps/replace with flash lights and watch man, etc. Another element of noise reduction is traffic routing. Supply trucks must follow a route that avoids cities and villages as much as possible. Often, a general waiting area is assigned away from the rig, e.g. a parking place at a highway. Trucks will be called to the rig site when needed. This practice avoids traffic congestion and running engines at the drilling site.

OTHER MEASURES

The measures to silence rig equipment, to increase crew awareness and to manage transport are often insufficient due to the proximity of dwellings. Additional measures are required to break the propagation of sound, such as sound walls around the drilling site, or occasionally a so-called cocoon structure.

SOUND WALLS

Sound walls are used to shield nearby dwellings from noise. They consist of sound insulation panels that measure 2.5 m wide and 10 m high and are connected to a concrete counterweight block of 10 tons. Each panel is held upright by a diagonal beam; the wall can survive a wind force of 12 Beaufort. A location may require up to 400 m of sound wall. At times, sections of sound wall have been used within the location to shield off mud pumps, shakers and generators. Erected a week prior to rig arrival, the walls include large rolling gates for transport and escape doors for personnel. The panels have to fit exactly to avoid air gaps that can become a noise source itself.

The noise reduction is approximately 25 dB immediately behind the wall; far field it is about 5 dB, measured at a height of 5 m above grass level (the first floor of a house). This is due to the bending of sound waves over the sound wall.

CASE STUDY

1. Acoustic and Engineering Consultants (AEC) Ltd. were commissioned by British Drilling and Freezing Co. Ltd. to measure noise levels during drilling operations using Rig 25 at Bevercotes, Nottinghamshire.
2. This report details the measured rig noise levels and the major sources of noise, based on the existing rig layout, of Rig 25.
3. The survey was conducted on 23/24 September 2008.

Description of Drilling Rig 25

Site layout

1. BDF Rig 25 was generally arranged in a layout as shown in Fig. 11.2.
2. In summary, the rig consists of a drill platform, with a diesel engine powered draw-works (Failling 2500SE), located at high level, together with an adjacent mud tank and shale shaker system. The diesel engine to the draw-works is located within an acoustic enclosure and is fitted with a substantial exhaust silencer.

3. The rig was fitted with two main generators, of which only one was used at any one time. These were both located at ground level and contained within acoustically treated enclosures with silencers fitted to air intake and exhaust vents. The generators provided all the electrical power to the rig and site in general during operations.
4. The site was also equipped with two mud pumps. The diesel engines powering these mud pumps were also contained with acoustically treated enclosures with silencers fitted to air intake and exhaust vents. The pump units themselves were open units.
5. There are also various ancillary containers and equipment around the site, including storage tanks and office buildings.

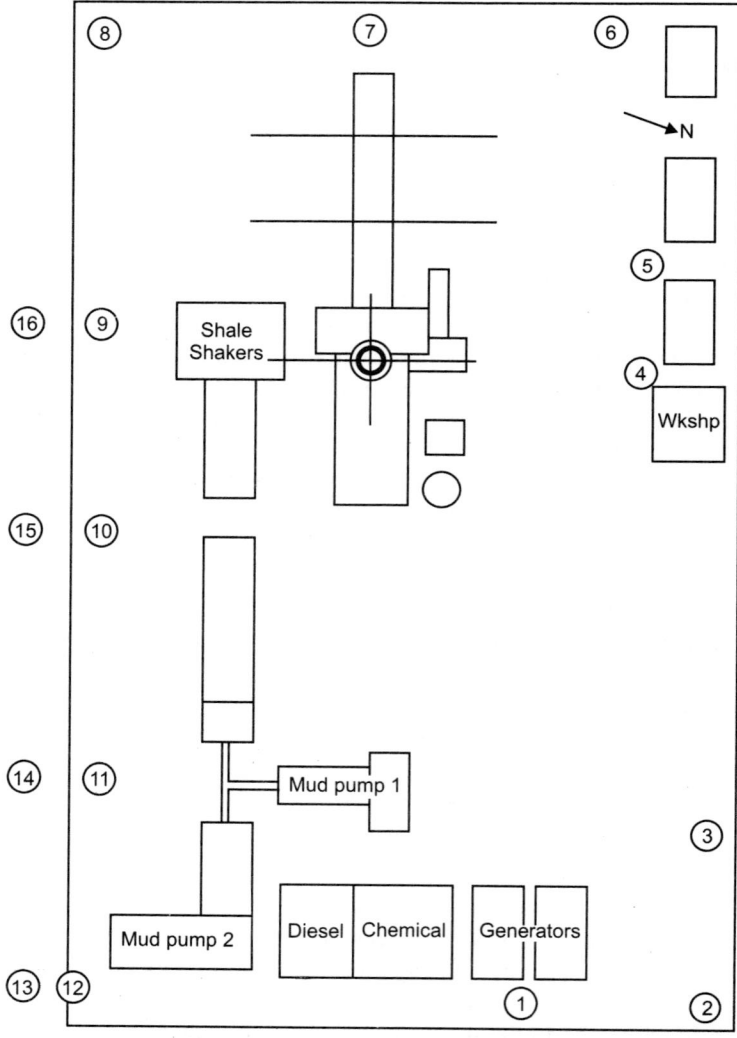

Fig. 11.2. General site layout (showing noise measurement locations).

Local topography

1. Drilling Rig 25 at Bevercotes is located on a relatively flat area, approximately 0.75 mile from the main A1 dual carriageway. The nearest residential property was at least 0.5 mile away, in the direction of the A1.
2. The area is rural, although noise from the A1 dominates the surrounding area, to within at least 25–30 m of the site boundary.

Noise Survey

Noise measurements

1. Noise levels were measured around the perimeter of the site during drilling ahead at a depth of approximately 200 m. As the drilling rate was only 0.5 m per hour, the drilling operation was continuous throughout the duration of the survey.
2. Noise levels were also monitored in the immediate vicinity of operational plant.
3. The survey particulars are as follows:

Engineer:	K A Worthington, AEC Ltd.
Date:	Monday/Tuesday 23/24 September 2008.

4. The instrumentation used for measurement was as follows:

Bruel and Kjaer	2260 Modular precision sound analyser	Type 2260
Bruel and Kjaer	1/2″ Prepolarised condenser microphone	Type 4189
Bruel and Kjaer	Electronic calibrator	Type 4231
Bruel and Kjaer	Windshield	
Kestrel	Anemometer/Thermometer	

5. At all environmental measurement locations the microphone was place 1.5 m above the ground and at least 3 m from the nearest reflecting surface.
6. The sound level analyser, which conforms to BS EN 60651:1998 'Specification for sound level meters' and BS EN 60804:2005 'Integrating-averaging sound level meters', was in calibration and check calibrated before and after the measurement periods using a Bruel and Kjaer type 4231 (94 dB) calibrator. There was no significant drift of calibration.

Measured Noise Levels

Drilling noise levels

1. Noise levels measured around the perimeter of the site during the various drilling operations are presented in Tables 11.1 and 11.2.
2. Noise levels in the immediate vicinity of operational plant are presented in Table 11.3.

Table 11.1. Noise levels around drilling Rig 25 perimeter during drilling.

Location number	Measured noise level dB(A), L_{eq}	Distance to bore hole (m)	Noise level, dB(A) (normalised to 1 m)
1	62	46	95
2	61	53	95
3	66	28	95

(Contd...)

Location number	Measured noise level dB(A), L_{eq}	Distance to bore hole (m)	Noise level, dB(A) (normalised to 1 m)
4	65	15	89
5	64	15	88
6	62	34	93
7	65	24	93
8	65	26	93
9	72	16	96
10	71	20	97
11	74	28	103
12	70	37	101
		Energy average level	96.9

Table 11.2. Octave band measurements around site perimeter—surface drilling.

Location	Direction	dB(A)	Measured noise level L_{eq} dB							
			Octave band center frequency, Hz							
			63	125	250	500	1k	2k	4k	8k
1	East	62	77	72	63	58	55	52	50	46
2	NE	61	78	72	61	56	53	49	46	41
3	–	66	85	78	64	56	58	55	49	43
4	N	65	80	75	62	59	59	54	49	41
5	–	64	79	74	60	57	59	54	48	41
6	NW	62	77	70	57	59	56	53	49	41
7	West	65	79	71	59	64	58	56	55	45
8	SW	65	76	72	62	62	58	56	51	44
9	–	72	84	73	67	70	66	63	61	55
10	South	71	85	78	69	68	65	62	58	52
11	SE	74	90	84	74	71	68	63	58	52
12	SE	70	80	81	72	65	63	60	55	51

Table 11.3. Noise levels at specific plant during drilling.

Location	L_{eq}, dB(A)
Shale shaker – 3 metre from front	76.6
Mud pump unit 2–3 metres to side enclosure	80.4
Mud pump unit 2–3 meters to rear of enclosure	73.3
Mud pump unit 2–8 m pump	73.6
Mud pump unit 1–3 m side of enclosure	80.5
Generator – 3 m corner	71.8
Generator – 3 m side	73.7

Discussion and Results

Sound power level

1. The sound power level of drilling Rig 25 during drilling ahead at a rate of 0.5 m per hour is calculated, from the site perimeter measured noise levels, to be 105 dB(A).
2. Due to the nature of the site, and the dominance of traffic noise from the A1, no measurements were taken at any significant distance from the rig. Hence, it is not possible to present noise contours for the rig from the data obtained at this site.

Noise control measures

1. Significant noise control measures have been implemented on this rig, including enclosure of generators and mud pumps units, together with exhaust silencing. The mud pump units themselves been enclosed. It is considered that enclosure of these would provide further reduction in overall noise levels.
2. There was no evidence of brake squeal during drilling, although this may have been due to the 'hard' nature of the ground being drilled and the relatively slow progress rate. The operator should be retested under drilling with a considerably more rapid progress rate, and/or surface drilling.

CONCLUSION

1. Noise levels have been measured at locations around the perimeter of the site and at varying distances, wherever practicable.
2. Based on the calculations from the measured sound pressure levels, the sound power level of drilling Rig 25 during drilling ahead operations at a progress rate of 0.5 m per hour was calculated to be 105 dB(A).

Chapter 12

Hydraulic Pneumatics

INTRODUCTION

Many designers have long accepted leaks as inherent to hydraulic systems, even though advances in technology should have eliminated hydraulic leakage a long time ago. Hydraulics suffers a similar identity crisis when it comes to noise. Noise certainly cannot be eliminated, but a number of products and techniques exist to at least bring noise down to an acceptable level. The problem is that noise reduction is a complex subject, and investing a great deal of time, effort, and money may produce only modest improvements.

SOURCES OF NOISE

A hydraulic system's greatest contributor to noise is the power unit. Noise not only emanates directly from the electric motor and pump, but also is caused by pressure fluctuations in the hydraulic fluid and by components—either resulting from these pressure fluctuations or from physical vibration. Transmitting vibration of the pump-motor assembly to the reservoir can transform this physical vibration into sound— in the same way a loudspeaker transforms electromagnetic vibrations into sound.

Electric-motor noise comes from bearings, the rotor and stator assembly (the characteristic hum), and, especially, the fan. A standard electric motor contains a fan with blades designed to provide cooling whether the motor shaft rotates clockwise or counter-clockwise. A fan designed for rotation in only one direction will generate less noise, so the expense of this option may be warranted if the application demands quiet operation.

Pump noise stems from rolling and sliding of bearings and pumping elements (vanes, pistons, rotors, gears, etc.), plus pressure fluctuations that result from the cyclical nature of the pumping process. Metal housings, whether part of the hydraulic pump or an electric motor, do little to prevent noise from being transmitted to the surrounding environment. Moreover, because the pump generally is coupled to an electric motor (and the coupling itself is a source of noise), noise control often involves treating the pump-motor combination as a unit. This design technique has produced power units where the pump-motor combination is submersed in oil or where the entirepower unit is submerged in the reservoir. This technique uses liquid to dampen sound waves by acting as a buffer between the pump-motor housing and the surrounding atmosphere.

Valve noise has occurred in cabs of construction and other mobile equipment for years. Often, a high-frequency, random noise occurs when fluid, travelling at high velocity through the valve, undergoes a rapid and severe drop in pressure. This causes air dissolved in the fluid to form bubbles which, when they collapse, generate noise. Other types of noise—such as chattering, squealing or buzzing—is generated when poppet-type valves do not seat properly.

Fortunately, most of these problems can be eliminated through better system design or by incorporating cushioning features into valves. A current trend replaces direct-operated valves with joystick-controlled remote electrohydraulic valves. This process of removing the hydraulics from the equipment cab offers other advantages beyond providing a quieter workplace environment.

Fluid conductors (tubing, hose, fittings), often are overlooked as noise sources. However, pressure pulsations in plumbing can distribute noise over a large area. Pressure pulsations can shake hose and tubing, causing rattling and eventual leakage.

Although reducing fluid-borne noise can be complicated, many manufacturers suggest rules of thumb to help reduce noise. For example, terminating a long run of metal tubing with a section of hose at each end helps isolate noise sources (Fig. 12.1). One might be tempted to simplify the design by instead specifying a single section of hose. Hose, however, is very sensitive to pressure pulsations, so in long sections it can be a greater source of noise than metal tubing or pipe.

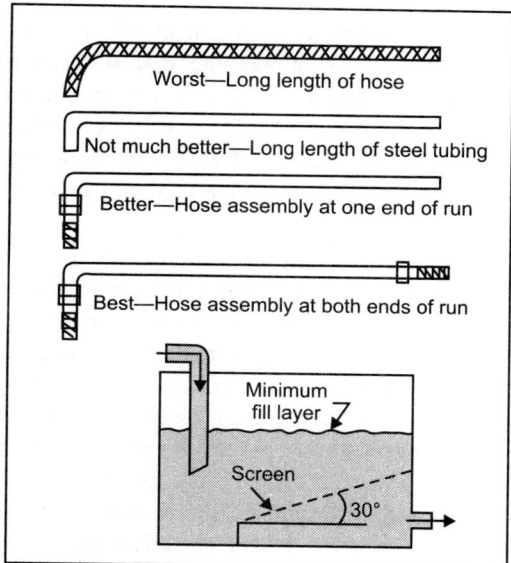

Fig. 12.1. Many simple techniques can be employed to quiet a noisy hydraulic systems. Among these are incorporating hose into long tubing assemblies, tiop, and using a 60 mesh screen positioned 30° from horizontal.

Securing tubing to framework with resilient clamps eliminates rattling and banging noise. However, care should be taken not to confine tubing too tightly, because lines may need to undergo thermal expansion. On the other hand, allowing a tube to fit too loosely could cause wear as the tube constantly rubs against a metal clamp surface. Likewise, resilient grommets should be used when a hose or tube passes through a hole in framework, covers, etc. Actuators, especially hydraulic motors, also generate noise. Hydraulic motors sometimes are considered to generate noise equivalent to that of pumps. However, hydraulic motors often operate at relatively slow speeds, so motors generally operate much quieter than pumps do.

PREVENTION AND CURE OF NOISE

The power unit generally holds potential for the greatest reduction in noise for a given amount of time, effort, and expense exerted. As mentioned, an optional cooling fan may reduce noise from the motor. Also, using a motor that operates at 1200 instead of the usual 1800 rpm may reduce noise. However, expect a 1200-rpm motor to be larger, heavier, and more expensive.

Pump noise may be reduced by running a large pump at a lower than normal speed (which can also increase pump life) or specifying four or five small pumps for a power unit instead of the usual one or two large pumps. Size and the type of pump (piston, vane, gear, etc.), number of pumping cycles per rotation, system pressure, and, especially, pump speed all influence noise. Check with the manufacturer for assistance in determining what parameters will best suit your application. In addition to specifying quiet pumps and motors, you can also reduce noise by:

1. Using vibration-damping mounts to mount the pump and the motor to a subframe.
2. Mounting the subframe to the power unit frame using vibration-damping mounts.
3. Installing a flexible coupling between the motor and pump (and aligning it properly before startup).
4. Using hose sections between tubing and components that are mounted to framework.
5. As a last resort, treating noise as a symptom rather than at its cause may be the only recourse for some applications. Installing sound-damping materials around the motor-pump or power unit not only adds expense and complexity to the system, but complicates maintenance and may hinder air circulation for cooling. Acoustic filters, which use internal reflections and resonant frequencies to cancel out noise, may also be effective. However, they must be tailored to the application and tend to be expensive.

Not allowing air to dissolve in hydraulic fluid goes a long way toward preventing cavitation, both in the pump and in downstream components. Cavitation usually causes noise when air bubble suddenly collapse as fluid becomes pressurised in the pump. Air can be removed most effectively when fluid is in the reservoir. Given enough time, air will separate from the fluid, so the path from the return line to the pump inlet should be as long and with as little turbulence as possible. In addition, incorporating a fine-mesh screen promotes removal of air. Furthermore, tests have shown that positioning a 60-mesh screen 30° from horizontal may remove as much as 90 per cent of entrained air, Fig. 12.1.

Another method of quieting the power unit is to reduce pressure pulsations. Accumulators often are specified for this purpose, but their effectiveness is limited because they dampen pressure pulsations within a range of frequencies for a given size and precharge pressure. Moreover, accumulator calculations are complicated, and several accumulators may be required to dampen the full range of pulsation frequencies experienced by a system.

An alternative is to mount an in-line surge suppressor to dampen pulsations over a wide range of frequencies. One such a suppressor consists of a housing containing an annular area that holds a pressurised charge of nitrogen, a cylindrical membrane, and a perforated tube. Under normal operation, fluid simply passes through the suppressor by entering one end of the tube and exiting the other. However, if pressure increases—from pump pulsation, for example—the fluid passes radially outward through the tube perforations, overcomes the nitrogen charge pressure, and expands the diaphragm outward. Allowing pressure fluctuations to act against the pressurised nitrogen cushions the vibration, so output pressure is much smoother and, therefore, pump operation is quieter. Moreover, sizing is simple, because the suppressor is selected according to the size of the pump discharge line.

Reducing Fluid-borne Noise

While fluid-borne noise attributable to pressure pulsation can be minimised through hydraulic pump design, it cannot be completely eliminated. In large hydraulic systems or noise-sensitive applications, the propagation of fluid-borne noise can be reduced by the installation of a silencer. The simplest type of silencer used in hydraulic applications is the reflection silencer, which eliminates sound waves by superimposing a second sound wave of the same amplitude and frequency at a 180° phase angle to the first.

Reducing Structure-borne Noise

The propagation of structure-borne noise created by the vibrating mass of the power unit (the hydraulic pump and its prime mover) can be minimised through the elimination of sound bridges between the power unit and tank, and the power unit and valves. This is normally achieved through the use of flexible connections i.e. rubber mounting blocks and flexible hoses, but in some situations it is necessary to introduce additional mass, the inertia of which reduces the transmission of vibration at bridging points.

Reducing Airborne Noise

The magnitude of noise radiation from an object is proportional to its area and inversely proportional to its mass. Reducing an object's surface area or increasing its mass can therefore reduce its noise radiation. For example, constructing the hydraulic reservoir from thicker plate (increases mass) will reduce its noise radiation.

Airborne noise can be reduced by mounting the hydraulic pump inside the tank. For full effectiveness, a clearance of half a meter between the pump and the sides of tank is required. The mounting arrangement must also incorporate decoupling between the power unit and tank to insulate against structure-borne noise. The obvious disadvantage to this is the access for maintenance and adjustment is restricted.

Energy Storage in Hydraulic Fluid

Another source of noise in hydraulic systems derives from the storage and subsequent release of energy in the hydraulic fluid. Hydraulic fluid is not perfectly rigid, and the compression of the fluid results in energy storage, similar to the potential energy stored in a compressed spring. Like a compressed spring, compressed fluid has the ability to perform beneficial work. If decompression is not controlled, the stored energy dissipates instantaneously. This sudden release of energy accelerates the fluid, which affects anything in its path. Uncontrolled decompression creates noise and stresses conductors, and can cause pressure transients that damage system components.

Bulk Modulus and Decompression

The ratio of a fluid's decrease in volume as a result of a pressure increase is given by its bulk modulus of elasticity. The bulk modulus for hydrocarbon-based hydraulic fluids is approximately 2,50,000 PSI, (17,240 bar) which results in a volume change of around 0.4 per cent per 1000 PSI (70 bar).

As a general rule, when the change in volume exceeds 10 cubic inches (160 cubic centimetres), decompression must be controlled. Decompression control is essential in presses or other applications that have large volume cylinders operating at high pressures. Although hydrocarbon-based hydraulic fluids compress 0.4 to 0.5 per cent by volume per 1000 PSI, in an actual applications compression should be calculated at 1 per cent per 1000 PSI. This compensates for the elasticity of the cylinder and conductors and variations in the volume of air entrained in the fluid.

If, for example, the combined captive volume of the cylinder and conductors on a press were 10 gallons and operating pressure was 5,000 PSI, the volume of compressed fluid would be half a gallon ($10 \times 0.01 \times 5 = 0.5$). This equates to a potential energy of around 33,000 watt-seconds. If the release of this amount of energy is not controlled, a big bang will be heard throughout the plant! Decompression is controlled by converting the potential energy of the compressed fluid into heat. This is achieved by metering the compressed volume of fluid across an orifice.

Water Hammer

Storage and release of energy in the fluid also occurs during a phenomenon know as 'water hammer'. Water hammer is the term used to describe the effect that occurs when the velocity of the fluid moving through a pipe suddenly changes. This change causes a pressure wave to propagate within the pipe. Under certain conditions, it can create a banging noise, similar to the noise made when beating a pipe ·with a hammer, hence the term water hammer. Not surprisingly, common symptoms of this problem are high noise levels, vibration and broken pipes.

When a moving column of fluid hits a solid boundary, for example when a directional control valve closes suddenly, its velocity drops to zero and the fluid column deforms within the confines of the rigid cross-sectional area of the pipe to absorb the (kinetic) energy associated with its motion. This is similar to a car hitting a concrete wall. However, unlike a car, the fluid is almost incompressible; therefore the deformation is small and energy accumulates in the fluid—similar to compression of a spring. The magnitude of the pressure rise that results from the subsequent release of this stored energy can be mathematically expressed as follows:

$$Pf = P + upc$$

where, P is initial pressure, u and p are initial fluid velocity and density respectively, and c is the speed of sound through the fluid.

When attempting to control this situation, accumulators are sometimes installed. Unfortunately, accumulators address only the symptoms. The significance of the pressure rise equation shown above indicates that fluid velocity is the only variable that can be altered when addressing the root cause of this problem. In other words, reducing the velocity of the fluid column that hits the solid boundary reduces the magnitude of stored energy and the subsequent noise and pressure rise caused by its release. Returning to the traffic crash analogy—the slower the car is travelling upon hitting the wall, the less damage that occurs. In hydraulics, the most efficient way to do this, on paper at least, is to increase the diameter of the pipe, which reduces fluid velocity for a given flow rate. The alternative is to control deceleration of the fluid column by choking the valve switching time to the point where the pump's pressure compensator and/or system relief valve reacts fast enough to reduce flow rate through the pipe and therefore the velocity of the fluid column.

KEEPING PNEUMATICS QUIET

Exhaust mufflers have taken on greater importance with the recent enactment of OSHA (Occupational Safety and Health Administration) Standard 1910.1000. In part, the standard mandates that noise may not exceed 90 dBA during an 8-hr day of a 40-hr week. In addition, compressed air exhaust may not exceed 4.32 ppm of oil mist contamination in any 8-hr work shift of a 40-hr work week. In addition, OSHA regulations spell out specific noise and emissions limits and time periods in which these limits are allowed. For example, exhaust noise from a pneumatic system as high as 115 dBA is acceptable. However, workers may only be exposed to this noise for 15 min. within an 8-hr shift. So if workers will be near the exhaust of a pneumatic system for an entire 8-hr shift, noise to only 90 dBA is allowed.

For decades, air exhaust mufflers have been used to reduce noise and emissions of compressed air exhausts. Now, however, specific guidelines exist. Internal geometry to reduce air velocity and baffles for audio damping take care of noise; filtration takes care of the oil. But not just any filter-muffler will do. A standard filter-muffler has a porous element to trap any solids that may have been entrained in the compressed air stream. Porous elements, however, are not designed to trap vapours or liquids, such as oil. So unless the pneumatic system uses an oil-free air compressor and no lubricators, exhaust air should be routed through a coalescing muffler.

A coalescing muffler operates on the same principles as a coalescing filter. As air flows through the coalescing element, oil particles are captured by three different mechanisms: direct interception, inertial impaction, and diffusion. In direct interception, oil particles simply collide with and are trapped by filter fibres. With inertial impaction, the element's turbulent air stream throws oil particles against fibres, which trap the oil. Diffusion causes the smallest particles to vibrate and collide with each other— and eventually the element's fibres—which traps the oil.

FINDING OUT MORE

However you decide to make the hydraulic systems you design run quieter, component manufacturers prove an invaluable resource. Not only can they provide specifications on components, but they may also have useful literature containing more information on noise control of hydraulic systems. Engineering service laboratories who specialise in design and testing of hydraulic systems may also provide solutions. Whether affiliated with major component manufacturers or engineering laboratories, application engineers possess a wealth of knowledge that may include solutions to noise problems very similar to those experienced by your applications.

But resources do not end there. Dozens of books, technical reports, and papers exist to help you learn more about controlling noise in hydraulic systems. Calling on these resources may not make you an expert on the subject, but you'll certainly be more able to decide which solutions are most practical for your applications.

Chapter 13

Power Presses and Metal Cutting Saws

INTRODUCTION

This chapter discusses noise control measures to reduce noise in power presses, metal cutting saws, metal fabrications and water pipes.

POWER PRESSES

Power presses are inherently noisy, and numbers of them are often used in a press shop at the same time, all contributing to a cumulative noise exposure for operators and for nearby workers. Noise levels as high as 95–115 dB(A) are typical in many press shops. The potential for hearing loss is well known. Some press users have taken some steps to reduce noise at source, but many have not taken all reasonably practicable measures and still rely too heavily on the use of hearing protection by employees.

Control Strategy

The Noise at Work Regulations require noise assessments to be made if noise levels exceed a daily personal noise exposure ($L_{EP,d}$) of 85 dB(A), and noise reduction measures to be taken where these are reasonably practicable. The use of ear protectors is a last resort where control at source is not reasonably practicable, or may be a temporary measure to be used until controls can be implemented.

The key to understanding whether there is a problem is the noise assessment. The assessment should be more than a series of noise measurements. It should give clear recommendations on what needs to be done to comply fully with the Noise at Work Regulations. In particular, it should either give clear recommendations on what measures can be taken to control noise, including control at source, or should indicate where such help is available. Where noise comes from a range of sources, as will normally be the case in press shops, the assessment should suggest priorities.

On the basis of the assessment, managers should produce an implementation plan to tackle the problems identified. It may be reasonably practicable to make some improvements quickly, i.e. in a few weeks; others may take longer. What is important is that there is a realistic and effective plan to deal with the known problems; that the plan is followed, and that where necessary suitable hearing protection is provided and used in the meantime. It will rarely be possible to control noise completely by means of a single control measure. Just as total noise exposure is often the cumulative effect of many noise sources, satisfactory noise reduction may depend on the cumulative effects of a number of separate control measures. Do not underestimate the effect of an apparently small noise reduction. It is often not appreciated that the decibel scale is logarithmic and that a reduction of 'only' 3 dB(A) halves the noise energy. It may not be reasonably practicable to reduce noise in many press shops to levels where hearing protection is no longer required, but all noise reduction measures will help to reduce the overall risk of hearing damage.

New Presses

The noise at work regulations require suppliers to provide noise test data for power presses unless a particular machine is unlikely to cause noise exposures of 85 dB(A) or more. The Supply of Machinery (Safety) Regulations require noise emission data to be provided where the equivalent continuous A-weighted sound pressure level at workstations exceeds 70 dB(A). They also require that machinery be so designed and constructed that risks resulting from emission of airborne noise are reduced to the lowest level, taking account of technical progress and the available means of reducing noise, in particular at source. Users considering buying new presses are advised to ensure that adequate data and noise reduction measures are included with any new power presses offered for sale. Care is needed in interpreting noise emission data because standard test conditions may be very different from actual conditions of use, which themselves may be very variable. Further information on noise reduction at new power presses can be found in BS EN 692:1997.

Control Measures

The information which follows is not a design guide but is intended to give press shop managers an outline of the main noise reduction measures they may need to consider. Not all of these measures will necessarily be appropriate at every press or in every press shop. Information has been included on case studies and sources of further information. The sources of power press noise identified by a good assessment are likely to be either pneumatic or mechanical:
1. Pneumatic: compressed air is used in many presses, either for the control system or for ejecting parts from the tools or both.
2. Mechanical: from the press action itself, i.e. tool/workpiece interaction; clutch/gear/flywheel mechanisms; and from workpiece impacts with discharge chutes, collection bins, etc.

Technical Measures

Various technical measures may be taken to reduce noise from these sources.

Pneumatic exhausts

Pneumatic exhausts should either be ducted away from work positions or fitted with silencers. Care needs to be taken to ensure that excessive back pressures do not adversely affect operating cycle times or control systems, particularly those operating press guards.

Workpiece air ejection nozzles

Noise from ejection air jets can be reduced by any of the following measures or by a combination of them:
1. Change a continuous jet to an intermittent supply.
2. Fit acoustic quiet nozzles.
3. Reduce the duration of the jet.
4. Reduce the air pressure.
These measures should also result in savings of compressed air.

Machine design

A good noise assessment will have identified which parts of each press are generating noise. Discussion with press suppliers may reveal modifications which can be made.

Transmission of noise

The flywheels of power presses radiate noise due to vibrations caused both by the operation of the clutch and by the impact of the tooling. Analysis of the vibration patterns and modes of vibration can allow dynamic absorbers to be designed to bond to the flywheel.

A sheet of metal with felt padding bolted to the outside surface of a flywheel, like a finger touching a bell to damp down its ringing, may have a similar effect and an acoustic cover around the flywheel and gear wheel can further reduce noise from this source.

The transmission of mechanical vibration from a press frame through its supporting legs may result in noise radiating from the legs and floor. Suitable anti-vibration mounts fitted between the press base and the floor may reduce this problem.

Tool design

It has been found that including a shear or skew cut in blanking tools (a technique adopted for extending the work range of a power press) can also reduce noise.

Press enclosure

The provision of acoustic enclosures on handfed and automatic, strip-fed presses will significantly reduce noise emission levels in many instances. Partial acoustic enclosures closely fitted around the tools or an existing guard modified to act as a partial acoustic enclosure may also offer significant noise reduction. In other circumstances it may be appropriate to provide an acoustic enclosure around a complete press or whole press line.

An existing guard could be modified to act as a partial acoustic enclosure by covering the guard panels with a lead sheet sandwiched between acoustically absorbent foam, inserting polycarbonate windows where necessary. Such acoustic panels are quite easy to fix using plastic band fixing strips, making sure that all gaps are sealed. Low voltage lighting could be fitted inside the guard to increase visibility.

Enclosures made by suspending strips of appropriate flexible PVC have been used around individual presses to reduce noise.

Absorption panels

In many press shops, most of the noise to which operators are exposed radiates directly from the nearest presses, but reflected noise from walls and ceilings can contribute to total noise and may sometimes be particularly significant. Sound absorbing panels suspended from ceilings, fitted as wall linings or provided as mobile screens may reduce noise exposures. Depending on the area to be treated, the expense of this type of control measure means that careful thought has to be given to its suitability (i.e. cost versus benefit) and correct installation.

Discharge chutes and collection bins

Press components are often discharged automatically or dropped manually into metal bins, often via metal chutes. Metal to metal impacts often contribute significantly to noise exposures. Bins and chutes can be lined with suitable resilient material like rubber, PVC, etc. of appropriate hardness; the under or outsides can be coated with sound-deadening compound. Alternatively, acoustic enclosures can be provided around bins.

Press maintenance

Virtually all machinery can be expected to work better, and often more quietly, if it is properly maintained.

CONTROL OF NOISE AT METAL-CUTTING SAWS

Metal-cutting circular saws, particularly those operating at high speeds, cutting nonferrous material such as aluminium, are known to produce high noise levels. These noise levels can typically exceed 100 dB(A). High-speed, pivoting head and up-stroking saws, for example, are commonly used for high volume cut-to-length and mitre-cutting operations. These machines process a wide variety of bar, strip and extruded materials. Less commonly, they are also used to cut lighter steel sections.

A free-running saw can also produce noise levels of around 90–95 dB(A). These periods of free running can have a significant impact on the daily noise exposure of the saw operator and those in its immediate vicinity.

Sources of Noise

Noise from circular saws can come from a number of sources including:

1. Free-running noise generally associated with the aerodynamic disturbance in the vicinity of the blade (producing a hissing sound). When this frequency coincides with the blade resonance frequency, it gets amplified and produces a high-intensity scream or whistle.
2. Cutting noise due to the impact between the saw blade and the work-piece and radiated by both.
3. Blade and work-piece vibration noise (sometimes at resonance, producing a ringing noise) which also depends on the feed rate of the cut.

Duties of Manufacturers and Suppliers

The Supply of Machinery (Safety) Regulations 1992 (as amended) require suppliers (who may or may not also be the manufacturers) to take measures to reduce risks from noise.

Control of noise at source by engineering means is the required option if the means are available and is generally best achieved at the design stage by careful consideration of the noise-generating mechanisms. A number of manufacturers already supply machines provided with full acoustic enclosures (usually fully automatic machines).

Manufacturers/suppliers are also required to supply data on noise emissions, measured according to relevant standards and an appropriate test code.

Duties of Machine Users

The Noise at Work Regulations 1989 require employers using noisy equipment or processes in their businesses to adopt measures to prevent their employees suffering hearing damage. Regulation 6 requires the risk of hearing damage to be reduced to the lowest level reasonably practicable. If the noise levels exceed a daily personal exposure level of 90 dB(A), reduction of the risk has to be achieved by means other than the use of ear protectors. In other words, businesses using metal-cutting saws should adopt the kind of engineering controls or other suitable measures, where it is reasonably practicable to do so.

Purchasing New Machines

Purchasers should use manufacturers' noise data to select quieter machines from the outset. It is also helpful to seek further recommendations, from the manufacturer, on any additional noise control measures that should be applied under the intended installation and operating conditions.

Engineering Controls

For existing saws, where reasonably practicable, steps should also be taken to reduce noise at source. Practical methods of noise reduction need to be considered as part of the noise assessment required by the Noise at Work Regulations. By careful measurement of the noise levels at each stage of the machining cycle, e.g. loading, idling, cutting, etc. the dominant noise sources can be identified. This information can then be used to help select the most appropriate noise reduction techniques.

Some Practical Methods of Noise Reduction

1. Select the correct saw blades for the intended operation.
2. Keep the blade sharp to maintain optimum cutting performance.
3. Provide correct clamping of the work-piece to reduce radiated noise. Additional clamping arrangements may be used, such as suitably positioned toggle clamps. Pads fitted to the clamps damp vibration transmission and prevent damage to work-surface finish.
4. Reduce radial and lateral imbalance possibly due to worn bearings, imbalance in saw blade or its associated arbour collar, inadequate maintenance or excessive duty cycles on a machine not designed for the service conditions.
5. Use damped saw blades (seek saw manufacturer's advice).
6. Use noise/vibration absorbing material on the surface of the feed table.
7. Damp the machine subframe or panels using proprietary damping compounds.
8. Enclose or partly enclose the cutting area using suitable sound-absorbing material.
9. Switch the saw off when not required.
10. Locate the saw in a separate room, to limit the noise exposure of other workers.

Damping the Saw Blade

Damped blades can significantly reduce noise, especially from those blades which exhibit 'resonance' while idling. Various methods of damping have been tried with varying degrees of success. These include laminated blades, damping discs or plates and resin-filled, laser-cut slots in the blade (elongated 'S' shape). On a pendulum crosscut saw, for example, a suitable foam can be added inside the existing top guard to absorb sound and at the same time produce a damping effect on the saw blade.

The overall noise reduction achievable in specific cases using these methods is difficult to predict but treatment has reduced the noise of some typical operations by up to 6 dB(A). Treatment may be applied to new blades and to used blades sent for refurbishment.

Maintenance

The control measures applied to reduce noise should be subject to periodic inspection and necessary maintenance to ensure that they continue to be effective.

Training

The Noise at Work Regulations require employers to give appropriate information, instruction and training to employees. This will include information on noise control measures fitted to saws and on how such controls are to be used and, where appropriate, maintained. The noise assessment should identify the specific matters which need to be covered.

METAL FABRICATION

Fabrication as an industrial term refers to building metal structures by cutting, bending, and assembling. The cutting part of fabrication is via sawing, shearing, or chiselling (all with manual and powered variants) and via CNC cutters (using a laser, plasma torch, or water jet). The bending is via hammering (manual or powered) or via press brakes and similar tools. The assembling (joining of the pieces) is via welding, binding with adhesives, riveting, threaded fasteners, or even yet more bending in the form of a crimped seam. Structural steel and sheet metal are the usual starting materials for fabrication, along with the welding wire, flux, and fasteners that will join the cut pieces. As with other manufacturing processes, both human labour and automation are commonly used. The product resulting from (the process of) fabrication may be called a fabrication. Shops that specialise in this type of metal work are called fab shops. The end products of other common types of metalworking, such as machining, metal stamping, forging, and casting, may be similar in shape and function, but those processes are not classified as fabrication.

In the metal fabrication industry, people are generally subjected to noise of a varying nature. High noise levels, particularly those of short duration such as impulse or impact noise, are present in many metal fabrication workshops and are capable of causing damage to hearing. The metal fabrication industry in Australia is responsible for a large proportion of workers compensation payouts for noise induced hearing loss of its workers. In metal fabrication workshops noise levels can be expected to range between about 80 and 125 dB(A). Most common noise sources are from electric angle grinders, metal presses, cutting saws and hammering and banging on metal objects.

Further sources may be found from welding and gouging which cause high noise levels to be emitted. Because these types of noise emission are generally of short duration, for example, a few belts with a sledge hammer on a metal plate or using an angle grinder to clean up a weld, the use of personal hearing protectors is often ignored or simply not even considered.

However, these relatively short duration exposures happen many times per shift and may therefore pose a serious hazard to hearing. Further, impact noise, such as from bashing on metal plates, are potentially more hazardous to hearing than noise such as that from machines.

Typical Noise Sources

Some typical noise sources at operator ear level in the metal fabrication include:

Hammering on metal objects	115–120 dB(A)
Punch press	102–107 dB(A)
Nine inch angle grinder	97–106 dB(A)
Gouging	97–99 dB(A)

Noise Control Measures

As with all risk exposures in the workplace, risk management must be applied through a hierarchy of control measures, that is elimination, substitution, engineering and/or administrative controls and, as an interim measure, reliance on protective equipment. Noise exposure in the metal fabrication industry should primarily be controlled through engineering and/or administrative noise control measures.

Examples of engineering noise control measures include:

1. Locating noisy equipment such as air compressors and power presses in separate enclosures.
2. Providing a sound proof enclosure for operators.
3. Using mobile enclosures where noisy work has to be carried out.

An example of an administrative noise control measure is the scheduling of noisy work from various fabrication activities into a particular part of the shift, such as early morning or late afternoon, with quieter work activities for the remainder of the shift. Where noise control cannot be achieved through these measures an employer must provide suitable personal hearing protectors as well as proper instruction in their use so that exposed workers can perform their work in a manner which is safe and without risks to their health and safety.

QUIETING NOISY WATER PIPES

Water flowing in pipes can cause all kinds of weird noises. We all know what water running through a pipe sounds like, but what about some of those other plumbing sounds—like creaks or cracking sounds, rattling, whistling and the most annoying or scary of them all, that loud banging noise? Let's look at what causes those sounds and how you can fix them.

Creaks or a Cracking Sound

These are usually caused by the expansion and contraction of the water pipes themselves. As hot water runs through a pipe, it naturally heats the pipe, causing it to expand slightly. Once the water stops flowing, the pipe cools and the metal contracts, resulting in the creaking or cracking sound. The easiest way to fix this is to put some insulation around the pipe, or if the pipe is running through a tight fitting hole in the wood framing, cut a notch in the framing so the pipe can expand and then contract without that creaking sound.

Rattling

The cause and remedy are similar to the cracking sound. The rattling sound comes from the pressure of water running through a loosely attached pipe, causing it to vibrate slightly. When a loose pipe vibrates against something solid, like framing or the strapping designed to hold it tightly, you hear the rattling sound. Stopping the pipe from vibrating will fix the rattle. Put some cushioning around the pipe or fasten the strapping more securely so the pipe won't vibrate.

Whistling

This is usually caused by water flowing through a restricted section of the plumbing. The restriction can be sediment in the pipe or a defective washer or valve. If the whistling only occurs when a particular faucet is turned on, that's likely where the problem is, and replacing the washer or repairing the valve seat should fix it. However, if the whistling sound occurs when any faucet is turned on, the problem is more likely in the main water supply valve itself. Adjusting the water pressure at the main water valve may dislodge the impediment, or the change in water pressure itself could get rid of the whistling sound. If that does not eliminate the problem, you may have to get the water valve replaced.

Banging

The loud banging sound when you shut off the water flow is actually called 'water hammer', and is a fairly common complaint in older homes. The flow of water through the pipes contains energy, and when the flow is abruptly stopped, this energy causes the loud banging sound. Initially a home's plumbing system was built with short pieces of pipe that filled with air and acted as air cushions to absorb the water's energy when the flow was stopped abruptly. However, over time the air has leaked out, meaning there is no air cushion left to absorb the water energy.

You can put that air cushion back into your plumbing system by turning off the main water supply and opening all the faucets in the house to drain the system. Next, turn the water supply back on and work your way up through the house, turning off the faucets as water flows through them. This should trap some air in the air chamber so it will once again provide the cushioning effect. If this does not work, you can buy a 'water hammer arrester' that attaches directly to the water supply pipe where the water hammer originates.

If your plumbing is making any of these sounds, you should be listening because it's telling you it needs some maintenance. The sound is caused by something not working properly in your system and if you ignore it, over time, that small sound could lead to larger problems.

Chapter 14

Issues in Combustion Noise

INTRODUCTION

This chapter discusses the inclusion of the 'boundary conditions' imposed by turbomachinery which is an essential ingredient of a rational approach to the physics and calculation of combustion noise. The calculation is carried to the noise output that may be expected at the downstream end of the turbine. The proposed analysis is linear and hence the noise output is proportional to the fluctuating heat release. The calculation includes details of all the waves produced fore and aft of the plane of fluctuating heat release prior to the calculation of noise (pressure wave) produced aft of the turbine. The boundary conditions needed to 'close' the analytical problem and to solve it are indicated along with assessments of their validity. The fact that interest in combustion noise is in low frequencies, i.e. in relatively long wavelength phenomena, is used to simplify the calculation of 'turbine transfer functions'.

If trends indicated by these calculations are actually representative of what might be the outcome of a more comprehensive approach, it could have significant impact on issues concerning combustion noise of aircraft engines, including issues such as experimental evaluation of combustion noise based on combustor component tests as well as the issue of approaches to the reduction of combustion noise.

It was argued that under conditions of inhomogeneous, steady heat release, the resulting inhomogeneous steady temperature distribution is a significant source of generation of entropy waves (much along the lines of how regions of high velocity gradients produce turbulence) and the so-called 'indirect noise' due to passage of such entropy waves through the multi-stage turbine seemed to be the predominant source of combustor noise as opposed to the 'direct noise' usually associated with the generation of pressure waves by unsteady combustion. Presumably a significant if not predominant reason for the generation of pressure waves by unsteady combustion would be the unsteady heat release produced by unsteady combustion. There could in turn be many reasons for unsteady combustion, both kinetic and fluid mechanical. A source of unsteady combustion followed by unsteady heat release which is fluctuations of mixture fraction can be estimated by 3D combustor codes even in the limit of equilibrium or fast chemistry models. It was this level of combustor computational fluid dynamics (CFD) that was used to derive estimates of the level of entropy waves, albeit the emphasis in that study was very much on the case of entropy wave generation by steady temperature gradients and not by unsteady heat release.

The indirect experimental data presented in support of entropy wave interaction with the (downstream) turbine as the dominant source of combustor noise, though (in those studies) this conclusion regarding entropy wave interaction with the (downstream) turbine being the dominant source of combustor noise was argued only in the case of spatially inhomogeneous steady heat release. The specific motivation of that study was the observation that dual annular combustors (DAC) relative to single annular combustors (SAC) seemed to have significantly more combustor noise at part load operation (approach condition). At such part load conditions the DAC produce a much more inhomogeneous steady temperature profile than equivalent SAC due to the difference in mode of burning. Figures 14.1 and 14.2 describe these results and Fig. 14.3 illustrates the difference between SAC and DAC. In Fig. 14.2 (full power condition) there is virtually no difference in the noise of the SAC and the DAC. In terms of Fig. 14.3, the DAC operates with only the pilot dome 'lit' at approach, whereas both main and pilot domes are lit at the full power condition.

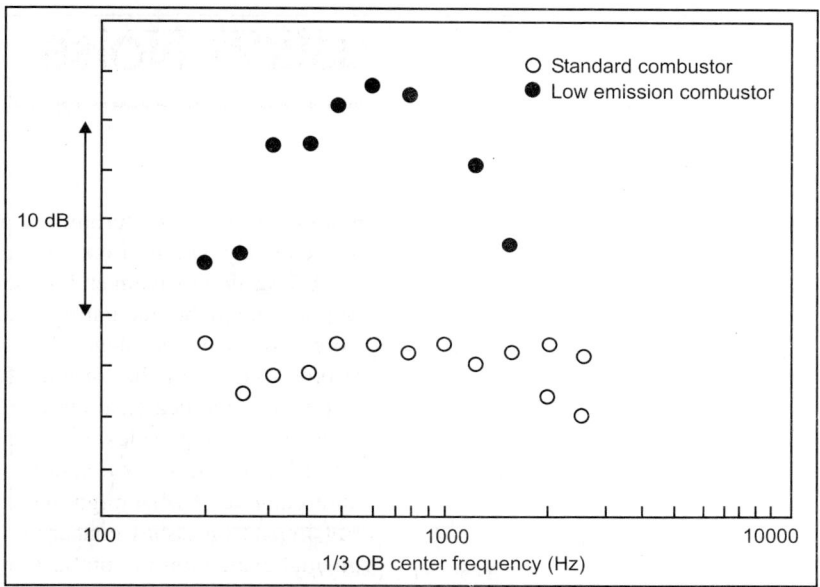

Fig. 14.1. Approach power peak angle combustor noise spectra. SPL at 120° to inlet axis.

In a pioneering study, Cumpsty forcefully advocates the predominance of 'indirect' noise for the case of noise produced by fluctuating heat release rather than due to spatially inhomogeneous steady heat release. He also presents two uncoupled calculations (the first a calculation of sound, shear and entropy waves produced by unsteady hear release from a flame sheet and the second, of 'indirect noise' produced by the passage of entropy waves through a multi-stage turbine), and then examines the product of the two solutions for a turbine and combustor typical of the RB211 at take-off conditions. His concluding remarks are quoted below:

> 'It suffices to say that the method predicts the noise below 1 kHz to be overwhelmingly generated by the interaction of entropy fluctuations with the turbine ('indirect' combustion noise). The fluctuating addition of heat to flow of gas produces pressure and entropy perturbations and in two (or three) dimensions, vorticity perturbations as well. The generation of all three types of disturbances are inextricably connected. In particular, it is inaccurate to associate noise from gas turbines with the pressure disturbance from the combustor in isolation: the interaction of entropy (and vorticity) with downstream components should also be considered. A simplified one-dimensional calculation of the interaction between a combustor and turbine indicates that it is the entropy interacting with the turbine (indirect noise) which predominates.'

Figures 14.1 and 14.2 illustrate the noise differences found (by test) at approach and full power conditions between a DAC design and a (standard) SAC design for a GE aircraft engine. Figure 14.3 illustrates the differences between a DAC and a SAC. It should be noted that most combustor codes for aircraft engines used in industry are 'incompressible' in that only density and pressure changes due to combustion are allowed for, not those due to compressibility. This is justifiable due to the low Mach numbers involved in aircraft engine combustors but renders these codes of poor utility in directly estimating the level of pressure fluctuations.

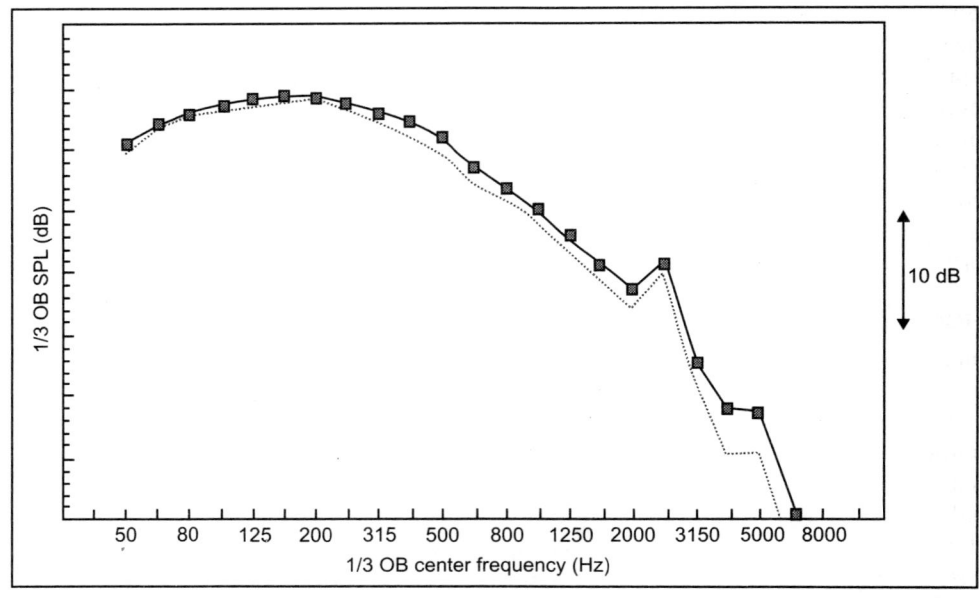

Fig. 14.2. Peak angle noise spectra. SPL at 120° to inlet axis. Full power condition.

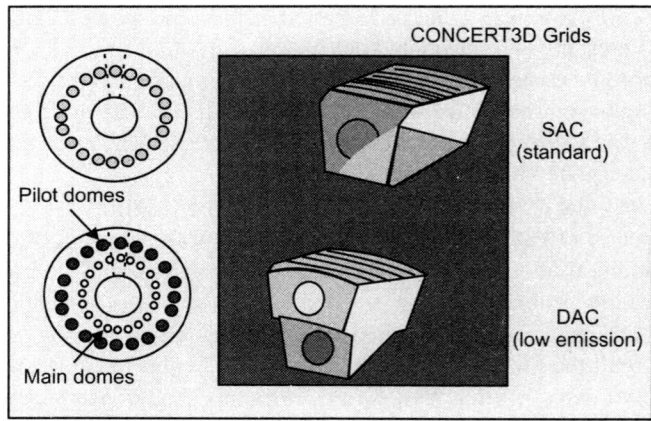

Fig. 14.3. Single annular combustor (SAC) and double annular combustor (DAC).

A multi-blade row actuator disk method similar to Cumpsty and Marble was used in the present study to estimate multi-stage 'turbine transfer functions'. This is justifiable at least as a first step, given the low frequency nature of combustion noise and in view of the great simplicity of multi-blade row actuator disk methods. Interest in combustion noise focuses on the pressure waves present downstream of the turbine. This has been addressed in Cumpsty but in an uncoupled fashion, i.e. the mix of aft-generated pressure, vorticity and entropy waves first calculated from a flame sheet model in 'isolation' followed by calculations of pressure waves and entropy waves incident on a multi-stage turbine assuming reflection free boundary conditions upstream and downstream of the turbine. The second calculation in Cumpsty was based on multi-blade row actuator disk (AD) methods. AD theory with its basic assumption that the wavelength of the waves involved substantially exceeds dimensions of the various length scales in the problem considered (e.g. turbine blade chord, spacing between adjacent blades, etc.) is justifiable. However, there is one exception: that at entrance to the inlet nozzle Mach numbers/flow velocities are low and hence the wavelength of entropy waves (which are propagated at flow velocity and not at the speed of sound) could be too short at the higher frequencies to justify applicability of AD methods. Both choked and unchoked blade row situations can be handled by AD methods, though Mani noted some unique aspects of solution procedure needed for the multi-stage case when one or more choked blade rows are involved.

METHOD OF ANALYSIS

Uncoupled combustor turbine calculations in Cumpsty indicate that unsteady heat release is more a source of entropy waves aft of it than of downstream travelling pressure waves and that the 'transfer function' for the pressure wave downstream of the turbine is similar in level for both incident pressure and entropy waves. Mani showed that there is no significant change in these conclusions when a fully coupled calculation is carried out. The conclusion is that even in the case of unsteady heat release induced combustion noise, the predominant chain of events is 'unsteady heat release $\rightarrow$ entropy waves aft of flame $\rightarrow$ sound waves aft of the turbine' upon interaction with the multi-stage turbine. Considerable doubt is cast on the conventionally assumed view of the chain of events of 'unsteady heat release $\rightarrow$ pressure waves aft of the flame $\rightarrow$ transmission loss' through the turbine. This result has a significant impact on the technical community concerning many aspects of combustion noise—physics of generation, relation of combustor component tests (typically executed without the turbine at the aft end of the combustor) to the engine case, strategies for combustor noise reduction, etc. Figure 14.4 defines the present project in schematic form. A linearised, two dimensional analysis (neglecting variations in the spanwise or radial direction) has been carried out with the compressor exit at the left end, the steady combustion process modelled as a discontinuous change of properties created by 'constant area' heat addition in a flame sheet and the turbine leading edge defining the right boundary.

The flame related inputs needed are upstream pressure, density, axial velocity (entry into the combustor is assumed to be axial), stagnation temperature ratio across the flame and an estimate of axial location of the 'flame sheet' relative to the compressor trailing edge plane and to the turbine leading edge plane. The magnitude or phase of the unsteady heat release is not needed for the analysis reported on, as all predictions of complex amplitudes of the various waves are made as multiples of this unsteady heat release. At compressor inlet we can assume that P_t and T_t are constant (independent of time). A 'quasi steady' relation between the P_t ratio across the compressor due to variations of mass flow rate at the compressor exit is assumed and used to yield a boundary condition at the cold end for the variation of linearised P_t at the cold end in terms of variations of linearised ρu. A similar exercise based on the T_t

versus ρu relation could be used to derive a relation for linearised T_t variation at the cold end. However it is simpler to assume that the low frequency unsteady behaviour of the compressor is isentropic and hence that the linearised $P_t' - T_t'$ relation at the cold end is also isentropic. The consequence of isentropic low frequency unsteady behaviour of the compressor (since there are no entropy waves at compressor inlet) is also that there are no entropy waves between the compressor trailing edge and the flame in Fig. 14.4.

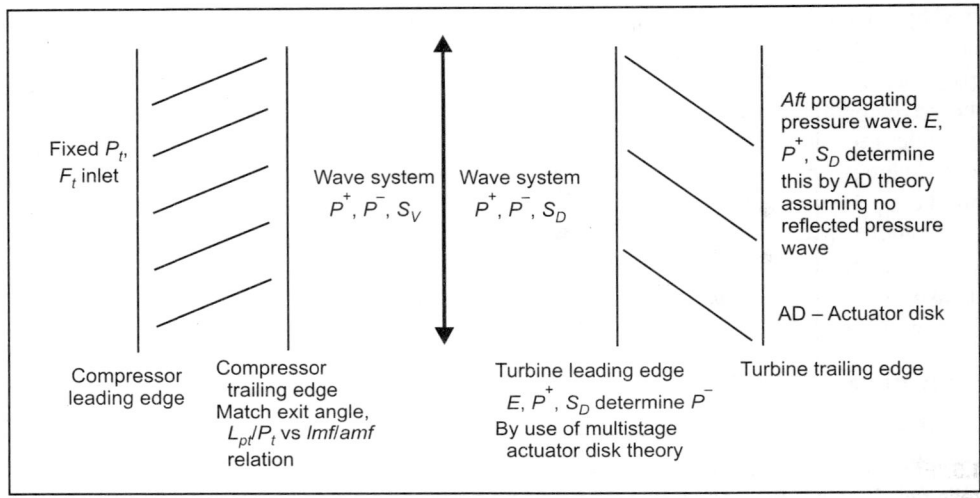

Fig. 14.4. Schematic of analysis 'model' used in project. Fluctuating heat release plane assumed moving in space such that burning velocity is unchanged. Match ρu, x– and y– momentum, T_t relative to flame across it.

Finally the least controversial additional boundary condition to be used at the cold end is that the compressor outlet guide vanes impose a constant leading angle condition there, which we will assume to be the axial direction. Summarising, the three boundary conditions at the cold end are: (i) absence of entropy waves, (ii) axial flow direction, and (iii) a relation between the linearised total pressure fluctuation (P_t') and linearised mass flow perturbation (linearised ρu). At the turbine end, using multi-stage AD methods and the assumption of reflection free termination at the downstream end of the turbine, a relation can be obtained between the incident entropy, pressure and shear waves and the reflected pressure wave. Across the flame are matching conditions of linearised ρu, linearised axial and tangential momentum and linearised stagnation enthalpy. In the absence of solid object flame holding devices as in augmentors, matching of linearised axial and tangential momentum imply matching of linearised $p + \rho u^2$ and of v across the flame where, v is the tangential velocity. In the present project, the flame is assumed to move such that there is no change in 'burning velocity' due to the heat release, and hence $u_f = u_1$. Matching of ρu and of $p + \rho u^2$ need to reflect the fact that the necessary u here is relative to the flame. The matching of linearised stagnation enthalpy across the flame will serve to bring into focus that all final amplitudes of the various wave systems will be in terms of the unsteady heat release.

Calculations are carried out for prescribed frequency and prescribed tangential wave number k_y. In the results shown in Figs 14.5 and 14.6, the tangential wave number range considered is limited by 2D cut-off considerations and k/k_y is a convenient way of accounting for this since $k/k_y > 1$ corresponds roughly to cut on waves. In Fig. 14.4, consider first the region bounded by the compressor trailing edge and the flame. Due to absence of entropy waves, there are three unknown wave amplitudes here—the shear wave SU, upstream and downstream propagating sound waves P^- and P^+. Downstream of the flame and upstream of the turbine leading edge there are similarly four unknown wave amplitudes the extra wave type being an entropy wave E.

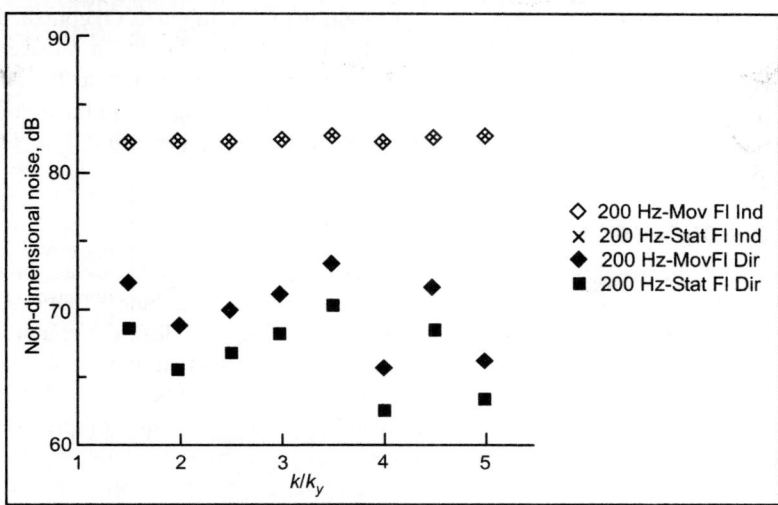

Fig. 14.5. Effect of matching conditions: moving vs. stationary flame-sheet : 90″ inlet diameter, 3600 rpm, 20:1 press ratio, turbine. Assume no change in burning velocity in moving flame case.

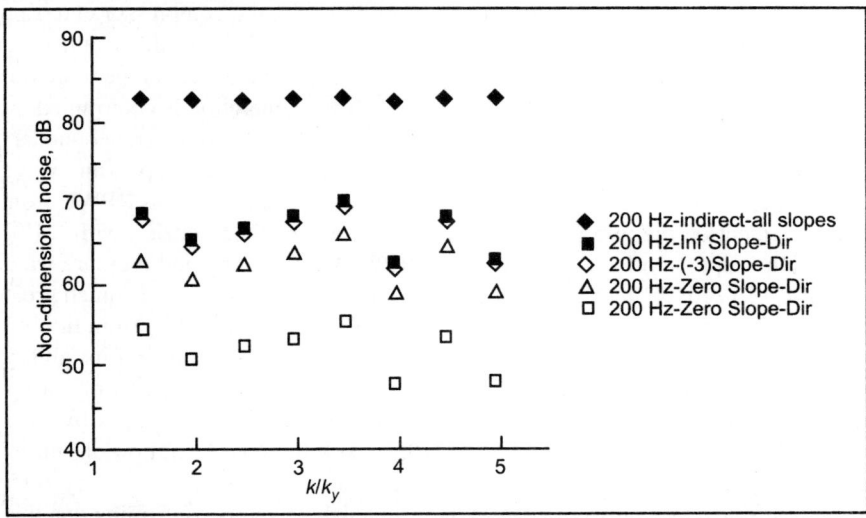

Fig. 14.6. Effect of boundary conditions at the compressor trailing edge stationary flame: 90″ inlet diameter, 3600 rpm, 20:1 press ratio, turbine.

These seven unknown wave amplitudes can now be determined by seven boundary and matching conditions—two at compressor trailing edge, four across the flame and one at turbine leading edge all of which have been discussed above and the seven wave amplitudes will be known in terms of the unsteady heat release. We will only be interested in the acoustic power radiated *aft* of the turbine and this is determined by knowledge of E, S_D and P^+. While the power in principle is related to the sum of the effects of E, S_D and P^+ (added with proper attention to phase relations), the separate contributions of E, S_D and P^+ were calculated and as expected the contribution of E dominates. Of course, the downstream power is known only in terms of the (unknown) amplitude of the unsteady heat release but clearly this does not prevent the ability to allocate the power to contributions from E, S_D and P^+. The results of the study should provide a reasonably 'first principles' based understanding of combustion noise generated by unsteady heat release in a combustor bounded by turbomachinery though several simplifying assumptions have been made.

RESULTS AND CONCLUSION

Many calculations were carried out but in summary, the key relevant results are as shown in Figs 14.5 and 14.6. Figure 14.5 contrasts the 'moving flame sheet' versus 'stationary flame sheet' cases and Fig. 14.6 shows the impact (for a 'stationary flame sheet' case) for four different compressor exit 'quasi steady' P_t versus ρu relations. The principal conclusion of the present study is as follows:

The conclusion concerning the predominance of indirect noise is not affected by consideration of 'moving' flame matching conditions or by use of boundary conditions at the compressor end other than zero mass flux. The ratio of indirect to direct noise is affected by consideration of 'moving' flame matching conditions or by use of different boundary conditions at the compressor end other than zero mass flux, while the indirect noise is almost unaffected by choice of flame matching conditions or compressor end boundary conditions, the direct noise is affected. But in view of the predominance of indirect noise, the predicted total noise is not affected much at all by consideration of 'moving' flame matching conditions or by use of different boundary conditions at the compressor end. Entropy waves 'participate' very little in the mass flux or momentum flux balances when the Mach numbers approaching the turbine are low (as is the case in gas turbine combustors) but the stagnation temperature flux associated with entropy waves is relatively large and hence entropy wave generation is determined mostly by the fluctuating heat release with little effect of flame matching conditions or by different boundary conditions at the compressor end. Not shown in Mani or in present report is the reflected pressure wave coefficient (with appropriate normalisation of entropy and pressure waves) abbreviated as RPWC from entropy or pressure waves incident on the turbine. For a one dimensional choked nozzle , with 'M' denoting the Mach number of the approach flow, to O(M), these RPWCs for entropy and pressure waves are –M/2 and [1 – (? – 1)M] respectively indicating that the RPWC for pressure waves is much greater than that for entropy waves. This result is also true for reflected pressure waves (RPW) from turbines. Since the RPW from incident waves are important for providing feedback in combustion instability, in case of combustion instability analysis, it would be erroneous to say that the generation of entropy waves by fluctuating heat release is dominant relative to the generation of pressure waves by fluctuating heat release. In case of combustion noise however it can be said that the generation of entropy waves by fluctuating heat release is dominant relative to the generation of pressure waves by fluctuating heat release. The present study has complemented Mani in a valuable way by confirming that this conclusion regarding the predominance of the generation of entropy waves by fluctuating heat release relative to the generation of pressure waves by fluctuating heat release as far as noise generated downstream of the turbine is not affected by use of a moving flame sheet model or by the choice of upstream boundary conditions other than zero mass flux.

Chapter 15

Hydraulic Pump

INTRODUCTION

Every hydraulic system needs a pump. Hydraulic systems are known for adjustable, controllable power transmission to a variety of actuators. But they require careful design attention to keep noise to acceptable levels. Fortunately, new pump technologies are making this task easier. Variable-displacement axial-piston pumps are increasingly used for hydraulic systems because they provide better efficiency, higher power density and increased versatility. While these areas are important in pump selection, there is also added emphasis on reducing pump and system sound levels. In addition to sound generated from fluid velocities, the pump and system pressure ripple contribute to the overall system sound levels.

There are two pump noise-generating mechanisms:

1. Alternating forces inside the pump body, considered the primary mechanism.
2. Noise emitted by all downstream components resulting from pulsating flow delivered by the pump.

PRIMARY PUMP NOISE

In an axial piston pump with swashplate, pistons rotate together with the cylinder block according to the drive speed of the shaft (nine pistons are shown in the Fig. 15.1). The valve plate ports fluid to and from the pistons: half a rotation with the pressure side and the other 180° with the suction side. Pistons connected to the pressure port may be either four or five as shown in Fig. 15.1. At an odd number of pistons (9 pistons, pitch angle 40°) this situation changes every 20° rotation angle. At an input speed of 1500 rpm, the total resulting force of the pressurised piston changes with a frequency of 225 Hz. For a pump operating at 350 bar, with a displacement of 46 cm³/rev (piston diameter 17 mm) that means a change of the internal forces from 3.2 to 4 tons. For the largest pump in Parker's PV-plus family, the force steps from 10.5 to 13.2 tons and back. In such an axial piston pump, a force of several tons stresses the structure by pulsating with a high frequency. Such high alternating forces require careful design of the pump body.

Under these high forces the pump case will be deformed. And, because of the rapid load changes—it takes less than 1 ms to change from the high to low force level or back—higher harmonic vibrations will be induced in the structure.

The deformations need to be minimised to reduce the effects on the surrounding air and the natural frequencies of the structure must be made as low as possible to make the noise level more tolerable.

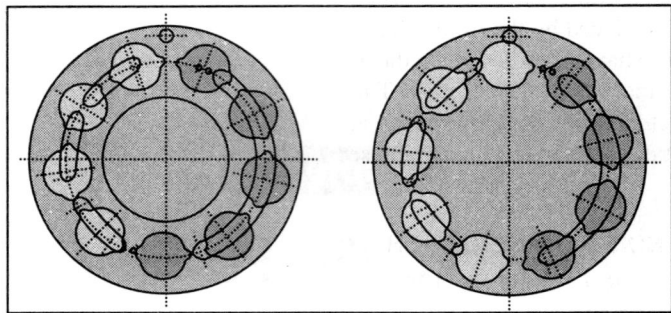

Fig. 15.1. Cylinder block over valve plate at 0° and 20° rotation angles.

Pulsation Reduction with a Precompression Chamber

One new pump design uses a precompression (ripple) chamber design in the pump to reduce the output flow pulsation. In the first part of the compression phase, an internal chamber—filled with fluid under pump output pressure—is connected to the piston coming from the low-pressure side. Fluid now starts to flow back into the chamber to refill it (Fig. 15.2).

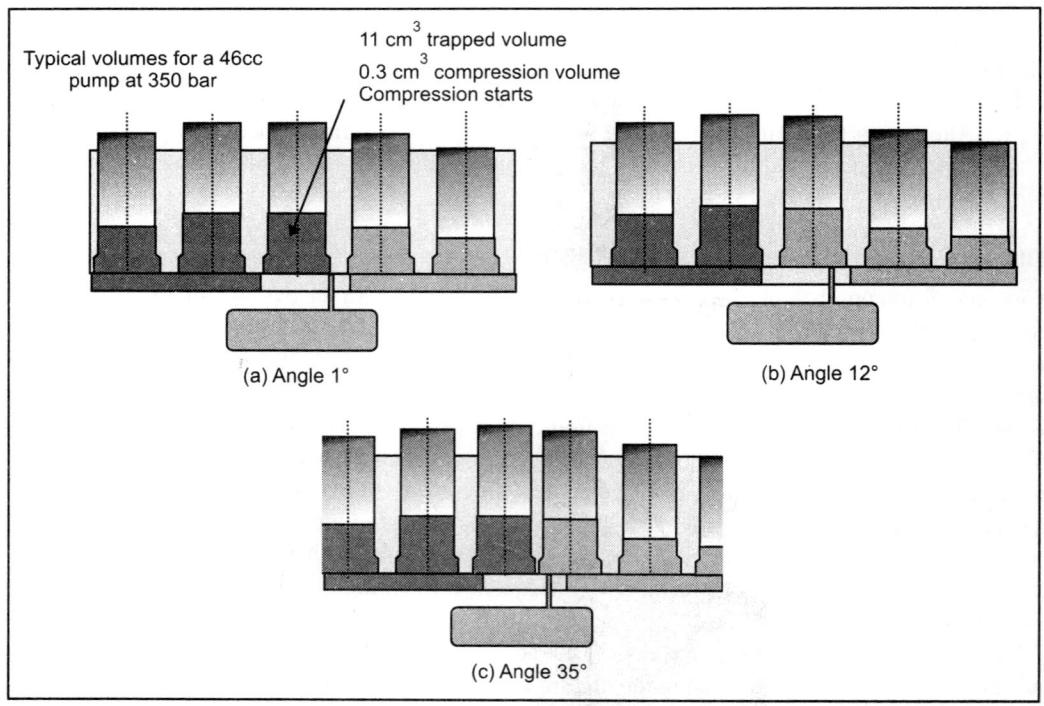

Fig. 15.2. (a) Pulsation reduction with a pre-compression (ripple) chamber at 1° of rotation where compression starts, (b) then advances to 12° rotation where compression ends and chamber refill starts, and (c) continues to 35° where chamber refill ends.

Chamber refill ends with refill time covering 25° of rotation. As the volume to bring the piston chamber to 350 bar is reduced by the ripple chamber input, now a lower average flow is required from the pressure chamber to have it refilled when the next piston gets connected. As a result, the peaks in the flow are reduced by more than 50 per cent. The whole system is less excited by the reduced flow pulsation and is quieter. Built to a power unit and incorporated into a hydraulic system, the lower pulsation causes a total reduction in primary and secondary noise by 3 dB(A) typical, and in some cases even up to 5 dB(A).

CASE STUDY OF NOISE TRANSMISSION AND REDUCTION FROM A HYDRAULIC PUMP

On the 10th Floor of the R. L. Smith ME-EM Building resides a laboratory for biomechanics research. The principle machine in this particular lab is a Hydraulic Tensile Test Machine (Fig. 15.3). To power the tensile test machine a three gallon per minute electric hydraulic pump provides 2000 psi pressure. The pump (Fig. 15.4) was acquired several years ago from the Civil Engineering Department and is approximately 35 years old.

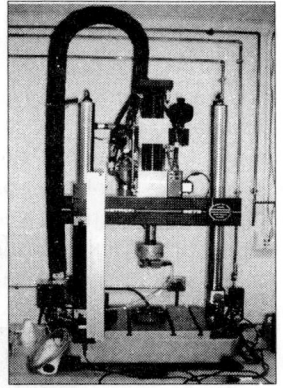

Fig. 15.3. Hydraulic tensile test machine.

Fig. 15.4. Hydraulic pump.

The pump is currently producing excessive amount of noise, which over periods of 3–5 hours can make for an uncomfortable work environment in the lab. Previous attempts have been made to reduce the overall noise levels in the lab caused by the pump. An old enclosure in use over the last several years was ineffective at noise reduction and overheated the pump. The objective of this project is to create a low cost solution that reduces the overall noise emitted from the pump but does not effect pump performance. Ultimately the goal is to create a more pleasant working environment for the occupants of the biomechanics lab.

Requirements

To complete this project's objective the final design must account for the following factors and design requirements:

1. Improve working environment of biomechanics lab by reducing noise.
2. No decrease in pump performance.
3. Accessibility to On/Off power switch.

The main causes of noise produced by the pump were two fold. Pressurised hydraulic lines and valves produced a very high frequency air borne noise. This was caused by hydraulic fluid passing through the orifice of the pressure regulator. The 3-phase motor produced a low frequency air borne noise. Structure borne noise was insignificant, but steps were taken to reduce its impact on neighbouring labs.

Due to the fact that very little could be done to reduce the sound levels of the pump itself, and the budget constraints, a new enclosure was selected as the best alternative. Comprised of a barrier with acoustical foam inside, its purpose would be to control and absorb sound before it could reach the occupants of the lab. To prevent the pump from over heating, an acoustical louver was designed into the enclosure. This louver would allow heat to exit the enclosure while containing the noise. An AC fan was also installed to supply forced convection through the box. A routine clean out of the heat exchanger also helped to remove heat from the unit.

Selection of Materials for Acoustic Enclosure

The materials used for the acoustic enclosure were selected on the basis of three criteria. The first was the applicability of the material, the material's accessibility, and on the basis of cost. As this project was privately funded, cost was a major concern.

The barrier was the first item considered on the project. Rugged construction was an important factor when considering materials for the enclosure. The enclosure would also be used as a work surface in the lab. Standard particle board (Fig. 15.5) was selected. This material was readily available at a local store and its consistent structure, no voids or cracks, made it a better choice then plywood.

A frame of 1″ × 1″ pine boards were used on the inside of the box for strength and rigidity. The acoustical foam used on the project was already selected for us. Due to budget reasons the foam used on the previous enclosure was recycled into the new one. The last component of the box, in the terms of acoustics, was the choice of a gasket material to be placed on the door, floor, and joints. These gasket materials would provide tight sealing and a complete enclosure around the pump. Neoprene weather stripping was purchased for use around the access door on the front of the enclosure. This material was easily acquired from the local hardware store. Around the base of the box ½″ Neoprene padding was used (Fig. 15.5). This material provided an excellent seal between the bottom of the enclosure and the floor to minimise cracks and therefore increasing overall transmission loss of the enclosure. For sealing the joints of the box, a standard silicon construction caulk was used. Also vibration isolators would be added to the bottom of the pump to alleviate structure borne noise. This would also help to eliminate vibrations affecting other labs, especially the micromachining laboratory next door.

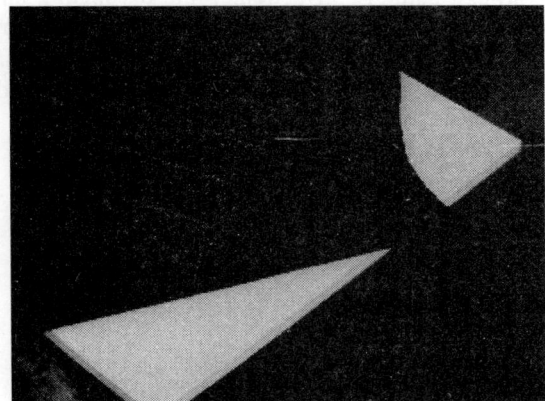

Fig. 15.5. Acoustic foam (top), neoprene (middle, and particleboard (bottom).

Acoustic Enclosure's Features

Before any fabrication of the acoustical enclosure began, some critical dimensions were taken of the pump and the surrounding environment. The dimensions of the pump were 44.5″ long, 36″ wide and 28″ high. One obstacle that needed to be considered was that the door opening to the lab is 33.5″ wide. It was decided that the dimensions of the acoustical enclosure would be 44.5″ × 36″ × 32.5″ (Fig. 15.6). This maximises the height of the enclosure and ensures it fits through the door.

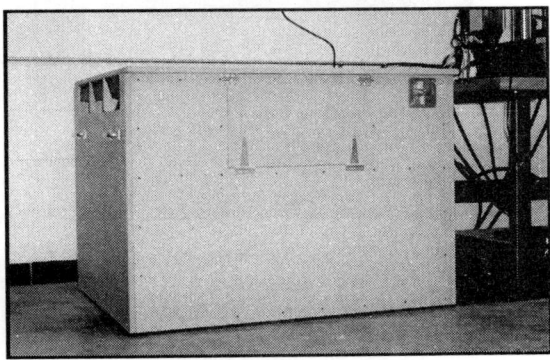

Fig. 15.6. Dimensioning of enclosure.

A 12″ × 24″ hinged door was installed to allow access to the control panel and power switch. To ensure a very tight seal and reduce the chance of noise loss, the door was fitted with 1/8″ Neoprene weather stripping at the seams (Fig. 15.7).

The entire interior of the enclosure was lined with acoustic foam to help dissipate the noise emitted from the pump. All five interior surfaces were fitted with the acoustic foam. Special attention was placed on making sure that there were no gaps in the foam layer (Fig. 15.8).

Overheating the pump was an important constraint. An acoustical louver (Fig. 15.9) lined top and bottom with acoustical foam was added to the side of the enclosure. The louver allows for heat to escape from within the enclosure through natural convection, while dissipating noise through the acoustic foam (Fig. 15.10). It was placed on the side of the enclosure facing away from the operator. An AC 110V fan with 100 cubic feet per minute flow was installed of the operator's side of the enclosure. This location was chosen because it was the farthest area away from the motor and hydraulic valves.

Fig. 15.7. Access door.

Fig. 15.8. Interior of enclosure lined with acoustic foam.

Fig. 15.9. Acoustical louvers.

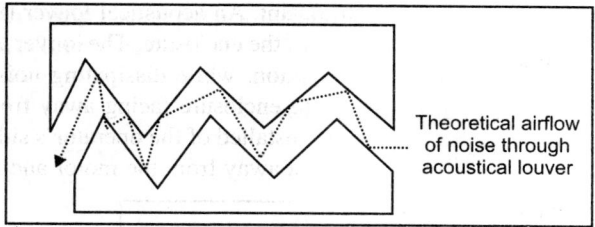

Fig. 15.10. Louver design.

The fan was designed to pull cool air into the enclosure to promote cooling and force convection of heat through the acoustic louver. Illustrated in Figs 15.8 and 15.11, a rectangular cut out at the right hand side of the enclosure allows the hydraulic lines and power cables to run into the box without leaving complete open spaces for sound to escape. Finally, ½″ Neoprene material was cut and stapled to the bottom of the box to act as a seal for the bottom of the box (Fig. 15.8).

Data and Acquisition

The initial testing included gathering third octave sound pressure readings of the background noise pump noise without using the old box, and pump noise using the old box. All of the sound pressure measurements were taken in the same fashion (illustrated in Fig. 15.11). An audio signal was created from three areas each scanned for 30 seconds. Measurement area 1 was 2′ wide by 3′ high on the left side of the pump. Area 2 was a section 4′ × 3′ high in front of the pump. The last measurement point simulates the noise heard by an operator. Thus it was taken from a point approximately 5′ high in front of the testing machine. The three audio signals were averaged using a logarithmic combination process in the dBFa32 program.

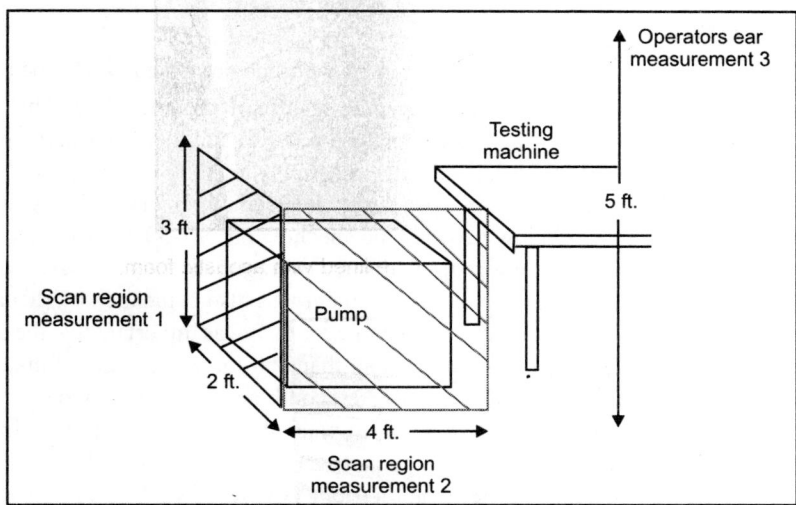

Fig. 15.11. Measurement techniques.

The combined signal was then post-processed using the dBFa32 software. A broadband sound pressure level plot was calculated by the system along with the Zwicker loudness and the overall sound pressure level. The first measurements taken were background noise and noise produced by the hydraulic pump while it is on. This was done to get a feel for what frequencies were contributing most to the overall sound pressure level in the room. Figure 15.12 indicates the sound pressure level of both the background and the noise produced by the pump in 1/3 octave bands.

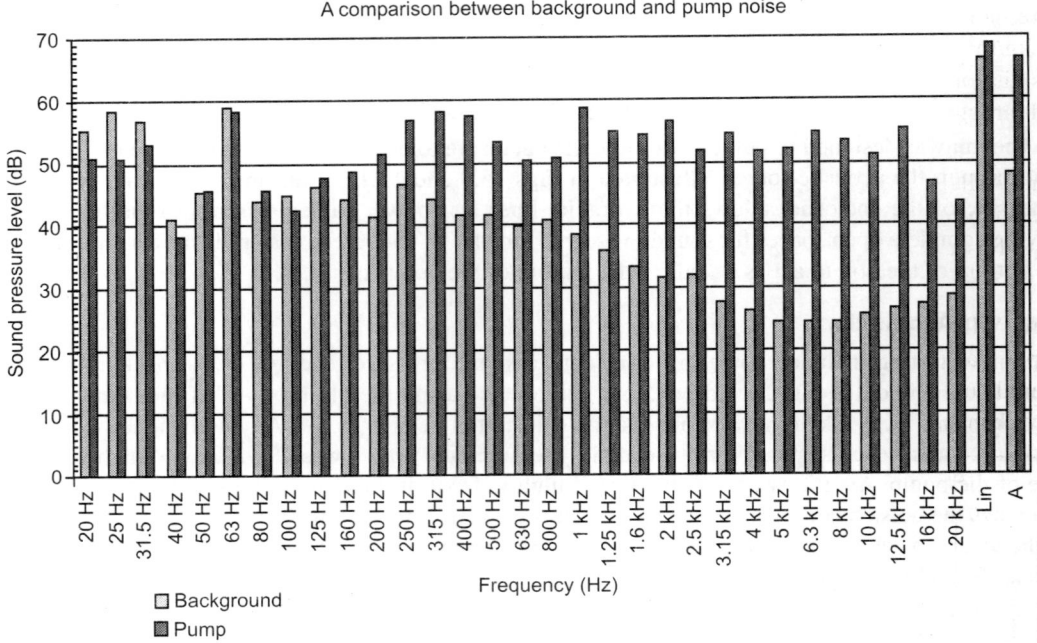

A comparison between background and pump noise

☐ Background
▨ Pump

Fig. 15.12. A graph that compares the background noise in the lab to the noise produced by the hydraulic pump.

The next step was to get a rough estimate expected results of the enclosure. This was done by creating 26″ × 26″ mock-up of the new acoustical enclosure's wall and then measurements were taken to find the noise reduction and transmission loss of the material using the SAE J1400 testing procedure. The procedure called for a homogeneous limp mass panel (1/16 inch thick aluminium plate was used in this instance) to be installed in the window between the anechoic on. Six 30-second audio signals were taken at certain points in the reverberation room and were averaged in the same fashion as the initial testing data. A broadband sound pressure level measurement was also calculated by dBFa32. The next step was to calculate the sound pressure inside the anechoic room. It was important that the measurements were not taken too close to the test sample but not farther than half the wavelength of major interest. An area inside this measurement window was scanned for 30 seconds and another broadband sound pressure level measurement was calculated. The next step was to place the test sample in the window between the anechoic and reverberation rooms. The side of the test sample that resembled the inside of the new enclosure was placed facing towards the reverberation room. Another six 30 second audio signals were taken at the same points in the reverberation room as were taken using the homogeneous limp mass panel. These audio signals were then averaged and a broadband sound pressure level measurement was calculated. The same area behind the sample material in the anechoic room was scanned for 30 seconds and a broadband sound pressure level measurement was calculated. Transmission loss (TL) was calculated for each of the frequencies on the 1/3 octave broadband spectrum using the equation provided by the SAE J1400 testing procedure listed below:

$$\text{TL (dB)} = 20 \log w + 20 \log f - 47$$

where, w is the surface weight of the homogeneous limp mass panel in kilograms per meter squared (2.8 kg/m^2 in this instance), and f is each of the frequencies in the 1/3 octave band in Hertz. This transmission loss was then used to find the correlation factor (CF) at every frequency in the 1/3 octave band through the following equation:

$$\text{CF} = \text{MNR} - \text{TL}$$

where, 'MNR' is the measured noise reduction between the reverberation and anechoic room. The correlation factor was used in the following equation to find the transmission loss in the test material (TL$_{TS}$) for each 1/3 octave band frequency.

$$\text{TL}_{TS} = \text{MNR}_{TS} - \text{CF}$$

where, 'MNR$_{TS}$' is the measured noise reduction of the test sample. The transmission loss for the test sample was calculated to be approximately 30 to 50 decibels at middle to high frequencies for the materials used in the test sample. Figure 15.13 illustrates the transmission loss broken down into 1/3 octave bands.

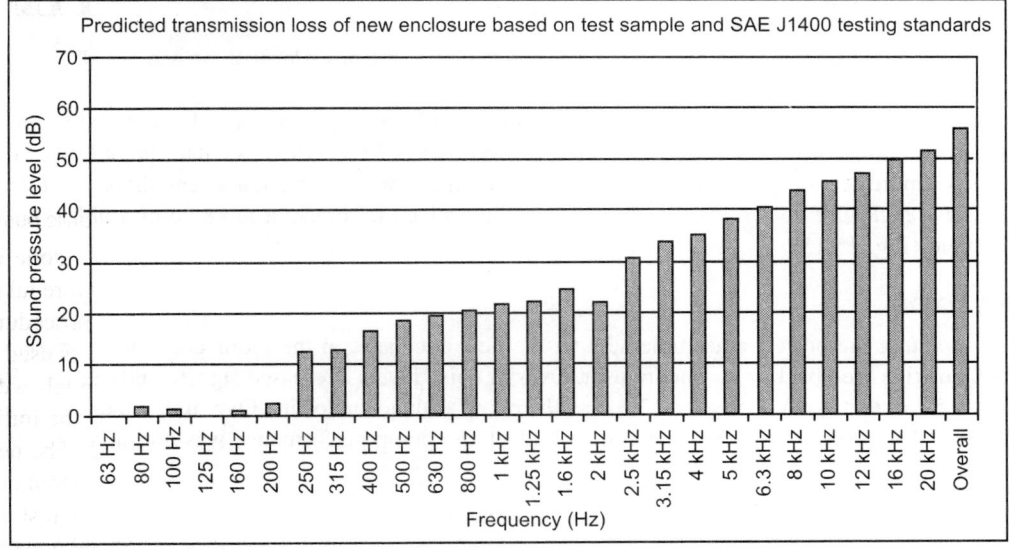

Fig. 15.13. A graph that illustrates predicted transmission loss of acoustical enclosure found through SAE J1400 testing standards.

From this predicted transmission loss the acoustical enclosure should efficiently absorb higher frequency noise. After the box was fabricated a similar process to the initial testing was followed. The background and the noise level outside the new enclosure were tested in the same fashion as the initial testing phase (Fig. 15.14).

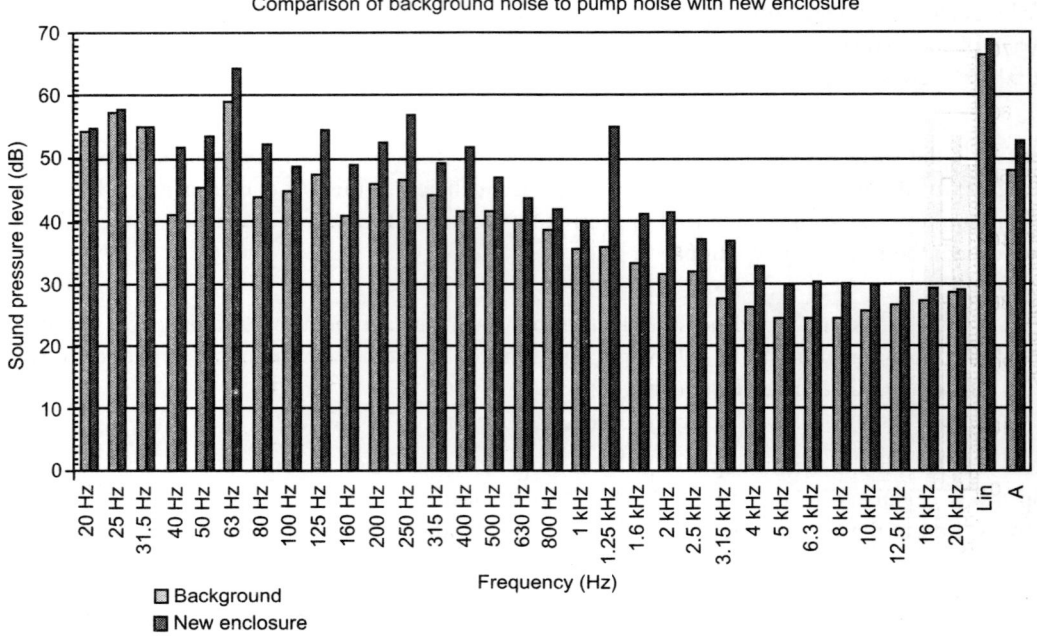

Fig. 15.14. A grpah that compares background noise in the lab to the noise level of the acoustically enclosed pump.

The new acoustic enclosure actually has a higher overall sound pressure level compared to the old box. However, the acoustic enclosure is 7.7 dBA lower and is 4.64 sones less than the old box. Since loudness is a more accurate measurement of annoyance experienced by the human ear, this data indicates that our new enclosure does meet the objective of the project, by create a more comfortable working environment (Fig. 15.15).

CONCLUSION

After completing the project and implementing the new enclosure in the biomechanics laboratory we received positive feed back from our sponsor. Overall noise levels decreased significantly in lab. From the Table 15.1 overall dBA decreased 12.6 over the pump with no enclosure. Overall loudness decreased 11.43 sones. The project was completed under budget and pump performance was not hindered by the enclosure.

Table 15.1. Comparison of data before and after new enclosure.

	Sound pressure level		Zwicker loudness
	dB	dBA	
Background	66.5	48.3	8.05 Sones
No enclosure	68.9	66.5	23.65 Sones
Old box	66.3	61.6	17.86 Sones
New enclosure	68.6	53.9	12.22 Sones

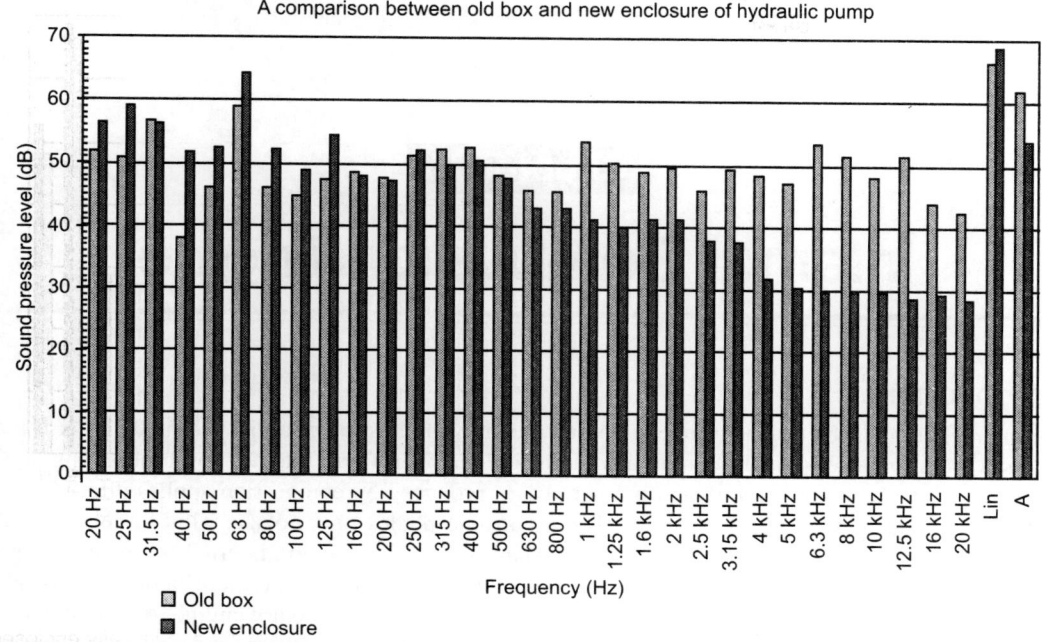

Fig. 15.15. A graph illustrating the performance of the old box compared to the new acoustic enclosure

This project was a great experience and incorporated many aspects of noise and vibration control. It was a great opportunity to learn hands on and incorporate classroom principles to an actual physical project.

Chapter 16

Internal Combustion Engine

INTRODUCTION

The internal combustion engine is an engine in which the combustion of a fuel (normally a fossil fuel) occurs with an oxidiser (usually air) in a combustion chamber. In an internal combustion engine the expansion of the high-temperature and pressure gases produced by combustion applies direct force to some component of the engine, such as pistons, turbine blades or a nozzle. This force moves the component over a distance, generating useful mechanical energy. The term internal combustion engine usually refers to an engine in which combustion is intermittent, such as the more familiar four-stroke and two-stroke piston engines, along with variants, such as the Wankel rotary engine. A second class of internal combustion engines use continuous combustion: gas turbines, jet engines and most rocket engines, each of which are internal combustion engines on the same principle.

Internal combustion engines are most commonly used for mobile propulsion in vehicles and portable machinery. In mobile equipment, internal combustion is advantageous since it can provide high power-to-weight ratios together with excellent fuel energy density. Generally using fossil fuel (mainly petroleum), these engines have appeared in transport in almost all vehicles (automobiles, trucks, motorcycles, boats, and in a wide variety of aircraft and locomotives). Where very high power-to-weight ratios are required, internal combustion engines appear in the form of gas turbines. These applications include jet aircraft, helicopters, large ships and electric generators.

All internal combustion engines depend on the exothermic chemical process of combustion: the reaction of a fuel, typically with oxygen from the air (though it is possible to inject nitrous oxide in order to do more of the same thing and gain a power boost). The combustion process typically results in the production of a great quantity of heat, as well as the production of steam and carbon dioxide and other chemicals at very high temperature; the temperature reached is determined by the chemical make up of the fuel and oxidisers, as well as by the compression and other factors.

Noise pollution: Significant contributions to noise pollution are made by internal combustion engines. Automobile and truck traffic operating on highways and street systems produce noise, as do aircraft flights due to jet noise, particularly supersonic-capable aircraft. Rocket engines create the most intense noise.

DEVELOPMENT OF AN ACTIVE EXHAUST SILENCER FOR INTERNAL COMBUSTION ENGINES

A silencer to attenuate engine exhaust noise using active control methods is developed. The device consists of an electrically driven valve, combined with a buffer volume, which is connected to the exhaust outlet. Using the mean flow through the valve and the pressure fluctuations in the volume, the valve regulates the flow in such a way that only the mean flow passes through the exhaust outlet. The fluctuations of the flow are temporally buffered in the volume.

To carry out optimisation and validation experiments, a cold engine simulator is developed. This device generates realistic exhaust noise and the matching gas flow using compressed air. The simulator allows quick and reliable acoustic and fluid dynamic experiments on exhaust prototypes. The active silencer is capable to reduce the exhaust noise from 91 dBA to 78 dBA after the tail pipe outlet, with a back pressure of 3 kPa to the engine. The reduction of external noise is now-a-days one of the important issues in car development. The legislators are lowering the noise emission standards continuously in the different countries. For example, the pass by noise level for passenger cars is reduced from maximum 77 dBA to 74 dBA in 2005. Also, in engine development, multi valve technology is becoming common to increase the engine efficiency by lowering in and outlet valve resistance. As consequence, exhaust system manufacturers must reach higher attenuation levels in combination with lower flow resistance. Much research on active noise cancellation in ducts is carried out in the recent years. Loudspeaker systems are successfully applied and commercial available in ventilation channels. A loudspeaker setup is developed for stationary six-cylinder diesel engines by Detroit Diesel Corporation. A problem using loudspeakers is the low sound generating efficiency and reliability in the extreme conditions of an engine exhaust. To increase sound generation performance, a loudspeaker with high diaphragm displacement, suitable for the low frequency range, is under development at the technical university of Dresden. If compressed air is available, an electro-pneumatic loudspeaker is also an option. Applying a controllable valve in the exhaust duct is a more robust concept. Good results are achieved mainly in the low frequency range, mostly on fan setups. On internal combustion engines, the sound power generated by the control sound source remains a problem.

In this section, the concept of a controllable valve in the exhaust is used. A solution to the sound power problem is achieved by mounting a buffer volume between the engine exhaust and the control valve. Idealised, the engine behaves as a volume velocity source. It is not possible to control the flow of a volume velocity source, if no capacitive elements are present between the source outlet and the active valve. The capacity of a duct is low, consequently the performance of the valve must be high. By adding an additional volume between exhaust and valve, the required noise generation performance of the valve is reduced. By balancing the volume and the valve performance, it is always possible to realise active noise attenuation for a combustion engine.

The dimensioning of the active silencer is based on acoustic simulations using electrical analogies. An electrical equivalent circuit is developed which consists of partial circuits for the engine, the active exhaust system and the controller. The simulation predicts a noise level reduction of approximately 27 dB, directly after the controlled valve. A main problem when conducting experiments is the availability of a representative sound source. Directly testing on an internal combustion engine is difficult because it is necessary to take precautions against the hot corrosive exhaust gases. Carrying out experiments on loudspeaker and fan setups results in unreliable data, because the acoustical circumstances on these setups differ considerably from these of an operating engine. Therefore, a cold engine simulator has been developed which generates the engine sound with the matching gas flow in a realistic way using compressed air. More generally, the simulator is applicable for the study of sound behaviour and flow phenomena in any kind of exhaust system.

The simulator has been validated by comparing experimental data with numerical simulations and experiments on a real combustion engine. These comparisons demonstrate that the cold engine simulator approximates the exhaust noise of an internal combustion engine very well. The experimental setup consists of an electrically driven valve, in series with a volume, which is connected to the exhaust outlet of the cold engine simulator.

The control signal for the valve is calculated from the pressure in the buffer volume and the flow through the exhaust duct. The valve is controlled such that only the mean flow passes the valve opening. The flow fluctuations are temporally stored in the volume. The control algorithm is a feed forward 'inverse plant' algorithm, without adaptation. This results in a fast algorithm, but it requires the active silencer to follow the theoretical proposed plant without major deviations in the desired frequency range. This approach results in a noise reduction from 91 dBA to 78 dBA, measured 10 cm out of the axis of the tail pipe opening, nevertheless, the 'inverse plant' approach limits the noise reduction capability considerably. In the near future, feed back and adaptive control algorithms will be implemented on the controller and tested.

Active Silencer

To control the flow in the exhaust duct, a valve is placed in the flow. The resistance of the valve is continuously variable by applying an external signal. It is assumed that the valve is purely resistive, it is not capable to store energy from the gas flow. The principle of using a valve to control duct flow is illustrated in Fig. 16.1.

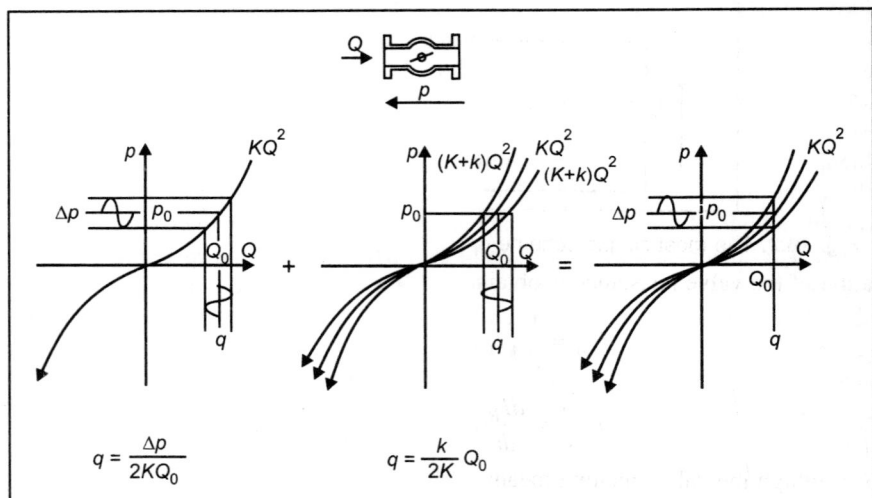

Fig. 16.1. Basic principle of using a valve controlling duct flow. The meaning of the symbols in this figure is: Q_0 is mean fluid flow, q is flow fluctuation, p_0 is mean pressure, Δp is the pressure fluctuation, K is the valve constant which expresses the relation between pressure and flow, k is the variation on K.

A valve placed in a volume flow causes a pressure drop over the valve. The principle of the active noise cancellation valve is, to obtain a constant volume flow despite a fluctuating pressure, by varying the valve resistance.

In Fig. 16.1, the first graph demonstrates how sound pressure influences the fluid flow via the valve resistance characteristic. The second graph demonstrates how the opposite fluctuating flow is generated from the mean pressure drop over the valve by varying the valve resistance. Superposition of both effects results in a constant volume flow, shown in the third graph. Mounting an active valve in the flow of a volume velocity source, like a combustion engine, has no effect if no capacitive elements are present between the source and the valve. The volume velocity source forces a prescribed flow through the system, whatever the pressure becomes in the system. Capacitive elements can be introduced using ducts or volumes. The most simple system for active control of a volume velocity source is presented in Fig. 16.2. The engine acts as a volume velocity source. At the exhaust, a volume with capacity C and a regulating valve with variable resistance $R(t)$ is connected. The translation of the physical system results in the electrical equivalent circuit shown in Fig. 16.2. The flow from the source will split over the capacity and the time dependent resistor. Now, the controller has to vary in time the valve resistance, such that the fluctuation flow through the resistance becomes zero.

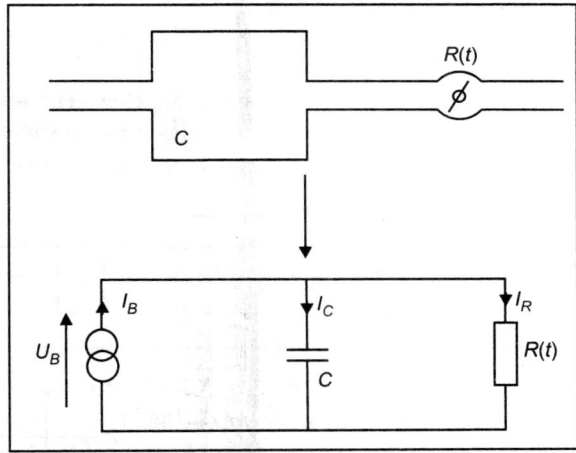

Fig. 16.2. The most simple scheme of active control of an engine using a valve.

The variation of the valve resistance is obtained from the electrical equivalent circuit:

$$U_B = \frac{1}{C}\int I_C dt = I_R R(t)$$

$$I_C \frac{1}{C} = \frac{dI_R}{dt} R(t) + I_R \frac{dR(t)}{dt}$$

Constant flow through the valve opening means:

$$\frac{dI_R}{dt} = 0$$

The controller has to vary the valve resistance during time according:

$$R(t) = R_0 + \int \frac{1}{C}\frac{I_C}{I_R} dt \qquad \qquad ...\,(16.1)$$

Two important conclusions follows from this simple consideration: First, by balancing the volume-valve combination, it is always possible to control the flow of a volume velocity source. Second, the resistance R_0 can be chosen freely with the only restriction that the resistance $R(t)$ remains always positive. The resistance R_0 must be optimised to obtain minimum energy loss of the engine.

Expanded Electrical Equivalent Model

The model looks not very realistic for an engine exhaust system, therefore the model will be expanded. The volume velocity source will be replaced by an engine model, and an exhaust duct is connected between the engine and the active silencer. The controller uses Eq. 16.1 as control algorithm. A scheme of this system is presented in Fig. 16.3.

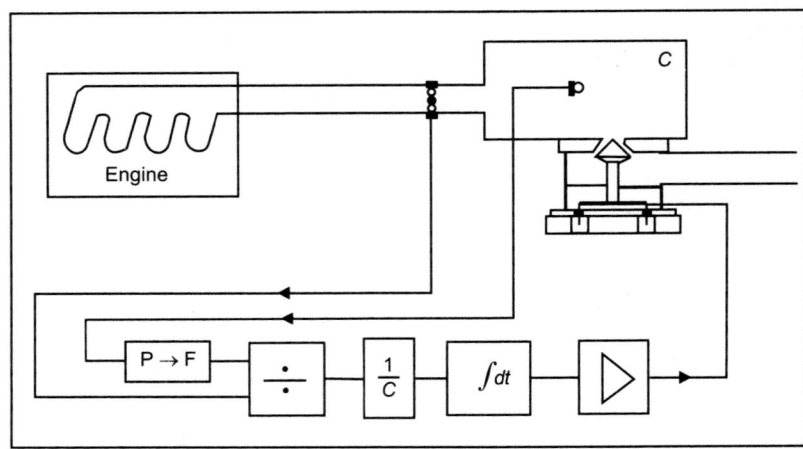

Fig. 16.3. Scheme of an internal combustion engine with an active exhaust noise cancellation device.

The electrical equivalent circuit of the system shown in Fig. 16.3 is presented in Fig. 16.4. The exhaust of the engine is considered as a series of discharges of constant volumes. When the outlet valve of a combustion engine opens, the remaining cylinder pressure drops in a few milliseconds to atmospheric pressure. The piston is at its lower dead point and the cylinder volume changes only 10 á 15 per cent during the escape time of the pressure pulse.

In the electrical equivalent model, the four capacitors represent the engine cylinders. The voltage of the source U_B equals the remaining pressure at the end of the expansion stroke of the engine. The capacitors are charged via the upper set of switch-resistor combinations, which represent the inlet valves of the engine. The capacitors are discharged via the lower set of switch-resistor combinations, which represent the outlet valves. The switches are operated in the same sequence as the cam shaft operates the engine valves.

The pulses enter the exhaust duct, represented by the transmission line T. The capacitor C and the variable resistor $R(t)$ forms the equivalent of the active silencer. The controller uses the current through the transmission line and the voltage over the capacitor to generate the control signal for the valve $R(t)$, using Eq. 16.1 as control algorithm. The inductor resistor combination, representing the acoustic impedance of free air, closes the circuit.

The simulation is carried out in time domain using a 1000 cc engine, a duct of 1 m length and $10^{-3} m^2$ cross-section, a buffer volume of $12 \cdot 10^{-3} m^3$ and a control valve resistance which can vary between 50 kΩ and 300 kΩ.

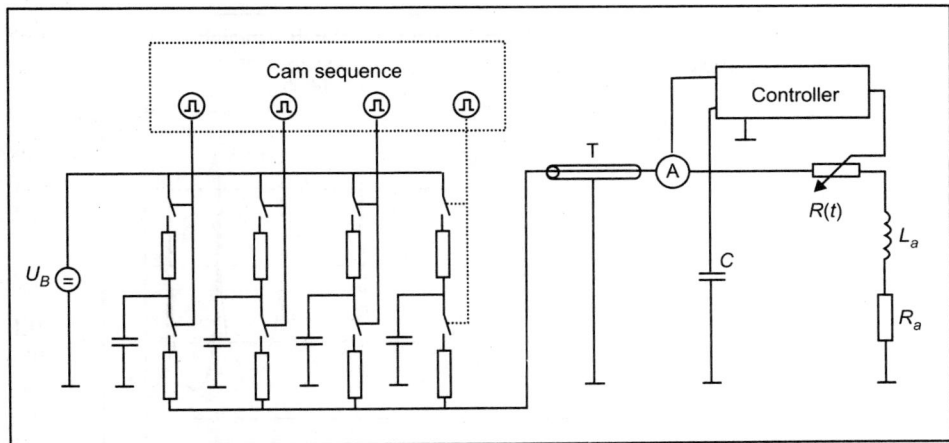

Fig. 16.4. Electrical analog circuit for a four cylinder engine with an active noise cancellation device.

The simulation result is presented in Fig. 16.5. Both lines represent the pressure directly after the control valve, before the outlet of the exhaust tail pipe. Before activating the controller, the pressure, represented by 'line 1', corresponds with a sound pressure level of 153 dB in the tail pipe. When the controller is activated, the pressure reduces to a sound pressure level of 126 dB.

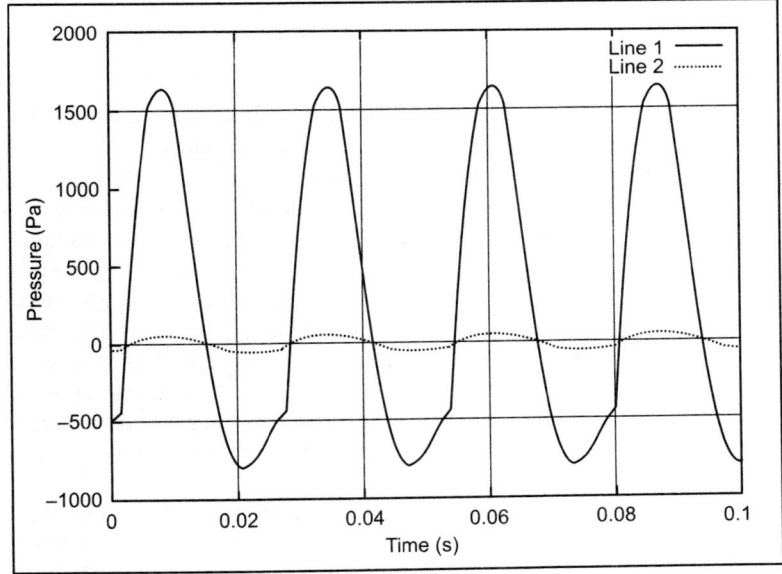

Fig. 16.5. Simulation result of the circuit presented in Fig. 16.4. 'Line 1' is the pressure at the outlet opening without control, 'line 2' is the pressure with control.

In frequency domain, the active control results in a shift of the noise spectrum with 27 dB downwards. This is illustrated in Fig. 16.6.

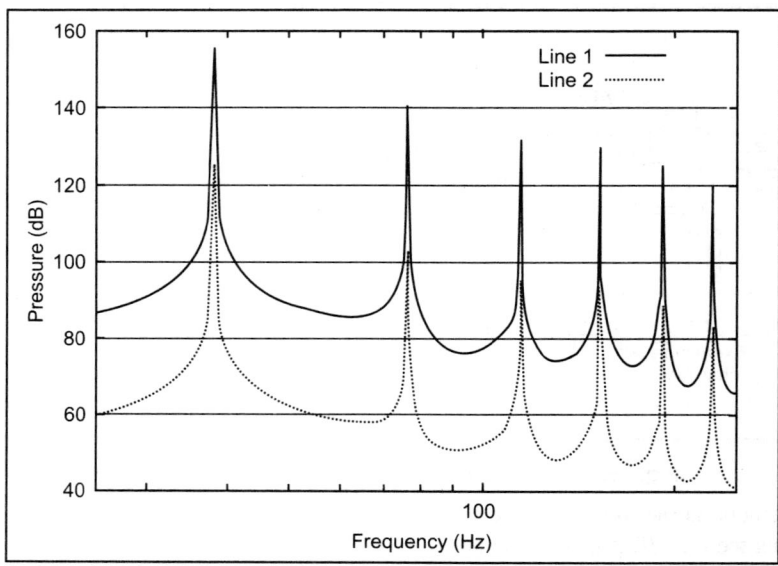

Fig. 16.6. Frequency spectrum of the time signal presented in Fig. 16.5. 'Line 1' is the pressure without control, 'line 2' is the pressure with control.

Experimental Results

Cold engine simulator

To carry out experiments with the active silencer, a cold engine simulator is developed which generates realistic engine noise and the matching gas flow using compressed air. It permits to experiment with new concepts of exhaust systems, without taking precautions against the hot corrosive gases of a real engine. The assumption that the pressure pulse can be considered as a discharge of a constant volume gas forms also the basis for the cold engine simulator. A scheme of the engine simulator is presented in Fig. 16.7. It consists of a regular engine block whose pistons are fixed at their lower dead points. The inlet collector is connected via an expansion vessel and a pressure reduction valve to a normal pressurised air supply network. The cam mechanism of the engine block is driven by an electric motor. The supplied pressure at the inlet collector is equal to the pressure in the cylinder of an operational combustion engine at the end of the expansion stage. During the inlet stage, the cylinder charges at the same pressure level as applied at the inlet. When the outlet opens, the cylinder discharges and the pressure pulse enters the exhaust. The discharge takes a few milliseconds. These pressure pulses are similar to these of a real combustion engine.

To validate the cold engine simulator, pressure measurements are carried out on the simulator and a real combustion engine. On the exhaust, a straight duct is connected of 7 m length and $10^{-3} m^2$ cross-section. The crank rotation speed is 300 rpm. The applied pressure at the inlet collector equals 200 kPa.

The −20 dB/decade decay is dominantly present. After 200 Hz, the higher frequencies show more damping as in the simulation. These damping effects are introduced by many causes. The most important ones are: the finite opening times of the outlet valves; the saturation of the flow in the valve opening; the difference in sound speed in the duct in forward and backward direction, due to the mean gas flow through the duct.

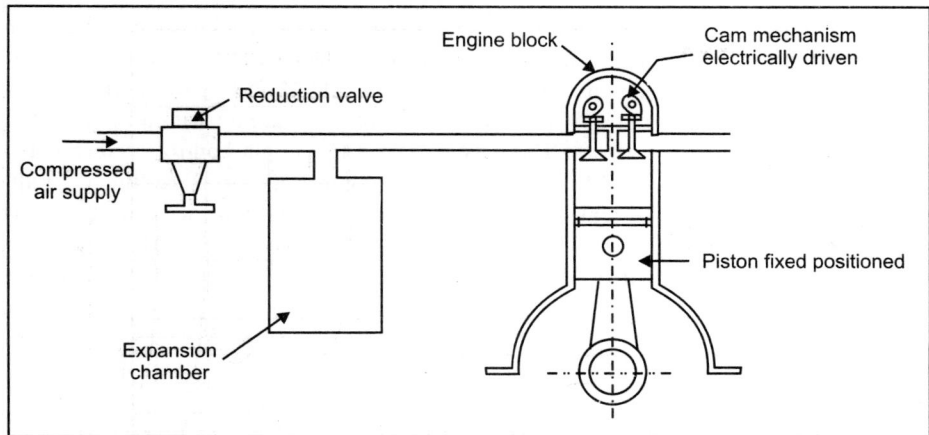

Fig. 16.7. Working scheme of the cold engine simulator.

The measurement, is the frequency spectrum of the manifold pressure of a real internal combustion engine. Here also, the –20 dB/decade slope is dominantly present. The cold engine simulator in Fig. 16.7 demonstrate that the cold engine simulator and the internal combustion engine exhibit a very similar behaviour. It demonstrates also the usefulness of the equivalent circuit for the engine.

Active Silencer

Experiments with the active silencer are carried out on the cold engine simulator. The electrically controllable valve is mounted on the buffer volume. As control signals, a piezo-resistive sensor measures the mean pressure in the volume, and is filtered by an analog 1 Hz second order low pass filter. The alternating (acoustic) pressure in the volume is measured by a piezo-capacitive sensor and is filtered by an analog 2000 Hz second order low pass filter. The mean flow through the outlet duct is measured by a small turbine and is also filtered by an analog 1 Hz second order low pass filter.

As Fig. 16.3 suggests, a voice coil in a magnet actuates the valve head to control the valve opening. The voice coil is powered by a current amplifier, with a maximum electrical power of 500 W. Before the current amplifier, a preamplifier buffers and filters the output signal from the controller. The filter is also an analog 2000 Hz second order low pass filter. By choosing these second order filters, the phase lag at 500 Hz is not higher than 45° after passing all in and output filters. As consequence, the valve should be able to attenuate pulses of at least 2 ms pulse duration. At low frequencies (below 17 Hz), the applied force on the voice coil is almost equal to the pressure drop over the valve times the valve opening area. At higher frequencies, the inertia of the valve motor becomes dominant and the force determines the acceleration of the valve head.

The controller is split in two parts. A low bandwidth PI-controller realises the set point resistance R_0, where around the active noise cancellation is carried out, using the piezo-resistive sensor. This controller uses the linear relationship between the applied force and the resulting pressure.

The active noise controller constructs the control signal using the flow through the duct and the pressure in the volume, based on Eq. 16.1. The input signals are taken from the piezo-capacitive sensor and the flow sensor. This controller takes the valve motor inertia into account to determine the resulting valve head displacement. This controller is not active in the low frequency range by implementing a 10 Hz first order high pass filter on the controller input signal. The sum of the signals of both controllers is applied to the preamplifier of the current amplifier. Both controllers are implemented in Z-domain on a C40 digital signal processor of Texas Instruments, using a sample frequency of 16 kHz.

In Fig. 16.8, the result of the control is presented. The pressures are measured in the tail pipe after the control valve using a pressure sensor. The pressure 'line 1' without control corresponds with the original sound pressure level of 151 dB. By activating the control, the pressure 'line 2' reduces to a sound pressure level of 138 dB. The realised noise reduction in the tail pipe amounts 13 dB. The same reduction is measured with a sound level meter 10 cm out of the axis of the outlet of the exhaust tail pipe. The radiated noise reduces with 13 dBA from 91 dBA to 78 dBA by activating the controller.

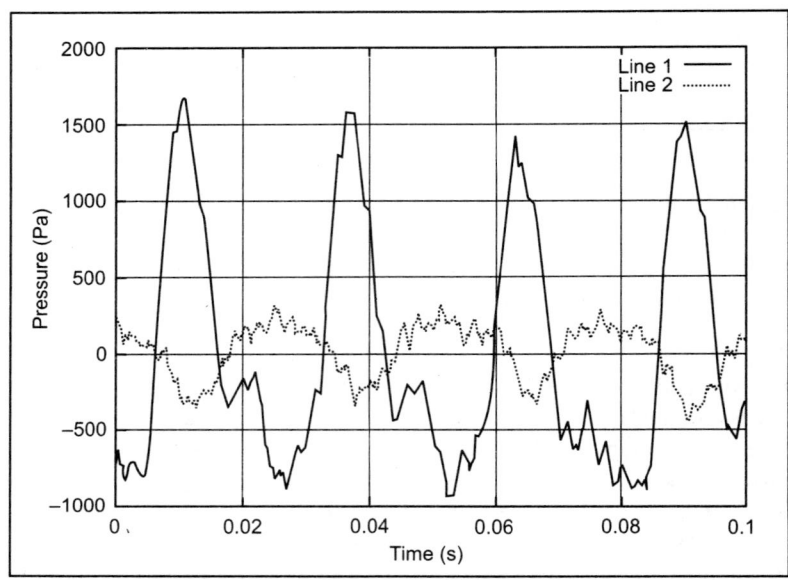

Fig. 16.8. Measurement of the pressure after the control valve in the outlet duct. 'Line 1' is the pressure without control, 'line 2' is the pressure with control.

The frequency spectrum in Fig. 16.9 of the pressure in the tail pipe presented in Fig. 16.8 shifts 13 dB downwards, by activating the controller.

The controller uses a linearised model of the valve resistance as function of the valve opening. The valve resistance depends quadraticly as function of the valve opening and this effect results in a distortion of the pressure drop over the valve. This limits the sound reduction performance of the active muffler. In the near future, feed back and adaptive feed forward control strategies will be implemented and tested, to minimise the influence of the quadratic valve resistance.

Future Investigations

Before creating an active silencer suitable for a real internal combustion engine, energy consumption and dimensions have to be optimised. The dimensions of the valve can be reduced considerably. This results also in a lower electrical energy consumption. Reducing the buffer volume dimensions results in a higher back pressure for the engine and consequently in a higher energy consumption from the exhaust gas flow. The buffer volume and the valve have to be balanced to obtain minimum energy consumption in an active silencer in realistic dimensions, suitable for passenger car combustion engines.

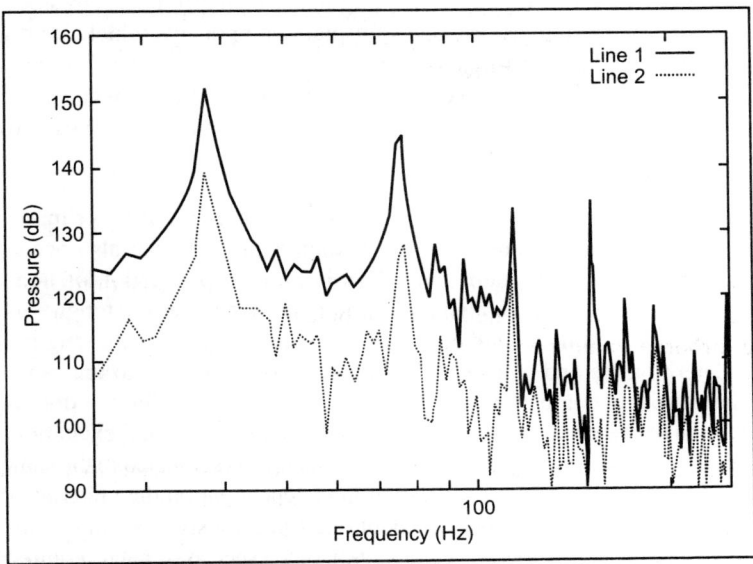

Fig. 16.9. Frequency spectrum of the time signals displayed in Fig. 16.8. 'Line 1' is the pressure without control, 'line 2' is the pressure with control.

CASE STUDY: FORD MOTOR

Ford Motor Company have a Sound Quality Lab where they try to find out what customers likes and dislikes are when it comes to the total acoustic environment in the car. If we have the capability to artificially create sounds (such as that from a running engine) we can present these to customers to determine what they desire in their vehicles. This section deals with engine sounds. Many aspects of engine noise generation parallel that of human speech production. The noise from an internal combustion engine is typically experienced from the passenger compartment of the vehicle. The engine noise is a complex signal produced by a number of sources in the engine compartment. These sources can be broadly classified as either deterministic (predictable) or stochastic (random) in nature. An analogy can be made between these two classes, and voiced and unvoiced speech. Unlike speech, however; both deterministic and stochastic components occur simultaneously for engine noise.

Deterministic Component

A large part of the deterministic component is due to the combustion of fuel in the engine cylinders. This produces pressure pulses not unlike those produced by the glottis in voiced speech. In this section we will discuss some of the components in speech production and compare these to that of sound production of a multicylinder internal combustion engine. The engine produces these pulses at a rate proportional to engine speed. For a four stroke engine, fuel is ignited once every two engine revolutions for each cylinder. Thus, an eight cylinder engine produces four pressure pulses per crankshaft revolution. Pressure pulse frequency is like that found in pitch extraction for speech in which the frequency of the glottal pulses are determined. As with speech, the source manifests itself spectrally as a frequency component at pulse frequency and its harmonics. Other deterministic sources include engine accessories such as the alternator, AC compressor, engine cooling fan (which also produces stochastic noise due to turbulent air flow), etc. These sources produce sinusoidal components which are highly correlated to the engine combustion since they are driven by the engine crankshaft. Depending on pulley ratios, these components do not necessarily represent multiples of the pressure pulse frequency. This is unlike voiced speech production in which only the pressure pulse frequency and its harmonics exist.

Stochastic Component

There is also a stochastic sound component produced by the engine. This is primarily due to turbulent air flow from the air intake and exhaust systems. Turbulent air flow can also occur from accessories such as the engine cooling and alternator fans. This is like the production of non-voiced sounds in speech in which coloured noise is generated by turbulent air flow over the tongue and lips.

If only the airborne component of the engine noise (coming from the engine compartment) is considered, both deterministic and stochastic components are subjected to the same transfer function before reaching the passenger compartment. For voiced speech this transfer function would be analogous to the transfer function of the vocal tract which spectrally shapes the glottal excitation. Unlike unvoiced speech production, the stochastic component of the engine noise is subjected to the same transfer function as the deterministic component. This transfer function is dependent on the materials and structure used to isolate the passenger from the engine. As mentioned previously, speech production is generally classified as voiced or unvoiced. Therefore, segmentation of the speech signal is required before traditional synthesis techniques can be implemented. This is not a concern for engine noise synthesis since both components occur simultaneously.

The preceding analysis gives us reason to believe that there may be speech synthesis techniques which can be applied to the problem of engine noise synthesis. We first briefly examine regular or periodic pulse excitation (pulses occur at a constant rate and height, and are spectrally shaped by a filter) and discuss some of the pitfalls of this approach for engine noise generation. Next, we introduce a new suboptimal multipulse (pulses occur at varying intervals and heights and are spectrally shaped by a filter) technique for the purpose of engine noise production. In this approach the deterministic components, as measured in the vehicle passenger compartment, are extracted from the signal. The remaining stochastic component is then synthesised by a multipulse excited time series model (the time series model is actually just a filter). Together, the deterministic and stochastic components are added to generate the synthesised sound. The resulting synthetic sound actually requires much less storage (2 per cent of the original) space. This is an added advantage when trying to store or transmit long sound files.

NOISE CONTROL APPARATUS FOR INTERNAL COMBUSTION ENGINE

A noise control apparatus has a thermal-type air flow meter for detecting engine load, an engine speed sensor for detecting engine speed, and an intake temperature sensor for detecting intake temperature. The engine load is detected based on the surging components of the signal from the air flow meter. An intake pipe is provided with a speaker that produces a noise control wave in accordance with a control signal from a controller. The controller has a memory that stores map data for noise control waves that are equal in sound pressure but opposite in phase with respect to intake noise. The map data regarding sound pressure and phase correspond to the engine load and speed on the basis of a reference temperature. A CPU of the controller computes a map-reading engine speed value based on a wavelength of intake noise that is determined based on intake temperature and engine speed, such that the map-reading engine speed value provides at the reference temperature substantially the same wavelength as that of the intake noise. The CPU generates a noise control wave signal based on the map-reading engine speed value and engine load information.

KEY STEPS AND METHODS IN THE DEVELOPMENT OF LOW NOISE ENGINES

The next generation of automotive engines will have to meet Euro 4 or similar exhaust emission limits, ideally with improved fuel economy and with noise emission which is at least 3 dBA below current levels. Using both simulation and experimental analysis, these challenging requirements can only be fulfilled by clearly defining all key steps in Noise, Vibration, and Harshness (NVH) development and by applying suitable technological methods. This section describes those phases of the entire development process which are decisive for engine NVH: the concept and design phase, the combustion and mechanical development phase and the NVH development and refinement phase. Related methodical approaches are discussed such as the close interaction between simulation and test bed work or the enhanced consideration of noise quality aspects using appropriate noise annoyance descriptors. The boundary conditions of engine NVH (Noise, Vibration and Harshness) development work are as simple as they are difficult to achieve. Relevant data must be available at the right time with an accuracy which is sufficient for making reliable design decisions. An important aspect in improving this process is a thorough integration of simulation and experimental analysis. This does not necessarily mean the use of a highly integrated analysis system, it is far more important to use consistent evaluation schemes and effective model updating procedures instead of unidirectional verification processes. A major characteristic of NVH development work is the continuous change of critical aspects and the balance of influencing factors with each modification or step of improvement. The risk of spending effort on ineffective measures is high and therefore simultaneous engineering is definitely required. Organisation structures can help a lot by defining NVH teams with full responsibility throughout the development process. Examples for the continuous interaction of worksteps are given in the following paragraphs.

Development Plan

A schematic diagram which outlines the appropriate steps and methods for a total NVH development plan up to the start of production (SOP) of the total vehicle or related machinery is shown in Fig. 16.10. Even if one is only concerned with a single automotive component, consideration of the final vehicle application from the very beginning of the development process is a must. Noise quality targets, packaging and installation considerations form the basis for numerous specifications. In view of pass-by noise regulations, a reliable prediction of noise radiation patterns in the final vehicle is very helpful.

Concept and Design Phase

Concept design decisions must be based on compiling and prioritising all target specifications. Knowledge about expected noise characteristics is essential for this work task and also for the specification of appropriate noise and vibration targets for subsystems. This is especially true when several system suppliers are involved and are responsible for the target specs. Decision making can only be based on experience and prediction. This becomes especially challenging when future concepts have to be developed and when only a very small empirical database is available. Such circumstances support all efforts to develop methods for early prediction.

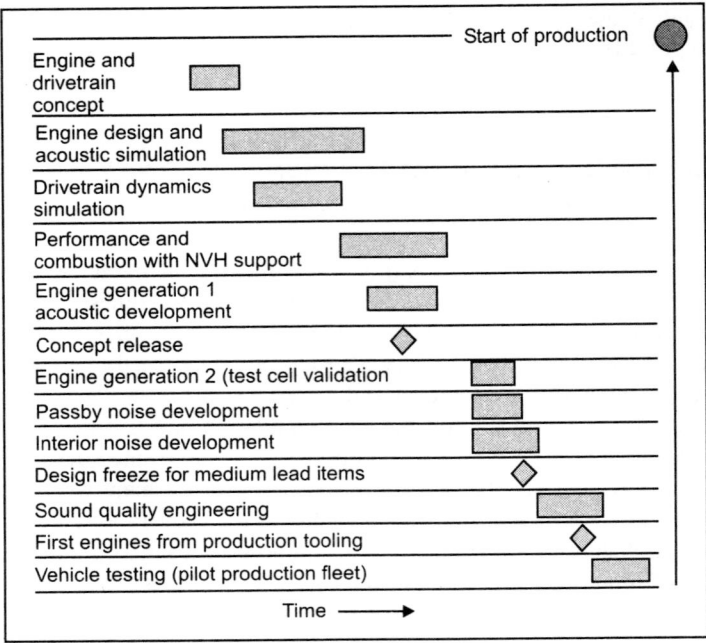

Fig. 16.10. Overall vehicle NVH development plan.

Objective of noise and vibration simulations

The current development trend for automotive engines is directed towards increased specific output. Additionally, reduced engine weight, manufacturing cost and, last but not least, reduced engine noise are important requirements. These requirements can be most effectively balanced by applying extensive simulation techniques in the early design stages. Performing structural vibration calculations in parallel with design studies offers substantial advantages. On the one hand, conceptual decisions can be supported by creating the structural prerequisites for low noise engines. On the other hand, the noise behaviour of an engine structure can be optimised early in the design stage through vibration simulation. Taking this approach ensures that first generation engine hardware is already structurally optimised for noise.

Simulation approach

Finite element methods (FEM) are utilised for simulation of structural vibrations. Natural frequencies and vibration modes may serve for verification purposes. The focus, however, is on forced vibrations of the engine structure with loads and boundary conditions as realistic as possible.

Simulation to assist conceptual decisions

The design concept of an engine is influenced by many parameters. Performance and emission requirements lead to peak firing pressure (PFP) levels of 160 bar or higher. Stringent pass-by noise regulations as well as interior noise considerations call for reduced engine noise. The achievable engine noise level is largely dependent on the design concept. The purpose of calculations in this stage is to check the effect of different design concepts on engine structural vibrations. For vibration simulation within the concept phase, simplified models based on conceptual drawings may be used. These simplified models are suitable for determining the influence of different designs on structural vibrations. However, it is important to accept that only trend predictions are possible in this stage and careful cross-checking with empirical data is required.

Simulation during layout and detail design

Once the design concept of an engine is fixed, layout design adds further detail to the Computer Aided Design (CAD) models. In this stage the complete engine is modelled, comprising all engine components which are relevant for engine structure dynamics. Evaluation of forced vibration calculation results clearly shows critical areas where high noise radiation is to be expected. Design modifications are then proposed to locally reinforce the structure and reduce vibration levels. The target of the simulation is to finally arrive at a structure showing evenly distributed vibration levels over the outer surface.

The use of automeshing methods will be standard operating procedure in the future. Together with 3-dimensional CAD data, it permits simulations to be performed in parallel with the design. In the generation of full-size engine structure acoustic models, however, automeshing typically results in very fine meshes, with high demands on the computer hardware and certain limitations in necessary modelling of modifications.

In current methodology it may therefore be more effective to apply automeshing only to certain subsystems of the structure such as crankshaft or brackets. In any case, the quality of the model is primarily determined by the definition of boundary conditions, parameters and load cases.

The result of this concurrent engineering approach is an engine structure that is already optimised for noise at the end of the layout design stage. Accordingly, first generation engine prototypes going into noise development should not require extensive structural changes to achieve noise targets.

Simulation models

The model used for vibration simulation consists of the vibration relevant components of the engine, that is:
1. Crankcase.
2. Cranktrain.
3. Cylinder head.
4. Oilpan, covers, manifolds.
5. Brackets and relevant ancillaries.

Excitation mechanisms and input data

Nonlinear forced vibrations are calculated considering structural responses to gas pressure as well as mass forces of the cranktrain. Piston secondary movement is simulated based on the crankcase and piston structure and calculated impact forces are then introduced into the forced vibration calculation. Additional excitations resulting from gear train and valve train forces are also considered, ideally based on specialised software tools. However, it is essential for the reliability of the results to ensure consistent operating and boundary conditions for all models. The simulation is performed in the time domain. Therefore, special attention has to be paid to the fact that the input data must represent realistic energy distribution over the whole frequency range of interest (e.g. requiring signal enhancement of cylinder pressure). Otherwise, there is a high risk of drawing wrong conclusions.

Simulation results

Calculation results are available as vibration accelerations, surface velocities as well as averaged levels for certain engine surface areas. Typically the surface vibration velocity levels are evaluated in octave or third octave bands. From these results, critical areas can clearly be detected and recommended design modifications may be derived. The optimised structure then offers the possibility to start engine noise development at a level where no extensive casting changes are necessary. Therefore, costly pattern modifications are also avoided—ideally one whole engine generation can be skipped in the development process. For this phase, evaluation of the vibration pattern is usually sufficient for the specification of design modifications. It should be noted, however, that sound radiation prediction will be an important step in a later phase. It should also be mentioned that acoustic post processing closely tied to test data processing is absolutely essential for several reasons:

1. Direct comparability of results supporting the same conclusions among all involved engineers.
2. Combined conclusions for deeper insight into the structure dynamics.

Development Phase

Due to the high level and reliability of the design supporting simulation work, the noise development phase can be started with acoustically preoptimised hardware. However, in spite of this fact, the noise development phase is still considered a one loop process. Modification and refinement of the hardware are based on experimental analysis of engine NVH. Using preoptimised hardware for the development phase shortens the entire process. In view of the short overall development schedules of engines, it is a must to run the different parts of an experimental engine development process simultaneously on several engines resulting, for instance, in a 'performance/emission engine', 'noise engine', 'mechanical engine', etc. Basically, noise development runs parallel to the performance and emission development of the combustion system as shown in Fig. 16.11.

The actual start of the noise development program, however, is shifted by a few months for two reasons. First, it is important to start the noise development with an advanced combustion development status. This ensures that the noise excitation from cylinder pressure remains substantially the same from the beginning of the noise development. After completion of combustion development, the combustion system of the noise engine has to be updated in order to represent the final status in view of performance, emission and noise excitation. Secondly, a late beginning of the noise development phase extends the procurement phase between end of design and beginning of noise development. This allows implementation of the finite element-derived modifications to the hardware of the noise engine providing a more authentic situation for the engine structure response and mechanical noise excitation.

In fact, the experimental part of the noise development and optimisation work do not only start with the noise engine but begin on the performance/emission engine with combustion development. As one of the key steps, the control of combustion noise can only be done in the context of performance, emission and economy requirements. Therefore, combustion noise meters have become an indispensable tool for combustion development. A typical criterion for combustion noise is shown in Fig. 16.12. In addition, it shows an example of a criterion for structural vibration to be applied to the rated condition of an engine. Such criteria provide essential guides for development engineers.

Note that the combustion noise level of Fig. 16.12 is in terms of the mean free field response (MFFR). This represents the use of a weighting function applied to the cylinder pressure measurements to obtain the equivalent combustion excitation in the cylinder and its transfer to a microphone position at a distance of 1 m from the engine.

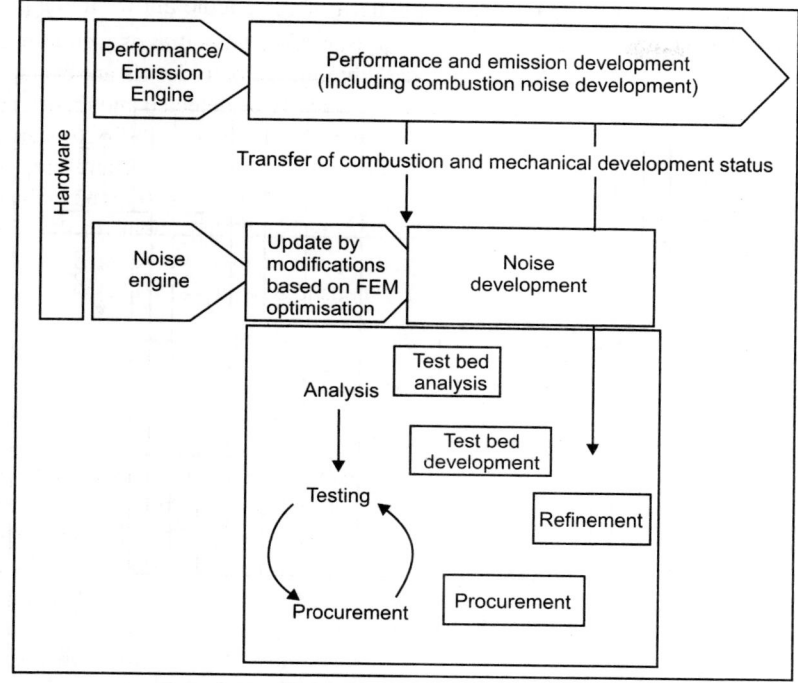

Fig. 16.11. Noise control scheme for series 100 engines.

In this phase of combustion development which has to consider combustion noise excitation under transient engine operations, the main task of noise specialists is to provide advisory support to the combustion development engineers. Any potential for combustion noise control not achieved during the combustion development is difficult to compensate for afterwards since it considerably worsens the initial status of the NVH development work on the 'noise' engine. The utmost significance of low-noise combustion on overall engine NVH can be seen in the example of Fig. 16.13. The reduction of combustion noise as a first development step yielded a decrease of engine noise level by 3.5 dBA. The gain in noise quality is two thirds of the overall quality improvement.

As previously mentioned, typical noise development work is a one loop process consisting of the four phases: analysis, testing and development of modifications, procurement of modified hardware, final testing and refinement. The essential NVH target parameters to be considered are the 1 m engine noise level, the noise quality and the low and mid frequency vibration behaviour at the engine mount positions (up to 800 Hz). The most significant operating conditions which have to be taken into account are: rated condition; full and low load conditions at mid speed (e.g. speed of maximum torque); low idle; and transient operation (pass-by test). For transient operations, it is important to consider not only a certain speed increase but to include a sudden change of load (to full load) at the beginning of the acceleration.

Experimental noise analysis

Experimental analysis is a key element in the development due to two aspects. First, measurements of the target parameters define the amount of noise reduction required to meet target specifications. Secondly, detailed analysis of the noise source and radiation pattern of the engine allows the appropriate noise reduction strategy to be established.

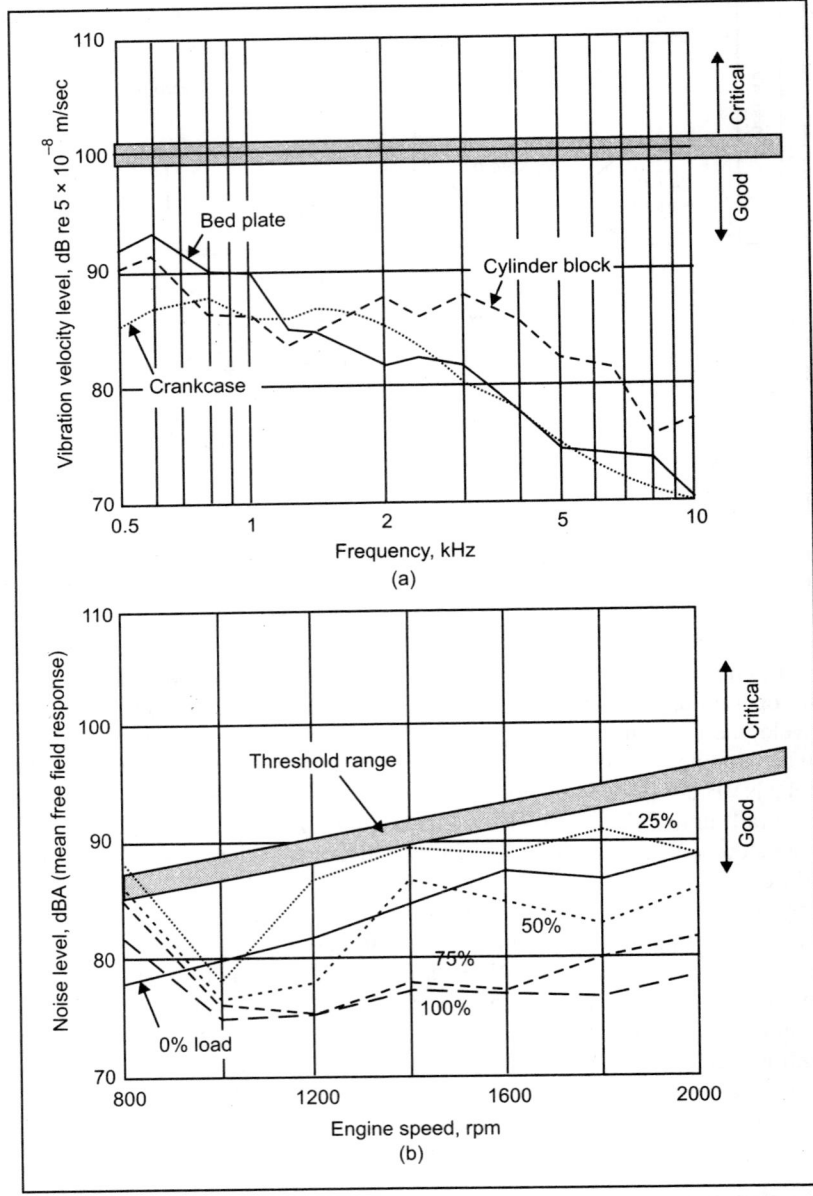

Fig. 16.12. Development criteria for vibration and combustion noise levels: (a) block vibration at full load, 2000 rpm; (b) combustion noise.

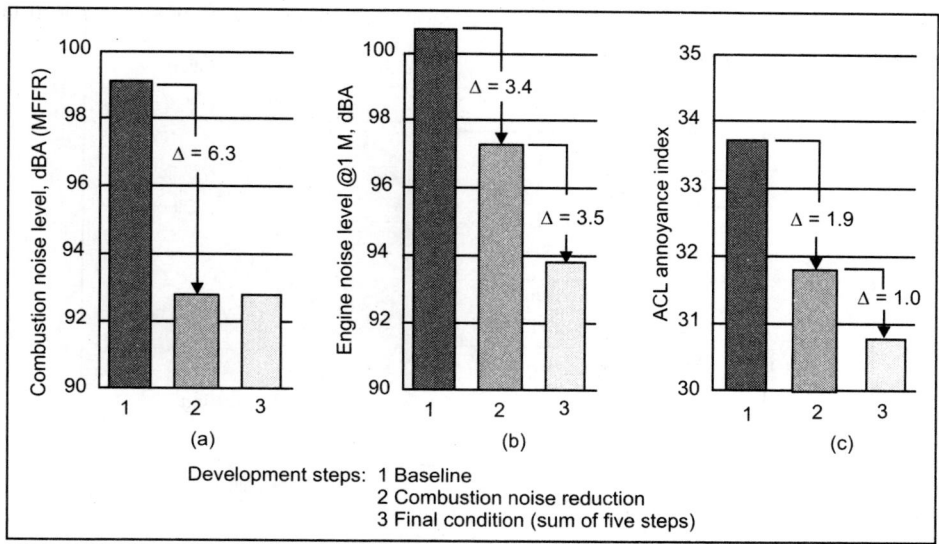

Fig. 16.13. Effect of combustion noise reduction on noise level and noise quality of a DI Diesel engine at rated operation: (a) combustion noise level; (b) engine noise level; (c) engine noise quality.

For the first part, precise measurements of the 1 m engine noise (according to test procedures DIN 45635 or SAE J1074 under anechoic conditions) and other target parameters are needed. The more challenging task is the second part, the detailed analysis. First, the noise radiation pattern has to be determined for critical operating conditions. An appropriate method is the well-known nearfield sound intensity analysis which identifies the engine surface and subsurface parts of highest noise radiation ranked as shown in Fig. 16.14. In a recent development, the identification of noise sources has become more efficient through 3D colour visualisation. This offers the advantage of a spatial resolution much finer than the subsurfaces that are conventionally averaged.

The results of the sound intensity analysis have the advantage that they are based on spatial averages, hence they are highly representative and not local in occurrence. On the other hand, since sound intensity is measured at a certain distance from the engine surface, the results cannot always be clearly assigned to a certain engine component. Therefore, the interpretation and understanding of sound intensity results need to be supported by the results of surface vibration analysis which are directly related to engine surface parts and their locations.

Conventional point-by-point measurement of surface vibration (e.g. by accelerometers) and the use of a scanning laser vibrometer system allow quick analysis of whole engine sides providing a good overview of global vibration behaviour. In addition, results obtained from a scanning analysis are very suitable for comparison with simulation results. For this evaluation, the measured vibration data are directly assigned to the FE-geometry via a specific data interface in order to allow identical post-processing. The graphs show very good correlation even in terms of absolute levels. Differences arise from the fact that the laser scanned engine surface includes corresponding surface areas of parts which are attached to the engine block structure such as oil filter, fuel filter, oil cooler, etc. On the other hand, the simulation results describe the engine block surface which can partly be hidden behind such attached parts.

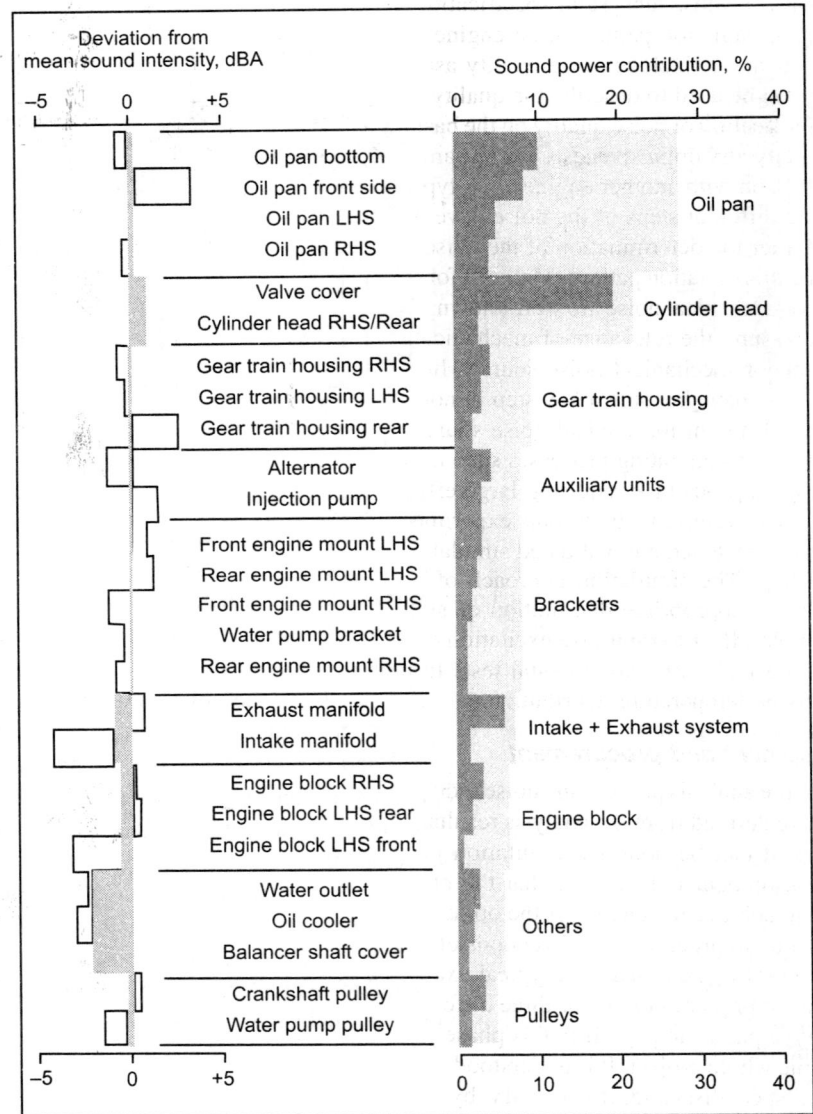

Fig. 16.14. Noise source ranking in terms of sound intensity and sound power.

Such comparisons between simulation and experimental analysis results are indispensable for the tuning or validation of simulation work and beyond. This helps targeting of a closed loop gain and regain of reliable results, experience and empirical data. For example, the closed loop procedure allows structural and system damping factors to be well matched over the significant frequency range. Correct assumptions are very important for the identification of critical frequency ranges and related noise sources. After several integration loops, an empirical data base can be built which again enhances the quality of concept decisions, subsystem specifications and so on, even on different engine types. The analysis of the noise radiation pattern of an engine must not only be confined to objective noise and vibration levels but has to include noise quality aspects. Noise quality indicators like the AVL noise annoyance index can be used to describe the quality of the radiated engine noise. The AVL Annoyance Index allows quantification of noise quality on the basis of the four psychoacoustic parameters: loudness, sharpness, periodicity and impulsiveness. These parameters as well as the overall annoyance index can be used for comparison with another engine, with typical ranges as derived from a wide data base or for comparison during different steps of the noise development.

The next step after the determination of the noise radiation pattern is the analysis of the main noise sources generating this radiation pattern. Methods for the analysis of cylinder pressure and the separation of combustion and mechanical noise are well-known. With the trend towards increased injection pressure and peak firing pressure, the relevance of mechanical noise contributions increases too. Therefore, the investigation of major mechanical noise sources like injection system, gear train, piston slap, valve train, crank train, etc. have become a key step in noise development. A pure experimental approach to the analysis of the different mechanical noise sources is very difficult in spite of advanced methods. Direct analysis of noise generating processes such as piston slap, gear impacts, crankshaft dynamics or valve impacts is an arduous task requiring large efforts, even if it is possible. Therefore, a combined approach is generally preferred which joins experimental methods with analytical simulation. This, of course, requires well matched and validated simulation models and tools. Important examples are gear train and piston slap. The simulation approach offers the possibility of developing noise reduction modifications like an appropriate distribution of gear moments of inertia within complex gear train systems or piston pin offset to minimise excitation of the cylinder liner. Of course, optimum piston pin offset has to be checked with experimental tests to account for wear and to assess the influence of combustion on piston temperature distribution.

Testbed development and procurement

The conclusion of the analysis phase is the noise reduction strategy represented by a set of noise reduction measures which are derived from the analysis results. If the effectiveness of the recommended measures has to be proven, it can be done by simulation or by experimental testing. Therefore, the aim of experimental development is to ensure that the effectiveness of the modifications recommended is sufficiently high to achieve the targets. If the optimum version of a modification cannot yet be defined, it may be worthwhile to procure several versions of a modification varying a critical parameter in order to save time in the refinement phase. A typical example is the elastomer components of a decoupling system which can be procured in two or three different steps of rubber hardness. Also, variants of more expensive prototype parts can pay off in this phase, since reaching the development target is absolutely time-critical for the whole project. It is understood that the modifications need to be commonly accepted by the production specialists and, if necessary, by subsuppliers.

Refinement and final testing

This last step of the prototype NVH development aims at several objectives:
1. Determination of the effectiveness of procured hardware.
2. Refinement of modifications.
3. Approval of target achievement.
4. Documentation of final development condition.
5. NVH sensitivity analysis in view of production tolerances.

Besides the fulfilment of development targets, the consideration of production tolerances is an essential task. Therefore, sensitivity analysis determines the effect of parameter variations within the expected production tolerances on engine NVH. Typical parameters are: clearances in the main and connecting rod bearings, piston-to-liner clearance, backlashes in the gear train and injection timing. As an example, Fig. 16.15 shows the negligible effect of injection timing variation by ±1° crank angle on the engine noise level. If the effect on noise is significant and jeopardises the achievement of the target level, additional refinement work or a redefinition of production tolerances might be necessary.

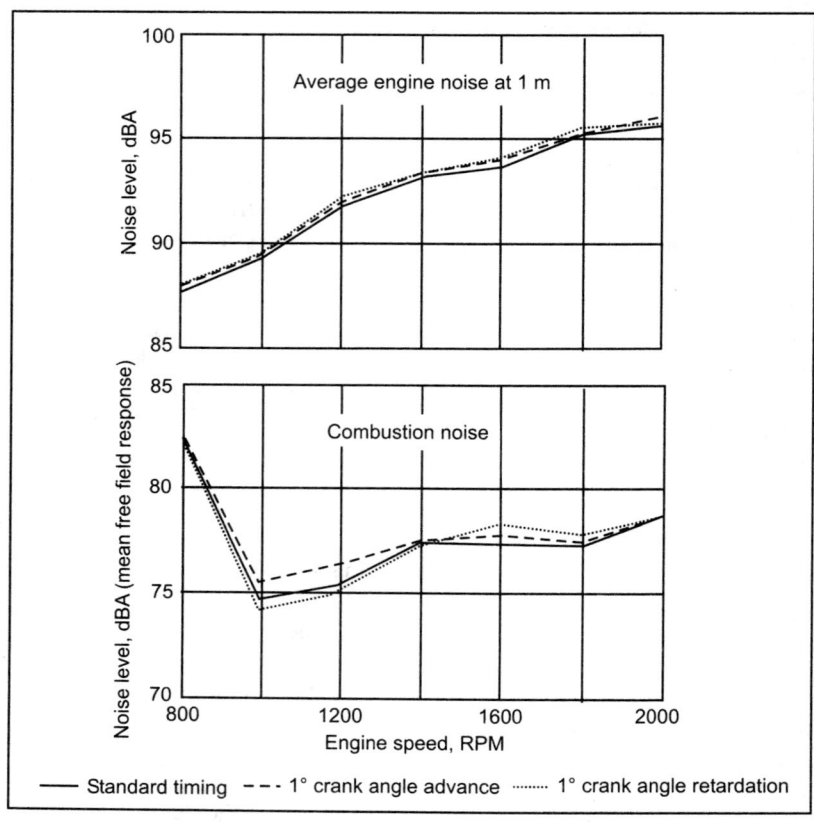

Fig. 16.15. Sensitivity analysis of noise as a function of timing tolerances.

Thus, by carefully analysing the engine NVH development process, time and quality critical steps can be determined. In order to effectively support the concept design phase, early availability of simulation results is required. Together with empirical data, a reliable prediction of engine noise characteristics can be achieved and also allow for evaluation of future concepts. Consequently, this leads to a two-stage simulation process. Whereas in the first stage the focus is on trend prediction, the second stage during the design phase means simulation with the consideration of all significant noise excitation mechanisms. This becomes more important the lower the target noise levels and requires a closed loop interaction of simulation and test bed work with comparable boundary conditions and evaluation schemes.

Chapter 17

Bearing and Valve

INTRODUCTION .

A bearing is a device to allow constrained relative motion between two or more parts, typically rotation or linear movement. Bearings may be classified broadly according to the motions they allow and according to their principle of operation as well as by the directions of applied loads they can handle.

Plain bearings use surfaces in rubbing contact, often with a lubricant such as oil or graphite. A plain bearing may or may not be a discrete device. It may be nothing more than the bearing surface of a hole with a shaft passing through it or of a planar surface that bears another (in these cases, not a discrete device), or it may be a layer of bearing metal either fused to the substrate (semi-discrete) or in the form of a separable sleeve (discrete). With suitable lubrication, plain bearings often give entirely acceptable accuracy, life, and friction at minimal cost. Therefore, they are very widely used. Wheel bearing is a device which allows a wheel to rotate freely around an axle.

WHEEL BEARING NOISE

A rumbling or cyclic noise of any kind is heard with bearings having a problem in their functioning. Even if the driver senses slight problem in the vehicle, he should immediately go for a wheel bearing noise diagnosis. The problem if not addressed in a timely manner, there could be a possibility of dangerous consequences.

Identifying Wheel Bearing Noise

The common trick of identifying a wheel bearing problem is to see whether the noise remains same while accelerating or decelerating. Wheel bearing noise does not change with changing speeds. The noise however, rises or lowers while turning. Thus, one can say that a noise heard at regular intervals while driving at a steady speed is a sign of bearing problem. Wheel bearing noise should not be confused with the noise produced by a damaged constant velocity (CV) joint; it is a completely different problem and should be dealt with separately. The noise produced by bearings is due to lack of lubrication or because of damage. The friction generated in dry areas of the bearing thus, results in a humming sound which repeats itself at regular intervals. The sound can also be called as bad wheel bearing noise.

To check wheel bearing noise, one should have a mechanism to lift the car for spinning the tyres. While spinning the wheel, one should place the other hand on the strut assembly. If there is any problem in the wheel bearing, one can hear a noise. Lifting the vehicle is necessary to determine the side from which noise is coming. Wiggling the tyre back and forth also helps in checking whether the tyre is loose.

Front Wheel Bearing Noise

The front wheel should be checked for noise and lubrication. Servicing of the front wheel bearing should be done every 30,000 miles. While driving, one should hear properly as to where the noise is coming from. Noise coming from the floorboard indicates that problem is not present in the wheel bearing but in other parts of the car. Sound coming from the windows gives a better idea about the correct problem.

Rear Wheel Bearing Noise

The noise produced by both the rear and front wheel bearings is the same. Therefore, no difference is present in the sounds produced by the rear and front wheels. A method of diagnosing rear wheel bearing noise, apart from the above mentioned, is to drive the vehicle at a constant speed, with slight variations in the speed being made. Slight variations are required because some cars make noise only at certain speeds. Here the point is that wheel bearing noise danger shouldn't go undetected, and should be dealt with or repaired as soon as possible.

During the diagnosis of the wheel bearing problem, one should check whether there are any cracks on the bearing surface. Even small cracks may lead to bigger problems. To detect minute cracks, one has to check the bearing in bright light. Cracked, pitted or worn out bearings need to be replaced. Parts such as the bearing hub bore and spindles should also be checked for damage. During inspection, one should check if there is any damage caused to the bearing hub bore; its fitting should also be checked. The bearing hub bore, if does not fit properly, it calls for the replacement of car parts such as drum, rotor or hub. It should be checked whether spindles are aligned properly. They may not be in a straight line; proper diagnosis of the problem is necessary. The steering knuckle should be replaced if any problem is found in the spindles.

COOLING FAN NOISE—SLEEVE BEARING VS BALL BEARING

Quite often in low-flow electronic cooling fans for computers, where the aerodynamic noise is low, the predominant noise emitted by the fan is generated in the bearing system. This type of fan usually has either an oil-impregnated sleeve bearing or ball bearings. Both bearing systems create noise that can be very different in both frequency content and amplitude. It is generally accepted in the computer and business machine industries that ball bearing fans are noisier than sleeve bearings fans. Furthermore, ball bearing fans can become noisier by mishandling or long-term running. This section will present a brief discussion of the various noise-source mechanisms in each bearing system and will show acoustical data, comparing the two systems against various factors such as life, shock, vibration, etc.

Discussion

Before discussing the acoustical aspects of sleeve and ball bearings, a brief discussion on the basic operation of each bearing system is in order. Figure 17.1(a) shows a typical sleeve bearing assembly for a small cooling fan. In this particular case the shaft rotates while the bearing is stationary. It is also possible to use a stationary shaft with the bearing rotating. In either case, lubricating oil is impregnated into the porous bronze bearing and is fed to the shaft via the small porosity openings in the bearing bore. Rotation causes a wedge or film of oil to build up on which the shaft rides. If a perfect bearing system could be built, this oil film would prevent metal-to-metal contact and thereby eliminate almost all-bearing noise. Because the shaft and bearing have rough surfaces, at least on a microscopic level, the bearing can create a scraping or grinding sound. Also due to forces such as unbalance and motor driving frequencies, vibration can cause the shaft to rock in the bearing and make contact at the bearing ends. This type of contact causes a knocking or rattling sound. Another source of noise is from the thrust washers which must slide relative to each other thereby creating a rubbing sound. At any rate, noise from a sleeve bearing is usually broad band in nature and somewhat intermittent. However, most good sleeve bearing designs are very quiet and stay quiet until they begin to run out of oil.

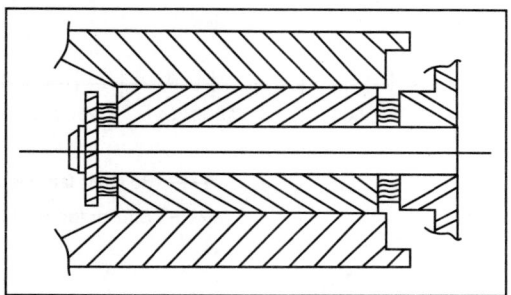

Fig. 17.1(a). A sleeve bearing assembly.

A typical ball bearing system is shown in Fig. 17.1(b). In contrast to a sleeve bearing system, the ball bearing system is relatively noisy to start with compared to a sleeve bearing and tends to get noisier over time. Because of a ball bearings construction, consisting of an inner and outer race, a series of balls and a cage to support the balls, there are a multitude of possible noise sources. Again surface finish, roundness, alignment, grease, etc. play an important part in noise from a ball bearing. Also ball bearings are easily damaged, particularly in the form of brinnelling, which is a denting of the race following a shock load. This brinnelling, although it has no major effect on life, at comparatively light loads, causes a great increase in noise.

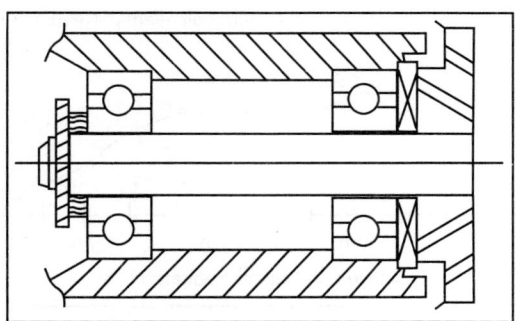

Fig. 17.1(b). Ball bearing assembly.

Noise from a ball bearing system is both broad band and pure tone in nature and is generally in the higher frequency ranges. Now looking at some data (Fig. 17.2) that compares sleeve to ball bearing fans one can see that the ball bearing fans are somewhat noisier than sleeve bearing fans. (Note: data presented is total airborne fan noise including aerodynamic noise).

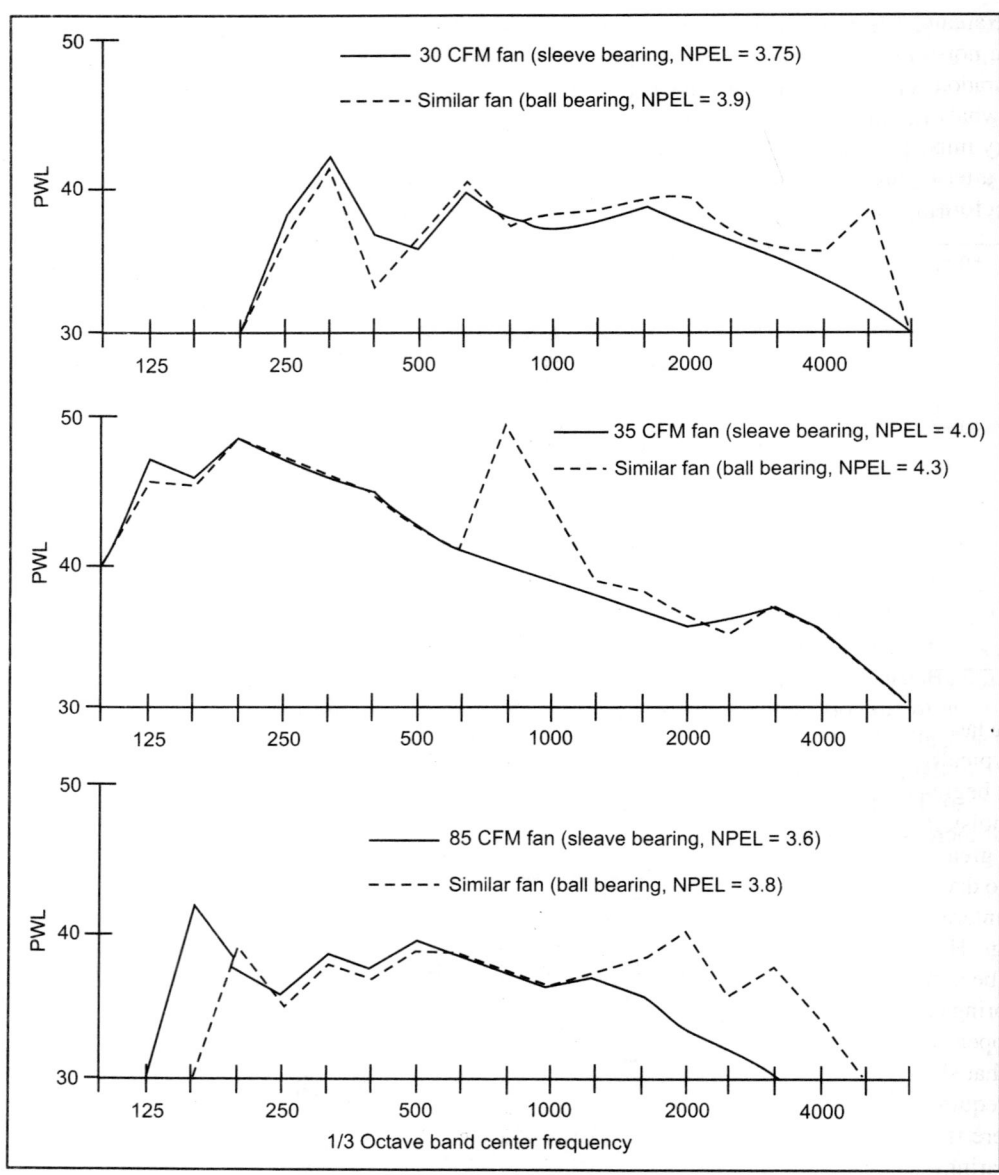

Fig. 17.2. Sleave bearing vs ball bearing fan noise.

This data supports the fact that, all else being equal, most ball bearing fans are noisier by 1 to 3 dBA over their counterpart sleeve-bearing fan. Also the additional noise is somewhat puretone in nature. Therefore, the annoyance level is considerably higher than with the sleeve-bearing fan. This higher noise level is also in the higher frequency ranges, which makes it even more annoying to the human.

One area that is important to fan noise is the ability of the bearing system to endure a shock and not become noisier. Sleeve bearing fans, generally speaking, can easily sustain multiple shocks of 80 g's with duration of 11 msec without impacting noise at all. This is not true for ball bearing fans. Figure 17.3 shows what can happen to ball bearing fan noise if the fan is subjected to 40 g's (11 msec duration). This is a very important factor since the equipment manufacturer has no control over how this equipment is treated after the fan is installed, particularly in shipment. It is quite common for a ball bearing fan to be noisy before it is even used just from the handling of the equipment it is installed in.

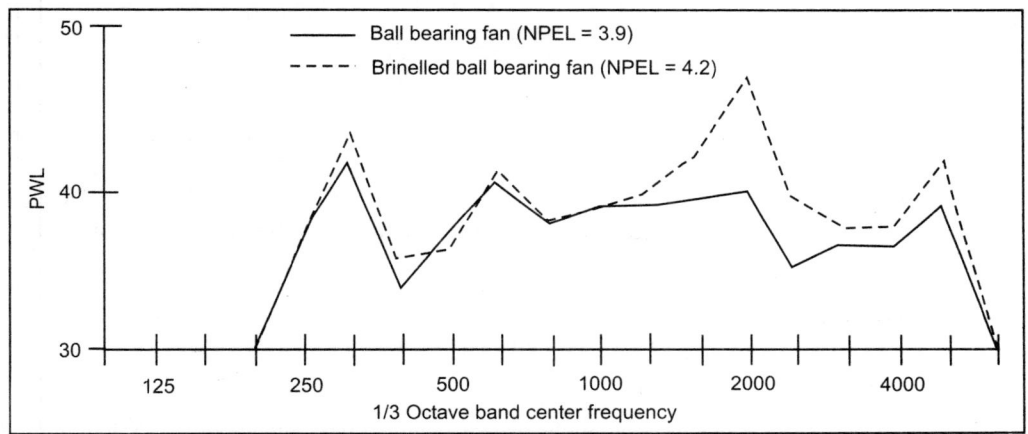

Fig. 17.3. Fan noise due to damaged ball bearing.

One last important point is what happens to both bearing systems noise level versus running time or life. Typically, sleeve bearing fan noise does not increase due to life. This remains true up until the system begins to fail due to loss of oil however; as can be seen in Fig. 17.4, ball bearing fans can begin to get noisy in a very short time. This increase in noise is due to many facts, such as grease channelling, loss of grease, damaged bearing camouflaged by the grease, etc. Also, as time goes on, the grease may begin to dry out which allows for a very noisy fan, but it will continue to run for a long time. This brings up an interesting point, the reason for the use of ball bearing fans is to extend the fan life past sleeve bearings. However, if usable life were defined to end when the fan became noisy it is quite possible the sleeve bearing fan would out live the ball bearing one. Thus, the data presented here, though somewhat brief, brings forth a point that fan users should consider before selecting a fan. Even though ball bearing fans appear to have longer life, they most likely will be noisier and cause far more complaints about noise that sleeve bearings fans. The point made here is that in those fan installations that have critical noise requirements, sleeve bearing fans will most likely meet the needs better than ball bearing fans.

There is one additional point to add and that is at low operating temperatures, sleeve bearings and ball bearings have similar life expectancies. At higher temperatures, sleeve bearings have a much lower life expectancy than the ball bearings. At higher operating temperatures, you must always consider the differences in life expectancy as well as acoustical advantages in the different bearings.

VALVE NOISE

Valve is mechanical or electromechanical device by which the flow of a gas, liquid, slurry, or loose dry material can be started, stopped, diverted, and/or regulated. The main types of valves include: (i) isolation valve: on/off valve that typically operates in two positions; the fully open and fully closed position, (ii) switching valve: usually a 3-way valve that converges and divert fluid flow in a piping system, and (iii) control valve: regulate the fluid flow in a piping system.

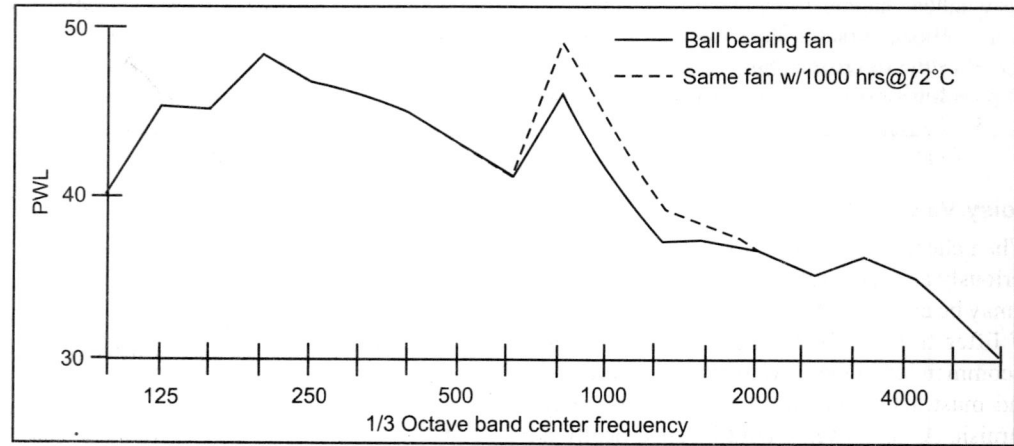

Fig. 17.4. Fan noise vs life.

Noisy Valve Train

The noise level of the valve mechanism cannot be properly judged where the engine is below operating temperature when the hood is raise, or when the valve rocker arm covers are removed. Before attempting to judge valve noise level, the engine must be thoroughly warmed up (at least 20 minutes of operation at 1200 to 1500 rpm) to stabilise oil and coolant temperatures and bring all engine parts to a normal state of expansion. When the engine is warmed up, listen for engine noise while sitting in the drivers seat with the hood close. Turn the engine at idle and at various higher speeds. It is advisable to observe the noise level in several engines that have been properly broken in, in order to develop good judgment for checking the noise level in any given engine. If the preceding check indicates the valve mechanism is abnormally noisy, remove the rocker arm covers so that the various conditions that cause noise may be checked. A piece of heater hose of convenient length may be used to pick out the particular valves or valve train components that are causing abnormal noise. With the engine running at a speed where the noise is pronounced, hold one end or hose to an ear and hold other end about 1/2" from point of contact between rocker arm and valve stem. Mark or record the noisy valves for investigation of following causes:

1. Excessive oil in crankcase: Crankcase oil level high enough to allow the crankshaft to churn the oil will cause air bubbles in the lubricating system. Air bubbles entering the hydraulic lifters will cause erratic operation resulting in excessive lash in the valve train. Locate and correct cause of high oil level, then run engine long enough to expel air from system.
2. Sticking, warped or eccentric valves, worn guides: Sticking valves will cause irregular engine operation or missing on a low speed pull and will usually cause intermittent noise. Pour penetrating oil over the valve spring cap and allow it to drain down the valve stem. Apply pressure to the one side of the valve spring and then the other, and then rotate the valve spring about 1/2 turn. If these operations affect the valve noise, it may be assumed that valves should be reconditioned.

3. Worn or scored parts in the valve train. Inspect rocker arms, push rod ends for scoring. Check for bent push rods. Check valve lifters and camshaft surfaces for scoring. Replace faulty parts.
4. Valves and seats cut down excessively. Noisy and improper valve action will result if a valve and its seat have been refinished enough to raise the end of the valve stem approximately 0.050″ above normal position. In this case it will be necessary to grind off the end of the valve stem or replace parts. The normal height of the valve stem above the valve spring seat is 1.933″, for 350 cu. in. engines and 2.082 inches for 455 cu. in. engines.
5. Faulty hydraulic valve lifters: If the preceding suggestions do not reveal the cause of noisy valve action, check operation of valve lifters.

Noisy Valve Lifters

When checking hydraulic valve lifters, remember that grit, sludge, varnish or other foreign matter will seriously affect operation of these lifters. If any foreign substance is found in the lifters or engine where it may be circulated by the lubrication system, a thorough cleaning job must be done to avoid a repetition of lifter trouble. To help prevent lifter trouble, the engine oil and oil filter must be changed as recommended in the service manual. The engine oil must be heavy-duty type (MS marked on container) and must also conform to GM Specification 6041-M to avoid detrimental formation of sludge and varnish. A car owner should be specifically advised of these requirements when the car is delivered. Faulty valve lifter operation usually appears under one of the following conditions:

1. Rapping noise only when the engine is started. When engine is stopped, any lifter on a camshaft lobe is under pressure of the valve spring; therefore, leak down or escape of oil from the lower chamber can occur. When the engine is started a few seconds may be required to fill the lifter, particularly in cold weather. If noise occurs only occasionally, it may be considered normal requiring no correction. If noise occurs daily, however, check for: (i) oil too heavy for prevailing temperatures, and (ii) excessive varnish in lifter.
2. Intermittent rapping noise: An intermittent rapping noise that appears and disappears every few seconds indicates leakage at check ball seat due to foreign particles, varnish, or defective surface of check ball or seat. Recondition, clean, and/or replace lifters as necessary.
3. Noise at idle and low speed: If one or more valve lifters are noisy or idle at up to approximately 25 mph but quiet at higher speeds, it indicates excessive leak down or faulty check ball seat on plunger. With engine idling, lifters with excessive leak down rate may be spotted by pressing down on each rocker arm above the push rod with equal pressure. Recondition or replace noisy lifters.
4. Generally noisy at all speeds: Check for high oil level in crankcase. See subparagraph (1) above. With engine idling, strike each rocker arm above push rod several sharp blows with a mallet; if noise disappears, it indicates that foreign material was keeping check ball from seating. Stop engine and place lifters on camshaft base circle. If there is lash clearance in any valve train, it indicates a stuck lifter plunger, worn lifter body lower end, or worn camshaft lobe.
5. Loud noise at normal operating temperature only: If a lifter develops a loud noise when engine is at normal operating temperature, but is quiet when engine is below normal temperature, it indicates an excessively fast leak down rate or scored lifter plunger. Recondition or replace lifter.

Chapter 18

Boilers

INTRODUCTION

Wall gas boilers allow to control a single apartment heating system independently from the other apartments ones. Wall gas boilers are usually characterised by high fluidodynamic performances and safety conditions. Thus, this kind of boilers is largely spread as civil houses heating system. Wall gas boilers are often installed indoors. For these reason, wall gas boiler noise emissions may cause annoyance, especially during the night, when human noise sensibility is very high.

Noise emissions reduction methods are here studied regarding a custom wall gas boiler. An intensimetric and vibration measurements campaign has been led in order to individuate wall gas boiler noise characteristics. Measurements results have shown the boiler fan as the main noise source. Thus, some vibration and acoustic insulation solutions have been proposed and realised on precombustion chamber and exhaust fan surrounding area. Numerical simulations by means of a volume finite code (fluent) and a measurement campaign have shown that the proposed noise reduction solutions do not affect the boiler thermofluidodynamic performances.

An intensimetric and vibration measurement campaign has been carried out after the realisation of the noise reduction solutions. Measurement results have allowed to individuate the optimum noise insulation solution which introduces a 3.5 dBA A-weighted power level reduction and a 5.0 dB vibration level reduction.

WALL GAS BOILER NOISE CHARACTERISATION

An intensimetric measurements campaign has been led in order to individuate the characteristics of the noise emitted by the wall gas boiler.

Measurement room characteristics are:

1. Approximately 65 m² absorbing units.
2. Dimensions equal to 4.5 m × 4.5 m × 3 m.

The wall the boiler is installed on is characterised by high absorbing coefficient, equal to 0.7. Measurements have been carried out for three boiler working conditions:

1. Only boiler pump is on.
2. Only boiler fan is on.
3. All noise sources are on.

First (1) condition measurements have been led in 12 points, placed on a fictitious parallelepiped surface which surrounds the noise source. Surface dimensions are $x = 1.2$ m, $y = 1.2$ m, $z = 0.6$ m (x–y

plane is the wall the boiler is installed on). Second (2) and Third (3) conditions measurements have been led on 20 points, placed on a fictitious parallelepiped surface which surrounds the noise source. Surface dimensions are $x = 1.8$ m, $y = 1.8$ m, $z = 1.8$ m. Measurement points number and positions have been chosen in according to ISO 9614–1/93. Measured sound pressure levels have been processed by the data acquisition system in order to calculate noise power spectrum. Figures 18.4 and 18.5 show the measured power level spectra relative to third (3) condition (indicated as before condition). Measured power levels have suggested the following considerations:

1. For first (1) condition, main noise component frequency (MNCF) is 125 Hz.
2. For second (2) condition, MNCF is 200 Hz.
3. For third (3) condition, MNCF is 200 Hz.

The A-weighted power level produced by the 'EOLO 27 Maior @' wall gas boiler is:

1. 39.5 dBA for first (1) condition.
2. 55.0 dBA for second (2) condition.
3. 55.0 dBA for (3) condition.

Measurements results have shown that noise power spectra for second (2) and third (3) conditions are characterised by the same behaviour and the same MNCF (200 Hz). Therefore, the exhaust gas fan is the boiler main noise source in the nominal working conditions. A remarkable contribute to the global boiler noise is solid-borne noise. Thus, a vibration measurement campaign has been led in order to individuate possible correlations between acceleration level spectra and the previously measured power levels ones. Measurements have been carried out for third (3) condition (all noise sources are on). Ten measurements points have been chosen on the boiler external panels. Vibration levels have been measured in 12.5–8000 Hz frequency range. The acceleration level average spectrum is reported in Fig. 18.6 (indicated as before condition). Measurements results show that vibration main component frequency (200 Hz) is equal to the power spectrum one due to the boiler fan. Therefore, also the boiler solid-borne noise is mainly due to the exhaust fan.

Noise Reduction Methods

Boiler fan power spectrum characteristics are:

1. 80–1600 Hz frequency range.
2. A tonal component whose frequency value (200 Hz) is proportional to the fan blades rpm.

Two kinds of noise reduction methods have been proposed and realised:

1. Airborne noise reduction: Two different solutions have been proposed and compared:
    (a) 20 mm thick noise insulation panels have been installed on the boiler internal walls and the combustion chamber external walls (Fig. 18.1). The panels are constituted by two polyurethane layers (35 kg/m^3 density, 0.018 W/m K thermal conductivity, 1400 J/kg K specific heat at constant pressure) separated by a 1 mm thick lead layer (11300 kg/m^3 density, 35.3 W/m K thermal conductivity, 130 J/kg K specific heat at constant pressure). Material performances are guaranteed for working temperatures less than 120°C. This condition always occurs for the boiler nominal working conditions.
    (b) 40 mm thick glass-wool insulation panels (22 kg/m^3 density, 0.034 W/m K thermal conductivity, 850 J/kg K specific heat at constant pressure) have been installed on the boiler internal walls and the combustion chamber external walls. Panels have been compressed in order to reduce the volume they fill into the boiler. Material performances are guaranteed for boiler nominal working conditions.

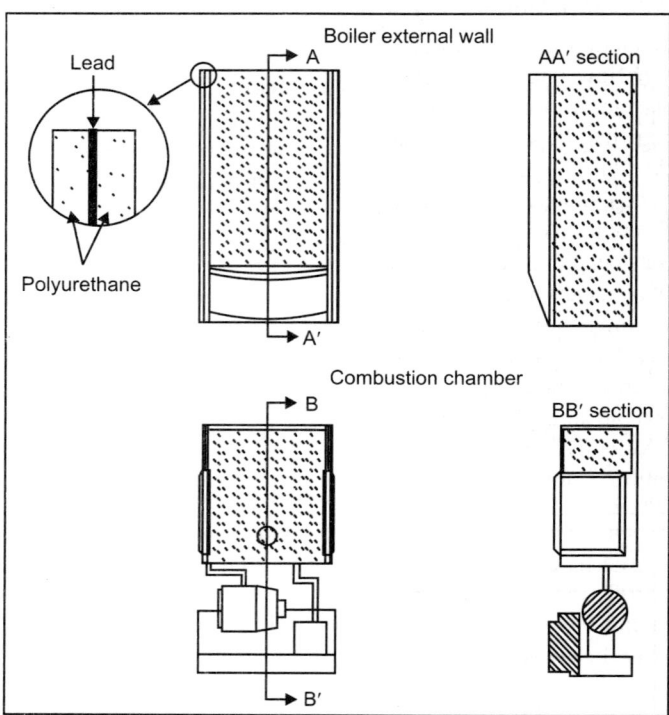

Fig. 18.1. Installation scheme of sound insulation multilayer panels.

2. Solid-borne noise reduction: exhaust fan has been insulated from the boiler other elements by means of a 3 mm thick polyurethane gasket. The gasket has been placed on the contact surface between the fan and the combustion chamber (Fig. 18.2). A polyurethane gasket has been also installed where the boiler fan is connected to the exhaust duct. Combustion chamber panels have been insulated by means of 1 mm thick Flexoid gaskets. Two 2 mm thick steel stirrups have been installed over the precombustion chamber internal walls in order to make them more rigid.

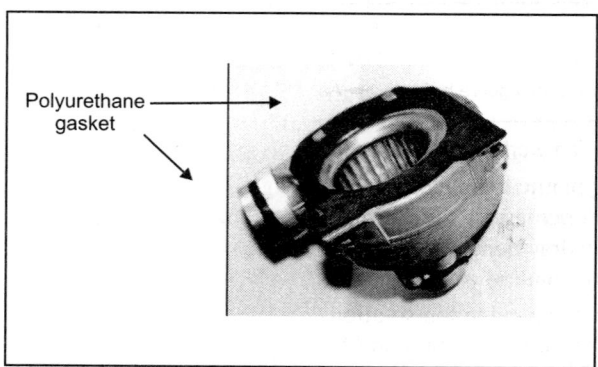

Fig. 18.2. Solid-borne noise reduction solution by means of polyurethane gaskets.

Noise Reduction Methods Compatibility with Boiler Thermofluidodynamic Performances

The proposed noise reduction methods (additional gaskets and noise insulation panels) may decrease the boiler aspiration air flow rate. This fact may induce a boiler thermofluidodynamic performances worsening: the exhaust gases flow rate may decrease and the internal temperature may increase. A measurement campaign has been carried out in order to evaluate the boiler thermofluidodynamic characteristics. In particular, the characteristics of the aeraulic circuit constituted by aspiration circuit, combustion chamber and exhaust gas duct have been determined. Air flow rate has been measured by varying the pressure difference (ΔP) between boiler inlet and outlet. Thus, the aeraulic circuit characteristic curve (ΔP-flow rate) has been evaluated (Fig. 18.3, before condition). At last, air temperature has been measured at boiler inlet by means of a sensor placed at the aspiration duct outlet; exhaust gases temperature has been measured at boiler outlet by means of a sensor placed at the exhaust duct inlet. Boiler fan working point is obtained as the intersection between the aeraulic circuit characteristic curve (ΔP-flow rate) and the fan characteristic curve (Fig. 18.3). Air and exhaust gases flow rates are equal to 115.5 m³/hr; this value corresponds to an air aspiration velocity equal to 4.7 m/s (aspiration duct diameter = 0.093 m) and an exhaust gases velocity equal to 5.2 m/s. Inlet temperature is the environmental temperature; exhaust gases temperature is about 125°C.

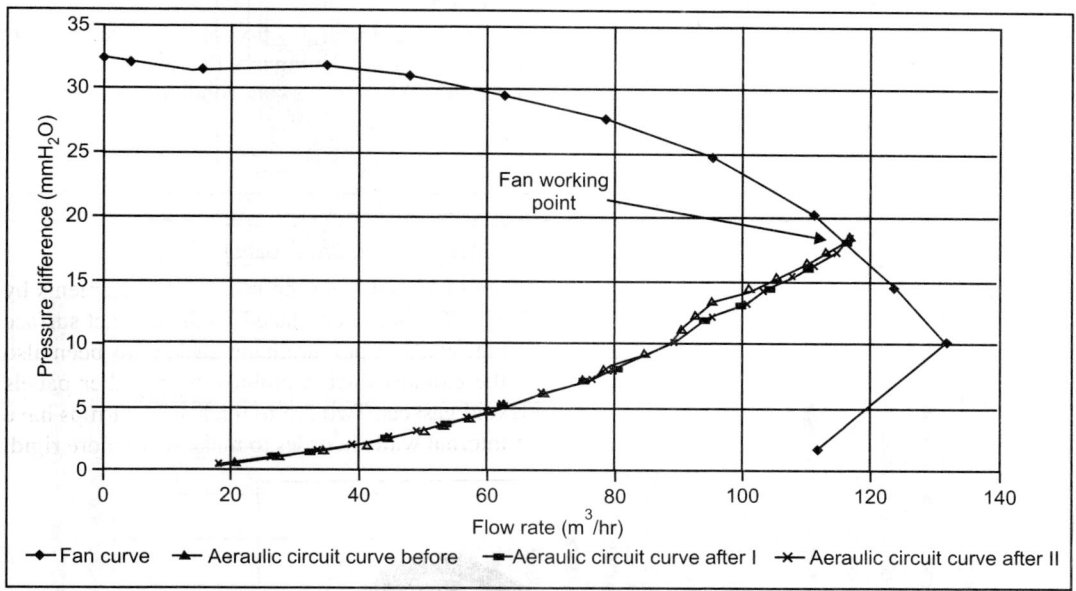

Fig. 18.3. Boiler fan working point after the realisation of noise reduction interventions.

Boiler thermofluidodynamic characteristics have been also studied by means of a volume finite numerical code (fluent). A numerical simulation has been led: boiler internal volume has been divided into about 5,00,000 tetrahedral elements. Boiler internal pressure, velocity and temperature have been simulated in the following conditions:

1. Boiler before the noise reduction solutions realisation (before condition).
2. Boiler after the realisation of 1(a) and (2) noise reduction solutions (after I condition).
3. Boiler after the realisation of 1(b) and (2) noise reduction solutions (after II condition).

Simulations have been carried out by applying the following boundary conditions: boiler inlet pressure equal to the boiler outlet one; pressure difference between exhaust fan inlet and outlet equal to the one individuated by means of the aeraulic measurements campaign; boiler inlet and outlet temperature equal to the measured ones (respectively 27°C and 125°C); 31.5 kW heat flux generated by the boiler burner; 28.5 kW heat flux absorbed by the boiler heat exchanger, placed in the combustion chamber upper side. Simulation results have shown no remarkable alterations in the boiler thermofluidodynamic properties due to the noise reduction solutions (both after I and after II).

Therefore, boiler thermofluidodynamic performances are not affected by the proposed noise reduction procedures. Thus, the proposed noise reduction methods have been applied to the boiler; a thermofluidodynamic measurement campaign has been carried out for after I and after II conditions. Figure 18.3 shows the comparison among the boiler aeraulic circuit characteristics curves relative to before, after I and after II conditions. Fan characteristic curve is reported too. It can be observed a negligible displacement of the aeraulic circuit curve after the realisation of the noise reduction solutions. Fan working point is not modified: thus, exhaust gases flow rate is 115.5 m³/hr for before, after I and after II conditions. Also aspirated air and exhaust gases temperatures are not modified by the proposed noise reduction solutions.

Individuation of the Optimum Noise Reduction Solution

The intensimetric measurements campaign has been repeated after the realisation of the noise reduction solutions. The campaign has been carried out in according to ISO 9614–1/93. Figures 18.4 and 18.5 show the comparison between the measured power level spectra for third (3) boiler working condition relative to:

1. Before and after I conditions (Fig. 18.4).
2. Before and after II conditions (Fig. 18.5).

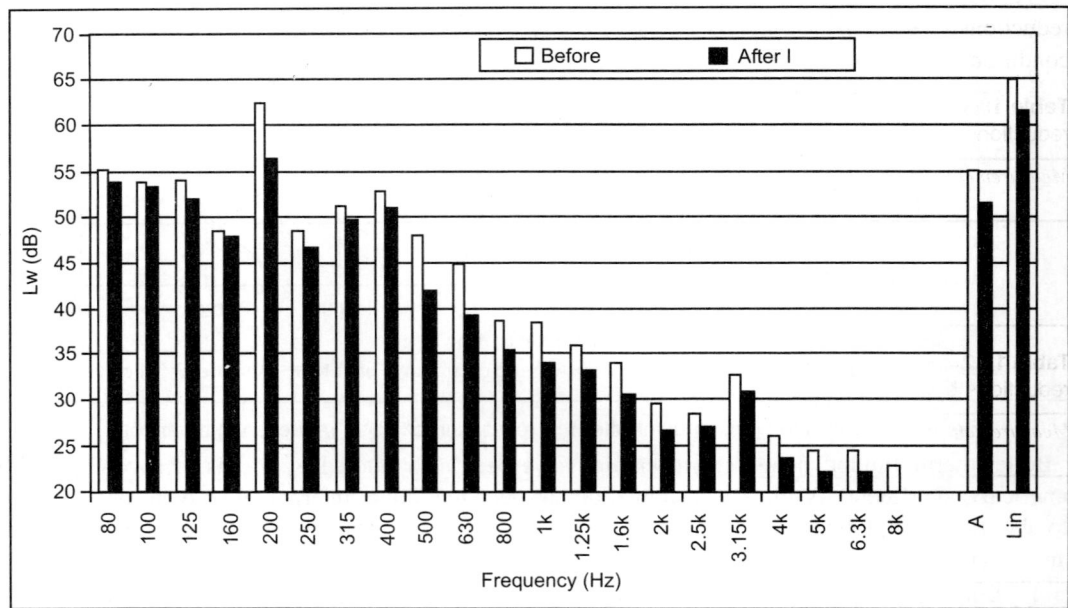

Fig. 18.4. Comparison between power level spectra relative to before and after I conditions third (3) boiler working condition.

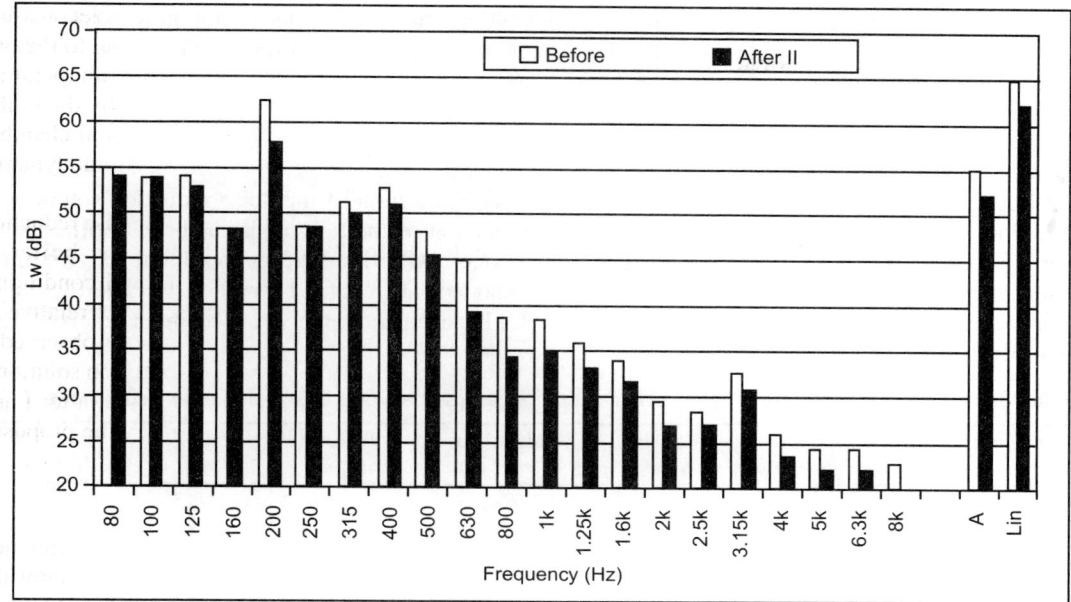

Fig. 18.5. Comparison between power level spectra relative to before and after II conditions third (3) boiler working condition.

A-weighted Lw_A power level reductions are reported in Table 18.1. A-weighted Lw_f power level reductions relative to the MNCF [125 Hz for first (1) condition, 200 Hz for second (2) and third (3) conditions] are reported in Table 18.2.

Table 18.1. Measured power levels relative to before, after I and after II conditions and obtained power level reductions by means of the proposed noise reduction methods.

Measurement condition	Lw_A (dBA) before	Lw_A (dBA) after I	Lw_A (dBA) after II	ΔLw_A (dBA) after I	ΔLw_A (dBA) after II
A	39.5	37.0	38.0	2.5	1.5
B	55.0	51.5	52.5	3.5	2.5
C	55.0	51.5	52.5	3.5	2.5

Table 18.2. Measured MNCF levels relative to before, after I and after II conditions and obtained MNCF level reductions by means of the proposed noise reduction methods.

Measurement condition	Lw_f (dB) before	Lw_f (dB) after I	Lw_f (dB) after II	ΔLw_f (dB) after I	ΔLw_f (dB) after II
A	50.5	46.0	47.5	4.5	3.0
B	62.5	56.5	58.5	6.0	4.0
C	62.5	56.5	58.5	6.0	4.0

Measurements results show after I solution allows to obtain more remarkable sound power level reductions than after II ones. In particular, results relative to the boiler nominal working conditions third (3) measurement condition) point out a 6.0 dB MNCF reduction for after I solution with respect to a 4.0 dB one for after II solution. Thus, after I is the optimum solution for airborne noise reduction. The obtained after I A-weighted power level reductions are:

1. 2.5 dBA for first (1) condition.
2. 3.5 dBA for second (2) condition.
3. 3.5 dBA for third (3) condition (boiler nominal working conditions).

Vibration measurements have been repeated after the realisation of the noise reduction solutions. Solid-borne noise reduction is mainly due to noise reduction method. Thus, it has been verified that after I obtained reductions are very similar to after II ones. After I solution is optimum for airborne noise reduction; thus, vibration measurement results are reported only for after I condition. In Table 18.3 La_T global acceleration level and main component (200 Hz) La_f acceleration level are reported relatively to each measurement point.

Table 18.3. Comparison between measured acceleration levels relative to before and after I conditions.

Measurement point	$La_T (dB)$ before	$La_T (dB)$ after I	$\Delta La_T (dB)$ after II	$Lw_f (dB)$ before	$Lw_f (dB)$ after I	$\Delta Lw_f (dB)$ after II
1	107.5	101.5	6.0	104.5	95.5	9.0
2	107.5	102.0	6.0	103.5	94.5	9.0
3	106.5	101.5	5.0	104.0	95.5	8.5
4	107.5	102.0	6.0	104.0	95.5	8.5
5	106.5	101.5	4.5	103.5	95.5	8.0
6	105.5	102.0	5.0	102.5	94.5	8.0
7	105.0	100.5	4.5	102.5	95.0	7.5
8	105.5	101.0	4.5	103.0	96.0	7.0
9	106.0	101.0	4.5	102.5	96.0	7.5
10	105.5	101.0	4.5	102.0	95.0	7.0

Figure 18.6 shows the comparison between acceleration levels spectra relative to before and after I conditions. Average acceleration level reduction is 5.0 dB. It is higher than 4.5 dB for each measurement point. Average main component reduction is 8.0 dB; maximum main component reduction is 9.0 dB. Maximum reductions correspond to the points near the boiler fan. In fact, solid-borne noise reduction solutions are intensified in that area.

Wall gas boilers noise problematics have been here investigated. A custom wall gas boiler has been studied in order to individuate its main noise sources. Intensimetric and vibration measurements results have allowed to identify the boiler fan as the main noise source, both solid-borne and airborne. Two noise insulation solutions have been proposed and compared: measurement results have shown that the highest noise reductions are obtained by means of polyurethane-lead multilayer panels (airborne noise reduction) and polyurethane and flexoid gaskets (solid-borne noise reduction). The proposed noise reduction methods does not modify fan working point, as a numerical simulation and aeraulic measurements results have shown.

The following noise and vibration levels reductions have been obtained by means of the optimum noise insulation solution:

1. 3.5 dBA A-weighted power level reduction.
2. 6.0 dB main noise component (200 Hz) power level reduction.

3. 5.0 dB global vibration level reduction.
4. 8.0 dB main noise component (200 Hz) vibration level reduction.

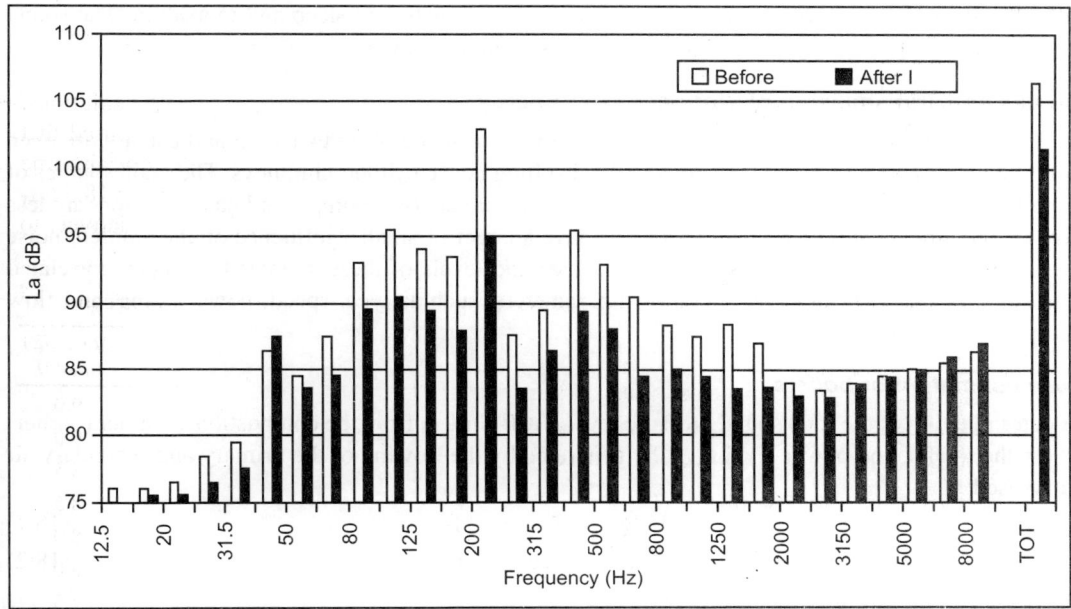

Fig. 18.6. Comparison between measured average acceleration levels spectra relative to before and after I conditions (all noise sources are on).

Symbols

Symbol	Description	Units
ΔP	Boiler inlet-outlet pressure difference	mmH$_2$O
ΔLa_f	Difference between MNCF acceleration levels before and after noise reduction	dB
ΔLa_T	Difference between global acceleration levels before and after noise reduction	dB
ΔLw_A	Difference between A-weighted sound power levels before and after noise reduction	dB
ΔLw_f	Difference between MNCF power levels before and after noise reduction	dB
La_f	MNCF acceleration level	dB
La_T	Global acceleration level	dB
Lw_A	A-weighted sound power level	dB
Lw_f	MNCF power level	dB

TRANSMISSION OF THE LOW FREQUENCY NOISE FROM BOILERS-ROOMS TO RESIDENCE

Low frequency sound may be generally defined as having a frequency below 160 Hz. Sound enters buildings through their structure, through open windows from boiler plants (rooms) which are situated directly in the residential area, or can be generated by boilers inside the building. Sound with a very long wavelength may be heard as noise (primary noise), cause the rattling of windows, doors or furniture (secondary noise), and may be difficult to distinguish from structural vibration. Both of these forms of noise can cause disturbance, particularly in the evening and at night. Low frequency noise can be more noticeable indoors, which is why it is often associated with disturbed sleep and relaxation. The sound pressure increases too for some special conditions—standing waves.

Sources and Transmission of Vibro-acoustic Energy

Boiler plants are usually located in living areas within a distance of up to 100 m and usually are even closer. Some of them are directly attached to the dwelling house by their chimneys. The main sources of the boiler rooms are boiler burners and ventilators. The air suctions, pumps, and gas regulators are less noisy technological equipment. All the sources have a larger or smaller influence on the ambient noise in residential and public buildings. The frequencies and levels of these sources have been correlated using some of the more obvious operational parameters, such as type, speed, power rating, and flow conditions.

Noise estimation of boilers

For determination of the total sound pressure level L (dB) in 1 m from the combustion chamber (burner) and for the total sound power level L_W (dB) generated by the flowing of the primary and secondary air can be used:

$$L = 44 \lg v + 17 \lg q_m - 146 \qquad \qquad ...(18.1)$$

$$L_W = 44 \lg v + 17 \lg q_m - 135 \qquad \qquad ...(18.2)$$

where, v is speed of air flow (m.s^{-1}).

q_m –mass flow rate (kg.s^{-1}).

The data obtained by measuring permit the calculation of sound pressure levels in octave bands at the plants with installed noisy equipment.

$$L = 10 \lg \left(\sum_{i=1}^{n} \frac{10^{0,1 L_{Wi}} Q_i}{\Omega_i r_i^2} + \frac{4}{A} \sum_{i=1}^{m} 10^{0,1 L_{Wi}} \right) \qquad \qquad ...(18.3)$$

where, L_W is sound power level in octave for i^{th}-source (dB).
Q factor of noise source direction.
$\hat{U}$ spatial angle of source radiation (rad).
r distance from acoustic centre of noise source up to rated point (m).
n number of noise sources.
A acoustic constant in the building (m^2).
m total number of noise sources in the building.

The character of noise generated during the combustion process of gas is aerodynamic and can be defined as a continuous spectrum which changes amplitude in the low frequency domain. This phenomenon can be proved during the measurements, and changes in amplitude reach approximately 6 dB. By using a narrow band filter, a spectrum characterised by a slow drop of the sound pressure level with growing frequency can be measured. The significant amplitude of noise for a boiler at the middle frequency 1/3 band is 112 Hz. This amplitude is increased in front of a protected dwelling house, identified as a tonal frequency, and this frequency usually developed the standing waves in the protected area. In most cases producers or suppliers of boiler products do not usually offer data about the noise emitted from the boiler to the boiler room and to the uptakes. Therefore, if the measurements for the boilers are not accessible, for determination of sound power level can be used diagrams in Figs 18.7 and 18.8 respectively.

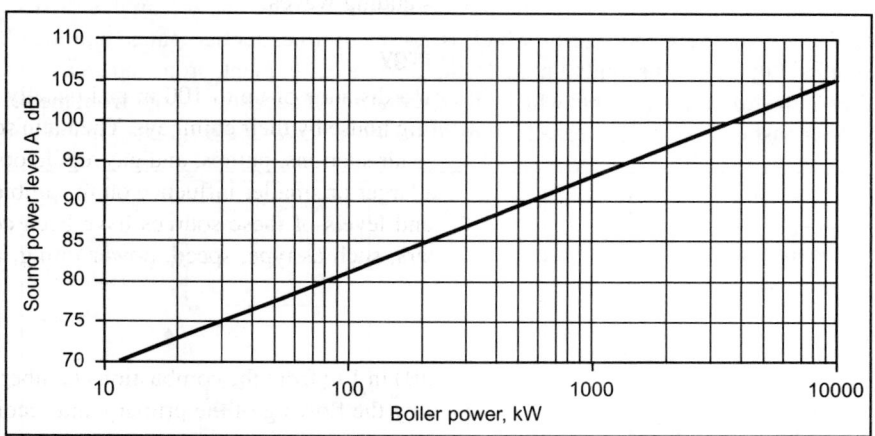

Fig. 18.7. Propagation of noise from combustion equipment into the boiler room.

Transmission and reduction of sound energy

The classical material structures are very good conductors of vibro-acoustic energy. The frequency domain in which the human ear is most sensitive to any sounds is the most important domain for sound reduction. Any attenuation of vibration levels within this frequency domain has a positive impact on the reduction of sound intensity emitted into the environment and on the sound exposure of humans.

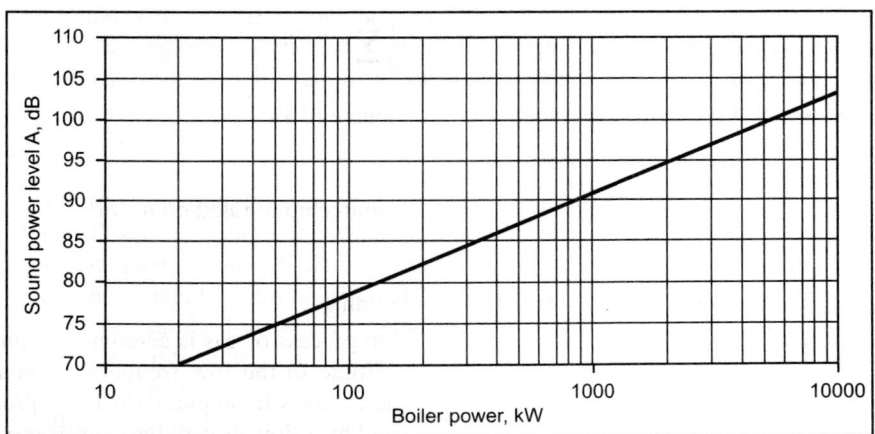

Fig. 18.8. Propagation of noise from combustion equipment into the uptake.

The transmission of sound power can be limited by reducing the amplitude of the vibration, reducing the emitting surface, reducing the time of emission, increasing the mechanical and acoustic impedance of the environment and increasing the loss coefficient and altering the Young's modulus of material. The reduction of the sound energy can be successfully solved by a dissipative silencer in combination with reflex principle. Vibro-acoustic energy is transmitted from the source into the point of exposition of a protected space via fluids and structures. Noise from boiler plants situated outside a protected area is transmitted by air. Acoustic energy is emitted from the noise source by a chimney orifice, chimney structure, ventilation windows, the size of which is not negligible, service gates, and also by certain structural elements. Figure 18.9 presents a model situation of a boiler plant, protected area and exposition places, together with structural elements which allow radiation of acoustic energy from the boiler plant. The radiation of acoustic energy is increased if partial standing waves are originated in the boiler house.

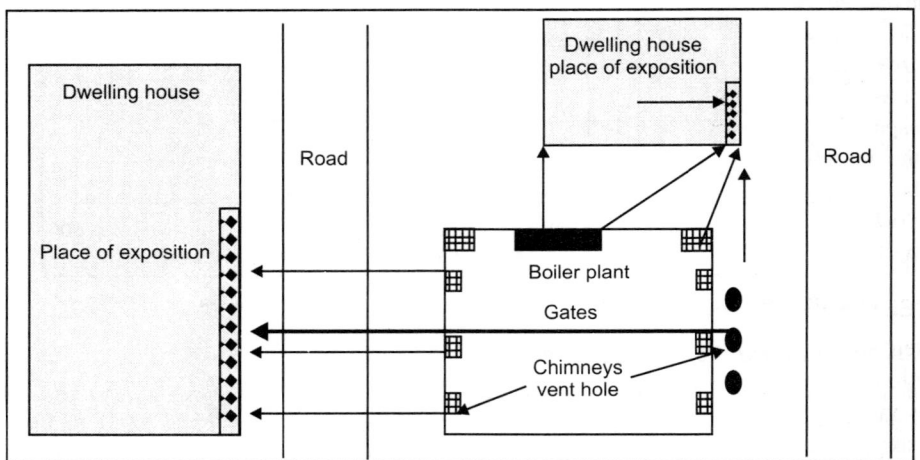

Fig. 18.9. Model situation of radiation sound energy from boiler room to projected objects.

Under certain circumstances acoustic energy excites vibration (rattling) in structural elements within a protected area. This phenomenon is an important element of the disturbance caused by a low frequency sound; rattling, as a factor related to strong low frequency and impulse noise, can be observed in structural elements located in dwelling houses as well.

Assessment of Low Frequency Noise

There are a large number of ways to measure and evaluate noise, each normally resulting in different noise measures, descriptors or scales. The different measures and descriptors result mainly from the different sources and the different researches involved in producing them. From these measures and descriptors, criteria have been developed to decide on the acceptability of the noise levels for different activities. These criteria are useful in determining whether noise control efforts are warranted in order to improve speech communication, reduce annoyance, and lessen sleep interference.

The sound spectrum of boilers is perceived to have a 'rumbly' quality with low-frequency sound energy. A rumbly spectrum is characterised as one with octave band sound levels that exceed the determined RC curve by more than 5 dB at and below 500 Hz. Acoustically induced perceptible vibration indicates sound pressure levels in the 16 Hz to 63 Hz octave bands at which perceptible vibration in building walls and ceilings can occur. Investigations have shown that the perception and effects of sounds differ considerably at low frequencies as compared to mid or high frequencies. The main reasons for these differences are as follows: a weakening of pitch sensation as the frequency of the sound decreases below 60 Hz; perception of sounds as pulsations and fluctuations; a much more rapid increase in loudness and annoyance with increasing sound pressure levels at low frequencies than at mid or high frequencies; complaints about feelings of ear pressure; annoyance caused by secondary effects like rattling of buildings elements, windows, and doors; less building sound transmission loss at low frequencies than at mid or high frequencies.

Therefore, for the assessment of sounds with strong low-frequency content, the rating procedures should be modified. The measurement location may be changed and the frequency weighting affected, since sounds with strong low-frequency content engender greater annoyance than is predicted by the A-weighted sound pressure level. In the assessment of low frequency noise the main factors are as follows:

1. The frequency range of interest appears to be about 5 Hz to about 160 Hz.
2. One of the issues in low-frequency noise assessment is that room resonances at low frequencies can create situations that may be hard to predict from outdoor measurements. This can be especially important in evaluating specific residences. However, for the purposes of estimating the prevalence of high annoyance in a large community population, outdoor measurements may be sufficient.
3. Sound-induced rattles in building elements are important determinants of the annoyance caused by low-frequency sound.

SOURCES OF NOISE IN HEATING SYSTEMS

The present contribution deals with the generation of noise in heating systems and with their propagation. This section analyses noise propagation from heat sources, circulation-water pumps and thermostatic valves by vibrations and turbulent heating water flow. The conclusion gives the results of measurements on parts and indicates potential methods of the reduction of noise based on the knowledge gained.

Primarily it has to be noted that from many arbitration procedures, technical consultations, discussions at conference, etc. it is apparent that many project engineers and persons responsible for the design and operation of heating equipment are not acquainted, from the viewpoint of generated noise, with acoustic terminology and have a tendency to confuse quantities used in technical mechanics. For this reason the first part of the present section gives a summary of basic and most frequently used quantities.

Noise in workplaces, flats, schools, hospitals, in the outdoor environment is rated by a quantity called sound pressure level A L_{pA} (dB). This is the reading from a sound level meter with a switched-on weight filter A, which inside the sound level meter simulates the sensitivity of the human ear.

The result is a single value rating of the exposure of man to noise according to Governmental Degree No. 502/2000 Sb 'On the Protection of Health Against Harmful Effects of Noise'. For noise the level of which is time variable a quantity is used called equivalent sound pressure level A Lp_{Aeq} (dB).

Sources of Noise

In recent years proprietors of residential buildings were able to refurbish their heating equipment with an aim to improve its efficiency, reduce the emission of pollutants into the outdoor environment and to reduce consumption of thermal energy for heating. Various alternatives are available of refurbishment of boilers, radiators, etc. One of the most frequently recommended measures is the use of thermostatic valves. Experience with these valves however, shows that this alternative is accompanied by unusual growth of noise generated by the heating equipment.

Boiler for Supply of Heat

Problems around propagation of noise from boiler rooms into indoor protected space are relatively well mastered and usually depend on the level of cooperation between the project engineer of the boiler room equipment and the project engineer of the boiler house, since what must be guaranteed is the sufficient air soundproofness of the separating construction elements of the boiler house. Noise inside boiler houses is a great extent affected by the noise of torches the noise parameters of which are given by sound pressure level A measured at a distance of 1 m from the front of the torch.

From numerous measurements of noise in boiler rooms the above sound pressure levels A have been interrelated with heat outputs of torches after being reduce to sound power levels A. The measured level is obviously affected by the space of the boiler room, configuration of boilers and sound absorptivity of walls of the boiler house. This is why sound pressure levels measured *in situ* 1 m from the front of the torch differ from those presented by the producers. A qualified estimate of noise generated in a boiler room can be made using the diagram in Fig. 18.7, which gives evidence of the sound power of gas torches and boilers (corrected by filter A) and heat output of the equipment.

The inherent source of noise in boiler rooms designed at present is the combustion process inside the boiler. Boilers with atmospheric torches belong to less significant sources of noise namely with respect to the outdoor environment. Problems are often encountered with pressure gas (or fuel oil) torches. In this case noise has to be distinguished between that produced by the fan supplying air into the combustion chamber and that produced by the combustion process inside the combustion chamber. The characters of noise generated during the process of combustion of gas or fuel oil is aerodynamic and is marked by a continuous spectrum which can be proved experimentally by pointing the flame of the torch into free space. Using a narrow band filter a spectrum can be measured characterised by a slow drop of the sound pressure level with growing frequency. If the torch is connected to the boiler they will interact, i.e. the boiler and the torch will mutually affect one another. By checking the spectrum of the sound pressure level or power in the exit branch of the boiler it can be found that the noise spectrum contains resonance frequencies which to a significant extent alter the frequency distribution of noise emitted into the chimney (stack). In literature data on noise emitted from within the boiler to the uptakes are very scarce and producers and/or suppliers of this type of equipment in most cases do not offer such information. The purpose of the results present below was identify *in situ* the sound properties of a type of boiler most frequently used in practical applications and to generalise the gained knowledge, if feasible, even to other cases.

Model tests in a certain scale cannot be performed with sufficient accuracy. Exhaust of combustion products by boiler uptakes and chimneys (stacks) into the outdoor environment, namely from higher output boilers, is a questionable issue already for a long time.

For the design of noise silencers for boiler uptakes it is absolutely necessary to know the spectrum of the sound power level in the exit branch of the boiler uptake. In this respect a series of experiments and theoretical considerations have been performed. It was established that the heat-exchanging interior parts of the boiler (combustion chamber, boiler tubes, separating space and space behind the tubes including the exit branch of the boiler uptake) behave like acoustic elements each with their won resonance and antiresonance frequencies. Theoretical considerations and experiments were performed on CKD Dukla KDVE 16 and BUDERUS SK 425-size 170 boilers which gave good agreement in the determination of frequencies. This indicates that in future boiler uptakes and chimneys (stacks) could be designed taking these facts into account. It is quite obvious that resonance effects of the whole combustion system, i.e. torch, boiler tubes and chimney (stack) mutually affect one another. For the successful design of such equipment a principle will necessarily have to be observed that series arranged acoustical elements (combustion chamber, boiler tubes, separating space and space behind the tubes including the exit branch) must not have identical resonance frequencies. This problem must be solved in the future namely for high output boilers where resonance frequencies reach as far as into the zone of infrasound.

Boilers can be consider as resonance dampers which exhibit considerably varying noise damping in the dependence on the frequency. The behaviour of insertion and transmission damping determined by various methods of a BUDERUS SK 425 boiler. This case shows that the design of noise silencer mounted in a boiler uptake cannot be performed by classical procedures and that for the successful solution for a particular boiler full account must be taken of the natural frequencies of the boiler-uptake-chimney (stack) system.

Circulating Pumps

Circulating pumps can be also considered as a significant source of noise in warm water heating systems. Provided the project engineer has not omitted or economised on separating the pump from the steel piping by rubber compensators on the suction and delivery side of the pump, more pronounced transmission of vibrations and noise from the pump via the piping system into the protected space can be definitely eliminated. In the boiler room, however, noise namely from powerful circulating pumps is much more apparent. In identifying sources of noise transmitted into the protected space of buildings namely during frequency analysis, noise from circulating pumps shows discrete frequencies which correspond to the basic frequency of rotation of the pump. After using rubber components for separating pumps as mentioned above another way how noise propagates is via the seating of the pump in the structure of the building. Lighter pumps can be fixed only by the rubber compensators. Heavier pumps should be resilient mounted on the structure of the building identically as fans, i.e. by means of rubber or steel springs (Fig. 18.10).

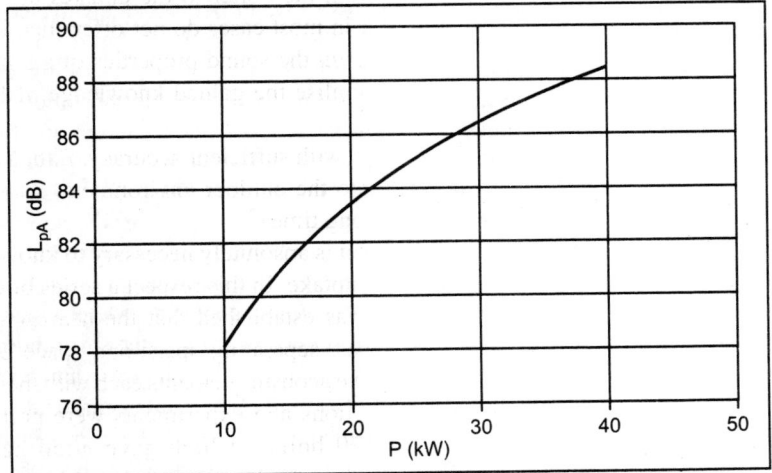

Fig. 18.10. Sound pressure level of circulating pumps at a distance of 1 m from the contour of the machine.

Thermostatic Radiator Valves

Control fittings are mounted in the close vicinity of radiators. At present thermostatic valves are preferably used and in operation appear to be elements with an increased dynamic pressure loss.

Aerodynamic (hydrodynamic) sources of noise (vibrations) can be characterised by a general dependence of the generated sound power W (W) on the velocity of the flowing fluid:

$$W \sim Kw^6 \qquad \qquad \text{... (18.4)}$$

where, K is a proportionality constant expressing the influence of dimensions and shape.

w is the velocity of the flowing fluid (m/s).

By doubling the fluid flow velocity the above relation leads to a theoretical 64 fold increase of sound power. In the logarithmic scale used in technical acoustics this is equivalent to an increase of the sound power level by 18 dB which follows from relation:

$$Lw \sim 60 \log(w) \qquad \qquad \text{... (18.5)}$$

It has to be realised that the above sound power is generated in water inside the piping system. Due to the relatively small dimensions of thermostatic valves emission of noise into the atmosphere usually proceeds in the presence of larger surfaces capable of transmission of vibrations (noise) into the surrounding environment. This is obviously provided by radiators in the closest vicinity of which thermostatic valves are mounted. Basically this is the case of secondary emission of sound power from water via the steel structure into the atmosphere. This obviously complicates the whole problem not only from the viewpoint of accurate determination of conditions of generation of noise into the atmosphere but also from the viewpoint of accurate determination of conditions of generation of noise into the atmosphere but also from the viewpoint of performance of check tests. The same thermostatic valves mounted on radiators of different sizes need not necessarily generate identical noise. And various alternatives of fixing of pipes and radiators to the structure of the building also contribute to its level.

The design of thermostatic valves is strongly influenced by a requirement to react accurately to volume changes of an active element (dilation due to temperature variations of ambient air) in the thermostatic head. This leads to very slow motion of the valve cone towards the valve seat—within the range of tenths of millimetres. The velocities of the flow of water between the valve cone and the valve seat achieve considerable values immediately before the water flow is closed down. In appropriate design of the heating system and therefore also setting of thermostatic valves to higher closing grades result in undesirable generation of noise. The hydraulic resistance of thermostatic valves in projects usually ranges from 1 to 10 kPa and an increase of this value leads to the generation of the above noise. Figure 18.11 (Danfoss) shows a typical diagram of the design of a thermostatic valve with a line of the constant sound pressure level A plotted in the diagram. This information on the noise of the given valve could be easily queried if the producer has not stated under which conditions the noise of the valve was determined. The design parameters, i.e. temperature gradient, ambient temperature, etc. must be maintained during the measurements. In this respect extensive experimental work is at present under way at the laboratories of the Department of Environmental Engineering. Determined is the effect of the magnitude of the heating surface on the radiation of sound energy into the surrounding environment.

The levels of sound pressure A originate by correcting the actual sound spectrum with a weight filter A, which approximately corresponds with the sensitivity range of the human ear to clear tones. Because this sensitive is low at low frequencies, e.g. for a frequency of 31.5 Hz the corresponding decrease is 39.6 dB, when evaluating noise the effect of the low-frequency section of the spectrum decreases. Therefore, relations for the dependence of the level of the sound power A on the water flow exhibit a lower value of the exponent. For the above type of valve the following relation would hold for the dependence of the level of the sound pressure A − L_{pA} on the water flow Q (l/hr):

$$L_{pA} = 41 \log (Q) - 78 \qquad \qquad ... (18.6)$$

provided the adjustment of the valve corresponds to the value of maximum number of preadjustment. If the preadjustment is set to 4, the constant before the logarithm in relation 12.6 drops 39.6. This is easily understood since the same pressure loss is reached in more steps, i.e. at a lower velocity between the valve cone and the valve seat and therefore also at lower noise.

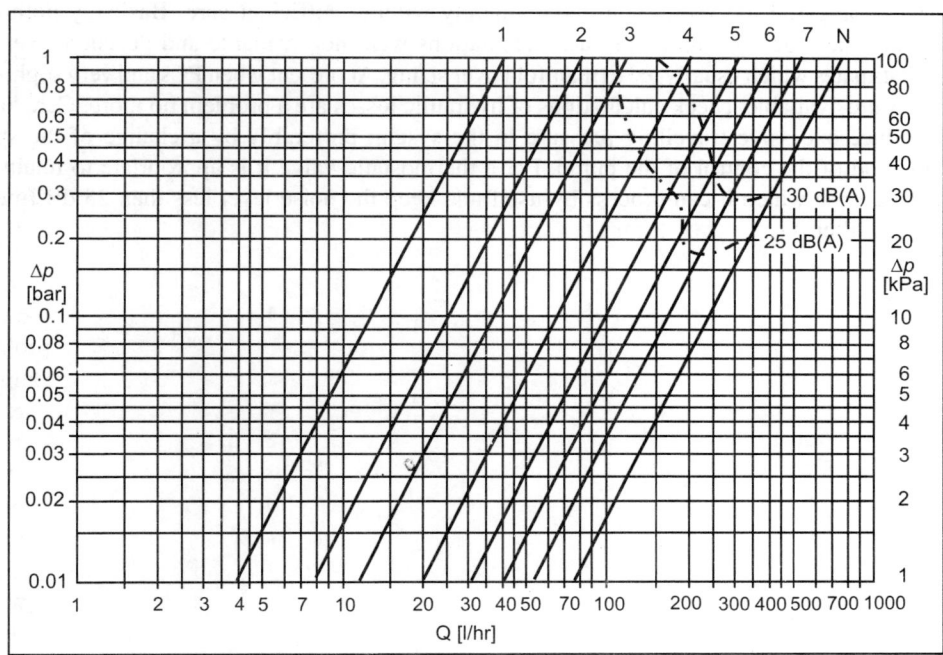

Fig. 18.11. Noise of thermostatic radiator valves in dependence on parameters of heating water flow.

In appropriate design of the heating system and therefore also setting of the thermostatic valve to a higher closing grade result in undesirable generation of noise. Cases are known when the consequences of an oversized piping network are solved by inserting pressure difference controllers or relief valves into the foots of risers. These elements are again a potential source of noise and it is indeed very difficult to decide which noise—that from thermostatic valves or from the elements previously mentioned—is the prevailing source of noise in the protected room. Noise as it is generally known propagates along the length of the steel piping with minimum losses of the sound power.

To sum up, troubles with the noise of boilers are central heating must be dealt with by producers and in this respect pressure must be exerted on them both by project engineers and investors. Due to the complexity of the entire system and since producers themselves release insufficient information on their products, it is an absolutely unrealistic task for the project engineer of the boiler house to design noise silencers in boiler uptakes.

For many decades piping systems for central heating of buildings are designed in various ways. Calculation procedures usually take no account of the fact that from the viewpoint of pressure losses heating systems are non-linear systems which follow modified Kirchhoff laws. This leads to simplifications which often obscure to designers the nature of the problem in question. Necessary pressure conditions in piping systems are predominantly adjusted by choking and bypassing. This, however, is only another sources of noise in heating systems.

A frequently asked question is how the designer should proceed in such cases. Primarily he should deal in detail with hydraulic losses along the entire central heating system and not only rely on data available from the original project. This will disapproved by some of the older generation, but most projects were prepared too rapidly and consequently without sufficient care. Basically nothing was properly calculated, because tools for such calculations were not available and decisions were made chiefly by estimates which usually led to manifold over sizing. At present when PCs and very sophisticated programmes are available, check calculations of hydraulic losses are a problem no more. Care has to be taken to reduce excessive travelling pressure in the system preferably by a change of the speed or another method of the control of the pump. For a thermostatic valve it is appropriate to retain such a pressure difference which it can process by itself and keep the noise level less than 25 dB (measured with a weight filter A).

Chapter 19

Smelter Industries

INTRODUCTION

Smelting is a form of extractive metallurgy; its main use is to produce a metal from its ore. This includes iron extraction (for the production of steel) from iron ore, and copper extraction and other base metals from their ores. Smelting uses heat and a chemical reducing agent to change the oxidation state of the metal ore; the reducing agent is commonly a source of carbon such as coke, or in earlier times charcoal. The carbon or carbon monoxide derived from it removes oxygen from the ore to leave the metal. The carbon is thus oxidised in two stages, producing first carbon monoxide and then carbon dioxide. As most ores are impure, it is often necessary to use flux, such as limestone, to remove the accompanying rock gangue as slag. Plants for the electrolytic reduction of aluminium, while not using carbon, are also generally referred to as smelters.

Smelting involves more than just 'melting the metal out of its ore'. Most ores are a chemical compound of the metal with other elements, such as oxygen (as an oxide), sulphur (as a sulphide) or carbon and oxygen together (as a carbonate). To produce the metal, these compounds have to undergo a chemical reaction. Smelting therefore consists of using suitable reducing substances that will combine with those oxidising elements to free the metal.

COPPER SMELTING

Copper can be produced either pyrometallurgically or hydrometallurgically. The hydrometallurgical route is used only for a very limited amount of the world's copper production and is normally only considered in connection with *in situ* leaching of copper ores; from an environmental point of view, this is a questionable production route. Several different processes can be used for copper production. The traditional process is based on roasting, smelting in reverbatory furnaces (or electric furnaces for more complex ores), producing matte (copper-iron sulphide), and converting for production of blister copper, which is further refined to cathode copper. This route for production of cathode copper requires large amounts of energy per ton of copper: 30–40 million British thermal units (Btu) per ton cathode copper. It also produces furnace gases with low sulphur dioxide (SO_2) concentrations from which the production of sulphuric acid or other products is less efficient. The sulphur dioxide concentration in the exhaust gas from a reverbatory furnace is about 0.5–1.5 per cent; that from an electric furnace is about 2–4 per cent. So-called flash smelting techniques have therefore been developed that utilise the energy released during oxidation of the sulphur in the ore.

Ambient Noise

Noise abatement measures should achieve either the following levels or a maximum increase in background levels of 3 dB(A). Measurements are to be taken at noise receptors located outside the project property boundary (Table 19.1).

Table 19.1. Ambient noise.

Receptor	Maximum allowable L_{eq} (hourly), in dB(A)	
	Day time	*Night time*
	07:00–22:00	*22:00–07:00*
Residential, institutional, educational	55	45
Industrial, commercial	70	70

The emission requirements given here can be consistently achieved by well-designed, well-operated and well-maintained pollution control systems.

LEAD AND ZINC SMELTING

Lead and zinc can be produced pyrometallurgically or hydrometallurgically, depending on the type of ore used as a charge. In the pyrometallurgical process, ore concentrate containing lead, zinc or both is fed, in some cases after sintering, into a primary smelter. Lead concentrations can be 50–70 per cent, and the sulphur content of sulphidic ores is in the range of 15–20 per cent. Zinc concentration is in the range of 40–60 per cent, with sulphur content in sulphidic ores in the range of 26–34 per cent. Ores with a mixture of lead and zinc concentrate usually have lower respective metal concentrations. During sintering, a blast of hot air or oxygen is used to oxidise the sulphur present in the feed to sulphur dioxide (SO_2). Blast furnaces are used in conventional processes for reduction and refining of lead compounds to produce lead.

Ambient Noise

Noise abatement measures should achieve either the levels given below or a maximum increase in background levels of 3 decibels (measured on the A scale) [dB(A)]. Measurements are to be taken at noise receptors located outside the project property boundary (Table 19.2).

Table 19.2. Ambient noise.

Receptor	Maximum allowable log equivalent (hourly measurements), in dB(A)	
	Day time	*Night time*
	07:00–22:00	*22:00–07:00*
Residential, institutional, educational	55	45
Industrial, commercial	70	70

NICKEL SMELTING AND REFINING

Primary nickel is produced from two very different ores, lateritic and sulphidic. Lateritic ores are normally found in tropical climates where weathering, with time, extracts and deposits the ore in layers at varying depths below the surface. Lateritic ores are excavated using large earth-moving equipment and are screened to remove boulders. Sulphidic ores, often found in conjunction with copper-bearing ores, are mined from underground.

Ambient Noise

Noise abatement measures should achieve either the following levels or a maximum increase in background levels of 3 dB(A). Measurements are to be taken at noise receptors located outside the project property boundary (Table 19.3).

Table 19.3. Ambient noise.

Receptor	Maximum allowable L_{eq} (hourly), in dB(A)	
	Day time 07:00–22:00	*Night time* 22:00–07:00
Residential, institutional, educational	55	45
Industrial, commercial	70	70

The emission requirements given here can be consistently achieved by well-designed, well-operated and well-maintained pollution control systems.

ALUMINIUM SMELTING

The basis for all modern primary aluminium smelting plants is the Hall-Héroult Process, invented in 1886. Alumina is dissolved in an electrolytic bath of molten cryolite (sodium aluminium fluoride) within a large carbon or graphite lined steel container known as a 'pot'. An electric current is passed through the electrolyte at low voltage, but very high current, typically 1,50,000 amperes. The electric current flows between a carbon anode (positive), made of petroleum coke and pitch, and a cathode (negative), formed by the thick carbon or graphite lining of the pot.

Molten aluminium is deposited at the bottom of the pot and is siphoned off periodically, taken to a holding furnace, often but not always blended to an alloy specification, cleaned and then generally cast. A typical aluminium smelter consists of around 300 pots. These will produce some 1,25,000 tonnes of aluminium annually. However, some of the latest generation of smelters are in the 350–4,00,000 tonne range. On average, around the world, it takes some 15.7 kWh of electricity to produce one kilogram of aluminium from alumina

Nuisances and Technological Risk

Noise

As with many industrial activities, noise is one of the possible nuisances that can have an impact on public health. Noise can be produced by fixed sources (e.g. ventilators, compressors, generators, and electrical transmission lines) or mobile sources (e.g. trucks and trains). Without mitigation measures, an aluminium smelter can produce an average noise level (L_{eq} 1 hour) of 55 dB at a distance of 2 km and 45 dB (L_{eq} 1 hour) at more than 3 km. With noise mitigation measures, however, intensity is reduced to 35 dB at 2 km and 25 dB at 4 km.

Aluminium smelter plants are extremely large and aesthetically unappealing facilities. Their related infrastructure (high-voltage power lines, railway lines, and roads) can also disrupt the agricultural, forest, and urban environments. A fairly broad buffer zone must be planned around a plant site, in anticipation of fluorine and PAH fallout. Aluminium plants and their infrastructure can affect local quality of life and lower residential property values. Notwithstanding the economic benefits of aluminium plants (jobs, commercial businesses, etc.), their negative social impacts should not be overlooked.

Technological risk

The potential for industrial accidents is generally confined to the plant itself and poses little risk to public health. Possible events that must be considered public health risks are:
1. A break in a gas or oil pipeline resulting in an explosion, which can cause injuries up to a distance of 100 m.
2. An ammonia cloud caused by dross or pot lining coming into contact with water (effects may extend as far as about 200 m from the point of incident).

CASE STUDY—ALUMINIUM SMELTER

Recently, an aluminium smelter underwent expansion adding a new potline to the existing operational potlines. The expansion was designed to massively increase the production capacity of the plant. To accommodate this increase in production capability, other sections of the plant had to be modified and/or expanded.

A principle environmental noise requirement of the project was that the new potline and expanded facilities were not to cause an increase in the existing background noise level in the surrounding community, and that the new development was not to hinder the ongoing noise reduction program of the existing potlines and associated facilities. A second requirement was that within the new potline and upgraded facilities, the internal noise levels were not to exceed the proposed Occupational Health and Safety noise limits of 85 dB(A) when in full production.

Project Management

The acoustic brief covered all phases of project/acoustic interactions. The construction company was supplied with extensive engineering advice on the building constructions, machinery designs, machinery layouts, and machinery selections. The acoustic requirements were most economically incorporated into their designs and specifications.

Separate acoustic specifications were prepared and issued to equipment suppliers, detailing information requirements and noise limits. The details included information format and measurement procedures to be used by suppliers in providing the necessary information. Where an equipment supplier was unable to supply specification details, the supplier's own noise information was used to modify the overall design to accommodate these variations. As designs and constructions progressed, ongoing changes were introduced to accommodate variations which occurred within large projects. Each subcontractor was required to submit noise guarantees. Following installation of equipment, the contractors had to provide certification confirming that they had achieved the guaranteed noise limits in accordance with the standard test procedures set down in the specifications. All this information had to be submitted for verification by the accredited acoustic engineers.

Environmental Noise Monitoring

At the same time, the smelter environmental group commissioned 'before and after' computerised noise emission simulations of the Smelter, and an assessment of the likely noise intrusions into the surrounding residential areas. All equipment and buildings on the site were measured and the determined sound power levels were entered into a computerised acoustic model of the site and surrounding areas. This exercise was repeated and included the proposed new expansion which at that stage had not been constructed.

Typical noise values obtained from the existing plant and from construction information was used to model the expansion. This exercise permitted the smelter noise contributions to be extracted from the existing background levels as part of the smelter operator's ongoing commitment of complying with the requirements of the Environment Protection Authority. In conjunction with the computerised modelling, an extensive automated field noise monitoring program was undertaken over a large area and continued over an extended period of time. Following the processing of the automated noise monitoring results to remove extraneous noise data, the computerised acoustic simulations were found to be accurate to within ±1 dB(A) at all residential locations monitored.

The computerised acoustic model permitted the clear identification of both meteorological conditions and noise sources responsible for the noise levels measured in the local community. The computerised acoustic model provided a full ranking of noise sources within the site by component contribution to each monitoring location. This approach has proven to be a valued tool in the smelter operator's long term strategic planning for noise reduction to meet the legal requirement of Environment Protection Authority.

Chapter 20

Blasting Operations

INTRODUCTION

Noise, vibration and air blast are unavoidable fallouts of mining operations, which involve using large mobile equipment, fixed plant and blasting. Noise, vibration and airblast are among the most significant issues for communities located near mining projects.

The adverse impacts due to noise, vibration and air blast emissions should be contained by the following three stage approach:

1. Noise, vibration and air blast impact assessment.
2. Developing and implementing a noise, vibration and air blast management plan.
3. A monitoring and audit program.

Blasting operations may cause excessive noise and vibrations impacts on the community. Excessive levels of structural vibration caused by ground vibration from blasting can result in damage to or failure of a structure. People are able to detect vibration at levels much lower than those required to cause even superficial damage to the most susceptible structures.

The criteria set out in this guideline assist in minimising annoyance and discomfort that may be caused by blasting at activities such as mining, quarrying, construction and other operations which involve the use of explosives for fragmenting rock.

Further, all blasting must be carried out in a proper manner by a competent person in accordance with best practice environmental management to minimise the likelihood of adverse effects being caused by the impact of airblast overpressure and ground-borne vibration in noise-sensitive places and people living in or using the surrounding area.

PROCEDURE

Noise Criteria

Blasting activities must be carried out in such a manner that if blasting noise should propagate to a noise-sensitive place, then

1. The airblast overpressure must be not more than 115 dB (linear) peak for nine out of any 10 consecutive blasts initiated, regardless of the interval between blasts.
2. The airblast overpressure must not exceed 120 dB (linear) peak for any blast.

Vibration Criteria

Blasting operations must be carried out in such a manner that if ground vibration should propagate to a noise-sensitive place:

1. The ground-borne vibration must not exceed a peak particle velocity of 5 mm per second for nine out of any 10 consecutive blasts initiated, regardless of the interval between blasts.
2. The ground-borne vibration must not exceed a peak particle velocity of 10 mm per second for any blast.

Measurement

Times of blasting

Blasting should generally only be permitted during the hours of 9 am to 3 pm, Monday to Friday, and from 9 am to 1 pm on Saturdays. Blasting should not generally take place on Sundays or public holidays. Blasting outside these recommended times should be approved only where:

1. Blasting during the preferred times is clearly impracticable (in such situations blasts should be limited in number and stricter airblast overpressure and ground vibration limits should apply).
2. There is no likelihood of persons in a noise-sensitive place being affected because of the remote location of the blast site.

Outdoor measurement of airblast overpressure

Measurement of airblast overpressure should be taken at an appropriate location that is exposed to the direction of blasting and at least 4 m from any noise-affected building or structure or within the boundary of a noise sensitive place, at a position between 1.2 m and 1.5 m above the ground.

Outdoor measurement of ground vibration

The ground-borne vibration transducer (or array) must be attached to a mass of at least 30 kg to ensure good coupling with the ground where the blast site and the measurement site cannot be shown to be on the same underlying strata. The mass shall be buried so that its uppermost surface is at the same level as the ground surface.

The ground-borne vibration transducer (or array) must be placed at a distance of at least the longest dimension of the foundations of a noise-affected building or structure away from such building or structure and be positioned between that building or structure and the blasting site.

Airblast overpressure and ground vibration monitoring

For the purposes of checking compliance with the airblast overpressure conditions and ground vibration conditions and for investigating complaints of noise and vibration annoyance, monitoring must be undertaken and at least the following descriptors, characteristics and conditions determined:

1. Maximum instantaneous charge (MIC) in kg.
2. Location of the blast within the quarry (including which bench level).
3. Airblast overpressure level, dB (linear) peak.
4. Peak particle velocity (mms^{-1}).
5. Location, date and time of recording.
6. Meteorological conditions (including temperature, relative humidity, temperature gradient, cloud cover, wind speed and direction).

7. Distance/s from the blast site to noise-affected building/s, structure/s or the boundary of any noise-sensitive place.

Where access to a noise-affected property for monitoring purposes is not feasible, the measurement may be undertaken at the appropriate property boundary and the results extrapolated to reflect the impact at the receptor premises.

Noise from blasting shall be measured using noise measurement equipment with a lower limiting frequency of 2 Hz (−3 dB response point of the measurement system) and a detector onset time of not greater than 100 microseconds as assessed in accordance with AS −1259.1 clauses 8.5 and 10.4.

Vibration instrumentation must be capable of measurement over the range 0.1 mms^{-1} to 300 mms^{-1} with an accuracy within 5 per cent and have a frequency response flat to within 5 per cent over the frequency range of 4.5 Hz to 250 Hz.

Weather effects

When a temperature inversion or a heavy low cloud cover is present, values of airblast overpressure will be higher than normal in surrounding areas. Accordingly, blasting should be avoided if predicted values of airblast overpressure in noise-sensitive places exceed acceptable levels. If this is not practicable, blasting should be scheduled to minimise noise annoyance. An appropriate period is generally between 11 am and 1 pm. Similarly, blasting should be avoided at times when strong winds are blowing from the blasting site towards noise sensitive places.

Quality assurance—airblast overpressure and ground vibration

The measurement and reporting of airblast overpressure and ground vibration levels must be undertaken by a person or organisation possessing both the qualifications and the experience appropriate to perform the required measurements and reporting.

Recording

Details of the measurement instrumentation, measurement procedure, location, date and time of recording and conditions prevailing during measurements must be recorded for each assessment. Records must be kept of the results of all airblast overpressure and ground vibration levels and other information required to be recorded in conjunction with such monitoring for a period of at least five years.

CASE STUDY 1: KUDREMUKH

Kudremukh produces high grade iron ore concentrate which is ideal for use as sinter feed and for pelletisation. The concentrate is being used in the steel plants in China and Japan for sintering and in Iran and China as a blend in pellet feed. The outstanding feature of Kudremukh ore is that it is very low in alumina, sulphur, phosphorous, vanadium and other deleterious elements. Magnetite content of the ore has an added advantage in that it requires relatively less energy for sintering and pelletisation when compared with other types of iron ore. Pellets: Similarly, Kudremukh pellets have excellent chemical, physical and reduction properties and are ideal feed for blast furnaces and direct reduction plants. Kudremukh blast furnace grade pellets have been used in blast furnaces of steel mills in Australia, China, Japan, Taiwan, Turkey and a host of other countries. The pellets have also been used in steel plants of Hungary, Yugoslavia, USA, West Germany, Poland, Czechoslovakia, Indonesia and in some of the direct reduction plants in India. In all an excellent material for steel production in blast furnaces and direct reduction plants.

Impact of Noise and Ground Vibration

No specific noise control measures were observed in the iron ore mining in the area, specifically in the smaller and manual mines. However, the existence of natural forests at places acts as a natural acoustic barrier for the local villages. The following steps are being taken for minimising the ground vibrations and air blast pressure due to blasting;

1. By proper blasting design and by selecting right explosives.
2. Reduction of charge weight per delay between the holes.
3. Adopting suitable delays (millisecond delays) and initiation.
4. Adopting 'non-electric delay initiation system'.
5. Ensuring a minimum stemming length of not less than 0.7 times of the burden.
6. Adopting muffle blasting and controlled blasting technique.
7. Discouraging practice of collar priming.
8. Avoidance of over fly confined charges and subgrade drilling.
9. Orientation of the quarry faces, where possible, so that they do not face directly towards residential areas.
10. Surveying fly rock distances for reference.

Though the bigger mines are practicing deep-hole drilling and blasting, the blast induced ground vibrations do affect the nearby residential villages, as apparent from the cracks developed in the residential units of the concerned mining company's townships. It also cannot be ruled out that the vibrations induced due to blasting do not affect the ground water aquifer system of the area. Thus, this have an important bearing on the existing forest in this area, as these plants/trees depend upon the ground water, which occur as a water table immediately below the earth surface. Further, these blasting operations produce impulsive noise, which may affect the wild life habitat existing in the nearby forest area. Also the noise produced by earth moving machinery and the mineral processing plants, to a major extent, effect the stillness for the forest area.

Impacts of Noise and Vibrations

Noise pollution in the Kudremukh is mainly due to the mining operations such as drilling, blasting, crushing, haulage etc. The noise levels in the nearby residential areas are being regularly monitored by the mine and shows lower noise levels. Further, the location of the concentrator has been so planned that the noise generated would be attenuated to an acceptable level by the time it traverses to the township and for the purpose, thick barriers of trees have been raised. Studies of blasting vibrations at Kudremukh mines conducted during 1998 by the Training and Safety Department of KIOCL revealed that the vibrations are well within the DGMS limits.

Moreover, no building (other than those owned by KIOCL) is located with in a distance of 2 km from the mine workings. The present workings are well away from the final slopes of the pits. However, when the area being worked out comes within 300 m of the final slopes, only controlled blasting will be carried out.

Control in this context, implies reduction of the maximum charge per delay, from the present level of over two tons to less than a ton, by judicious use of delays. Since blast vibrations are directly proportional to the charge, the already low vibration levels can be reduced to half or even one third of the present levels by such controlled blasting.

In the Donimalai, the locations of mines and plant are quite far away from the Township and other adjoining villages like Narasingpur, Ranjitpura, etc. the horizontal and vertical distances being more than 3.5 kms and 400 m respectively.

Due to the natural barrier like configuration of the hilly terrain, the noise levels in the mine and infrastructural areas do not affect the township and office buildings, etc. as the latter are situated far away in the flat terrain down the hill. Moreover, the noise produced from various locations in the mines is also not continuous and only intermittent.

CASE STUDY 2: IRON ORE PROJECT—SALEM (TAMIL NADU)

Kanjamalai in Salem and Kavuthimalai and Vediappanmalai in Thiruvannamalai Districts of Tamil Nadu have low grade iron ore deposits . Mine able reserves at Kanjamalai is estimated @75 million tonnes and at Kavuthimalai and Vediappanmalai @35 million tonnes. Tamil Nadu Iron Ore Mining Corporation Ltd. (TIMCO) a joint venture company of TIDCO and Jindal Vijayanagar Steel Limited is implementing the iron ore project at Kanjamalai , Salem District and at Kavuthimalai, Vediappanmalai at Thiruvannamalai Districts at a project cost of Rs 400 crores.

TIMCO is taking necessary steps to get the following approvals/clearances from State and Central Government for mining activities:

1. Mining lease under Mineral Concession Rule 1960 from Department of Mining, Government of India.
2. NOC under Air and Water Act/from Ministry of Environment and Forests, Government of India.
3. Clearance from Ministry of Forest under Section (2) of Forest (Conservation) Act 1980 for carrying out mining activity in the reserved forests.

Noise Environment

Noise level monitoring was carried out at all the AAQ locations and the noise levels were found to be within the prescribed standards. The minimum and maximum noise levels obtained were as follows:

Zone	Minimum dB(A)	Maximum dB(A)
Core zone	36.7	56.0
Buffer Zone	35.7	52.0

Noise pollution sources

1. Stationary mining equipment.
2. Mobile mining equipment.
3. Transportation vehicles.
4. Blasting operations.

The management plan for controlling noise pollution is as follows:

1. Selection of suitable machinery and equipment, proper mounting of equipment, providing noise insulation/padding wherever practicable and machinery fitted with properly designed silencers.
2. Proper maintenance of noise generating parts of the machine.
3. Provision of earmuffs to workers as a measure to protect their ears.
4. Thick plantation in and around the mine.

PROPOSED NOISE AND AIRBLAST STANDARDS

Noise Level Standards

The noise levels in the mining and other associated activities shall not exceed the following limits:

Parameter	Noise limits	
	Day time *(6.00 am to 10.00 pm*	*Night time* *(10.00 pm to 6.00 am)*
Noise level – L_{eq}	75 dB(A)	70 dB(A)

Noise levels shall be monitored both during day and night times on the same day while in operation. The noise measurements shall be taken outside the broken area, boundary of ore processing and material handling areas, which include mine site and general offices, statutory buildings, workshops, stores, etc.

In addition to this, occupational exposure limit of noise specified by the Director General of Mines Safety (DGMS) shall be complied with by the iron ore mines.

SECTION IV

Noise Reduction in Electrical, Electronics and Telecommunication Industries

Chapter 21

Transformer

INTRODUCTION

A transformer is an electrical device, which changes voltage levels and facilitate transmission, distribution and utilisation of electrical power in the most efficient and economic manner. The health of transformer Industry depends largely on the power generation and transmission system program. The major users of this product are the State Electricity Boards, Power Grid Corporation of India Ltd. and other Industries. Some special types of transformer are also manufactured which are used for the purpose of welding, traction and electrical furnaces, etc.

Noise is defined as unwanted sound. But, what is unwanted sound? A mellow sound to some, can be completely unacceptable to others. Attending rock concerts with noise levels at eardrum rattling levels is totally stimulating to many people. Put those same people in a different environment, possibly next to a transformer, and there will be wild protests. The difference then between noise and sound is in the 'ear of the hearer'. Since it is necessary to place electrical apparatus alongside a wide spectrum of people we have to accept the inevitable, that even under normal conditions, somebody will always complain. Transformer 'humming' has been known to soothe people (which makes it a sound) but generally it is thought to be a nuisance (which makes it a noise). The causes and reduction of transformer noise has been the subject of many articles for at least two generations. It has come to prominence again, mainly because transformers are placed closer to the populace — in high rise office buildings, apartments, shopping malls and in their gardens. It is becoming even more necessary to locate these units carefully and some planning, preferably ahead of time, is needed. The remedies we use to counter possible objections to transformer noises are varied and in some cases, expensive, because we cannot produce a blanket remedy to cover all situations. It is absolutely necessary to consider each case on its merits, to apply the general rules of acoustic technology and to be familiar with the causes of transformer noise. The techniques can be explained simply enough for anyone to understand and the rules are, in the main easy to apply. The best rule however is to plan ahead. Finding out you have a noise problem (or vibration problem) after the placement of the unit is costly, time consuming and frustrating.

TRANSFORMER'S HUM

Transformer noise is caused by a phenomenon called magnetostriction. In very simple terms this means that if a piece of magnetic sheet steel is magnetised it will extend itself. When the magnetisation is taken away. A transformer is magnetically excited by an alternating voltage and current so that it becomes extended and contracted twice during a full cycle of magnetisation.

This extension and contraction is not uniform, consequently the extension and contraction varies all over a sheet. A transformer core is made from many sheets of special steel. It is made this way to reduce losses and to reduce the consequent heating effect. If the extensions and contractions described above are taking place erratically all over a sheet, and each sheet is behaving erratically with respect to its neighbor, then you can get a picture of a moving, writhing construction when it is excited. Of course, these extensions are only small dimensionally, and therefore cannot usually be seen by the naked eye. They are, however, sufficient to cause a vibration, and as a result noise. The act of magnetisation by applying a voltage to a transformer produces a flux, or magnetic lines of force in the core. The degree of flux will determine the amount of magnetostriction (extensions and contractions) and hence, the noise level.

The obvious question is why not reduce the noise in the core by reducing the amount of flux. Why? Because it is not that simple. Transformer voltages are fixed by system requirements, and the amount of magnetisation, by the ratio of these voltages to the number of turns in the winding. The decision on what this ratio of voltage to turns will be, is made for reasons, mainly economic. It means that the amount of flux at the normal voltage is invariably fixed, thus setting the noise and vibration level. Also, increasing (or decreasing) magnetisation does not increase or decrease the magnetostriction by the same amount. In technical terms the relationship is not linear. Therefore, when we are asked, as we invariably are,– 'can you reduce the noise level at the source?'—the answer is that it can be done, at a cost and for not much improvement in noise level.

NOISE FREQUENCY

We have established that the transformer hum is caused by the extension and contraction of the core laminations when magnetised. Under alternating fluxes, we can expect this extension and contraction to take place twice during a normal voltage or current cycle. This means that the transformer is vibrating at twice the frequency of the supply, i.e. for 60 cycles per second supply frequency, the noise or vibration is moving at 120 cycles per second. This is called the fundamental noise frequency.

Nothing is this world is ever perfect and so it is with transformer cores. Since the core is not symmetrical and the magnetic effects do not behave in a simple way, the resultant noise is not pure in tone. That is the noise or vibration produced is not only composed of a 20c/s frequency, we find from practical work that transformer noise is made up of frequencies of odd multiples of the fundamental known as 1st, 3rd, 5th and 7th harmonics. This means we get noise frequencies of 120 (1st), 360 (3rd), 600 (5th), 840 (7th) cycles per second. They are not equally important for we find that the first and third harmonics predominate and produce most of the transformer sound (Fig. 21.1).

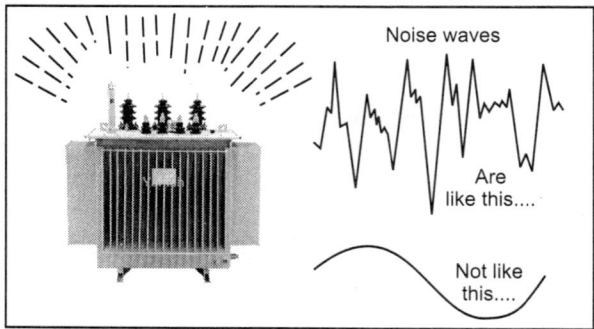

Fig. 21.1. Transformer waves.

It is important to know this because, with this knowledge, we can measure the amount of noise at these frequencies and determine whether amongst a number of other noises, we really are picking up a transformer noise.

TRANSFORMER ON LOAD

It is usually asked — 'what proportion of the transformer noise is contributed by the windings and does the noise increase as the load increases?'. There are, of course, mechanical forces existing between individual conductors in a winding when the transformer is excited. These forces will produce a vibration and a noise, but only one which is pure in tone, i.e. at twice the exciting frequency –120 cps. This, however, is swamped by the fundamental and harmonics produced by the core.

The difference between no load and full load, at constant flux density is usually no greater than 1 or 2dB. An exception to this is when special flux shields are placed inside a transformer tank to reduce stray flux effects. It has been explained that the noise from a transformer is caused by mechanical movement of the individual lamination of the core under magnetisation. The pulsation will cause not only air disturbances, thus producing noise, but also physical vibration of the core structure and everything attached to it. The vibration will have similar frequencies to those measured in the noise analysis. Reducing (attenuating) these mechanical pressure pulsations is vital to noise and vibration control and consequently, isolating the core and coils of a transformer, either in the tank or through a tank, or just as the core and coils, is important. Baffling transformer noise and forgetting to isolate the vibrations will only lead to a disappointing result and is something which should not be done. Remember noise is usually air borne. Vibration is ground borne. They are very much connected.

TRANSFORMER TYPES

We talk about dB's (decibels) but what do they really mean? In simple terms, we are trying to take what we hear and relate it to scientifically measurable terms. The decibel as used in acoustics is a measurement comparing the pressure generated by a noise against some standard level. These decibels will vary according to the frequency of the noise, but this is taken care of in the noise level meter.

We refer to dB. The 'A' part refers to a position on a sound level meter which more closely follows the human ear. It is important when taking measurements to specify if the noise level was taken on the 'A' weighted scale. Since the transformer is not necessarily symmetrical, we cannot take one reading of noise level from a sound level meter and call the noise level of the unit. It is necessary to take many readings around the transformer and to average them. The resultant will become the transformer noise level.

Standards are laid down on how this should be done. The main ones are ANSI Standard C57–12–90 or NEMA Standard TRI–2–068–1954.

What happens is that you imagine a string following the contours of the transformer. You step back 1 foot from that contour line with the unit excited at the normal voltage, and record a measurement. You take these measurements at 3 foot intervals along the imaginary string. The measurements are totalled and then averaged. The result is the transformer noise level. To measure amounts of noise in each frequency range you need a frequency analyser. This is a worthwhile acquisition (Fig. 21.2).

It is always necessary to measure the background (ambient) noise level before you start and when you finish the tests. There has to be a difference between the ambient reading and the average noise level of 7dB or better, for it to be valid, otherwise you could be increasing the actual reading of the transformer. This sometimes makes night owls of the testers!

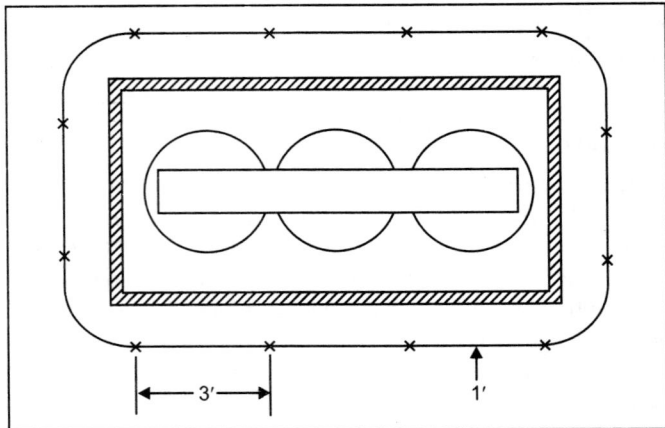

Fig. 21.2. Measuring of noise in each frequency range.

So now we know where the noise comes from and how to measure it. What can we do about it?

First of all, accept that there is a noise and you are stuck with it. We have to consider how to avoid making it a nuisance to people. The most obvious strategy is to place the transformer in a field miles away from habitation. The noise level drops away as the square of the distance from the noise, but even so, it would take a very large field to hide it. However, we invariably have to place transformers near people and we must face up to that fact.

We have both noise and vibration to worry about and as we have said noise is usually air born, vibration is structure born.

CUTTING AIR BORN NOISE

1. Put the object in room in which the walls, floor are massive enough to reduce the noise to a person listening on the other side. Noise is usually reduced (attenuated) as it tries to pass through a massive wall. Walls can be of brick, steel, concrete, lead, etc.

2. Put the object inside an enclosure which uses a limp wall technique. This is a method which uses two thin plates separated by viscous (rubbery) material. The noise hits the inner sheet — its energy (some) is used up inside the viscous material. The outer sheet should not vibrate.

3. Build a screen wall around the unit. This is cheaper than a full room. It will reduce the noise to those near the wall, but the noise will get over the screen and fall elsewhere (at a lower level). Screens have been made from wood, concrete, brick and with dense bushes (although the latter becomes psychological).

4. Do not make any reflecting surface coincident with half the wave length of the frequency. What does this mean? Well, every frequency has a wave length. To find the wave length in air, for instance, you divide the speed of sound, in air (generally reckoned as 1130 feet/second) by the frequency (Fig. 21.3).

If a noise hits a reflecting surface at these dimensions it will produce what is called a standing wave. Standing waves will cause reverberations (echoes) and an increase in the sound level. If you hit these dimensions and get echoes you have to apply absorbent materials to the offending walls (fiberglass, wool, etc.).

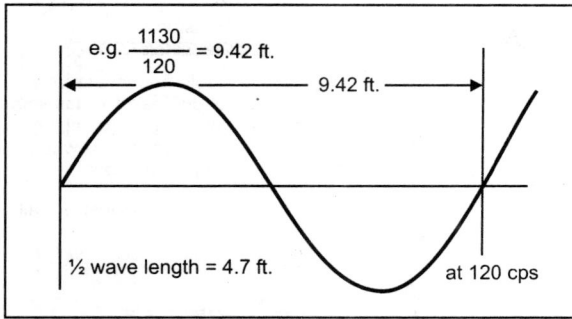

Fig. 21.3. Wave length in air.

CUTTING STRUCTURE VIBRATION

1. Isolate the core and coils of the transformer from the ground. In air cooled dry types this means to isolate the core and coil from its support on the ground. For an oil filled unit it means to isolate the core and coil from its tank base, and isolate its tank base from the supporting ground (Fig. 21.4).
2. Use isolating materials guaranteed to eliminate transformer frequencies (at 120 cps upwards). This is important. Not many materials can do this. Seek advice on the best anti-vibration pads to use.
3. Make sure all connections to a solid reflecting surface are flexible. This includes incoming cables, bus bars, stand off insulators, etc. Any solid connection from the vibrating transformer to a solid structure will transmit vibration (Fig. 21.5).
4. Make sure shipping bolts are removed so that they do not short circuit anti-vibration pads.
5. Additional information is given in ANSI C57.94.

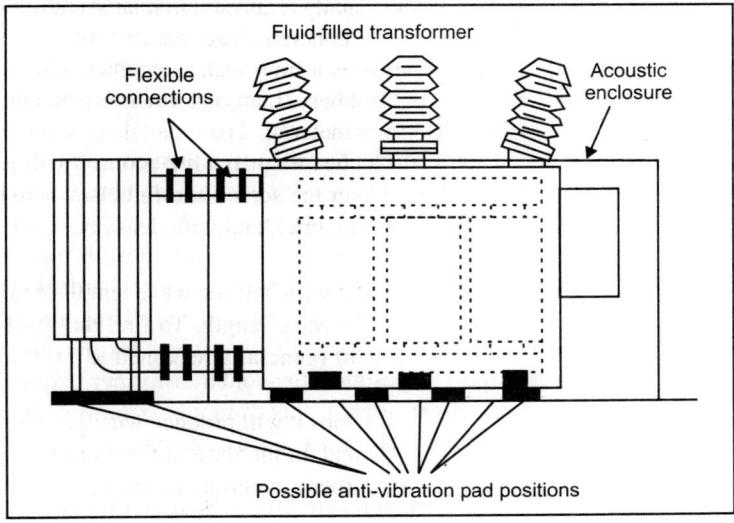

Fig. 21.4. Oil filled transformer.

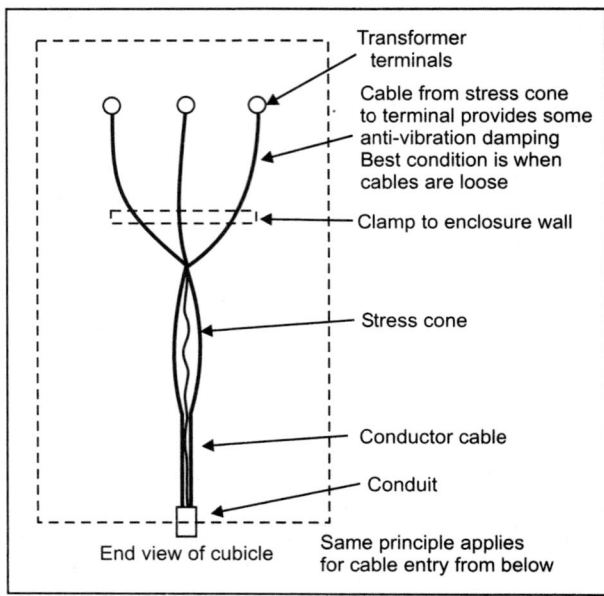

Fig. 21.5. Various terminals of transformer.

WHAT CAN THE MANUFACTURE DO?

The manufacturer must first insure that he achieves the noise level as specified by the appropriate specification. If something unusual is required by way of a very low noise level then there should be discussions and agreement between the manufacturer and the user, as to what steps to take. Remember the only course left to the manufacturer is usually to lower the flux density and this means increased cost. There have to be trade-offs between cost and noise annoyance or treatments.

If the manufacturer is only supplying core and coils, then what happens next is in the hands of the user, assuming all noise level requirements have been met. If the core and coil is mounted in a containing cabinet then the manufacturer has some precautions to take.

He must insure that the core and coils are correctly resiliently mounted, for it they are not, the noise level will increase. The stiffness of the mounts must be such that they do not weaken the installation by being too soft or spoil their attenuation properties by being too hard.

The choice of the resilient must be carefully considered. It has to absorb transformer frequencies which, by most commercial shock absorber systems are very low. 'Shore' hardness (resilience), ability to withstand the environment and stiffness sufficient to carry the unit, are all important design parameters.

Bus bar or other connections to the core must not transmit vibrations. Flexibility is the key. Ventilators must be carefully positioned. The core must be designed to avoid transformer frequencies of half wavelength dimensions, or multiples of these dimensions. If this cannot be achieved, then consideration of damping material applied to the case, is required. This is an added cost and must be part of the arrangement with the user.

Now comes the interface with the user. For shipment purposes, it is often necessary to 'block out' the core and coils against the case, to avoid shipping damage. This can include holding down bolts which if left in a fastened condition, can short circuit the anti-vibration effects of the resilient mounts. The manufacturer can draw attention to these bolts by painting them with a florescent paint, and advising his customer to remove all such marked bolts before use. All other blocking and wedging, not part of the design, should be carefully removed since these might interfere with the vibration isolation. The user should be made aware of any of these requirements. After this, it is up to the user!

The user thinking should start at the conceptual stage. If he can, he must consider if he has a noise problem before he specifies his transformer. If he does, a noise survey including frequency analysis, would be advisable. If for instance, a building is only in the conceptual stage, then a little thought beforehand will make sure that transformers are not placed in small reverberant rooms next to a proposed board room, sleeping areas, study areas or other occupied areas where the normal sound level is low. Closets under stairs seem, very popular for dry type transformers — but are usually acoustically bad. Some discussion with the manufacturer is useful at this point.

A word of warning here. The noise level as measured and given by the transformer manufacturer is usually for the core and coils inside a cubicle. There is no way that the manufacturer can assess the effect on the transformer noise level by the location in which the unit will be place. It is advisable that if a user wants to maintain a particular noise level in a particular environment they should work backwards.

First of all assess what level is tolerable (say 65dBA). Allow for the effect of the room (say 3dB). Allow for the efficiency of all the connections (say 2dB) and as a result ask for a transformer to meet 60dBA! This will ensure that the required noise level is met. Advice on how to assess these corrections is available within federal pacific.

The design of the room to house the transformer is the next consideration. Avoiding half wavelengths of transformer noise, or multiples thereof, is advisable. This includes dimensions in all directions, including the ceiling. If these dimensions concurrences are involved. This is a caution against using acoustical treatments which are only effective for speech frequencies. Choose damping materials for the noise frequencies to be damped.

Isolation of the transformer from the ground is vital. Installation instructions must ensure that nobody tightens down shipping bolts — but removes them. Connecting cables must be as flexible as possible. Ventilation ducts must be placed in positions where these are effective thermally without affecting the acoustic performance. After taking all the precautions, a noise survey after installation, with the transformer excited might be useful.

SALIENT POINTS

It might be useful to review the salient points and give some extra pointers:

1. Transformer noise is difficult to change at the source. Flux density reduction is the main thrust, but this means increased cost.
2. Transformer core constructions help to a degree. Reputable manufacturers will use good joints, flat steel, consistent thickness, good core supports, few bolts, etc.
3. Transformer current loading has little or no effect on the noise level.
4. Placing transformers in liquid (oil) does not help since oils are incompressible.
5. Vibration — isolating core and coils within a tank does assist vibration isolation although isolation of the whole tank is still needed.
6. Noise reduction by distance is the simplest form of attenuation. If it can be achieved without cost — excellent. Usually it cannot.

7. Noise reduction by screens, bushes, etc. is the next simplest but use should be made of the topography of the site. Remember the shadow effect means the noise could be heard outside the shadow of that screen.
8. Full enclosure is usually the only option left to a troublesome transformer.
9. Full enclosure can be made of any material with a high mass/weight ratio. Brick, concrete, steel have been used. Expect 25.30 dBA reductions.
10. Full enclosures using masonry products are not easily demountable. Prefabricated concrete block is the best for this application.
11. Steel, mass or limp panel techniques make good demountable enclosures. A 15–20 dBA reduction is possible with properly designed enclosures.
12. External cooling to the enclosure requires flexible treatments to the connecting pipe work.
13. Enclosure mounts should be separate from the transformer base or at least, isolated somehow.
14. All connections—cables, etc. to enclosed transformers should be flexible.
15. Remember bushings vibrate and losses (acoustic) are experienced through them. Flexible acoustic protection between enclosure and bushings are needed.
16. Bushings used in an enclosure might have to have a longer ground sleeve to accommodate the enclosure roof distances.
17. Pay close attention to access doors and removable covers on enclosures. Tight fits are essential.
18. Watch the dimensions of rooms in which units are mounted. Damp them if necessary, suitable for transformer frequencies.
19. Damping materials are needed if standing waves or reverberations are possible.
20. Choose damping materials compatible with transformer frequencies.
21. If steel plates are used for enclosures ensure that they are gasketed. Isolate the fastening down bolts.
22. Carry out sound surveys before and after installations. Remember to do a frequency analysis so that transformer noise can be differentiated.
23. Anticipate transformer noise problems when accommodating them inside a building-especially for dry types.
24. Pay careful attention to removing unnecessary bolting or stiffening used originally for shipping. Make sure the manufacturer identifies what can and cannot be used or removed.
25. Remember transformers need cooling air in rooms. Be careful (acoustically) when you position air ducts, ventilators and grilles, etc.
26. Pay attention to flexible connections inside rooms containing transformers.
27. Make sure the vibration isolators are correctly mounted and will accommodate transformer frequencies.
28. Select rooms which are not near potential complaint areas.
29. Check the voltage on the system. Increased flux density by having a higher than normal system voltage will raise the noise level.
30. When assessing the required noise level of a transformer work backwards from the required noise level at a location. Consider the inefficiencies of the site.
31. Consider very carefully where transformers will be mounted. Resilient structures such as wooden mezzanines might be harmful.

It has not been possible to give all the points and suggestions that might assist a user in producing a trouble free (noise) site.

SUMMARY

1. Transformer noise is produced by the core.
2. The amount of noise is generally fixed by the design of the transformer.
3. Adjustments to a design to reduce the noise level can be made at cost but don't expect a huge reduction in the noise level.
4. Loading a transformer has little effect on the noise level.
5. Vibrations are produced as well as noise and these are just as important as the noise. We have established that the core and coils of a transformer will, when magnetised, produce a hum (noise) and mechanical vibrations, but, the transformer category will also have an effect on what happens once the noise and vibration is produced.

There are three basic categories currently in use:

1. Those immersed in liquids—oils, silicones, etc.
2. Those immersed in vapours and gases—nitrogen, fluro-carbons, etc.
3. Those mounted in air.

A basic statement can be that irrespective of how transformer core and coils are surrounded, noise and vibration will still be transmitted. Oil is incompressible, and gas and air, we know, transmit sound very effectively. Until we put units in absolute vacuums, we have to accept that they will transmit sound almost as if all were in air.

However, each type requires special consideration and treatments, and it is important that these are understood. Transformer size, requirements, and it is important that these are understood. Transformer size, requirements and applications will determine more exactly where and how a transformer is placed, but there are certain treatments which are common to all type.

Chapter 22

Induction Motor

INTRODUCTION

An electric motor converts electrical energy into mechanical energy. Most electric motors operate through interacting magnetic fields and current-carrying conductors to generate force, although a few use electrostatic forces. The reverse process, producing electrical energy from mechanical energy, is accomplished by some type of generator such as an alternator or a dynamo. Many types of electric motors can be run as generators, and vice versa. For example a starter/generator for a gas turbine, or traction motors used on vehicles, often perform both tasks.

Electric motors are found in applications as diverse as industrial fans, blowers and pumps, machine tools, household appliances, power tools, and disk drives. They may be powered by direct current (e.g. a battery powered portable device or motor vehicle), or by alternating current from a central electrical distribution grid. The smallest motors may be found in electric wristwatches. Medium-size motors of highly standardised dimensions and characteristics provide convenient mechanical power for industrial uses. The very largest electric motors are used for propulsion of large ships, and for such purposes as pipeline compressors, with ratings in the millions of watts. Electric motors may be classified by the source of electric power, by their internal construction, by their application, or by the type of motion they give.

NOISE REDUCTION

Design Trends

Small electric motors are a ubiquitous source of noise in daily life. They occur in almost all home appliances and industrial machines. Current trends in product design are increasing the noise that motors make. Greater power densities produce higher levels of excitation. The use of the field (stator) as a structural component reduces the isolation between the motor and supporting structure. The expanding use of lightweight and stiff plastics for appliance structures increases both vibration transmission to noise radiating surfaces and the ability of those surfaces to radiate sound.

The most common motors in use for small and moderate sized appliances are induction motors and universal motors. In both, the primary noise mechanisms are usually imbalance, misalignment, and/or electromagnetic forces. Universal motors also have a contribution from brush/commutator interaction.

Harmonic Content

The noise from a universal motor is at multiples of rotation speed. These harmonics are 'modulated' at twice the line frequency, the frequency of magnetic attraction between the rotor and stator. This modulation

248

also produces sidebands on the motor speed harmonics. These harmonics and their sidebands are strong and highly audible. Although simple imbalance may generate the highest amplitude tone in the noise spectrum, the electromagnetic 'runout' is often the largest noise problem.

Induction motor noise also has tonal components at the rotation speed due to imbalance, but its electromagnetic excitation is also strong at a frequency that is the product of the number of rotor conducting bars and the 'slip frequency.' The slip frequency is the difference between the rate of rotation of the stator's magnetic field and of the rotor. In a small 'shaded pole' motor, this electro-magnetic induced frequency can be readily heard.

Redesign for Noise Reduction

Reducing the noise from these motors can be done at various levels. Magnetic circuit modelling and analysis can reveal the effects of variations in geometry, materials properties, and structural deformation on the magnetic runout. Changes in the design of magnetic materials, geometry, and structural components can then be made to determine the optimal configuration that is achievable at an acceptable cost. Such an exercise can have ancillary benefits apart from noise, such as improved starting torque or smoother operation.

Other Noise Reduction Options

Even if a redesign of the magnetics is not contemplated, modification may still be possible for a quieter design. For example, motors generally do not radiate much sound directly, since they are small and they may be enclosed within the product. But they do excite the outer structure mechanically, and the resulting vibration will radiate sound. Motor vibrations can be determined from their ODS's (operating deflection shapes), which in turn are determined from the forces due to imbalance and electromagnetics and from the vibrational mode shapes of the structure.

Redesigning the structure to minimise vibration transmission may therefore involve a combination of the following: experimental procedures (ODS's), mode shape determination from experiment (modal analysis), and calculation (finite element analysis or FEA), extraction of the forces from ODS and FEA information, and modal modifications to minimise the response of the structure to these forces.

Close Coordination Necessary

A close coordination of experimental procedures and analytical modelling is critical to the success of the above procedures. This combination of methods has application to many devices other than motors, such as gear trains, hydraulic valves and pumps, and actuators.

INDUCTION MOTOR NOISE

Induction motors are the most common motors in use today. Motor noise can be a challenge for most motor manufacturers. So what causes motor noise and what can be done to reduce the noise? In this chapter we will discuss where motor noise comes from and how it can be minimised.

Types of Motor Noise

The three basic types of motor noise are:
1. Magnetic noise.
2. Bearing noise.
3. Unbalance noise.

Magnetic noise

The main source for magnetic noise in induction motors comes from force waves created by the rotating magnetic field. Since this field is the vehicle by which power is transmitted from the stator to the rotor, elimination of this noise is not possible. However, magnetic noise can be minimised by utilising several techniques:

1. Using the correct stator and rotor slot combination.
2. Designing the stator and rotor lamination geometry.
3. Applying the proper coil pitch.
4. Formulating the appropriate rotor skew.

The proper selection of stator and rotor slots is one of the most important aspects of a quiet motor design. This selection also affects manufacturing costs. To keep manufacturing costs low, the number of stator slots should be the number of poles times the number of phases within a given frame size. This number of stator slots should be such that a balanced winding can be accommodated. There are hardly any slot combinations that are perfect; however, over the years there have been some combinations that have proven to be successful. The following general rules should be used as a guide when designing the laminations:

1. The number of rotor slots should not equal the number of stator slots. Acceptable performance has been attained when the number of rotor slots differs by the number of stator slots by 15 to 30 per cent.
2. To avoid dead points or cogging, the difference between stator and rotor slots should not equal +/– number of phases times number of poles.
3. To minimise noise and vibration the difference between the number of stator slots and rotor slots should not equal +/–1, +/–2, +/– (number of poles+/–1) and/or +/– (number of poles +/–2). For example: If you have a 6 pole motor and the stator lamination has 18 slots then the rotor laminations should not have the following number of slots: 17, 19, 16, 20, 13, 23, 14 or 26.

When designing the stator and rotor lamination geometry, many parameters must be reviewed. A closed rotor slot will cause less permenace variation, however, thickness of the bridge must be reviewed to ensure the lamination does not saturate. The width of the stator slot opening should be kept to a minimum, however, you still need it large enough to insert the coils. The length of the air gap also affects permeance variation such that the larger the gap the less the variation. But keep in mind, the air gap also is dependent on the other design criteria such as power factor and efficiency. The motor design engineer needs to take into account all requirements when designing laminations.

Utilising the proper stator coil pitch is additionally important. The stator coil pitch ratio should be selected to give small pitch factors for lower order of air gap harmonics. The motor designer will need to keep this in mind when designing the winding. Skewing of either rotor or stator slots is another way to reduce motor noise and provide smooth acceleration. The degree of skew also affects motor performance, therefore the skew is limited to performance parameters such as breakdown torque, starting torque and starting currents.

Bearing noise

Bearing noise can be influenced by several factors such as preload, shaft to bearing fits/tolerances, and bearing to bore fits/tolerances. The decision to which bearing to use should always depend on the application of the motor. It is highly recommended to involve bearing manufactures when choosing the bearing. Proper preloading is required to ensure the bearings run smooth. Too much preloading will damage the bearings. Same philosophy goes with the bearing fits to the motor shaft and housing and end bell bores. Bearing manufactures have recommended fits/tolerances to implement during the mechanical design of the motor.

Unbalance noise

Unbalance noise is created by an unequal weight distribution of the rotor assembly around the motor axis of rotation. It has been found that vacuum impregnation and/or encapsulation of motor stator windings has also helped in coil vibration and noise. There are many ways to reduce motor noise. All motor parameters must be taken into account when designing the unit.

Stepper Motor Noise

Stepper motors can run very quietly. They need not make any more noise when attached to the scope as when held in your hand. It's common star party etiquette to run a quiet operation, and you will make a good impression of your system. The noise of the switching stepper motor windings is greatly amplified when attached to metal and wood mounts. The mount acts as a drum.

To run them quietly, isolate the stepper from wood and other resonant materials by a thin piece of rubber or styrofoam or something similar such as a mouse pad (neoprene). Use nylon screws or nylon bolts to attach the stepper to its mounting plate, further isolating the screws with rubber grommets. Use a short piece of automobile vacuum hose to attach the stepper to the input drive shaft of the gear reducer, leaving a gap of a millimeter between the shafts. Another suggestion that receives rave reviews is to mount the motor with rubber standoffs, such as available from Small Parts.

Paul Shankland suggests using a PC wrist support made of a long strip of gel material. He goes on to say, 'Cut away the hard rubber base revealing the material (colour of vanilla pudding) underneath, with the cloth left on top — it's a tacky but cohesive firm gel — looks like some syntheic blubber, and leaves a faint tacky resiudue — very resiliant, absorbent, but of a resilience that with clamps, would not let the torque whip the steppers out of alignment.

The magnetically excited acoustic noise is calculated for three high-speed permanent-magnet motors with identical stators, but with different rotors. An analytical approach is used for the calculation of the magnetic radial forces. The components of the radial magnetic flux density in the middle of the air gap are determined with the help of finite element method (FEM) static calculations, which allow a decomposition of the air gap field into a Fourier series of flux density waves. The natural bending frequency is calculated using an analytical and alternatively a numerical calculation of a simplified mechanic model. The radial deformation of the stator yoke is calculated and the sound intensity level is calculated for the first 13 tonal frequencies and compared for the three motor configurations.

The noise of the rotating electrical motor can be strongly influenced by the change in the loading conditions. In extreme cases a noise component of electromagnetic origin may increase by 36 dB or decrease by 24 dB upon loading. Protection of a human being from the noise depends on the noise characteristics of the machine under operating conditions. The standards relevant to the noise qualification measurements permit the use of the sound power level determined under no-load conditions, if the noise of the machine does not vary with the change in the load. This is a concession that is made because it is difficult to separate the noise level of the loading machine from the noise of the electric motor being investigated. Several methods, and their analyses, where the sound power level emitted by a loaded electrical motor can be determined experimentally in an industrial environment. Wide ranging experimental work has been made in this field.

Chapter 23

Motor Vibration Problems

INTRODUCTION

Vibration problems in induction motors can be extremely frustrating and may lead to greatly reduced reliability. It is imperative, in all operations and manufacturing processes that down time is avoided or minimised. If a problem does occur the source of the problem is quickly identified and corrected. With proper knowledge and diagnostic procedures, it is normally possible to quickly pinpoint the cause of the vibration. All too often erroneous conclusions are reached as a consequence of not understanding the root cause of the vibration. This may result in trying to fix an incorrectly diagnosed problem, spending a significant amount of time and money in the process. By utilising the proper data collection and analysis techniques, the true source of the vibration can be discovered. This includes, but is not limited to:

1. Electrical imbalance.
2. Mechanical unbalance—motor, coupling, or driven equipment.
3. Mechanical effects—looseness, rubbing, bearings, etc.
4. External effects—base, driven equipment, misalignment, etc.
5. Resonance, critical speeds, reed critical, etc.

Once the electrical and mechanical interactions in a motor are understood, and the influence external components have on the apparent motor vibration, identification of the offending component is usually straightforward. This chapter provides an analytical approach for expeditiously understanding and solving these types of problems. Vibration problems can occur at anytime in the installation or operation of a motor. When they occur it is normally critical that one reacts quickly to solve the problem. If not solved quickly, one could either expect long-term damage to the motor or immediate failure, which would result in immediate loss of production. The loss of production is oftentimes the most critical concern. To solve a vibration problem one must differentiate between cause and effect. For this to happen, one must first understand the root cause of the vibration. In other words: where does the force come from? Is the vibratory force the cause of the high levels of vibration or is there a resonance that amplifies the vibratory response. Perhaps the support structure is just not stiff enough to minimise the displacement. In this chapter the various sources of electrical and mechanical forces will be explained. Additionally, how the motor reacts or transmits this force and how this force can be amplified or minimised will be explained as well. When a vibration problem occurs it is important that one use a good systematic, analytical approach in resolving the problem. This includes performing the proper diagnostic tests. The process starts by listing all the possible causes for the particular identified frequency of vibration and any variations under different operating conditions. Then eliminate the incorrect causes one by one until all that remains is the true source of the problem, and now this can be efficiently eliminated.

SOURCES OF VIBRATION

There are many electrical and mechanical forces present in induction motors that can cause vibrations. Additionally, interaction of these various forces make identification of the root cause elusive. In subsequent sections, the major mechanisms are discussed. For a more comprehensive list of electrically and mechanically induced vibrations Table 23.1 should be referenced. There are many different forces and interactions as a result of the power source and the interactions between the stator and rotor as seen in Fig. 23.1. The power source is a sinusoidal voltage that varies from positive to negative peak voltage in each cycle. Many different problems either electrical or mechanical in nature can cause vibration at the same or similar frequencies. One must look closely to differentiate between the true sources of vibration.

A power supply produces an electromagnetic attracting force between the stator and rotor which is at a maximum when the magnetising current flowing in the stator is at a maximum either positive or negative at that instant in time. As a result there will be 2 peak forces during each cycle of the voltage or current wave reducing to zero at the point in time when the current and fundamental flux wave pass through zero as demonstrated in Fig. 23.2. This will result in a frequency of vibration equal to 2 times the frequency of the power source (twice line frequency vibration). This particular vibration is extremely sensitive to the motor's foot flatness, frame and base stiffness and how consistent the air gap is between the stator and rotor, around the stator bore. It is also influenced by the eccentricity of the rotor.

Some people are inaccurately under the premise that twice line frequency vibration varies with load. This misconception comes from the belief that twice line frequency vibration excitation is due to a magnetic field generated by the current in the stator coil which varies with load and creates a magnetic force which varies with the load current squared. In reality the ampere-turns of the stator and rotor tend to balance one another except for the excitation ampere-turns. To explain this to those not familiar with motor electrical theory, the excitation ampere-turns are created by the motor no load current. This establishes the magnetic field in the motor necessary to generate a back EMF approximately equal to the applied voltage. As load is applied to the motor, both stator and rotor currents increase together and balance one another, therefore, there are no significant changes in flux. This means that the basic magnetic forces are independent of load current and are nearly the same at no load or full load. Therefore the main component of twice line frequency vibration which is created by an unbalanced magnetic pull due to air gap dissymmetry and does not change with load.

On 2 pole motors, the twice line frequency vibration level will appear to modulate over time due to its close relationship with 2 times rotational vibration. Problems in a motor such as a rub, loose parts, a bent shaft extension or elliptical bearing journals can cause vibration at 2 times rotational frequency. Due to its closeness in frequency to twice line frequency vibration the two levels will add when they are in phase and subtract when they are out of phase and then add again when they return to being in phase. This modulation will repeat at a frequency of 2 times the slip on 2 pole motors. Even at no-load, twice rotation vibration on 2 pole motors will vary from 7200 cpm (120 Hz) due to slip. Since there is some slip on induction motors, although small at no-load, it may take 5 to 15 minutes to slip one rotation. For those of you not familiar with the term slip, there is a rotating field around the stator that the rotor is trying to stay in phase with, but the rotor will fall behind the stator field a certain number of revolutions per minute depending upon the load. The greater the load the greater the slip. Slip is typically 1 per cent of rated speed at full load, and decreases to near 0 slip at no-load. Since vibration levels are not constant, to measure vibration, many times it is necessary to perform what is referred to as a modulation test. In a modulation vibration test the motor is allowed to run for a period of typically 10 or 15 minutes, and vibration is recorded continuously to allow the maximum and minimum to be established.

Table 23.1. Electric motor diagnostic chart.

Cause	Frequency of vibration	Phase angle	Amplitude response	Power cut	Comments
Misalignment: Bearing	Primarily 2 × Some 1 × Radial high at DE and axial	Phase angle can be erratic	Steady	Drops slowly with speed	2 × can dominate during coast-down. 2 × is more prevalent with higher misalignment
Misalignment: Coupling	Primarily 1 × Some 2 × Radial high at DE and axial	Drive 180° out Phase with NDE	Steady	Level drops slowly with speed	Parallel causes radial forces and angular causes axial. Load dependent
Rub- Seal/or bearing	1/4×, 1/3×, 1/2× or 10–20× can be seen Primarily 2 × Some 1 ×. Radial.	Erratic	Erratic depending upon severity	Disappears suddenly at some lower speed	Full rubs tend to be 10 to 20× higher. Bearing misalignment can give rub symptoms. Severe pounding.
Rotor	1/4×, 1/3×, 1/2×, and 1× with slip freq side bands. Radial	Erratic	High		
Looseness: Bearing (non-rotating)	2 × 3 × may be seen Radial	Steady	Fluctuates	Disappear at some lower speed	Bearing seat looseness. Looseness at bearing split
Rotor core (rotating)	1–10× with 1, 2, and 3 predominant Radial	Can exist relative to type of looseness. General core loose gives erratic symptom	Erratic, high amplitude	Drops with speed Can disappear suddenly	End plates loose. Core ID loose
Pedestals (non-rotating)	1–10× with 2 and 3 predominant Radial and axial	Steady	Fluctuates	Disappears at some lower speed	
External fans	1 and 3 × Radial and axial – OE (fan end)	N/A	Fluctuates	Drops with speed Can disappear suddenly	
Unbalance rotor	1× rotor speed Radial	NDE and DE in phase	Steady	Level drops slowly	Rotor has unbalance can be due to thermal problems

(Contd...)

Cause	Frequency of vibration	Phase angle	Amplitude response	Power cut	Comments
		Couple gives out of phase condition			
Unbalance of external fan	1× Radial high at NDE (fan end) 1× Axial with high at fan end	Couple DE 180° out of phase with EO	Steady	Level drops slowly.	
Coupling unbalance	1 × Radial and higher on drive end		Steady	Level drops slowly	Unbalance due to coupling or key
Bent shaft extension	2 × Primarily 1 × may be seen Axial	EO 180° out of phase with DE	Steady	Level drops slowly	DE runout should give higher 2 × axial at that end. Normal runout on core - 1–2 mil.
Eccentric air gap	Strong 120 Hz Radial	N/A	Steady	Immediately drops	Difference between max. and min. air gap divided by ave. should be less than 10%
Soft foot					
Eccentric rotor.	1× Primarily Some 60 and 120 Hz Radial	Unsteady	Modulates in amplitude with slip	Immediately drops	Eccentricity limit 1–2 mil. Slip beat changes with speed/load
Loose stator core	120 Hz Axial and radial	Frame and bearing brackets in phase at 120 Hz	Steady	Immediately drops	Look for relative motion of core with respect to housing
Rotor bow (thermal bow)	1× Primary Some 120 Hz may be seen May have Modulators on 1X and 2X vib.-Radial	Unsteady	Changes with temperature Time or load related Varies at Freq. slip × poles	Some drop but high level would come down with speed	Heat related Examine rotor stack for uneven stack tightness or looseness Shorted rotor iron Check bar looseness
Broken rotor bars	1× and modulates at slip × # poles May have high stator slot frequencies On slower speed motors	Dependent upon where broken bars are located	Strong beat possible Varies @ Freq. Slip × poles Amplitude increased with load	Immediately drops	Sparking in the air gap may be seen Long term variation in stator slot frequencies can be indicator of bar problems Broken bars cause holes in magnetic field

(Contd...)

Cause	Frequency of vibration	Phase angle	Amplitude response	Power cut	Comments
Loose bars	1 × Possible balance effect with thermal sensitivity. Radial Stator slot freq. plus sidebands @ ±(# Poles*Slip)	1 × vibration will be steady Stator slot freq. will modulate causing a fluctuation in phase angle on overall vibration	Steady	Stator slot freq. will immediately disappear Imbalance effect can suddenly disappear at some lower speed	Large current fluctuations Current analysis shows slip frequency side bands. Excessive looseness can cause balance problems in high speed motors
Interphase fault	60 and 120 Hz Radial	N/A	Steady and possible beat	Immediately disappears	
Ground fault	60 Hz and 120 Hz slot freq. Radial	N/A	Steady and possible beat	Immediately disappears	
Unbalanced line voltages	120 Hz Radial	N/A	Steady 120 Hz and possible beat	Immediately disappears	
Electrical noise vibration	(rpm × # of rotor slots)/60 +/-120, 240, etc. Radial	Due to modulation overall vibration will fluctuate	Steady	Immediately disappears	Increases with increasing load
System resonance	1 × rpm or other forcing frequency one plane—usually horizontal	Varies with load and speed	Varies	Disappears rapidly	Foundation may need stiffening—may involve other factors
Strain	1 × rpm		Steady		Caused by casing or foundation distortion from attached structure (piping)
Poorly shaped Journal	2× rotational usual	Erratic	Steady		May act like a rub
Oil film instability (oil whirl)	Approx (0.43–0.48) rotational	Unstable	Steady	May disappear at lower speed	
Anti-friction bearing problems	Various frequencies dependent on bearing design	Unstable	Steady		Four basic frequencies.
Resonant parts	At forcing frequency or multiples	N/A	Steady	Drops rapidly	May be adjacent parts
Top cover fit	120 Hz Radial	N/A	Steady	Disappears immediately	Magnification of 120 Hz electrical Top cover rests on basic core support

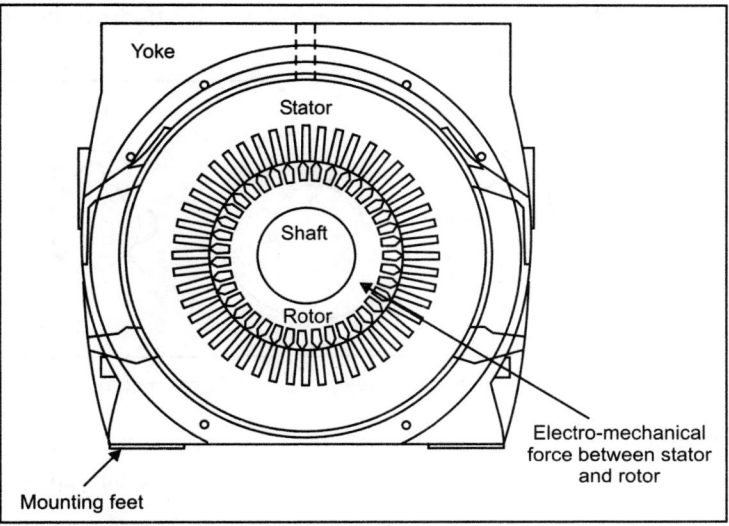

Fig. 23.1. Stator and rotor.

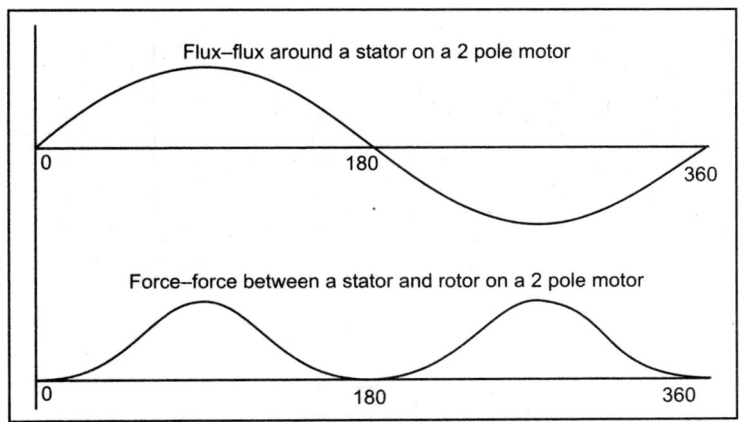

Fig. 23.2. One period flux wave and magnetic force wave.

Elliptical Stator due to Fundamental Flux

As can be seen in Fig. 23.3, for 2-pole motors the electromechanical force will attempt to deflect the stator into an elliptical shape. The primary resistance to movement is the strength of the core back iron and the stiffness of the housing around the stator core, which is restraining the core's movement. On 4 pole motors the distance between the nodes is only 45 mechanical degrees, ½ that seen on 2 pole motors, thereby making the 4 pole stator core much stiffer to movement resulting in much lower twice line frequency vibration. Calculations on a typical 1000 HP two pole motor at 60 Hz show 120 Hz vibration at the stator core OD of about 0.12 inches per second, peak, while values for a four pole motor of the same size are only about 0.02 to 0.03 inches per second, one sixth to one quarter of this value.

This twice line frequency vibration is transmitted through the motor frame to the bearing brackets where it is reduced somewhat in amplitude.

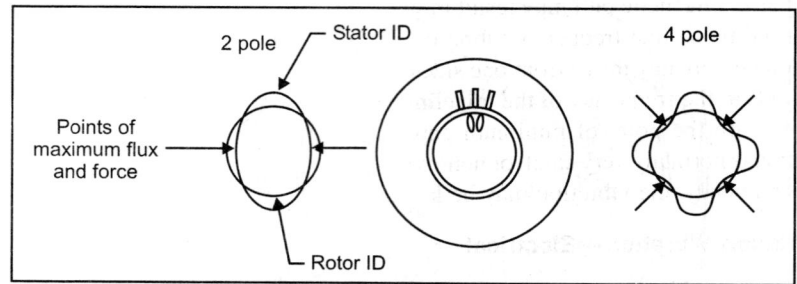

Fig. 23.3. Electromechanical force on 2 and 4 pole motors.

Nonsymmetrical Air-gap

Twice line frequency vibration levels can significantly increase when the air gap is not symmetrical between the stator and rotor as shown in Fig. 23.4.

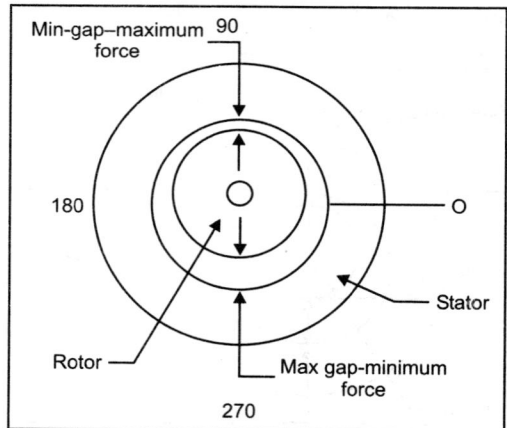

Fig. 23.4. Unsymmetrical air gap around rotor.

This particular condition will result in the force being greater in the direction of the smaller air gap. That is, an unbalanced magnetic pull will exist in the direction of the minimum air gap.

$$\text{Force} \approx B^2/d$$

where,

B = flux density.

d = distance across air gap.

Of interest here, not only is the stator pulled in one direction, but also the rotor is pulled in the opposite direction, to the side that has the minimum air gap. This causes higher shaft vibration, which is more detrimental to bearing life. Note that in Fig. 23.4 the rotor OD is concentric with the axis of rotation thereby causing the force to remain a maximum in the direction of minimum air gap.

One Times Line Frequency Vibration

Although not nearly as prevalent as twice line frequency vibration, one times line frequency vibration can exist. Unbalanced magnetic pull may result in vibration at line frequency (one times line frequency) as well as the usual twice line frequency vibration. If the rotor or stator moves from side to side, the point of minimum air gap may move from one side of the motor to the other. When the frequency of this motion corresponds to the frequency of the traveling flux wave, the unbalanced magnetic pull will shift from side to side with the point of minimum gap, resulting in vibration at line frequency. This line frequency vibration is normally very small or non-existent, but if the stator or rotor system has a resonance at, or near, line frequency, the vibration may be large.

One Times Rotation Vibration—Electrical

Eccentric rotor

An eccentric rotor, which means the rotor core OD is not concentric with the bearing journals, creates a point of minimum air gap which rotates with the rotor at one times rotational frequency. Associated with this there will be a net balanced magnetic force acting at the point of minimum air gap, since the force acting at the minimum gap is greater than the force at the maximum gap, as illustrated in Fig. 23.5. This net unbalance force will rotate at rotational frequency, with the minimum air gap, causing vibration at one time rotational frequency.

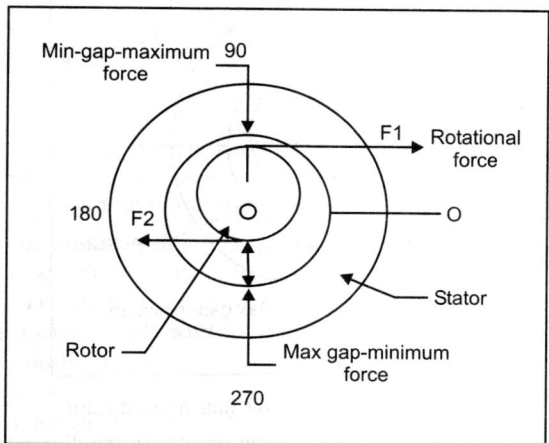

Fig. 23.5. Exagerated view of eccentric rotor.

The flux causing the magnetic force is the fundamental flux wave, which rotates around the stator at the synchronous speed of the motor. The rotor attempts to keep up with the rotating flux wave of the stator, but the rotor slips behind the stator field as needed to create the necessary torque for the load. When the high point of the rotor (point of minimum air gap) aligns with the high point (maximum) of the stator flux, the force will be a maximum, and then it will decrease, becoming small under a point of minimum flux. Thus, an unbalance force is created which rotates at rotational speed and changes in magnitude with slip. The end result is a one times rotational speed vibration, which modulates in amplitude with slip. This condition occurs at no load or full load. At no load, the frequency approaches synchronous speed and could have a modulation period of 5 to 15 minutes. At full load the frequency of modulation in CPM will equal the slip in rpm times the number of poles. The slip is equal to the synchronous speed minus the full load speed, typically 1 per cent of rated rpm. For example, a 2-pole motor with a full load speed of 3564 rpm at 60 Hz will have a slip of 3600–3564 = 36 cycles per minute (1 per cent slip) and will result in a modulation frequency of 2 × 36 = 72 cycles per minute.

Broken rotor bar

If a broken rotor bar or open braze joint exists, no current will flow in the rotor bar as shown in Fig. 23.6. As a result the field in the rotor around that particular bar will not exist. Therefore the force applied to that side of the rotor would be different from that on the other side of the rotor again creating an unbalanced magnetic force that rotates at one times rotational speed and modulates at a frequency equal to slip frequency times the number of poles.

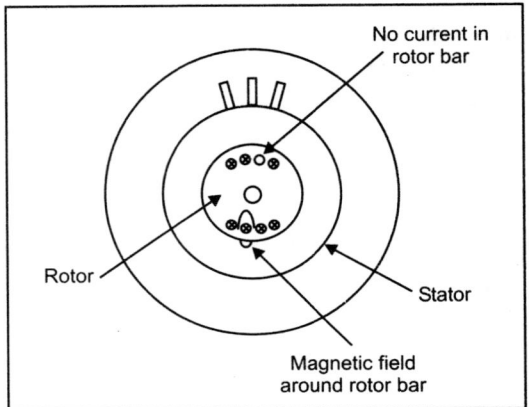

Fig. 23.6. Rotor with broken rotor bar.

If one of the rotor bars has a different resistivity a similar phenomenon (as in the case of a broken rotor bar) can exist. It should be noted that this is one of the few conditions that cannot be seen at no-load. But there is an additional phenomenon associated with this condition that can be seen at no-load after the motor is heated to full load temperature by any method that creates rotor current. These methods would include, coupled full load test, dual frequency heat run, multiple accelerations or heating by locking rotor and applying voltage. In addition, broken rotor bars or a variation in bar resistivity will cause a variation in heating around the rotor. This in turn can bow the rotor, creating an eccentric rotor, causing basic rotor unbalance and a greater unbalanced magnetic pull, thereby creating a high one times and some minimal twice line frequency vibration.

Rotor bar passing frequency vibration

High frequency, load-related magnetic vibration at or near rotor slot passing frequency is generated in the motor stator when current is induced into the rotor bars under load. The magnitude of this vibration varies with load, increasing as load increases. The electrical current in the bars creates a magnetic field around the bars that applies an attracting force to the stator teeth. These radial and tangential forces which are applied to the stator teeth, as seen in Fig. 23.7, create vibration of the stator core and teeth. This source of vibration is at a frequency which is much greater than frequencies normally measured during normal vibration tests. Due to the extremely high frequencies, even very low displacements can cause high velocities if the frequency range under test is opened up to include these frequencies. Though these levels and frequencies can be picked up on the motor frame and bearing housings, significant levels of vibration at these higher frequencies will not be seen between shaft and bearing housing where they could be damaging. For this reason vibration specification requirements normally do not require that these frequencies be included in overall vibration.

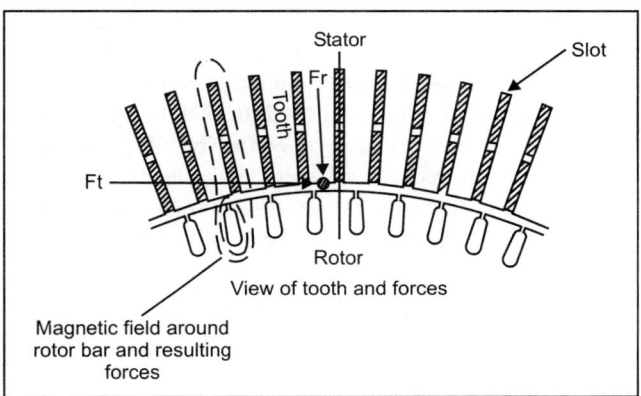

Fig. 23.7. Magnetic field around rotor bar and resulting force on stator teeth.

Since vibration at rotor bar passing frequency occurs at a high frequency, the vibration velocity level may be significant, but the effect on motor reliability is insignificant. Considering the stress that results in the motor as a consequence of the vibration makes this determination. For example, suppose a two pole motor exhibiting a vibration at 2800 Hz due to rotor bar passing frequency plus a 120 Hz side band:

Velocity, (IPS)	0.1	0.5
Displacement (mils)	0.011	0.057
Stress in stator core iron	30 psi	150 psi
Stress in stator tooth iron	50 psi	250 psi

The typical fatigue strength of the core iron is 35,000 psi. Similar low stress levels can be calculated for all parts of the motor (including the stator windings). In addition, the typical minimum oil film thickness ranges from 1.0 mils to 1.5 mils. Since only a small displacement such as 0.011 to 0.057 mils as mentioned above could be seen, this vibration will not have an adverse affect on bearing performance. The rotor slot and side band frequencies are in the frequency range normally related to noise rather than vibration performance, and are taken into account in noise predictions during motor design. In fact, these force components are the principal sources of high frequency noise in electrical machines, which has been for some time subject to noise regulations and limits. Experience has shown that where noise has been within normal or even high ranges, there has been no associated structural damage. The significance of these high frequency vibrations is distorted by taking measurements in velocity and then applying limits based on experience with lower frequency vibration.

Load Related Magnetic Force Frequencies and Mode Shapes

The frequencies of the load related magnetic forces applied to the stator teeth and core equal the passing frequency of the rotor bars plus side bands at + or − $2f$, $4f$, $6f$ and $8f$ Hz, where f is the line frequency.

A magnetic force is generated at the passing frequency of the rotor slot (FQR), which is motor speed in revolution per second times the number of rotor slots as shown in (Eq. 23.1).

$$FQR = RPM \times N_r/60, \text{ Hz} \qquad \qquad ... (23.1)$$

where,

N_r = number of rotor slots.

For the typical two pole 3570 rpm motor with 45 rotor slots in the example above, FQR = 2680 Hz. The side bands are created when the amplitude of this force is modulated at two times the frequency of the power source. On a 60 Hz system the 120 Hz modulation produces the side bands, giving excitation frequencies of FQR, FQR + 120, FQR – 120, FQR + 240, FQR – 240 Hz, etc.

The forces applied to the stator teeth are not evenly distributed to every tooth at any instant in time; they are applied with different magnitudes at different teeth, depending upon the relative rotor — and stator-tooth location. This results in force waves over the stator circumference. The mode shape of these magnetic force waves is a result of the difference between the number of rotor and stator slots as shown in (Eq. 23.2).

$$M = (N_s - N_r) +/- KP \qquad \qquad ... (23.2)$$

where,

N_s = number of stator slots.
N_r = number of rotor slots.
P = number of poles.
K = all integers 0, 1, 2, 3, etc.

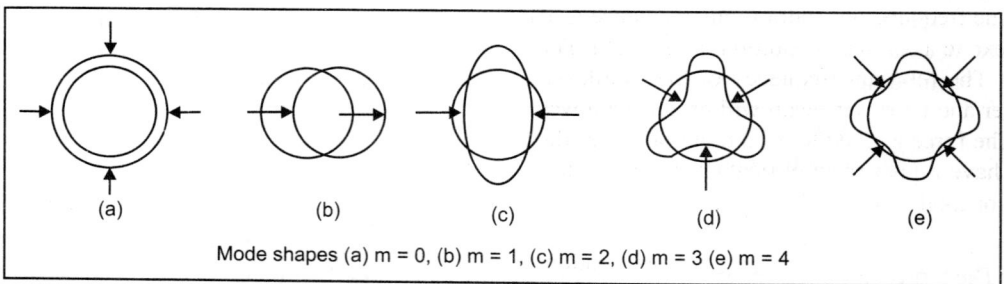

Mode shapes (a) m = 0, (b) m = 1, (c) m = 2, (d) m = 3 (e) m = 4

Mode shapes and natural frequencies of core vibration

Under the applied magnetic forces the stator core is set into vibration in the same manner that a ring of steel would respond if struck. Depending upon the modal pattern and frequencies of the exciting force, as described above, the stator would vibrate in one or more of its flexural modes m of vibration, as shown in Fig. 23.8. Each of the mode shapes has its associated natural frequency. The core may be somewhat influenced by the stator frame in actuality, but in analysis the frame is usually neglected, both due to complexity and because the effect on higher frequency modes is minimal.

To understand the resonant frequency of the core at a given mode of vibration, the core can be represented as a beam, which is simply supported on both ends and flexes between the ends due to forces applied on the beam. The length of the beam is equal to the circumferential length of the mean diameter of the stator core for one-half the mode wave length (Fig. 23.9).

$$L = \frac{\Pi D_s}{2M}$$

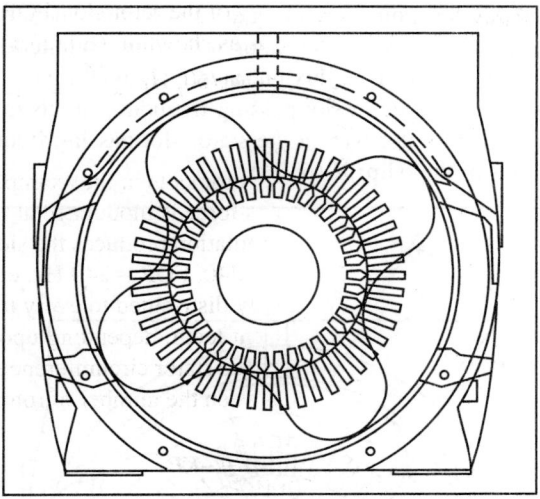

Fig. 23.8. Mode shapes.

If the resonant frequency of the core is close to the forcing frequency, a high level of vibration will result. The lower modes of vibration may produce resonant frequencies that are close to the primary forcing frequencies.

The frequency of stator tooth resonance is also a concern. The tangential forces applied to the teeth can excite a resonant condition in the tooth. The tooth is a cantilever beam supported at the root by the core. The resonant frequency of the cantilever beam is a function of the beam length and width. A longer and narrower beam will produce a lower resonant frequency (Fig. 23.10).

The force applied to each tooth produces displacement of the tooth and the core. The displacement will have a greater amplification the closer the forcing frequency is to the resonant frequency of the core or tooth (Eq. 23.3):

$$\text{Amplification factor} = \frac{1}{1 - \left(f/f_0 \right)^2} \qquad \text{... (23.3)}$$

where, f is the line frequency and f_0 is the natural frequency.

This vibration is sometimes incorrectly associated with loose rotor bars, but there are reasons why loose rotor bars won't create rotor slot passing frequency vibration.

First, on most larger motors the centrifugal forces are so great that the only time there could possibly be rotor bar movement is while the rotor is accelerating. This in itself could be a serious problem since the number one cause of rotor bar to end connector failure is rotor bar movement as a result of multiple restarts of a high inertia load. But, the only increase in vibration at speed due to loose rotor bars would be due to a shift in the rotor cage causing a one times rotational mechanical unbalance.

Secondly, looking at any one rotor bar, the bar itself is never subject to a force at the rotor slot passing. The bars are rotating at rotational speed. There is an alternating field in the rotor, which has a frequency close to 0 cycles per minute at, no-load, then increases to a frequency equal to the slip frequency times the number of poles at full load. On a 2 pole motor typically 2 (poles) times 36 rpm (typical slip) or 72 cycles per minute. To make this easier to understand consider one point or bar on the rotor of a 2 pole motor, and that this point is rotating at 3564 Rpms. There is a field around this bar at a very low frequency. It is applying a force to the stator at varying magnitudes depending on the level of flux in the rotor at that instant in time. This flux pulsates each time it passes by a stator slot. Note that the force that the rotor sees is at the stator slot passing frequency and is modulating at twice the slip. This will produce vibration of the rotor bars at the stator slot passing frequency plus and minus side band frequencies in multiples of the (slip) X (poles).

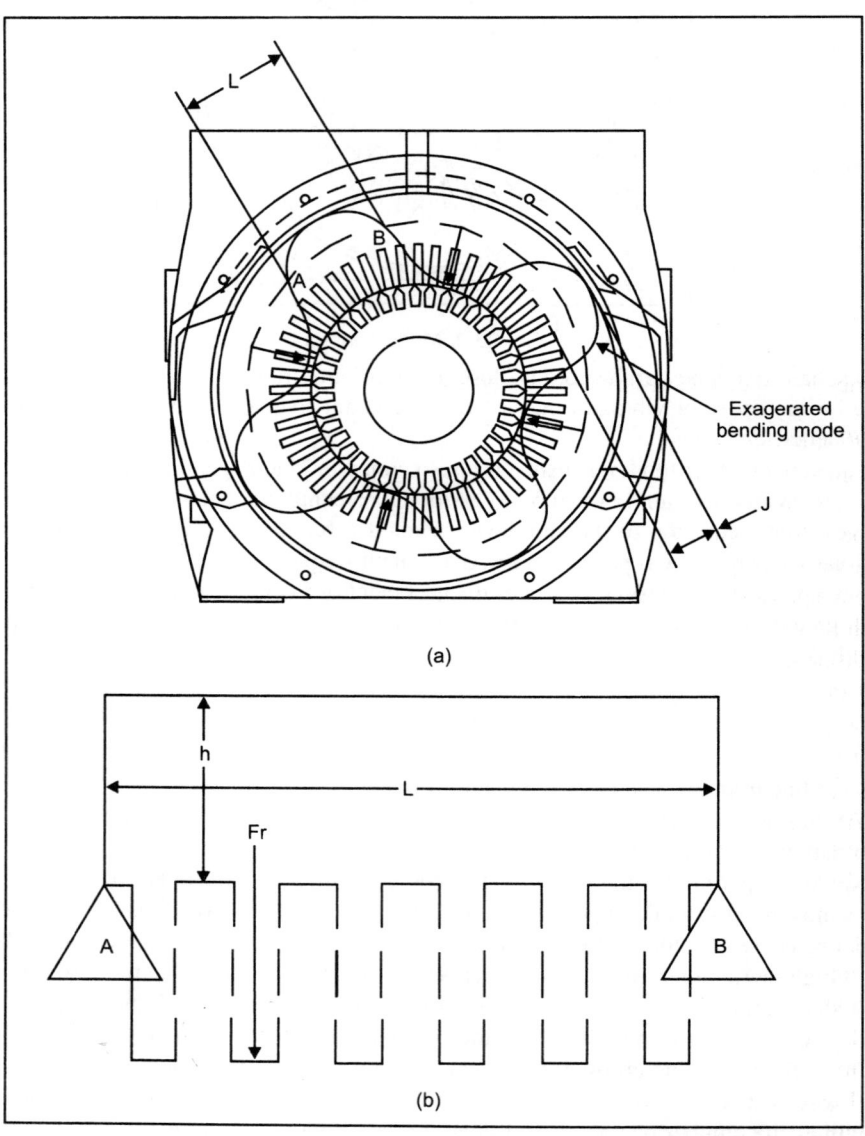

Fig. 23.9. (a) Fourth mode of vibration, (b) Linear representation of core for one-half wavelength of force.

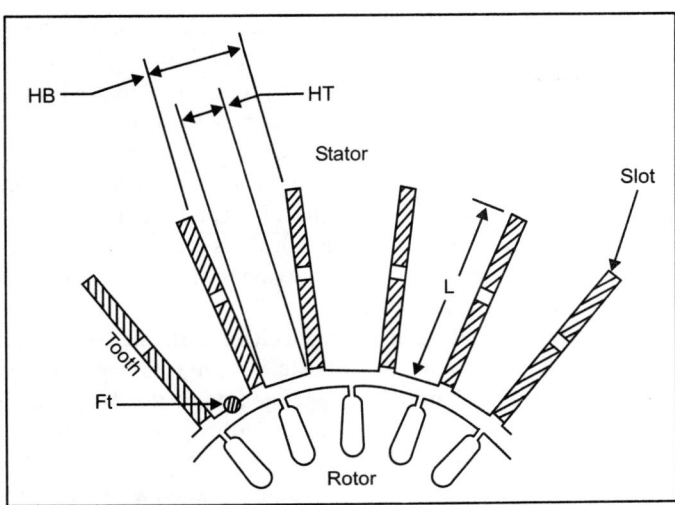

Fig. 23.10. Stator tooth forces.

One Times Rotation Vibration—Unbalance

Motor unbalance

Balancing is required on all types of rotating machinery, including motors, to obtain a smooth running machine. This is performed in the factory in a balance machine at a level of precision determined by the motor speed, size, and vibration requirements. The highest precision is required for two pole motors. Two pole and large four pole motors should be balanced at their operating speed in the balance machine. The assembled motors are then run in test to confirm that vibration requirements are met in operation. Although they do not usually concern the user directly, a few salient factors affecting factory balance will be discussed here. These mainly apply to two pole motors.

Most medium to large motors are used for constant speed applications, although there has been a recent increase in the number and size used for variable speed applications on adjustable speed drives. Constant speed motors need only be precision balanced at one speed, their operating speed. Variable speed applications require that good rotor balance be maintained throughout the operating speed range, which typically may be from 40 per cent to 100 per cent of synchronous speed. Rotor balance involves the entire rotor structure which is made up of a multitude of parts, including the shaft, rotor laminations, end heads, rotor bars, end connectors, retaining rings (where required) and fans. These many items must be controlled in design and manufacture to achieve stable precision balance.

Fundamental requirements for precision balance on any machine are:

1. Parts must be precision manufactured for close concentricities and minimal unbalance individually.
2. Looseness of parts, which can result in shifting during operation, causing a change in balance, must be avoided or minimised.
3. Balance correction weights should be added at or near the points of unbalance.

For motors, rotor punchings must be precision manufactured with close concentricities of all features and have a shrink fit on the shaft that is maintained at all operating speeds and temperatures. The punchings must be stacked square with the bore, uniformly pressed, and clamped in position when shrunk on the shaft to prevent movement with speed change. When end connectors require retaining rings, the rings are of high strength material designed with proper interference fit. Rotor bars are shimmed and/or swaged so they are tight in the slots. There are other methods to assure tight rotor bars, such as heating the core and chilling the bars, but these methods are not common. End connectors should be induction brazed symmetrically to the bars, which helps eliminate variations in balance due to thermal change. The shaft and assembled rotor are precision machined and ground to concentricities well within 0.001 inch. The rotor is prebalanced without fans, then the fans are assembled and final balanced on the rotor. The fans are individually balanced before assembly on the rotor. For motors with a heavy external fan, two plane balance of the fan may be required.

Constant speed applications are usually satisfied with either a stiff shaft design, for smaller machines, or a flexible shaft design for larger motors. A 'stiff shaft' design is one that operates below its first lateral critical speed, while a 'flexible shaft' design operates above the first lateral critical speed. When the rotor is precision designed and manufactured as described above, a two plane balance making weight corrections at the rotor ends, will usually suffice even for flexible rotors. Occasionally, however, a flexible rotor may require a three plane balance to limit vibration as the machine passes through its critical speed during runup or coastdown. This is accomplished by also making weight corrections at the rotor center plane as well as at the two ends.

Adjustable speed applications require a stiff shaft to prevent major balance changes with speed due to shaft deflection, such as may occur with a flexible shaft. In addition, however, the many other factors affecting balance in this complex structure, discussed above, must also be controlled to maintain good balance at varying speeds. In particular, any bar looseness will result in excessive change in balance with speed. This is prevented by rotor bar shimming and sometimes swaging as noted above. Shims around bars, such as used here allow the bars to be driven tightly into the slots without concern for having the laminations shear pieces of the bar off, causing bars to be loose. This design also prevents the bars from becoming loose over time in the field due to a similar phenomenon, which may occur during heating and cooling where the bars may not expand and contract at the same rate as the core.

During balancing and no load testing in the shop, the shaft extension keyway is completely filled with a crowned and contoured half key held in place by a machined sleeve to avoid any unbalance from this source. Load testing is carried out with the motor mounted on a massive, rigid base, accurately aligned to a dynamometer and coupled to the dyne with a precision balanced coupling and proper key.

Thermal unbalance

Thermal unbalance is a special form of unbalance. It is caused by uneven rotor heating, or uneven bending due to rotor heating. The proper solution is to determine the reason for uneven heating affecting shaft straightness, and fix the rotor. Before such major rework is performed, the severity of the thermal situation needs to be ascertained. All rotors will have some change in vibration in transitioning from a cold state to a hot one. API 541, 3rd edition allows 0.6 mils change in shaft vibration (at rotational frequency, 1X), and, 0.05 inches per second change in housing vibration. However, if the application is one of continuous duty, and, vibration levels are not excessive during startup (i.e. motor cold), it is permissible to allow more change cold to hot without any damage to the motor. In these situations if the lowest vibration levels are desired at operating conditions, a hot trim balancing procedure can be performed. To perform this procedure, run the motor until all conditions thermally stabilise, and quickly perform a trim balance. If necessary, run the motor again after the initial trial weights have been installed and let the motor thermally stabilise before taking additional vibration measurements for final weight correction.

Coupling unbalance

The coupling unbalance limit given in API 671 of 40W/N, when applied to a typical 1000 HP 3600 rpm 2 pole motor for example, gives a value equal to about one-third of the motor unbalance limit for one end. Analysis shows this would be about the correct value to have minimal effect on motor vibration. Comparing this to AGMA coupling unbalance limits commonly used in the industry, it is comparable to a Class 11 balance which requires a balanced coupling. It is considerably better than a Class 9 balance (by a factor of 3) which is not a balanced coupling. AGMA Class 9 balance couplings are sometimes used for 2 pole motors, but do not meet API 671 and can give vibration problems with precision motors. Use of a proper key and a balanced coupling leaves the machine alignment and mounting and the driven equipment balance as the remaining major factor in system vibration.

Oversize coupling

One consideration in coupling selection is coupling size. The coupling should be large enough to handle the application, including the required service factor, but should not be exceptionally large. Potential results of oversize couplings are:

1. Increased motor vibration due to increased coupling unbalance and/or a change in the critical speed or rotor response due to increased weight. This is particularly true for flexible shaft machines.
2. A greatly oversize coupling can result in greatly severe shaft bending, excessive vibration, and, heavy rubbing of seals, ultimately resulting in catastrophic shaft failure.

The predominant vibration frequency as a consequence of an oversized coupling would be at one times rotation, just like an unbalance condition. The concept of 'bigger is better' does not hold true here!

Driven machine unbalance

Under normal circumstances, the unbalance of the driven machine should not significantly affect the motor vibration. However, if the unbalance is severe, or if a rigid coupling is being used, then the unbalance of the driven machine may be transmitted to the motor.

Maintaining balance in the field

When a finely balanced high speed motor is installed in the field, its balance must be maintained when the motor is mated to the remainder of the system. In addition to using a balanced coupling, the proper key must be used.

One way to achieve a proper key is to have the shaft keyway completely filled, with a full key through the hub of the coupling and the entire key outside the coupling crowned to match the shaft diameter. A second approach is to use a rectangular key of just the right length so that the part extending beyond the coupling hub toward the motor just replaced the unbalance of the extended open keyway. This length can be calculated if the coupling hub length and keyway dimensions are known.

An improper key can result in a significant system unbalance, which can cause the vibration to be above acceptable limits. For example, calculations for a typical 1000 HP, 2 pole 3600 rpm motor show that an error in key length of 0.125 inches will give an unbalance of 0.7 oz.-in. This is about equal to the residual unbalance limit for each end of the rotor of 4W/N given in API 541 for motors, and exceeds by a factor of three the residual unbalance tolerance of a typical one-half coupling of 40W/N given in API 671 for couplings.

A problem occasionally arises in the field when a flexible shaft machine with a high speed balance is sent to a service shop for repair. If the rotor is rebalanced in a slow speed balance machine at the service shop, then this usually results in unbalance at operating speed, and the machine will run rough when tested or reinstalled. The solution, of course, is to not rebalance unless absolutely required by the nature of the repair. If rebalance is absolutely required, than it should be done at the operating speed of the rotor, otherwise, a trim balance may need to be performed after the motor is reassembled.

Forcing Frequency Response Vibration

Weak motor base

If the motor is sitting on a fabricated steel base, such as a slide base, then the possibility exists that the vibration which is measured at the motor is greatly influenced by a base which itself is vibrating. Ideally the base should be stiff enough to meet the 'Massive foundation' criteria defined by API 541. Essentially, this requires that support vibration near the motor feet to be less than 30 per cent of the vibration measured at the motor bearing. To test for a weak base, measure and plot horizontal vibration at ground level, at bottom, middle, and top of the base, and at the motor bearing. Plotted, this information would look like as shown in Fig. 23.11, for a motor sitting on a weak base. In this particular example, had the motor been on a rigid base, the vibration at the bearing would have been closer to 0.25 mils rather than the measured 2.50 mils.

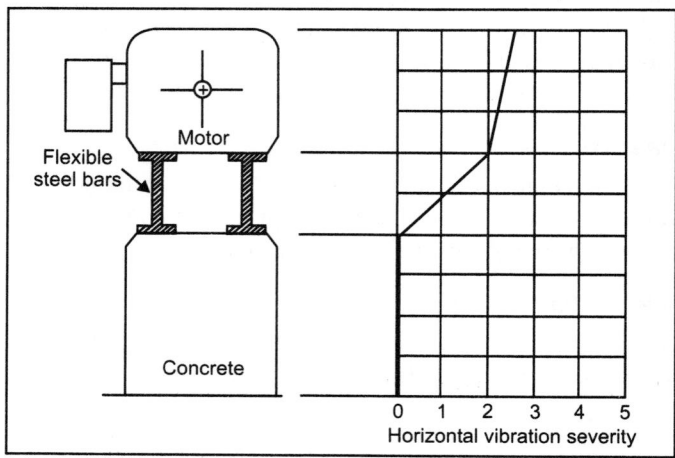

Fig. 23.11. Plot of Vibration (in mils) Vs. base/motor position.

A weak motor base usually results in high 1x vibration, usually in the horizontal direction as shown in Fig. 23.11. However, it may also result in high 2x (twice rotational frequency) or 2f (twice line frequency) vibration, which also is a common vibration frequency in motors. To determine the nature and source of this high 2x vibration requires vibration measurements be made at the motor feet in both the vertical and horizontal direction, taking phase as well as amplitude to determine a mode shape. The 'rocking mode' of the motor observed in a particular case is illustrated in Fig. 23.12. The horizontal component δ_{HV} due to the rocking adds to the inherent δ_{HM} of the motor alone to give a high total at the bearing housing, as shown by the equivalency below.

$$\delta_H = \delta_{HM} + \delta_{HV}$$

where,

δ_H = Actual motor horizontal vibration measured in the field.

δ_{HM} = Horizontal vibration of motor alone measured on a massive base in shop.

$\delta_{HV} = \dfrac{D}{E} VB$, calculated horizontal vibration component due to δ_{VB} , measured vertical vibration at each motor foot in the field.

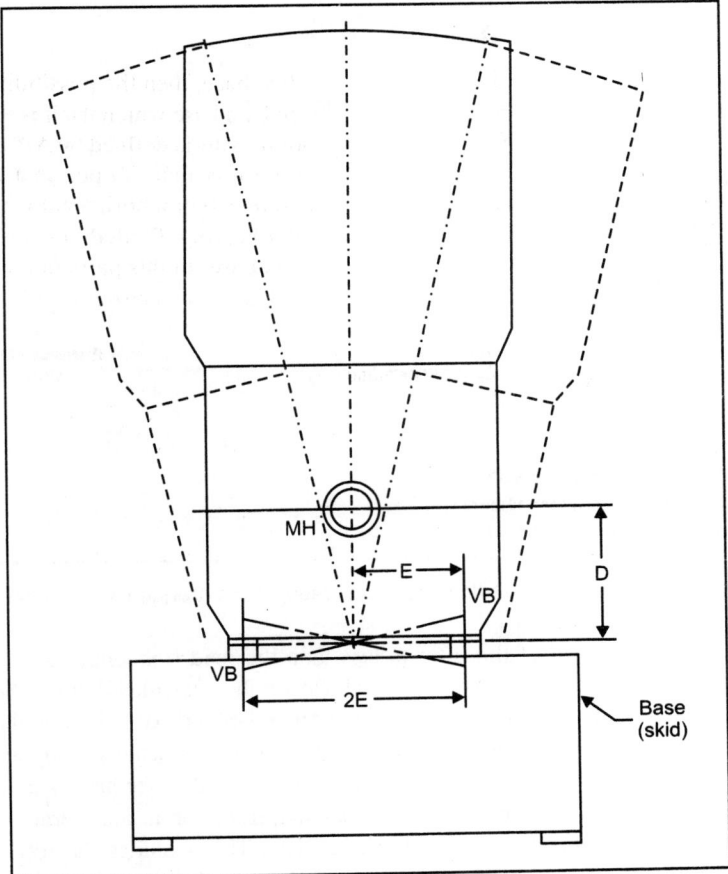

Fig. 23.12. Rocking mode due to weak base.

The recommended repair for the weak motor base illustrated is that the support posts be tied together and heavily stiffened with the intent to meet the criteria for a 'massive foundation'. Even where resonance of the base is not a factor, heavy stiffening of a light support structure can greatly reduce vibration.

Reed critical base issues

A vertical motor's reed critical frequency is a function of its mass, distribution of mass, and base geometry. The reed critical should not be confused with the motor rotor's lateral critical speed. However, in large vertical motors, the rotor lateral critical speed may be a determining factor in the reed critical frequency, particularly of the motor alone. The effect of the rotor may be determined by considering it as a separate mass and including rotor shaft flexibility in the reed frequency calculation. That is, consider the motor as a two mass, two degrees of freedom system as shown in Fig. 23.13, rather than a single degree of freedom system as described in NEMA MG 1–20.55. Figure 23.13 shows that the motor structure (a) is basically a two mass system which can be progressively simplified, first to a beam-mass structural schematic (b), then to an equivalent two mass, two spring system (c).

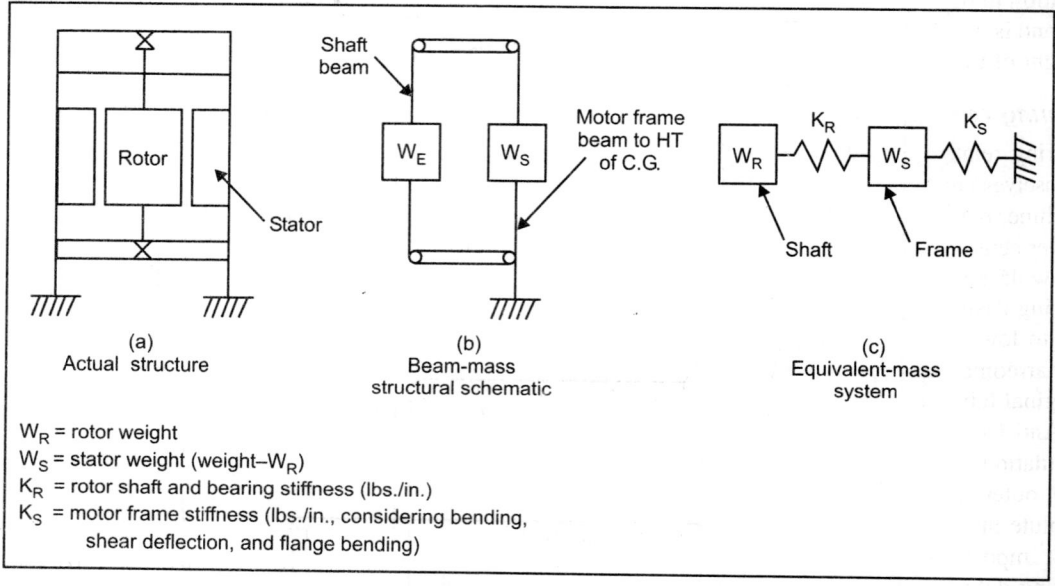

W_R = rotor weight
W_S = stator weight (weight–W_R)
K_R = rotor shaft and bearing stiffness (lbs./in.)
K_S = motor frame stiffness (lbs./in., considering bending, shear deflection, and flange bending)

Fig. 23.13. Structural representation of vertical motor for reed critical frequency calculation including rotor shaft flexibility.

Where the lateral critical speed of the rotor is less than the reed frequency calculated as a single degree of freedom system, the true reed frequency will be lower than calculated. It will be approximately equal to the rotor lateral critical speed. However, when mounted on a flexible base in the field, the rotor shaft effect will be less and a single degree of freedom calculation is usually adequate. Just as in the case of a lateral critical, if the motor's operating speed (or any other frequency at which a forcing function is present) coincides with the reed critical, great amplification in the vibration amplitude will occur. Motor manufacturers routinely issue reed critical data. This includes the reed critical that the motor alone would have if it were mounted on a rigid, seismic mass. In addition the motor manufacturer supplies the following information to aid in determining the system resonant frequency with the motor mounted on the user's base. Machine weight, center of gravity location, and static deflection. Bases found in typical installations are not as stiff, and correspondingly, the reed critical frequency will be lowered. If the reed critical drops into a frequency at which there is a forcing function present (most commonly the operational speed), the reed critical frequency will have to be changed. Usually, this is not difficult to do, and is most commonly accomplished by either changing the stiffness of the base, or by changing the weight of the base/motor. Where the reed critical drops below the operational speed to about 40 per cent to 50 per cent of running speed, this can result in subharmonic vibration at the system resonant speed in motors with sleeve guide bearings. This could be due to either oil whip effects or inadequate guide bearing oil film.

Resonant base

If the motor's operating speed (or any other frequency at which a forcing function is present) coincides with the base resonant frequency, great amplification in the vibration amplitude will occur. The only solution to this problem is to change the resonant frequency of the base. Usually, this is not difficult to do, and is most commonly accomplished by either changing the stiffness of the base, or by changing the weight of the base/motor.

Bearing related vibration

Bearing related vibrations are common to all types of rotating equipment, including motors, and in themselves encompass extensive fields of technology. They will be dealt with briefly here. Sleeve bearing machines may occasionally experience 'oil whirl' vibration, which occurs at a frequency of approximately 45 per cent of running speed. This may be quite large, particularly if there is a critical speed at or just below 45 per cent of running speed, which is referred to as an 'oil whip' condition. Other than basic bearing design considerations which will not be dealt with here, a common cause is high oil viscosity due to low oil temperature in flood lubricated motors operating in cold ambient conditions. Similar subharmonic vibration, but low in amplitude, may occur in ring lubricated bearings, probably due to marginal lubrication. Other causes of vibration are journal out of roundness or bearing misalignment.

Anti-friction bearings have four identifiable rotational defect frequencies for which formulas for calculation or tabulations of values are given in the literature. These defect frequencies are for the inner race, outer race, ball (or roller) spin, and cage fundamental train. Much research has proven that no absolute answer can be given to allowable amplitudes at bearing defect frequencies. Therefore, the most important thing to look for indicating significant bearing wear is the presence of a number of bearing defect frequency harmonics, particularly if they are surrounded by sidebands independent of amplitude. Tracking of vibration should be carried out starting at installation, observing these indicators to predict remaining bearing life.

IDENTIFICATION OF CAUSE OF VIBRATION PROBLEM

Now that the causes of vibration are understood it is time to establish a systematic approach to solve any problem that may arise.

Vibration Data Gathering/Analysis

Many of the details of rotor dynamics, vibration data gathering, and analysis have not been presented in detail in this chapter. Now one must keep in mind that all of the electrical sources of vibration and the mechanical sources of vibration are not necessarily at the same phase angle or exactly the same frequency. To make matters worse, the electrical vibration may modulate, and when superimposed on the mechanically induced vibration may result in an overall vibration signature that is unsteady in amplitude and phase. Through proper data collection, testing, and analysis, it is possible to identify the root cause of the vibration.

Vibration Units

Vibration can be measured in units of displacement (peak to peak, mils), units of velocity (zero to peak, inches per second), or units of acceleration (zero to peak, g's). Acceleration emphasises high frequencies, displacement emphasises low frequencies, and velocity gives equal emphasis to all frequencies. This relationship is better illustrated in Fig. 23.14. In this figure the vibration level is constant at 0.08 inches per second throughout the entire frequency range, with corresponding vibration levels shown in acceleration (in g's) and displacement (in mils). It is possible to convert from one unit of measurement to another at discrete frequencies of the vibration. To do so on an overall vibration measurement, complete knowledge of the entire spectral data is required (i.e. amplitude for each frequency band, for all the lines of resolution).

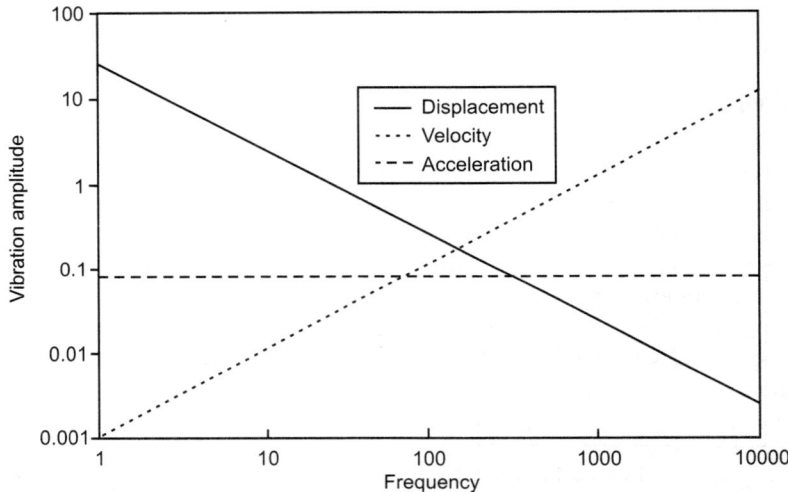

Fig. 23.14. Comparison of vibration amplitudes expressed in acceleration, velocity, and displacement.

Today, the most common units are displacement for shaft vibration measurement, and velocity for housing vibration measurement. The use of these units is further reflected in most current standards such as API and NEMA.

Direction of Measurement

Measurements should be made in three planes (vertical, horizontal, and axial) on both bearing housings, as shown in Fig. 23.15.

Shaft Vibration vs. Housing Vibration

The determination of obtaining shaft vibration data vs. housing vibration data is dependent upon the type of problem being experienced. Oftentimes it is advantageous to have both shaft and housing vibration data. If the problem originates in the rotor (unbalance or oil whirl for instance), then shaft vibration data is preferable. If the problem originates in the housings or motor frame (twice line frequency vibration for instance), then housing vibration data is preferable. Housing vibration is generally obtained with magnetically mounted accelerometers.

Shaft vibration can be obtained one of two ways: shaft stick or proximity probe. There is an important distinction between the two methods of obtaining shaft vibration data: the proximity probe will give vibration information of the shaft relative to the housings, whereas measurements obtained with a shaft stick yield vibration information with an absolute (i.e. inertial) reference. Housing vibration data is always obtained in terms of an absolute reference. If the motor has proximity probes then they should be used. If it does not, then proximity probes may be carefully set up with magnetic mounts. In this case it is important to have the tip of the proximity probe on a ground, uninterrupted surface. Even with this precaution taken, the electrical runout will be higher than in a motor specifically manufactured for use with proximity probes.

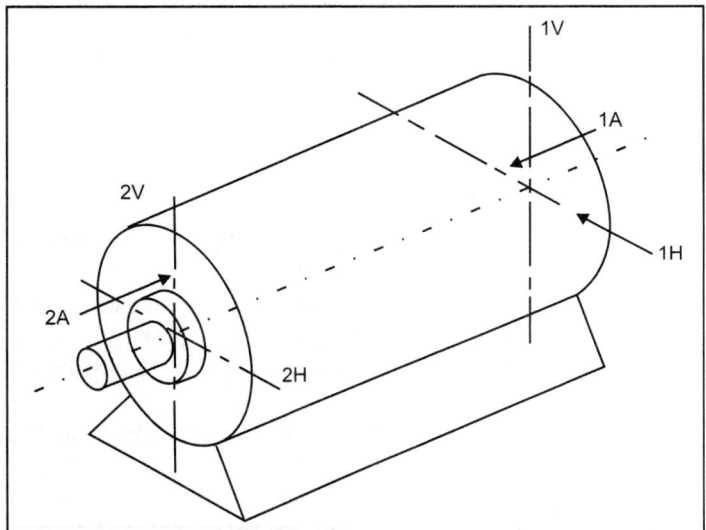

Fig. 23.15. Vibration measurement positions.

Modulation vs. Snapshot

A snapshot refers to obtaining spectral vibration data at an instant in time. Details of amplitude vs. frequency is readily available in this format. A modulation refers to collecting vibration data for a period of time (typically ten or fifteen minutes), so that the variation in vibration as a function of time can be analysed. Typically, the following frequencies are tracked when taking a modulation: 1/2X, 1X, 2X, and $1f$, $2f$, and overall vibration levels (i.e. unfiltered), where X corresponds to rotational frequency and f, line fequency. Additionally, the phase information should be tracked when taking the modulation, especially for the one times rotational frequency. This will make the identification and subsequent correction of various vibration problems possible.

It is sometimes desired to separate twice line frequency and twice rotational frequency vibration. Different methods are required to do this at no load and full load. Under full load the difference in frequency is large enough so that the separate components can each be measured directly with most vibration analysers.

However, at no load, the frequencies are so close together that this cannot be done, even using the zoom mode on a high resolution analyser, so that an indirect method is required. This can be accomplished by measuring the 2 × rpm value at reduced voltage (25 per cent) where the 2 × line component is negligible, and then subtracting this from the peak 2 × component in the modulation test which is the sum of 2 × line and 2 × rpm components. This is usually only possible at a motor manufacturer's facility or at a motor service shop.

Troubleshooting Procedure

If a vibration problem occurs there are various tests that should be performed. But first, the following maintenance items should be checked.

1. Maintenance items.
2. Check for loose bolts—mounting or other loose parts.
3. Keep motor clear of dirt or debris.
4. Check for proper cooling and inlet temperatures or obstructions such as rags, lint or other enclosures.
5. Check bearing and stator temperatures.
6. Lubricate as recommended.
7. Check proper oil levels.
8. Check vibration periodically and record.

The affected frequencies and other vibration characteristics are listed in Table 23.1.

1. Are all bolts tight? Has soft foot been eliminated?
2. Is hot alignment good? If it's not possible to verify hot alignment, has cold alignment been verified (with appropriate thermal compensation for cold to hot)?
3. Is any part, box top cover, piping vibrating excessively (i.e. are any parts attached to motor in resonance)?
4. Is the foundation or frame the motor is mounted to vibrating more than 25 per cent of motor vibration (i.e. is the motor base weak or resonant).
5. Is there any looseness of any parts on motor or shaft?
6. Integrity of fans and couplings—have any fan blades eroded/broken off, are any coupling bolts loose/missing, is coupling lubrication satisfactory?

If all of the above items check out satisfactorily, and vibration remains high, a thorough vibration analysis shall be required.

Essentially, there are only two steps in diagnosing a problem:

1. Obtain vibration data—not always clear cut because of noise, sidebands, combination of signals, modulation, etc.
2. Determine what conditions increase, decrease, or have no effect on vibration through different test conditions to help isolate root cause.

Ideally, vibration measurements should be obtained with the motor operating under the following conditions:

1. Loaded, coupled, full voltage, all conditions stabilised (i.e. normal operating conditions):
   (a) First measurement to be obtained.
   (b) Represents state of machine in actual operation.
   (c) May indicate which test should be taken next.
2. Unloaded, coupled, full voltage:
   (a) Removes load related vibration, while everything else remains the same.
   (b) Not always possible to get to zero load, but some reduced load is usually possible.
3. Unloaded, uncoupled, full voltage:

(a) Removes all effects of coupling and driven machine.

(b) Isolates motor/base system.

4. Unloaded, uncoupled, reduced voltage (25 per cent if possible):

(a) Effect of magnetic pullover forces minimised (most effective use is in comparison to vibration at full voltage.

(b) 25 per cent usually only possible at motor service shop or motor manufacturers facility. If motor is a Y–Δ connected motor, then Y connection is effectively 57 per cent voltage as compared to Δ connection at the same terminal voltage. A comparison of vibration under both connections will reveal voltage sensitivity of motor.

5. Unloaded, uncoupled, coast down:

(a) Will make any resonance/critical speed problem apparent for entire motor/base/driven equipment system.

(b) Observation of vibration change when the motor power is cut will give information similar to reduced voltage operation as illustrated by Fig. 23.16.

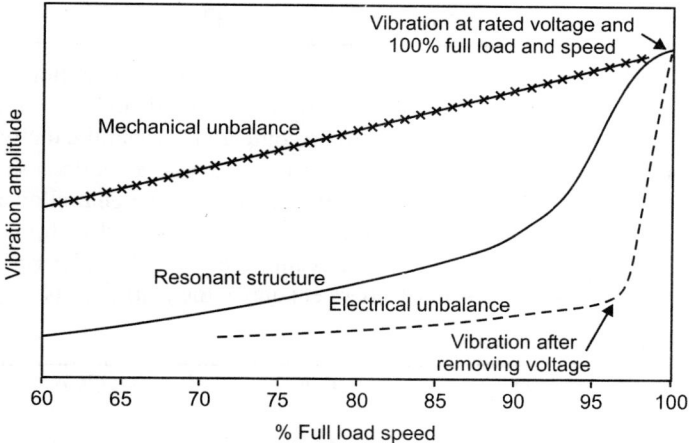

Fig. 23.16. Vibration after power is removed.

Vibration Limits

Many publications of 'vibration limits' exist. Table 23.2(a) and 23.2(b) lists various industry vibration limits. Both current revisions, as well as older revisions of these standards are listed, as these older revisions are commonly referenced. Furthermore, these motor vibration limits are applicable to a motor mounted on a seismic mass, and either uncoupled, or coupled to a piece of equipment in such a way that any vibration influence from the driven equipment is totally eliminated.

As a motor ages, the vibration levels may slowly increase. There may be a multitude of reasons of why the levels may increase over time:

1. Degradation of the bearings (sleeve bearings).

2. Loosening of rotor bars

3. Accumulation of debree in the oil guards, between rotor and stator, etc.

4. Changes in mounting conditions: deterioration of grouted base, changes in alignment/soft foot, etc.

5. Loosening of things mounted to the motor.

Table 23.2(a). Industry housing vibration limits.

	NEMA-1993	NEMA-old			API 541 3rd Ed	API 541 2nd Ed.			IEEE 841
	2, 4, 6 pole	2 pole	4 pole	6 pole	2, 4, 6 pole	2 pole	4 pole	6 pole	2, 4, 6 pole
Unfiltered (overall)	0.12 IPS	1 mil	2 mils	2.5 mils	0.1 IPS	0.8 mils	1.5 mils	1.5 mils	0.08 IPS
Filtered-1X	0.12 IPS				0.1 IPS	0.5 mils	1 mil	1 mil	
Filtered-2X	0.12 IPS				0.1 IPS				0.05 IPS
Filtered-2f					0.1 IPS				0.05 IPS

Table 23.2(b). Industry shaft vibration limits.

	NEMA–1993			API 541 3rd Ed.	API 541 2nd Ed.		
	2 pole	4 pole	6 pole	2, 4, 6 pole	2 pole	4 Pole	6 pole
Unfiltered (overall)	1 mil	2.0 mils	2.5 mils	1.5 mils	2.0 mils	2.5 mils	3.0 mils
Filtered-1X				1.2 mils	1.5 mils	2.0 mils	2.5 mils
Filtered-2X				0.5 mil	1.0 mils	1.5 mils	1.7 mils
Filtered-2f				0.5 mil			

Obviously, if conditions are identified which increase the motor's vibration level, they should be corrected. If for whatever reason it is not feasible to rectify the identified condition or identify the offending condition, the level of vibration needs to be compared to what the motor can safely tolerate. The appropriate vibration limits for a particular application are dependent upon several factors such as motor speed, size, design, and lastly, criticality of the process. In the end, allowable motor vibration limits depend greatly upon what the user is willing to tolerate, tempered with knowledge of what the motor can safely tolerate. In the absence of any other information, Table 23.3 can serve as a guide for alarm limits. Trip limits can be safely set at 10 per cent above the alarm limits.

Table 23.3. Motor vibration alarm limits.

Speed	3600	1800	1200	900
Housing IPS	0.2	0.2	0.2	0.2
Shaft mils	3.0	3.4	3.9	4.5

The factor limiting the vibration limits at these levels is the motor bearings. Generally, sleeve bearings (as compared to anti-friction bearing motors) are more restrictive in terms of vibration limits. Sleeve bearing motors can operated continually at one-half their diametrical bearing clearance, without any damage. They can operate at slightly higher levels for short periods of time as well, but these higher limits must be established with the motor manufacturers. If the motor is sitting on a weak base, higher housing vibration limits and shaft vibration limits (if measured by shaft stick, and not by a proximity probe) can be tolerated. Effectively, the vibration measured at the motor feet can be subtracted from the vibration measured at the bearing. Refer to Fig. 23.11, and the section on forcing frequency response vibration for further explanation.

Thus, vibration problems can vary from a mere nuisance to an indication of imminent motor failure. With solid knowledge of motor fundamentals and vibration analysis, it is possible to identify the root cause of the problem, and more significantly correct, or ascertain the impact of increased vibration on motor reliability and longevity.

Chapter 24

Electronic Noise

INTRODUCTION

Electronic noise is a random fluctuation in an electrical signal, a characteristic of all electronic circuits. Noise generated by electronic devices varies greatly, as it can be produced by several different effects. Thermal and shot noise are unavoidable and due to the laws of nature, rather than to the device exhibiting them, while other types depend mostly on manufacturing quality and semiconductor defects.

In communication systems, the noise is an error or undesired random disturbance of a useful information signal, introduced before or after the detector and decoder. The noise is a summation of unwanted or disturbing energy from natural and sometimes man-made sources. Noise is, however, typically distinguished from interference, (e.g. cross-talk, deliberate jamming or other unwanted electromagnetic interference from specific transmitters), for example in the signal-to-noise ratio (SNR), signal-to-interference ratio (SIR) and signal-to-noise plus interference ratio (SNIR) measures. Noise is also typically distinguished from distortion, which is an unwanted alteration of the signal waveform, for example in the signal-to-noise and distortion ratio (SINAD). In a carrier-modulated passband analog communication system, a certain carrier-to-noise ratio (CNR) at the radio receiver input would result in a certain signal-to-noise ratio in the detected message signal. In a digital communications system, a certain E_b/N_0 (normalised signal-to-noise ratio) would result in a certain bit error rate (BER).

While noise is generally unwanted, it can serve a useful purpose in some applications, such as random number generation or dithering.

TYPES OF ELECTRONIC NOISE

Thermal Noise

Johnson–Nyquist noise (sometimes thermal, Johnson or Nyquist noise) is unavoidable, and generated by the random thermal motion of charge carriers (usually electrons), inside an electrical conductor, which happens regardless of any applied voltage. Thermal noise is approximately white, meaning that its power spectral density is nearly equal throughout the frequency spectrum. The amplitude of the signal has very nearly a Gaussian probability density function. A communication system affected by thermal noise is often modelled as an additive white Gaussian noise (AWGN) channel.

The root mean square (RMS) voltage due to thermal noise v_n, generated in a resistance R (ohms) over bandwidth Δf (Hertz), is given by:

$$v_n = \sqrt{4k_B T R \Delta f}$$

where, k_B is Boltzmann's constant (joules per kelvin) and T is the resistor's absolute temperature (kelvin).

As the amount of thermal noise generated depends upon the temperature of the circuit, very sensitive circuits such as preamplifiers in radio telescopes are sometimes cooled in liquid nitrogen to reduce the noise level.

Shot Noise

Shot noise in electronic devices consists of unavoidable random statistical fluctuations of the electric current in an electrical conductor. Random fluctuations are inherent when current flows, as the current is a flow of discrete charges (electrons).

Flicker Noise

Flicker noise, also known as $1/f$ noise, is a signal or process with a frequency spectrum that falls off steadily into the higher frequencies, with a pink spectrum. It occurs in almost all electronic devices, and results from a variety of effects, though always related to a direct current.

Burst Noise

Burst noise consists of sudden step-like transitions between two or more levels (non-Gaussian), as high as several hundred millivolts, at random and unpredictable times. Each shift in offset voltage or current lasts for several milliseconds, and the intervals between pulses tend to be in the audio range (less than 100 Hz), leading to the term popcorn noise for the popping or crackling sounds it produces in audio circuits.

Avalanche Noise

A junction phenomenon in a semiconductor in which carriers in a high-voltage gradient develop sufficient energy to dislodge additional carriers through physical impact; this agitation creates ragged current flows which are indicated by noise. This is the noise produced when a junction diode is operated at the onset of avalanche breakdown.

QUANTIFICATION

The noise level in an electronic system is typically measured as an electrical power N in watts or dBm, a root mean square (RMS) voltage (identical to the noise standard deviation) in volts, dBμV or a mean squared error (MSE) in volts squared. Noise may also be characterised by its probability distribution and noise spectral density $N_0(f)$ in watts per Hertz.

A noise signal is typically considered as a linear addition to a useful information signal. Typical signal quality measures involving noise are signal-to-noise ratio (SNR or S/N), signal-to-quantisation noise ratio (SQNR) in analog-to-digital coversion and compression, peak signal-to-noise ratio (PSNR) in image and video coding, E_b/N_0 in digital transmission, carrier to noise ratio (CNR) before the detector in carrier-modulated systems, and noise figure in cascaded amplifiers.

Noise is a random process, characterised by stochastic properties such as its variance, distribution, and spectral density. The spectral distribution of noise can vary with frequency, so its power density is measured in watts per Hertz (W/Hz). Since the power in a resistive element is proportional to the square of the voltage across it, noise voltage (density) can be described by taking the square root of the noise power density, resulting in volts per root Hertz $\left(v/\sqrt{Hz} \right)$. Integrated circuit devices, such as operational amplifiers commonly quote equivalent input noise level in these terms (at room temperature).

Noise power is measured in Watts or decibels (dB) relative to a standard power, usually indicated by adding a suffix after dB. Examples of electrical noise-level measurement units are dBu, dBm0, dBrn, dBrnC, and $dBrn(f_1-f_2)$, dBrn(144-line).

Noise levels are usually viewed in opposition to signal levels and so are often seen as part of a signal-to-noise ratio (SNR). Telecommunication systems strive to increase the ratio of signal level to noise level in order to effectively transmit data. In practice, if the transmitted signal falls below the level of the noise (often designated as the noise floor) in the system, data can no longer be decoded at the receiver. Noise in telecommunication systems is a product of both internal and external sources to the system.

Dither

If the noise source is correlated with the signal, such as in the case of quantisation error, the intentional introduction of additional noise, called dither, can reduce overall noise in the bandwidth of interest. This technique allows retrieval of signals below the nominal detection threshold of an instrument. This is an example of stochastic resonance.

PHASE NOISE

Phase noise is the frequency domain representation of rapid, short-term, random fluctuations in the phase of a waveform, caused by time domain instabilities ('jitter'). Generally speaking, radio frequency engineers speak of the phase noise of an oscillator, whereas digital system engineers work with the jitter of a clock.

Historically there have been two conflicting yet widely used definitions for phase noise. The definition used by some authors defines phase noise to be the power spectral density (PSD) of a signal's phase, the other one is based on the PSD of the signal itself. Both definitions yield the same result at offset frequencies well removed from the carrier. At close-in offsets however, characterisation results strongly depends on the chosen definition. Recently, the IEEE changed its official definition to $L(f) = S_\phi/2$ where S_ϕ is the (single-sided) spectral density of a signal's phase fluctuations.

An ideal oscillator would generate a pure sine wave. In the frequency domain, this would be represented as a single pair of delta functions (positive and negative conjugates) at the oscillator's frequency, i.e. all the signal's power is at a single frequency. All real oscillators have phase modulated noise components. The phase noise components spread the power of a signal to adjacent frequencies, resulting in noise sidebands. Oscillator phase noise often includes low frequency flicker noise and may include white noise.

Consider the following noise free signal:

$$v(t) = A \cos(2\pi f_0 t).$$

Phase noise is added to this signal by adding a stochastic process represented by φ to the signal as follows:

$$v(t) = A \cos(2\pi f_0 t + \varphi(t)).$$

Phase noise is a type of cyclostationary noise and is closely related to jitter. A particularly important type of phase noise is that produced by oscillators.

Phase noise $(L(f))$ is typically expressed in units of dBc/Hz, representing the noise power relative to the carrier contained in a 1 Hz bandwidth centered at a certain offsets from the carrier. For example, a certain signal may have a phase noise of –80 dBc/Hz at an offset of 10 kHz and –95 dBc/Hz at an offset of 100 kHz. Phase noise can be measured and expressed as single sideband or double sideband values, but as noted earlier, the IEEE has adapted as its official definition, one-half the double sideband PSD.

Jitter Conversions

Phase noise is sometimes also measured and expressed as a power obtained by integrating $L(f)$ over a certain range of offset frequencies. For example, the phase noise may be –40 dBc integrated over the range of 1 kHz to 100 kHz. This Integrated phase noise (expressed in degrees) can be converted to jitter (expressed in seconds) using the following formula.

Jitter (seconds) = Phase error (degrees) / (360x Frequency (Hertz))

In the absence of $1/f$ noise in a region where the phase noise displays a –20 dBc/Hz slope, the rms cycle jitter can be related to the phase noise by:

$$\sigma_c^2 = \frac{f^2 L(f)}{f_{osc}^3}$$

Likewise:

$$L(f) = \frac{f^2 \sigma_c^2}{f_{osc}^3}$$

Measurement

Phase noise can be measured using a spectrum analyser if the phase noise of the device under test (DUT) is large with respect to the spectrum analyser's local oscillator. Care should be taken that observed values are due to the measured signal and not the Shape Factor of the spectrum analyser's filters. Spectrum analyser based measurement can show the phase-noise power over many decades of frequency, e.g. 1 Hz to 10 MHz. The slope with offset frequency in various offset frequency regions can provide clues as to the source of the noise, e.g. low frequency flicker noise decreasing at 30 dB per decade (=9 dB per octave).

Spectral purity

The sinewave output of an ideal oscillator is a single line in the frequency spectrum. Such perfect spectral purity is not achievable in a practical oscillator. Spreading of the spectrum line caused by phase noise must be minimised in the local oscillator for a superheterodyne receiver because it defeats the aim of restricting the receiver frequency range by filters in the IF (intermediate frequency) amplifier.

PHASE NOISE IN OSCILLATORS

As well known from oscillator theory, two conditions are required to make a feedback system oscillate: the open loop gain must be greater than unity; and total phase shift must be 360 degrees at the frequency of oscillation. An oscillator circuit can be a combination of an amplifier with gain A (jω) and a frequency dependent feedback loop H (jω) = βA. Oscillator has positive feedback loop at selected frequency.

1. Frequency stability is a measure of the degree to which an oscillator maintains the same value of frequency over a given time.
2. Phase noise can be described as short-term random frequency fluctuations of a signal; is measured in the frequency domain, and is expressed as a ratio of signal power to noise power measured in a 1 Hz bandwidth at a given offset from the desired signal.

3. Low oscillator phase noise is a necessity for many receiving and transmitter systems. Adjacent channel rejection as well as transmitter signal purity are dependent on the phase noise of the receiver local oscillator or transmit local oscillator.

4. The local oscillator phase noise will limit the ultimate signal-to-noise ratio (SNR) which can be achieved when listening to a frequency modulated (FM) or phasemodulated (PM) signal.

5. The oscillator phase noise is transferred to the carrier to which the receiver is tuned and is then demodulated by the FM discriminator. The phase noise results in a constant noise power output from the discriminator.

6. The performance of some types of AM detectors or SSB detectors may be degraded by the local oscillator phase noise. Reciprocal mixing may cause the receiver noise floor to increase when strong signals are near the receiver's tuned frequency; this limits the ability to recover weak signals. All of these effects are due to local oscillator Phase Noise, and can only be reduced by decreasing the phase noise.

7. Local oscillator Phase Noise will affect the Bit Error Rate (BER) performance of a phase-shift keyed (PSK) digital transmission system. A transmission error will occur any time if the local oscillator phase, due to its noise, becomes sufficiently large that the digital phase detection makes an incorrect decision as to the transmission phase. For instance, a QPSK transmission system (used in microwave links, CDMA, DVB, etc.) will make a transmission error if the instantaneous oscillator phase is offset by more than 45° since the phase detector will determine that baud to be in the incorrect quadrant. Digital transmission systems with smaller phase multiples are more sensitive to degradation due to local oscillator phase noise.

A variable controlled oscillator (VCO) part of a phase-locked loop (PLL), will always have some spurious signals present on its output. The amplitude and frequency of these spurious modulations may vary as the local oscillator is tuned.

1. Poor layout of the phase-locked loop oscillator circuitry may increase the amplitude and number of these spurious signals.

2. Oscillator phase noise has two components: Phase noise resulting from direct upconversion of white noise and flicker noise (1/f noise), and phase noise resulting from the changing phase of the noise sources modulating the oscillation frequency.

The Phase Noise of an oscillator is best described in the frequency domain where the spectral density is characterised by measuring the noise sidebands on either side of the output signal center frequency.

Single side band (SSB) phase noise is specified in dBc/Hz at a given frequency offset from the carrier (Fig. 24.1).

SSB phase noise places limit on receiver adjacent channel selectivity (ACS) and also affects the receiver signal to noise ratio.

A model for oscillator SSB phase noise was introduced by *David B. Leeson* in 1966.

$$\mathcal{L}_{PM} = \log\left[\frac{FkT}{A}\frac{1}{8Q_L^2}\left(\frac{f_o}{f_m}\right)^2\right]$$

$\mathcal{L}_{PM}$ = Single side band (SSB) phase noise density [dBc/Hz].
A = Oscillator output power [W].
F = Device noise factor at operating power level A (linear).
k = Boltzmann's constant, 1.38×10^{-23} [J/K].

T = Temperature [K].
Q_L = Loaded-Q [dimensionless].
f_o = Oscillator carrier frequency [Hz].
f_m = Frequency offset from the carrier [Hz].

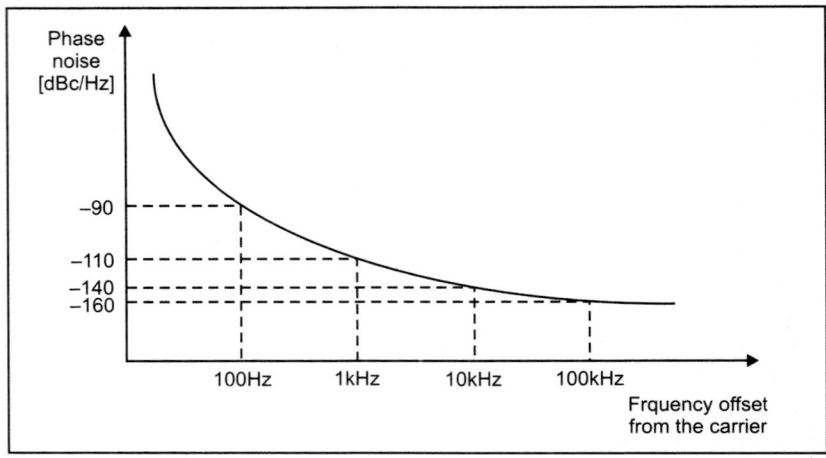

Fig. 24.1. Single side band noise.

Leeson's equation only applies between $1/f$ flicker noise transition frequency (f_1) and a frequency (f_2) where white noise (flat) dominates (Fig. 24.2).

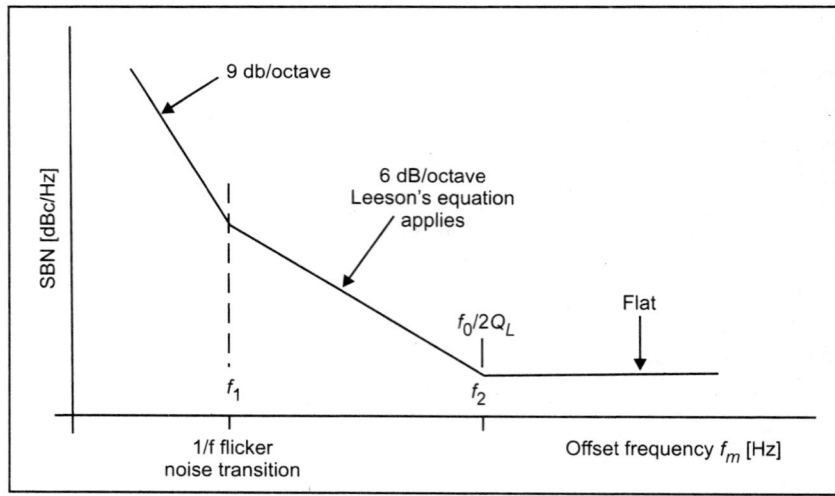

Fig. 24.2. Leeson's equation curve.

Leeson equation provides several insights about oscillator SSB Phase noise:
1. Doubling the Loaded-Q improves Phase Noise by 6dB
2. Doubling the operation frequency results 6dB Phase Noise degradation.

Unloaded-Q: Means the resonant circuit is not loaded by any external terminating impedance. In this case the *Q* is determined only by resonator losses.

Loaded-Q: Means the width of the resonance curve, or phase slope, including the effects of external components. In this case the *Q* is determined mostly by the external components.

In the Fig. 24.3 phase noise in dBc/Hz is plotted as a function of frequency offset (f_m), with the frequency axis on a log scale. Note that the actual curve is approximated by a number of regions, each having a slope of $1/f^x$, where $x = 0$ corresponds to the 'white' phase noise region (slope = 0 dB/decade), and $x = 1$ corresponds to the 'flicker $1/f$' phase noise region (slope = 20 dB/decade). There are also regions where $x = 2, 3, 4$, and these regions occur progressively closer to the carrier frequency (Fig. 24.3).

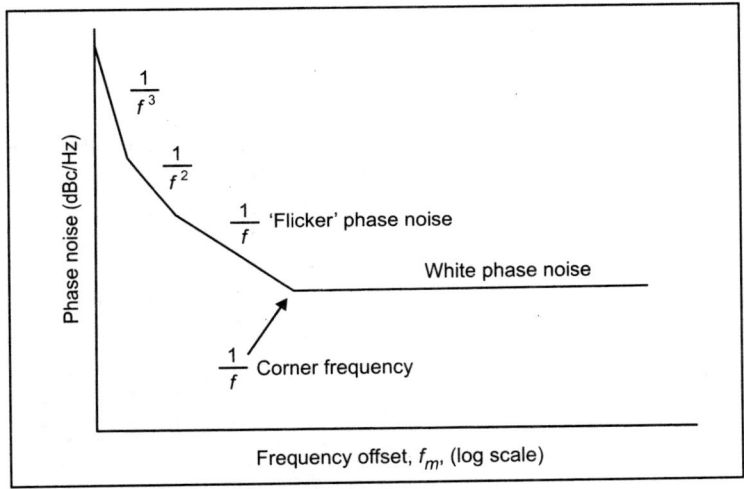

Fig. 24.3. Phase noise as function of frequency.

Whilst harmonics can be filtered out by a simple low pass-filter, the spurious levels close to the wanted signal can only be minimised by careful oscillator design:

1. Power supply (Vcc) and tuning voltage (Vtune) returns must be connected to the printed circuit board ground plane. VCO ground plane must be the same as that of the printed circuit board and therefore all VCO ground pins must be soldered direct to the printed circuit board ground plane.

2. Adequate RF grounding is required. Several chip decoupling capacitors must be provided between the Vcc supply and ground.

3. Good, low noise power supplies must be used to prevent AM noise. Ideally, DC batteries for both supply (VCC) and tuning (V tune) voltages will provide the best overall performance.

4. Output must be correctly terminated with good load impedance. It is also a good practice to use a resistive pad between the VCO and the external load.

5. Connections to the tuning port must be as short as possible and must be well screened, shielded, and decoupled to prevent the VCO from being modulated by external noise sources. A low noise power supply must be used for tuning voltage (Vtune) supply.

6. Avoid saturation of the active devices at all cost, and try to have either limiting or automatic gain control (AGC) without degradation of the *Q* of the resonator.

7. Using active components with low 1/f-noise. Flicker noise in BJTs is also known as 1/f noise because of the 1/f slope characteristics of the noise spectrum (the amplitude varies inversely with frequency). Mainly traps associated with contamination and crystal defects in the emitter-base depletion layer cause this noise. These traps capture and release carriers in a random fashion. The time constants associated with the process produce a noise signal at low frequencies.

In order to design an oscillator with low 1/f noise, the following are required:

1. A resonating circuit (Crystal, L, C or Varactor) with a high Q-factor.
2. Active components with low flicker noise or 1/f-noise.

To construct a resonant structure with a high Q-factor low losses are required in all of the constituent parts. The following points should therefore be carefully considered:

1. Q of resonator device itself.
2. Series resistance of capacitors.
3. Series resistance of tuning diode.
4. Loss of printed circuit board.

Low 1/f noise of the transistor in the oscillator is very important, because the 1/f noise appears as sideband noise around the carrier frequency of the oscillator output signal.

The basic rules to select the right transistor for an optimised design are:

1. The best oscillator transistor is a device with the lowest possible noise figure and lowest f_T. A commonly used criteria is: $f_T \leq 2 \times$ fosc.
2. The 1/f noise is directly related to the current density in the transistor. Transistors with high Icmax used at low currents have best 1/f performance. For low phase noise operation use a medium power transistor. If you need your output power to be achieved at 6–9 mA, select a transistor with Icmax of 60–90 mA. However, the ft of a transistor drops as current is decreased. Additionally, the parasitic capacitances of a high current transistor are higher due to the larger transistor structure required.
3. The effect of flicker noise can be reduced through RF feedback. An unbypassed emitter resistor of 10–30 Ω in a BJT circuit can improve the flicker noise by as much as 40 dB. The proper bias point of the active device is important.
4. Precautions should be taken to prevent modulation of the input and output dynamic capacitances of the active device; which will cause amplitude-to-phase conversion and therefore introduce noise.
5. Low noise figure combined with a small correlation coefficient.
6. Higher output power.
7. Low output conductance.
8. Reasonably high input impedance.
9. Meeting an impedance condition at the input of the active device, which can be achieved by optimisation of the feedback factor and which leads to optimum impedance noise matching.

In a PLL the design of the loop filter can affect the phase noise of the system:

1. Within the loop bandwidth, the phase noise of the oscillator will tend to cancel itself, leaving a phase noise essentially equal to the frequency multiplied phase noise of the crystal reference.
2. Multiplied phase noise of the crystal reference at particular frequency offset is equal with reference phase noise at the same frequency offset plus $20 \times \log(N_{VCO\_divider})$ plus 1dB (multiplication efficiency factor).
3. Outside the loop bandwidth, the phase noise of the oscillator is not cancelled, and will continue to decrease, until reaching its half bandwidth, $\omega_o/2Q$ or 1/f corner frequency. Since the Q of the

crystal reference is very large, its half bandwidth is very small, and its frequency multiplied phase noise will remain relatively flat down to very small frequency offsets. Further, at some moderate frequency offset, this multiplied phase noise power spectral-density will be crossed by the decreasing oscillator phase noise power spectral-density (Fig. 24.4).

4. The bandwidth of the loop should be chosen equal to the frequency offset of this crossover.

5. Although the phase-locked loop bandwidth is not a barrier frequency with a discontinuity on either side of the barrier, it can be approximated as such with the proviso that small errors around the offset frequency equal to the loop bandwidth are accepted.

6. The role of the loop filter, which is a low-pass filter inserted between the phase comparator and the VCO control voltage circuit, eliminates the high frequency component of the phase correction pulse generated by the phase comparator so that the only the DC component is provided to the VCO.

7. As a rule of thumb, the cut off frequency of the low-pass filter is chosen as equal or less than comparison frequency divided by ten ($F_{cutooff} < F_{comparision}/10$).

8. Usually the low-pass filter is an RC network. The analysis of the phase noise performance shows that the phase noise depends on the resistor value, part of the low-pass filter. The higher the resistor, the higher is its contribution to the phase noise.

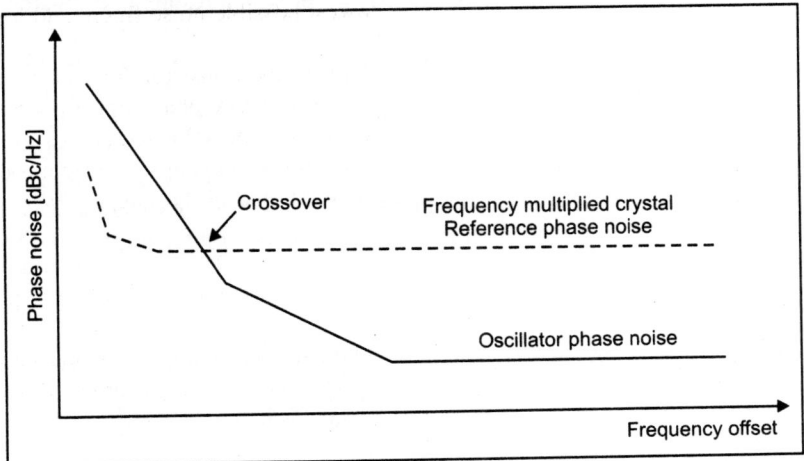

Fig. 24.4. Frequency multiply crystal.

Continuing with parameters that affect VCO phase noise we can see:

1. When frequency of the carrier increases, it is more difficult to achieve good phase noise.

2. It's easy to achieve good phase noise when the frequency range covered by VCO is narrow; the tuning bandwidth must be small. Generated energy should be coupled from the resonator rather than from another portion of the active device so that the resonator limits the bandwidth.

3. Increasing tuning sensitivity (measured in MHz/V) degrades phase noise.

4. For a given frequency it's easy to achieve good phase noise in VCO's using a wide tuning voltage range.

5. Temperature affects the phase noise. In a range of –55°C to +85°C the variation is +/–3 dB of the phase noise.

NOISE SPECTRAL DENSITY

In communications, noise spectral density N_o is the noise power per unit of bandwidth; that is, it is the power spectral density of the noise. It has dimension of power/frequency, whose SI coherent unit is watts per Hertz, which is equivalent to watt-seconds or joules. If the noise is white noise, i.e. constant with frequency, then the total noise power N in a bandwidth B is BN_o. This is utilised in signal-to-noise ratio calculations. The thermal noise density is given by $N_o = kT$, where, k is Boltzmann's constant in joules per kelvin, and T is the receiver system noise temperature in kelvins. N_o is commonly used in link budgets as the denominator of the important figure-of-merit ratios E_b/N_0 and E_s/N_0.

FLICKER NOISE

Flicker noise is a type of electronic noise with a $1/f$, or pink spectrum. It is therefore often referred to as $1/f$ noise or pink noise, though these terms have wider definitions. It occurs in almost all electronic devices, and results from a variety of effects, such as impurities in a conductive channel, generation and recombination noise in a transistor due to base current, and so on. It is always related to a direct current. Its origin is not well known. In electronic devices, it is a low-frequency phenomenon, as the higher frequencies are overshadowed by white noise from other sources. In oscillators, however, the low-frequency noise is mixed up to frequencies close to the carrier which results in oscillator phase noise.

Flicker noise is often characterised by the corner frequency f_c between the regions dominated by each type. MOSFETs have a higher f_c than JFETs or bipolar transistors which is usually below 2 kHz for the latter. The flicker noise voltage power in MOSFET can be expressed by $K/(Cox \bullet WLf)$, where K is the process-dependent constant, W and L are channel width and length respectively.

Flicker noise is found in carbon composition resistors, where it is referred to as excess noise, since it increases the overall noise level above the thermal noise level, which is present in all resistors. In contrast, wire-wound resistors have the least amount of flicker noise. Since flicker noise is related to the level of DC, if the current is kept low, thermal noise will be the predominant effect in the resistor, and the type of resistor used will not affect noise levels.

Measurement

For measurements the interest is in the 'drift' of a variable with respect to a measurement at a previous time. This is calculated by applying the signal time differencing:

$$1 - e^{-Td \cdot s}$$

to

$$\frac{a}{f}$$

where, Td is the time between measurements,

$$s = i \cdot 2\pi f$$

and a includes the contribution of both positive and negative frequency terms.

After some manipulation, the variance of the voltage difference is:

$$2a \int_0^{f_h} \frac{1 - \cos(Td \cdot 2\pi f)}{f} df = -2a \cdot \text{Cin}(Td \cdot 2\pi f_h)$$

where, Cin is the cosine integral:

$$\mathrm{Cin}(x) = \int_0^x \frac{1 - \cos t}{t} dt.$$

where, f_h is a brick-wall filter limiting the upper bandwidth during measurement (Fig. 24.5). Real measurements involve more complicated calculations.

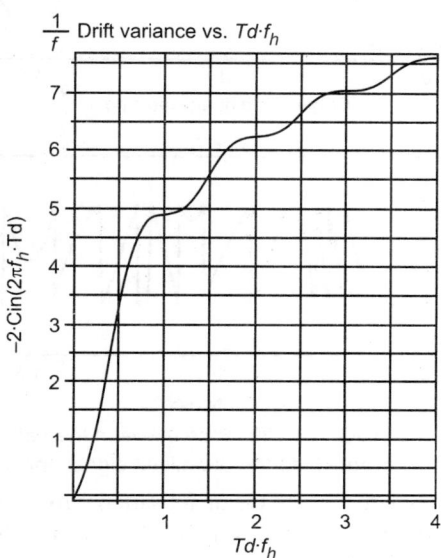

$\frac{1}{f}$ Drift variance vs. $Td \cdot f_h$

Fig. 24.5. Measurement of flicker noise.

Removal in Instrumentation and Measurements

For DC measurements 1/*f* noise can be particularly troublesome as it is very significant at low frequencies (tending to infinity with integration/averaging at DC.) One powerful technique involves moving the signal of interest to a higher frequency, and using a phase sensitive detector to measure it. For example the signal of interest can be chopped with a frequency. Generation-recombination noise or **g-r noise**, is a type of electrical signal noise caused statistically by the fluctuation of the generation and recombination of electrons in semiconductor-based photon detectors.

QUANTISATION ERROR

In source coding (analog-to-digital conversion and compression), the difference between the actual analog value and quantised digital value is called quantisation error. This error is either due to rounding or truncation. The error is sometimes considered as an additional random signal called quantisation noise.

Quantisation Noise Model of Quantisation Error

Quantisation noise is a model of quantisation error introduced by quantisation in the analog-to-digital conversion (ADC) in telecommunication systems and signal processing. It is a rounding error between the analog input voltage to the ADC and the output digitised value. The noise is non-linear and signal-dependent. It can be modelled in several different ways (Fig. 24.6).

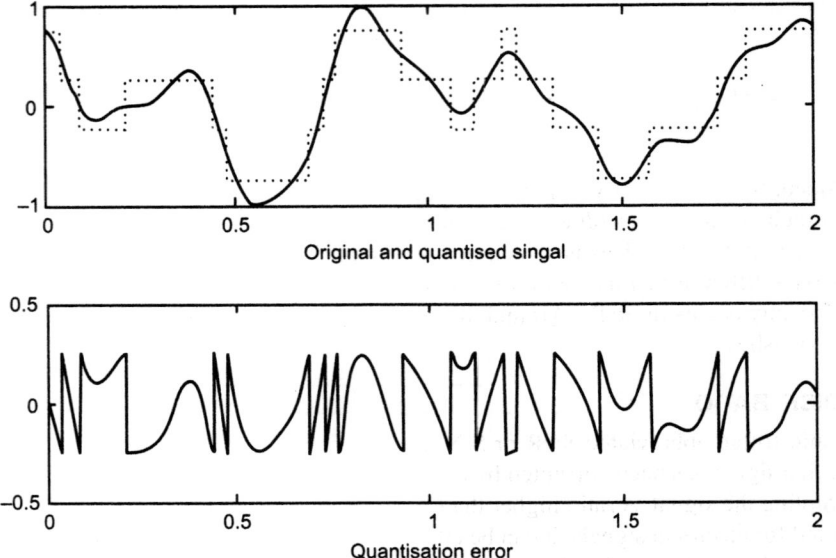

Fig. 24.6. Quantisation noise for a 2-bit ADC. The difference between the dark and dotted signals in the upper graph is the quantisation error, which is 'added' to the quantised signal and is the source of noise.

In an ideal analog-to-digital converter, where the quantisation error is uniformly distributed between −1/2 LSB and +1/2 LSB, and the signal has a uniform distribution covering all quantisation levels, the Signal-to-quantisation-noise ratio (SQNR) can be calculated from:

$$\text{SQNR} = 20 \log_{10} (2^Q) \approx 6.02 \cdot Q \text{ dB}$$

where, Q is the number of quantisation bits.

The most common test signals that fulfil this are full amplitude triangle waves and sawtooth waves. For example, a 16-bit ADC has a maximum signal-to-noise ratio of 6.02 × 16 = 96.3 dB.

When the input signal is a full-amplitude sine wave the distribution of the signal is no longer uniform, and the corresponding equation is instead:

$$\text{SQNR} \approx 1.761 + 6.02 \cdot Q \text{ dB}$$

Here, the quantisation noise is once again assumed to be uniformly distributed. When the input signal has a high amplitude and a wide frequency spectrum this is the case. In this case a 16-bit ADC has a maximum signal-to-noise ratio of 98.09 dB. The 1.761 difference in signal-to-noise only occurs due to the signal being a full-scale sine wave instead of a triangle/sawtooth. (Typical real-life values are worse than this theoretical minimum, due to the addition of dither to reduce the objectionable effects of quantisation, and to imperfections of the ADC circuitry. On the other hand, specifications often use A-weighted measurements to hide the inaudible effects of noise shaping, which improves the measurement.) For complex signals in high-resolution ADCs this is an accurate model. For low-resolution ADCs, low-level signals in high-resolution ADCs, and for simple waveforms the quantisation noise is not uniformly distributed, making this model inaccurate. In these cases the quantisation noise distribution is strongly affected by the exact amplitude of the signal. The calculations above, however, assume a completely filled input channel. If this is not the case—if the input signal is small—the relative quantisation distortion can be very large. To circumvent this issue, analog compressors and expanders can be used, but these introduce large amounts of distortion as well, especially if the compressor does not match the expander.

$$\text{Quantisation error} = \text{Quantised output} - \text{Analog input}$$

Other fields

Many physical quantities are actually quantised by physical entities. Examples of fields where this limitation applies include electronics (due to electrons), optics (due to photons), biology (due to DNA), and chemistry (due to molecules). This is sometimes known as the 'quantum noise limit' of systems in those fields. This is a different manifestation of 'quantisation error', in which theoretical models may be analog but physically occurs digitally. Around the quantum limit, the distinction between analog and digital quantities vanishes.

SIGNAL-TO-NOISE RATIO

Signal-to-noise ratio (often abbreviated SNR or S/N) is a measure used in science and engineering to quantify how much a signal has been corrupted by noise. It is defined as the ratio of signal power to the noise power corrupting the signal. A ratio higher than 1:1 indicates more signal than noise. While SNR is commonly quoted for electrical signals, it can be applied to any form of signal (such as isotope levels in an ice core or biochemical signaling between cells). In less technical terms, signal-to-noise ratio compares the level of a desired signal (such as music) to the level of background noise. The higher the ratio, the less obtrusive the background noise is. 'Signal-to-noise ratio' is sometimes used informally to refer to the ratio of useful information to false or irrelevant data in a conversation or exchange. For example, in online discussion forums and other online communities, off-topic posts and spam are regarded as 'noise' that interferes with the 'signal' of appropriate discussion.

Definition

Signal-to-noise ratio is defined as the power ratio between a signal (meaningful information) and the background noise (unwanted signal):

$$\text{SNR} = \frac{P_{\text{signal}}}{P_{\text{noise}}}$$

where, P is average power. Both signal and noise power must be measured at the same or equivalent points in a system, and within the same system bandwidth. If the signal and the noise are measured across the same impedance, then the SNR can be obtained by calculating the square of the amplitude ratio:

$$\text{SNR} = \frac{P_{\text{signal}}}{P_{\text{noise}}} = \left(\frac{A_{\text{signal}}}{A_{\text{noise}}} \right)^2$$

where, A is root mean square (RMS) amplitude (for example, RMS voltage). Because many signals have a very wide dynamic range, SNRs are often expressed using the logarithmic decibel scale. In decibels, the SNR is defined as:

$$\text{SNR}_{\text{dB}} = 10 \ \log_{10}\left(\frac{P_{\text{signal}}}{P_{\text{noise}}} \right) = P_{\text{signal, dB}} - P_{\text{noise, dB,}}$$

which may equivalently be written using amplitude ratios as:

$$\text{SNR}_{dB} = 10 \log_{10} \left(\frac{A_{\text{signal}}}{A_{\text{noise}}} \right)^2 = 20 \log_{10} \left(\frac{A_{\text{signal}}}{A_{\text{noise}}} \right).$$

The concepts of signal-to-noise ratio and dynamic range are closely related. Dynamic range measures the ratio between the strongest un-distorted signal on a channel and the minimum discernable signal, which for most purposes is the noise level. SNR measures the ratio between an arbitrary signal level (not necessarily the most powerful signal possible) and noise. Measuring signal-to-noise ratios requires the selection of a representative or reference signal. In audio engineering, the reference signal is usually a sine wave at a standardised nominal or alignment level, such as 1 kHz at +4 dBu (1.228 V_{RMS}).

SNR is usually taken to indicate an average signal-to-noise ratio, as it is possible that (near) instantaneous signal-to-noise ratios will be considerably different. The concept can be understood as normalising the noise level to 1 (0 dB) and measuring how far the signal 'stands out'.

Alternative Definition

An alternative definition of SNR is as the reciprocal of the coefficient of variation, i.e. the ratio of mean to standard deviation of a signal or measurement:

$$\text{SNR} = \frac{\mu}{\sigma}$$

where, μ is the signal mean or expected value and σ is the standard deviation of the noise, or an estimate thereof. Notice that such an alternative definition is only useful for variables that are always positive (such as photon counts and luminance). Thus it is commonly used in image processing, where the SNR of an image is usually calculated as the ratio of the mean pixel value to the standard deviation of the pixel values over a given neighbourhood. Sometimes SNR is defined as the square of the alternative definition above.

The Rose criterion (named after Albert Rose) states that an SNR of at least 5 is needed to be able to distinguish image features at 100 per cent certainty. An SNR less than 5 means less than 100 per cent certainty in identifying image details.

Yet another alternative, very specific and distinct definition of SNR is employed to characterise sensitivity of imaging systems. Related measures are the 'contrast ratio' and the 'contrast-to-noise ratio'.

Improving SNR in practice

All real measurements are disturbed by noise. This includes electronic noise, but can also include external events that affect the measured phenomenon — wind, vibrations, gravitational attraction of the moon, variations of temperature, variations of humidity, etc. depending on what is measured and of the sensitivity of the device. It is often possible to reduce the noise by controlling the environment. Otherwise, when the characteristics of the noise are known and are different from the signals, it is possible to filter it or to process the signal. When the signal is constant or periodic and the noise is random, it is possible to enhance the SNR by averaging the measurement.

Digital signals

When a measurement is digitised, the number of bits used to represent the measurement determines the maximum possible signal-to-noise ratio. This is because the minimum possible noise level is the error caused by the quantisation of the signal, sometimes called Quantisation noise. This noise level is non-linear and signal-dependent; different calculations exist for different signal models. Quantisation noise is modeled as an analog error signal summed with the signal before quantisation ('additive noise').

This theoretical maximum SNR assumes a perfect input signal. If the input signal is already noisy (as is usually the case), the signal's noise may be larger than the quantisation noise. Real analog-to-digital converters also have other sources of noise that further decrease the SNR compared to the theoretical maximum from the idealised quantisation noise, including the intentional addition of dither.

Although noise levels in a digital system can be expressed using SNR, it is more common to use E_b/N_o, the energy per bit per noise power spectral density. The modulation error ratio (MER) is a measure of the SNR in a digitally modulated signal.

Fixed point

For n-bit integers with equal distance between quantisation levels (uniform quantidation) the dynamic range (DR) is also determined.

Assuming a uniform distribution of input signal values, the quantisation noise is a uniformly-distributed random signal with a peak-to-peak amplitude of one quantisation level, making the amplitude ratio $2^n/1$. The formula is then:

$$DR_{dB} = SNR_{dB} = 20 \log_{10}(2^n) \approx 6.02 \cdot n$$

This relationship is the origin of statements like '16-bit audio has a dynamic range of 96 dB'. Each extra quantisation bit increases the dynamic range by roughly 6 dB.

Assuming a full-scale sine wave signal (that is, the quantiser is designed such that it has the same minimum and maximum values as the input signal), the quantisation noise approximates a sawtooth wave with peak-to-peak amplitude of one quantisation level and uniform distribution. In this case, the SNR is approximately:

$$SNR_{dB} \approx 20 \log_{10}\left(2^n \sqrt{3/2}\right) \approx 6.02 \cdot n + 1.761$$

Floating point

Floating-point numbers provide a way to trade off signal-to-noise ratio for an increase in dynamic range. For n bit floating-point numbers, with $n-m$ bits in the mantissa and m bits in the exponent:

$$DR_{dB} = 6.02 \cdot 2^m$$
$$SNR_{dB} = 6.02 \cdot (n - m)$$

Note that the dynamic range is much larger than fixed-point, but at a cost of a worse signal-to-noise ratio. This makes floating-point preferable in situations where the dynamic range is large or unpredictable. Fixed-point's simpler implementations can be used with no signal quality disadvantage in systems where dynamic range is less than 6.02 m. The very large dynamic range of floating-point can be a disadvantage, since it requires more forethought in designing algorithms.

Optical SNR

Optical signals have a carrier frequency, which is much higher than the modulation frequency (about 200 THz and more). This way the noise bandwidth covers a bandwidth which is much wider than the signal itself. The resulting signal influence relies mainly on the filtering of the noise. To describe the signal quality without taking the receiver into account the optical SNR (OSNR) is used. The OSNR is the ratio between the signal power and the noise power in as given bandwidth. Most commonly a reference bandwidth of 0.1 nm is used.

This bandwidth is independent from the modulation format, the frequency and the receiver. For instance a OSNR of 20dB/0.1nm could be given, even the signal of 40 GBit DPSK would not fit in this bandwidth. OSNR is measured with a optical spectrum analyser.

CONTRAST-TO-NOISE RATIO

Contrast-to-noise ratio (CNR) is a measure used to determine image quality. CNR is similar to the metric, signal-to-noise ratio (SNR), but subtracts off a term before taking the ratio. This is important when there is a significant bias in an image, such as from haze. The intensity is rather high even though the features of the image are washed out by the haze. Thus this image may have a high SNR metric, but will have a low CNR metric. One way to define contrast-to-noise ratio is:

$$C = \frac{|S_A - S_B|}{\sigma_o}$$

where, S_A and S_B are signal intensities for signal producing structures A and B in the region of interest and σ_o is the standard deviation of the pure image noise.

CARRIER-TO-NOISE RATIO

In telecommunications, the carrier-to-noise ratio, often written CNR or C/N, is the signal-to-noise ratio (SNR) of a modulated signal. The term is used to distinguish the CNR of the radio frequency passband signal from the SNR of an analogue base band message signal after demodulation, for example an audio frequency analogue message signal. If this distinction is not necessary, the term SNR is often used instead of CNR, with the same definition.

Digitally modulated signals (e.g. QAM or PSK) are basically made of two CW carriers (the I and Q components, which are out-of-phase carriers) . In fact, the information (bits or symbols) is carried by given combinations of phase and/or amplitude of the I and Q components. It is for this reason that, in the context of digital modulations, digitally modulated signals are usually referred to as carriers. Therefore, the term carrier-to-noise-ratio (CNR), instead of signal-to-noise-ratio (SNR) is preferred to express the signal quality when the signal has been digitally modulated. High C/N ratios provide good quality of reception, for example low bit error rate (BER) of a digital message signal, or high SNR of an analogue message signal.

Definition

The carrier-to-noise ratio is defined as the ratio of the received modulated carrier signal power C to the received noise power N after the receive filters:

$$CNR = \frac{C}{N}.$$

When both carrier and noise are measured across the same impedance, this ratio can equivalently be given as:

$$CNR = \left(\frac{V_C}{V_N}\right)^2,$$

where, V_C and V_N are the root mean square (RMS) voltage levels of the carrier signal and noise respectively. C/N ratios are often specified in decibels (dB):

$$\text{CNR}_{dB} = 10 \, \log_{10}\left(\frac{C}{N}\right) = C_{dB} - N_{dB}$$

or in term of voltage:

$$\text{CNR}_{dB} = 10 \, \log_{10}\left(\frac{V_C}{V_N}\right)^2 = 20 \, \log_{10}\left(\frac{V_C}{V_N}\right)$$

The C/N ratio is measured in a manner similar to the way the signal-to-noise ratio (S/N) is measured, and both specifications give an indication of the quality of a communications channel.

In the famous Shannon–Hartley theorem, the C/N ratio is equivalently to the S/N ratio. The C/N ratio resembles the carrier-to-interference ratio (C/I, CIR), and the carrier-to-noise-and-interference ratio, $C/(N + I)$ or CNIR.

NOISE MARGIN

In electrical engineering, noise margin is the amount by which a signal exceeds the minimum amount for proper operation. It is commonly used in at least two contexts:

1. In communications system engineering, noise margin is the ratio by which the signal exceeds the minimum acceptable amount. It is normally measured in decibels.

2. In a digital circuit, the noise margin is the amount by which the signal exceeds the threshold for a proper '0' or '1'. For example, a digital circuit might be designed to swing between 0.0 and 1.2 volts, with anything below 0.2 volts considered a '0', and anything above 1.0 volts considered a '1'. Then the noise margin for a '0' would be the amount that a signal is below 0.2 volts, and the noise margin for a '1' would be the amount by which a signal exceeds 1.0 volt. In this case noise margins are measured as an absolute voltage, not a ratio. Noise margins for CMOS chips is usually much greater than TTL because the $V_{OH\,min}$ is closer to the power supply voltage and $V_{OL\,max}$ is closer to zero.

Noise margins are generally defined so that positive values ensure proper operation, and negative margins result in compromised operation or perhaps outright failure.

Chapter 25

Semiconductor Manufacturing Facilities

INTRODUCTION

This chapter summarises in modelling and controlling environmental noise from more than 30 semiconductor manufacturing plants ('wafer fabs'). This type of facility typically has a large number of concentrated noise sources because of the unusually large amount of intake and circulation air required to maintain clean room conditions, the complex exhaust and pollution control equipment necessary to handle many specialty gases and chemical products, and high heating and cooling load requirements. Furthermore, these facilities are most often located in populated areas, making noise control important, as well as challenging. Modelling and noise control details for specific types of equipment (make-up air and exhaust fans, cooling towers, ventilated boiler and chiller rooms, piping, valves, etc.), as well as other considerations — such as site building layout with respect to sensitive residential areas — are discussed.

FAB NOISE SOURCES

The typical 'campus' for semiconductor manufacturing contains buildings of several different functions. There is necessarily one or more fab building, in which the primary manufacturing process is carried out. This building is typically three to four stories in height to accommodate and separate vibration sensitive processes on one level, and cleanroom air movement systems plena and other mechanical equipment at other levels. Due to the vibration sensitivity of the process, it is common to have the most energetic building services equipment (boilers, chillers, major process fluid pumps, etc.) located in a separate 'energy center' or 'central utility building' ('CUB'). Noise-generating equipment might also be located in yards outside the CUB. Many facilities have open-air air separation plants that produce process gases (nitrogen, oxygen, etc.). Finally, there is usually one or more administration buildings on site. In most respects, these are similar to typical office buildings, although they may also contain light manufacturing or laboratory facilities. Associated with these facilities are many environmental noise sources. These include:

Exhaust Fans

Large centrifugal fans are used for scrubbed acid exhaust, heat or 'general' exhaust, volatile organic compounds (VOC) exhaust, and for other gasses and functions. These fans are typically located on the roof of the fab building, but are also often found on CUB and office buildings. A typical campus may have 30 or more exhaust fans with a total flow of more than 2 million m³/hr. Noise is primarily radiated from the stack exhaust openings with a significant degree of directivity. In addition, noise can be radiated from the fan ductwork (breakout), and from the fan motors (this is only an environmental noise concern if these sources are located outside).

Make-up Air Units and other Air Handlers

Make-up air fans supply outside air to the cleanroom recirculation air systems. These are located typically on the roof of the fab building. It is not uncommon to have several of these units, which most often employ 'plug' (plenum) fans, supplying a flow of 90,000 to 1,80,000 m^3/h each. In addition, the campus will typically have several smaller units supplying air to the non-process levels of the fab, office, and other buildings.

Cooling Towers

A bank of cooling towers is typically located in an equipment yard near the CUB building. Occasionally these are located on rooftops, but this is usually not done because of their large size (it is common to have up to 20 units with a motor power of 20 to 80 kW each). Most often induced-draft propeller fan units are used, but occasionally forced-draft centrifugal units are used. Cooling towers produce both 'waterfall' and fan noise.

Boilers

Several large boilers with capacities on the order of 700 boiler horsepower are located inside the CUB building. Due to requirements for combustion air, one or more sides of the boiler room is typically louvered, allowing the noise from the boilers and associated pumps to radiate to the outside.

Compressed Air Vents

Air separation plants typically have several associated air vents, periodically producing mid- to high-frequency noise. These release excess gasses back to the atmosphere. One also finds vents outside the CUB building, associated with the oil-free air (OFA) compressors.

Emergency Generators and Continuous Power Supply (CPS) Units

These sources are typically located outside or in well-ventilated rooms inside the CUB. The generators themselves are relatively quiet in comparison with the diesel engines that power them. The diesel engines typically only operate in emergency situations, and are exercised periodically for test purposes. With 'standard' emergency generators there is no noise unless they are being exercised or during power outages, but with CPS units, the generator runs continuously and can generate significant noise from its cooling fan. Up to four of these units may be required for each wafer fab on the campus.

Pumps

Most process and building facility pumps are located within the buildings and are, therefore, not environmental noise contributors. However, certain pumps are typically located outside, including those used for condenser water supply to the cooling towers and chemical feed pumps associated with outdoor storage tanks.

Miscellaneous Sources

Other noise sources include tank piping valves, degasifier towers, delivery truck traffic, etc. Generally speaking, it is useful to characterise the nature of wafer fab noise. Nearly all fabrication facilities operate 24 hours a day, 365 days per year. Most of the noise sources associated with building services (air, water, process materials, etc.) produce continuous noise day and night. Exceptions may be cooling towers and boilers that operate in accordance with the varying heat load in the building, and emergency equipment, such as smoke exhaust fans and emergency electricity generators. Blow-off vents at air separation plants are periodic sources that occasionally 'punctuate' the continuous noise.

The continuous sources produce noise throughout the audible frequency range. In most cases, the periodic sources (vents and valves) tend to generate mid- to high-frequency noise. All of these sources are likely to be audible near the plant. But since the high-frequency portion of the continuous noise, and most of the venting and valve noise, tends to be well attenuated by air absorption at larger distances from the plant (say, over 300 meters), one tends to hear, primarily, continuous low-frequency noise. Thus, when receivers are located near the fab, many sources of several types may be the subject of noise control; for communities at relatively long distances from the fab, it is most often the dominant producers of low-frequency noise (e.g. large exhaust fans, cooling tower fans, and boilers) that are the subjects of noise control.

ENVIRONMENTAL NOISE CRITERIA

Due to the need for highly-trained and educated personnel, wafer fabrication plants are most often located near populated areas large enough to support institutions of higher education. Even when located remotely, it is not uncommon for residential areas to develop near the plant to house employees and people working in support industries. Combined with the large numbers of noise sources described in the previous section, this is the reason why control of environmental noise is a concern and a challenge.

Throughout the world, environmental noise criteria and regulations take many forms. Although we can provide some general characteristics, reference must be made to local, state, national, and international ordinances. Noise criteria may vary depending on whether the site has existing developments or is a previously undeveloped site. Well-written ordinances will refer to potential receivers of noise in terms of land use, e.g. residential, commercial, or industrial. Most ordinances regulate continuous noise; some also account for time-varying noise exposure [for example, by the use of the equivalent energy (L_{eq}) metric], and others explicitly acknowledge, with stricter regulation, that intermittent and tonal noise can cause greater annoyance. Other variations include: reference to a fixed criterion level versus the ambient noise level; the use of other indices such as statistical centile levels (e.g. L10, L50, and L90) and day-night noise levels (L_{dn}); and references to the property line of the noise generator versus that of the noise receivers. These details, along with the existing ambient noise levels on site, must be accounted for in any environmental noise control design.

A detail that is often omitted from environmental noise regulation, but which must be considered, lest the needs for noise control be underestimated, is the effect of meteorological conditions on the propagation of noise. In many instances, due to the large size of the fab campus, the sensitive receivers may be located at significant distance from the noise sources. When the receivers are located at distances of 100 meters or more from the primary noise sources, it is important to consider the effects of the variations in temperature, humidity, and wind speed and direction on the propagation of noise. Under certain conditions, meteorological effects can cause variations in noise level of 10 dB or more.

It is often difficult to obtain specific meteorological data for sites. While this is available from most airports and certain other authorities, it rarely contains wind and temperature data as a function of altitude, and, of course, it does not account for local variations at the site, caused by terrain, ground conditions, etc.

When the meteorological conditions under which noise control design is to be carried out are not specified by regulation, we typically take one of two approaches that must be agreed to in advance by the concerned parties. The first is to define noise control based on 'nominal' conditions: average temperature and humidity for the area, and average wind conditions. When wind direction varies such that time in opposite directions is nearly equal, the implication is that wind can be ignored, on the average. The second approach is to use arbitrary 'worst-case' conditions, where conditions most favourable to propagation towards the sensitive receivers are used. The word 'arbitrary' is used, because, due to the non-stationary random nature of weather conditions, there will always be a worse 'worst-case'. In this sense, it is pointless to use maximum conditions. It should also be borne in mind that the most extreme conditions are not suitable for evaluation of noise (such as in wind speeds greater than 5 m/s). Therefore, conditions are selected such that most of the time, the design noise levels are within the criterion. Under certain conditions, the International Standards Organisation recommends the use of the arbitrary wind speed of 1 to 5 m/s within ±45° of the direction of the receiver, as one example.

NOISE CONTROL

Layout

If noise control is addressed early enough in the project, the simplest solution is to control noise by advantageous layout of the campus buildings and external noise sources. The following are some general principles used to reduce noise propagation to sensitive areas:

1. Identify surrounding noise-sensitive areas (neighbourhoods, hospitals, parks, etc.)—including potential future developments of these areas. Also identify less-sensitive surroundings: roadways, railways, and industrial facilities. Arrange the campus buildings such that the noisy facilities (CUB, equipment yards, etc.) face the less noise-sensitive areas, and are shielded from the noise-sensitive areas by large, relatively quiet buildings (e.g. the office buildings).

2. Locate equipment yards and air separation plants inside a 'courtyard' formed by the CUB, fab, and other buildings, so that these buildings will act as noise barriers. Face oiler room combustion air intakes and other noisy CUB building services (compressor vents, etc.) inwards towards this courtyard.

3. Utilise building feature barriers whenever possible. On rooftops, locate sources at lower elevations than building parapets. For buildings without parapets, it is sometimes possible to utilise the roof edges as a barrier if receivers are located nearby and the sources are located on the roof as far from the receivers as possible.

General Modelling Considerations

It is rarely possible to control all semiconductor plant noise sources by advantageous layout. In many cases, the layout is dictated by zoning, access, aesthetic, economic, and other considerations. Control of individual or groups of sources is often necessary.

In most cases, we have found that semiconductor facilities are not well represented by simple models that depict only a few of the most significant sources. This is because there is often a large number of 'secondary' sources that contribute substantially to the overall noise level at receiver points. We therefore use a modelling package that can contain details about a large number of sources (between 40 and 80 in a typical fab campus), as well as a detailed representation of the building masses, ground contours and types, and meteorological conditions, since all of these play a major role in the prediction and control of noise in this situation.

Sources may be modelled differently depending on the nearness of the receivers. For distant receivers, the sound power of less significant and proximate (to each other) sources can be combined to reduce computation time (although this reasoning is becoming less relevant with fast computers). When the receivers are nearby, more detail about the radiation characteristics of sources (such as the various sources of noise from a cooling tower) must be modelled separately. In addition, certain forms of noise control, such as partial barriers, require more accurate definition of the radiation and directivity characteristics of sources.

General Philosophy

When excessive noise levels are predicted, the general philosophy is to find the most cost-effective solution out of several possible solutions. The best solution will depend on whether there are a few dominant sources, or many sources with similar noise levels. The determination must be made whether noise control is most cost-effectively applied to certain significant sources, or to groups of sources simultaneously. The most significant sources are identified using the noise model. In the model, noise control is applied to the worst-case sources in turn, until the noise goals are met.

Methods

The following discussion can apply to facilities in the design stage or in the operating stage. In the latter case, one has the advantage of being able to use measured sound power data in the noise model. For the former, equipment manufacturer's submittal data or standard prediction algorithms can be used to develop sound power data for the sources, but of course with less precision.

Exhaust fans (stack-radiated noise)

Especially in the case where receivers are nearby, it is important to apply a directivity correction to the sound power radiated from exhaust stacks. This index is a function of stack diameter, among other parameters. For control of this source, there are a number of different silencer configurations that can be used, including standard dissipative silencers, stack inserts, and active noise control. With dissipative silencers, care must be taken to ensure that fill materials are compatible with the chemical content of the emissions and the rigors of temperature and flow velocity. Sometimes the use of refurbishable or packless silencers is specified. Proper location of the silencer within the stack is important to avoid creating system effects that can reduce the effective insertion loss and increase pressure drop through the silencer.

In at least one case, where receivers were in very close proximity to a group of stacks of short height upon a tall roof, a barrier was effective in reducing the noise levels by several decibels as required. With this solution, one has to balance path-length increases that increase attenuation, with decreases in attenuation that may occur by placing the top of the barrier in higher noise regions dictated by the directivity of the source.

Exhaust fans (casing-radiated noise)

The most effective control of this source is by full enclosure. Partial enclosures or barriers can also be effective at reducing noise, especially when high frequency motor noise is a problem. Lagging the ductwork and fan casing can provide some benefit.

Make-up air units and other air handlers

Often, with air handlers, it is most efficient to apply noise control within the unit itself at the manufacturing stage. Manufacturers typically have several means of reducing the noise radiated from their units, including advantageous fan wheel sizing, the use of plenum liners, splitter silencers, and fan inlet flow straighteners. In other cases, external silencers and barriers can be applied as appropriate.

Cooling towers

For design of noise control for large cooling towers, it may be necessary to distinguish between fan and water noise in the model. The simplest control solution is often modification of tower operation modes. When a greater number of towers can be operated at lower fan speed, the total fan noise will be reduced. It may be possible to use barriers in some cases, but barrier design is often complicated in this case due to height and air flow requirements.

Boilers

As much as possible, isolate other significant CUB internal noise sources, such as chillers, pumps, and compressors from the boiler room, which will necessarily be open to the environment to a large degree due to combustion air requirements. Acoustic louvres can be applied to boiler room vents, although these are usually of limited effectiveness at the low frequencies characteristic of large boilers. Absorption can be applied within the boiler room to reduce the noise available for radiation; this should be designed to work well at low frequencies.

Compressed air vents

A number of manufacturers make dissipative/reactive-type silencers specifically for control of blow-off noise sources.

Emergency generators and CPS units

Use full acoustically-lined and sealed enclosures with intake and exhaust silencers in the most critical applications. Partial enclosures and barriers will have a significantly more limited effect, but may be acceptable in less critical situations.

Pumps

The most effective control of this source is by full (ventilated) enclosure. Partial enclosures or barriers will also work in less critical situations.

Valves and piping

Standard noise control for these sources is acoustical lagging. A more effective solution for valve noise is to construct a sealed enclosure.

Chapter 26

Computers

INTRODUCTION

A quiet PC is a personal computer that makes little noise. Common uses for quiet PCs include video editing, sound mixing, home servers, and home theater PCs. A typical quiet PC uses quiet cooling and storage devices and energy-efficient parts.

Like noise, the term 'quiet PC' is subjective and there is currently no standard definition for a 'quiet PC'. However, a general definition accepted by some is that the sound emitted by such PCs should not exceed 30 dB$_A$ when measured 1 meter away from the computer. In addition to the average sound pressure level, the frequency spectrum and dynamics of the sound are important in determining if the sound of the computer is noticed. Sounds with a smooth frequency spectrum (lacking audible tonal peaks), and little temporal variation are less likely to be noticed. The character and amount of other sound in the environment also affects how much sound will be noticed or masked, so a computer may be quiet with relation to a particular environment or set of users.

CAUSES OF NOISE

The main causes of PC noise are:
1. Mechanical friction noise generated by micro motors and fan bearings, as well as vibration noise from low quality chassis and improper assemblies.
2. Turbulence caused by obstructions in the flow of air, such as poorly designed fan grilles and heatsinks. A lack of clearance between rotating fan blades and nearby support struts and grilles will create an audible noise, similar to how a mechanical siren operates. As the fan blades spin, they produce a vortex of air that trails off the edge of the blade. When the vortexes path crosses an object it may produce noise. Fan blades can be designed to reduce this problem by using special notched shapes.
3. Noise generated by electrical coils or transformers used in power supplies, motherboards, video cards or LCD monitors.

Noise in personal computers has been increasing with rising computing power and number of transistors on a single die (integrated circuit). More transistors of a given size use more power, which releases more heat. Faster-rotating cooling fans are one common way to remove this heat. Also, motor rotation speed for hard disc drive (HDD) and Optical disc drive (ODD) has been rising for faster data processing with technical advances in micro motors. Faster rotation (usually measured in RPM) causes higher bearing friction, thus more noise, given the same bearing technology.

The noise issue had received widespread attention with AMD's early Athlon CPUs and Intel's Pentium 4 Prescott core CPU known for its excessive heat and bundled fan noise running on high RPM. With the introduction of Home Theatre PCs (HTPC), the excessive heat and noise problem, that had been mostly confined to the overclocking and quiet computing communities, came to the attention of the general public.

The main approaches to reducing noise problems from personal computers are:

1. Reduce heat generation by using energy efficient parts—nearly all the energy used by a computer is converted into heat.
2. Improve cooling—by using more efficient cooling parts and lower friction, quieter bearings.
3. Use soundproofing to reduce the effects of remaining noise sources.

NOISE REDUCTION METHODS

Common Noise Reduction Methods

1. Replace heat sinks with more efficient models. This often entails the use of larger copper or aluminum heat sinks which may incorporate heat pipes.
2. Replace fans with passive cooling solutions where possible, such as fans on motherboards and GPUs.
3. Replace fans with low-speed, large-diameter fans with low bearing and motor noise. Larger fans can move more air per revolution than smaller fans.
4. Replace the constant-speed fan in the power supply with a thermostatically controlled variable speed fan that runs at less than maximum speed, and thus runs quieter most of the time.
5. Replace the power supply with a quieter model. The main considerations, from a noise-reduction point of view, in choosing a power supply are fan quality, AC/DC conversion efficiency, and how good the thermal fan speed control is at keeping the fan running slow and steady. Efficiency is important because the less heat that is produced the less work the fan has to perform.
6. Replace hard drives with quieter models. Hard drives can also be replaced with laptop hard drives, with solid state devices like compact flash or networked file systems like NFS. This reduces or eliminates this source of noise and reduces system power and cooling requirements.
7. Place a damping material around hard drives (or other spinning drives) such as Sorbothane.
8. Cover the case with sound insulation material such as rubber, foam or fibre mat, although this method has limited effectiveness. The material can (because of its weight) dampen case resonance, and can also absorb some high-frequency sound. Care must be taken to be sure the soundproofing does not interfere with airflow and cooling.
9. In energy-hungry computers, water cooling may be necessary for quiet operation. Older water pumps sometimes can make systems noisier than air-cooled, low-power computers. However, recent advances in 12v DC pump technologies have resulted in highly reduced noise levels from many pumps. In a modern watercooling system geared towards silence rather than performance, the loudest component in the computer is often the Hard Drive or Optical disc drive when performing accesses to the media.

Low-Cost Methods

A number of methods exist for reducing computer noise at little or no added cost.

1. Reduce CPU supply voltage (undervolting). Many of today's CPUs can run stably at their stock speed, or even with a slight overclock, at a reduced voltage, which reduces heat output.

Underclocking can be done for the same effect, however this reduces performance and is not as effective as undervolting; all the same, underclocking may allow further undervolting. Power consumption is approximately proportional to $V^2 \cdot f$, that is, it varies linearly with the clock frequency and quadratically with the voltage. This means that even a small reduction in voltage can have a large effect in power consumption. Undervolting and underclocking can also be used with chipsets and GPUs.

2. Enable Cool'n'Quiet for AMD CPUs or EIST on Intel's CPUs.

3. Reduce fan speed. For newer computers, the speed of fans can be varied automatically, depending on how hot certain parts of the computer get. Lowering a DC fan motor's supply voltage will reduce its speed while making it quieter and lowering the amount of air the fan moves. Doing this arbitrarily could lead to components overheating; therefore, whenever performing hardware work it is advised to monitor the temperature of system components. Fans with Molex connectors can be modified easily. With 3-pin fans, either fixed inline resistors or diodes, or commercial fan controllers, such as the Zalman Fanmate, can be used. Software like speedfan may allow fan speed control. Many newer motherboards support Pulse-width modulation (PWM) control, allowing the fan speed to be set in the BIOS or with software.

4. Mount fans on anti-vibration mounts.

5. Remove restrictive fan grills to allow easier airflow, or replace noisy fan grills with quieter versions.

6. Use software such as Nero drivespeed to reduce the speed of optical drives.

7. Isolate hard disk noise, either by using anti-vibration mounts (generally rubber or silicone grommets), or by suspending the hard disk to fully decouple it from the computer chassis by mounting it in a 5.25 inch drive bay with viscoelastic polymer mounts.

8. Set the hard disk's AAM value to its lowest setting. This reduces the seek noise produced by the hard drive, but also reduces performance slightly.

9. Set operating system to spin down hard drives after a short time of inactivity. This may reduce a drive's life span and commonly conflicts with the OS and running programs, though it can still be useful for drives that are only used for data storage.

10. Defragment the hard drives to reduce the drive heads' need to search widely for data, also improves performance. This may not be practical or even possible with some operating systems, such as Linux, UNIX or free BSD.

11. Arrange components and cables to improve airflow. Wires hanging inside the computer can block the airflow, which can increase the temperature. They can be easily moved to the side of the case so that air can pass through more easily.

12. Remove dust from inside the computer. Dust on computer parts will retain more heat. Fans draw in dust along with outside air, it can build up quickly inside the computer. Dust can be removed with a vacuum cleaner, gas duster or compressed air. Special anti-static vacuum cleaners should be used however to prevent electrostatic discharge (ESD). Ideally, this would be done often enough to prevent a significant amount of dust from ever building up. How frequently this would need to be performed would depend entirely on the environment in which the computer is used.

SOUND POWER AND PRESSURE MEASUREMENT

Though standards do exist for measuring and reporting sound power output by such things as computer components, they are often ignored. Many manufacturers do not give sound power measurements.

Some report sound pressure measurements, but those that do often do not specify how sound pressure measurements were taken. Even such basic information as measurement distance is rarely reported. Without knowing how it was measured, it is not possible to verify these claims, and comparisons between such measurements (e.g. for product selection) are meaningless. Comparative reviews, which test several devices under the same conditions are more useful. Although even then, an average sound pressure level is only one factor in determining which components will be perceived as quieter.

INDIVIDUAL COMPONENTS IN A QUIET PC

The following are notes regarding individual components in quiet PCs. The motherboard, CPU and video card are major energy users in a computer. Components that need less power will be easier to cool quietly. The quiet power supply is selected to be efficient when providing enough power for the computer.

Motherboards

A motherboard based on a chipset that uses less energy may be easier to cool quietly. Many modern motherboard chipsets have hot northbridges (notably nForce4), which may come with active cooling, usually a small, noisy fan. Fanless heatsinks, such as the Zalman ZM-NB47J, ZM-NBF47 or the Thermalright HR-05, may be used to eliminate a noisy chipset fan. Some motherboard manufacturers have got rid of these fans by incorporating large heatsinks or heatpipe coolers, however they still require good case airflow to remove heat. Also, motherboard voltage regulators often have heatsinks and may need airflow to ensure adequate cooling.

Motherboards can also produce coil noise. Undervolting and underclocking generally require motherboard support. Some motherboards can control the fan speed using software like speedfan. Most recent motherboards have built in PWM based fan control for one or two fans.

CPUs

The heat output of a CPU can vary according to its brand and model—to be exact, its TDP. Intel's third revision Pentium 4, using the 'Prescott' core, was infamous for being one of the hottest-running CPUs on the market. By comparison, AMD's Athlon series and the new Intel Core 2 perform better at lower clock speeds, and thus produce less heat.

Modern CPUs often incorporate energy saving systems, such as Cool'n'Quiet, LongHaul, and SpeedStep. These reduce the CPU clock speed and core voltage when the processor is idle, thus reducing heat. The heat produced by CPUs can be further reduced by undervolting, underclocking or both. Most modern mainstream and value CPUs are made with a lower TDP to reduce heat, noise, and power consumption. Most of Intel's desktop Core 2 Duo processors have 65W TDPs, and AMD has newer processors with a TDP between 35W and 65W. Some processors come in special low power versions.

Modern low power CPUs

Maximum TDP:
1. Athlon 64 X2: 45W, 65W, 85W.
2. Intel Core and Core 2 series: 35W, 65W, 85W.
3. Intel Pentium M and Celeron M (lacks SpeedStep).
4. VIA C7: 12W-20W (Fanless).
5. Transmeta processors.

6. Geode (processor): 5W, 25W.
7. Intel atom: 2W, 4W, 8W.
8. Loongson: 1W, 5W, 7W (Fanless).

Video cards

Video cards can produce a significant amount of heat. A fast GPU may be the largest power consumer in a computer. For instance, 161 watts peak power consumption for an ATI Radeon HD 2900 XT 512. Because of space limitations, video card coolers often use small fans running at high speeds, making them noisy.

Display options for making a quiet computer include:

1. Use motherboard video output-typically motherboard video takes less power than an external video card, typically at the price of lower gaming or HD video decoding performance.
2. Select a video card that does not use a fan.
   (a) many of the more efficient GPUs are available in fanless models.
   (b) many older video cards used less power than more recent video cards, thus heatsinks or fans were often not required.
3. Replace the GPU cooler with a larger heatsink and possibly a larger, slower fan.
4. No video card—use a terminal, thin client, USB display or desktop sharing software if display required.

Power supplies

PSUs are made quieter through the use of higher efficiency (which reduces waste heat and need for airflow), quieter fans, more intelligent fan controllers (ones for which the correlation between temperature and fan speed is more complex than linear), more effective heatsinks and through designs which allow air to flow through with less resistance. For a given power supply size, more efficient supplies, such as those certified 80 plus, generate less heat.

Selecting a power supply of appropriate wattage for the computer is important for high efficiency and minimising heat. Power supplies are typically less efficient when lightly or heavily loaded. High wattage power supplies will typically be less efficient when lightly loaded, for instance when the computer is idle or sleeping. Most desktop computers spend most of their time lightly loaded. For example, most desktop PCs draw less than 250 watts at full load, and 200 watts or less is more typical.

Power supplies with thermally controlled fans can be made quieter by providing a cooler and/or less obstructed source of air. For instance, the power supply is in a separate compartment in the Antec P180 to keep the air supplied to the PSU cool. The fan in a power supply can be replaced with a quieter one, although there is a risk of electric shock when doing this, and it usually voids the warranty.

Fanless power supplies are available.

1. Some of them are equipped with large passive heat sinks and rely on convection or case airflow to dissipate heat. It is also imperative that such fanless power supplies be installed in a case with good ventilation.
2. There are also fanless DC to DC power supplies that operate like those in laptops, using an external power brick to supply DC power, which is then converted to appropriate voltages and regulated for use by the computer. These power supplies usually have lower wattage ratings.

The electrical coils in power supplies can produce noise which can become noticeable in a quiet PC.

Cases

Cases designed for low noise usually include reasonably quiet fans, and often come with a relatively quiet power supply. Some cases for quiet computers incorporate heatsinks to cool components passively.

Cases that provide lots of space make it easier to quiet a PC, both by allowing for airflow and by accommodating large coolers.

Case airflow

Noise optimised cases like the Antec P180 and Antec P150 often have ducting and partitioning within the case to optimise airflow and thermally isolate components. For example, the P180 has the PSU mounted in the bottom of the case in an isolated partition. This design feature allows cooler air to enter the PSU, reducing the necessary airflow and accordingly, the noise output of the fan. Apple has also employed this tactic in their G5 workstations in an effort to reduce noise. Antec's Sonata is often considered by the mainstream to be one of the quietest PC cases; however, it has since been surpassed by the P180 and other more-advanced cases. Vents and ducts may easily be added to regular cases.

More obstructive fan grills increase pressure drop and lower airflow, necessitating higher fan speeds and more noise output. They also increase the turbulence of the flow, which causes some noise of its own. Cases designed to be quiet typically have wire grills or honeycombed fan grills, which perform almost as well as wire grills; both are far superior to the old style of stamped grill.

Features that facilitate neat cable management, such as brackets and space to run cables behind motherboard tray, help increase cooling efficiency.

Air filtering

Air filters can help to prevent dust from coating heatsinks and surfaces, thereby impeding heat transfer, making fans spin faster. However the filter itself can increase noise if it restricts airflow too much, or is not cleaned, requiring a larger or faster fan to handle the pressure drop behind the filter.

Regular cleaning of air intake filters will help to keep airflow minimally restricted and keep the interior clean. In some cases a fine mesh intake screen is sufficient to stop most large dust particles from entering the system. These screens should be vacuumed or washed to remove dust.

Case soundproofing

The inside of a case can be lined with dampening materials to reduce noise by:
1. Attenuating the vibration of the case panels via extensional damping or constrained-layer damping.
2. Reducing the amplitude of the vibration of the case panels by increasing their mass.
3. Absorbing airborne noise, such as with foam.

Cooling systems

Heat sinks

Heat sinks that operate efficiently with little airflow are often used in quiet computers. Typically they are (relatively) large, and have larger spaces to allow freer airflow. Often heat pipes are used to help distribute heat. For instance, in 2007, the Scythe Ninja or the Thermalright Ultra-120 were frequently used as CPU heat sinks in quiet computers.

Fans

Bearing and motor noise vary between different fan models and often between different samples of the same model.

Quiet PCs typically use larger (e.g. 120 mm) low-speed fans. Although 140 mm fans are made by some manufacturers, such as Aerocool and Yate Loon, there are very few cases or heatsinks that can use them. Fan adapters, which allow larger fans to be used in place of smaller ones, and fan brackets, like the Zalman FB123, often help when replacing small fans.

Quiet fan manufacturers include Nexus, EBM-Papst, Yate Loon and Scythe. In situations where the resistance to flow is very low, like in free-air conditions, Noctua fans also perform very well. Extensive comparative surveys have been posted by Silent PC Review and MadShrimps.

Fan controllers can be used to slow down fans and to precisely choose fan speed. Fan controllers can produce a fixed fan speed using an inline resistor or diode, or a variable speed using a potentiometer or pulse width modulation (PWM). Resistor-based fan control feeds the fan a lower voltage, while PWM fan control rapidly cycles between feeding the fan full voltage and no voltage. PWM fan control reduces rotational speed, and is the easiest and most efficient option for motherboards which have PWM fan headers. This is because PWM fans in conjunction with the motherboard chipset obtain temperature data from Digital Temperature Sensors on the CPU itself. All PWM fans are four pin, and if plugged into a conventional three pin supply will operate at full speed just like a three pin fan.

Fans can also be plugged into the power supply's 5 volt line instead of the 12 volt line (or between the two for a potential difference of 7 volts, although this cripples the fan's speed sensing) to run them at a reduced speed. Most fans will run at 5 volts once they are spinning, but may not start reliably at less than 7 V. Some simple fan controllers will only vary the fans' supply voltage between 8 V and 12 V to avoid this problem entirely. Some fan controllers start the fan at 12 V, then drop the voltage after a few seconds.

Soft mounting fans (e.g. with rubber or silicone fan isolators) can help reduce transfer of fan vibrations to other components. Intel has recently developed a piezoelectric fan for use in desktop PCs, which is quieter than motor fans and consumes a fraction of the power.

Watercooling

Watercooling is a method of heat-dissipation by transferring the heat through a conductive material which is in contact with a liquid, most often demineralised water and an additive to prevent bacterial growth and provide cosmetic effects. This heated water travels in a loop which usually contains a reservoir, radiator and pump. Recent advances in 12 V DC pump technologies (for the first time specifically geared-for PC development) allow for new pumps to be both extremely powerful and extremely quiet. Loops can be made up of any combination of these items and some aren't required such as the radiator or reservoir if alternatives methods are used. The radiator often uses one or more fans to air cool the radiator fins and dissipate the majority of the heat at this point.

The most common loop order is reservoir to pump, radiator then the watercooling block and back to the reservoir. The radiator and fan efficiency has the greatest effect on the noise level and cooling efficiency but watercooling is currently the most effective and potentially quietest methods of cooling above ambient temperatures.

There is an inherent danger in the use of water around electrical equipment and leak testing the loop is always recommended before attaching any parts to the motherboard, after all loop connections have been made. The 12 V DC pump can be run using batteries or a power supply making sure no power is going to any other part of the system. Because of these risks and the use of water under pressure watercooling is a greater technical challenge to set up due to the number of components and case modification usually required.

Secondary storage

Hard drives

Previously, hard drives used ball bearing motors, but they got noisy when the rotational speed of the drive was increased to 5400 RPM or 7200 RPM. More recent desktop hard drives use fluid bearing motors. The first hard drive widely reputed to be quiet was the Seagate Barracuda ATA IV.

The smaller 2.5′ form-factor hard drives generally vibrate less, are quieter, and use less power than the traditional 3.5′ drives. On the other hand, they often have lower performance and less capacity, and cost more per gigabyte. To minimise vibrations from a hard drive being transferred to, and amplified by, the case, hard drives can be mounted with soft rubber studs, suspended with elastics or placed on soft foam or Sorbothane. Hard disk enclosures can also help reduce drive noise. Care must be taken to ensure that the drive gets adequate cooling. Hard disk temperatures can often be monitored by SMART software.

External components

Laptops

Laptop computers typically do not have power supply fans or video card fans, and they use smaller hard drives. They also use many lower power components. However, laptop CPU coolers are usually smaller, so may be noisier than their desktop counterparts. Limited space, limited access and proprietary components make silencing laptops more difficult.

A few laptops do not use cooling fans, for instance the Dell Latitude X1, Panasonic Toughbook W5 and T5, Fujitsu Lifebook P7120. Also, some netbooks, such as the Dell Mini 9, 10 and 12 do not have fans. The Mini 9 used SSD rather than a hard disk. The OLPC XO-1 has no internal moving parts.

Monitors

CRT monitors can produce coil noise, as can the external power supply for an LCD monitor or the voltage converter for the monitor's backlight. LCD monitors tend to produce the least noise (whine) from when at full brightness. Reducing brightness using the video card does not introduce whine, but may reduce colour accuracy. An LCD monitor with an external power supply tucked out of the way will produce less noticeable noise than one with the power supply built into the screen housing.

Printers

Dot matrix and daisy wheel printers are often noisy, and soundproofed boxes or cabinets can be used to reduce the noise. Another solution is to locate the printer away from the immediate work area or in another room, especially if it can be controlled through a local area network.

PARTS AND INSTALLATION STEPS CAN HELP BRING PEACE BACK TO YOUR WORKSPACE

Following parts and installation steps can reduce the noise. A wide range of PC-quieting products are available. They can be used singly or in combination.

Acoustic insulation: Sound-absorbing foam, installed inside the case, turns your PC into an almost-soundproof miniature room, though it can make your PC run a bit hotter.

Hard-drive enclosure: Dampen the whine of your PC's hard drive by enclosing it in an isolation chamber. Note: This requires a free 5.25-inch drive bay for installation.

Silent case: If you're building your own PC, consider a case designed with a quiet power supply, quiet fans, and large air vents.

Case fans: Specially designed and thermostatically controlled case cooling fans can move enough air to keep your system cool, while keeping noise to a minimum.

Quiet power supplies: The power supply may be the noisiest component of your PC. Quiet power supplies can dramatically reduce the clatter.

Gasket kits: Gaskets and insulated mounting washers for power supplies and fans can help reduce vibration.

CPU coolers: If your existing CPU cooling fan sounds like a miniature vacuum cleaner, consider fans with larger heat sinks and quiet rotary blades; these can virtually eliminate CPU cooling noise.

Move the PC

First, try a simple solution. If your system is sitting on the desk next to you, move it onto the floor (assuming that its case is designed to sit vertically rather than horizontally). The vibration from the PC often sets up a sympathetic vibration with the desk surface, turning the case into a soundboard. Moving the computer under your desk can eliminate this problem.

You'll get the best results if the PC sits on a carpeted surface. If you have a tile or hardwood floor, try putting the system on a carpet scrap or a piece of packing foam. For a really 'far out' solution, purchase keyboard, mouse, and monitor extension cables (available from any computer dealer) and move the PC farther away. This will, of course, make the CD or DVD drives harder to access.

Tighten up to reduce rattling

Sometimes, the biggest contributor of PC noise isn't so much the direct sound of fans and motors as it is the vibration of PC components. It can be as annoying as a rattle somewhere in your car. You can often solve these problems by unplugging your PC, removing its cover, and methodically tightening the mounting screws of parts such as the power supply, drives, motherboard, and cooling fans. Be careful, though: Overtightening screws is worse than leaving them too loose, since you could damage components. Some noise-reduction kits include screws with polymer or rubber washers to reduce vibrations further.

Install gaskets

Inexpensive polymer gasket kits can isolate vibrating fans and parts from the case, reducing noise. NoiseMagic's No-Vibes hard-drive suspension kit can do the same for your hard drives. Follow the manufacturer's instructions.

Install heat-sensitive case fans

Because of the heat generated by fast processors and other components, almost all PC cases have one or more fans dedicated to exhausting heat. The inexpensive fans that many manufacturers use can be noisy, but fans don't always have to work at full speed, since a PC produces much less heat when it's idling. Thermostatically controlled fans use temperature sensors that slow the fans down when the internal case temperature drops.

Alternatively, some manufacturers supply single-speed fans that are designed to be quiet. Installing them is usually simple, though you may need to remove drives or add-in boards to take out the old fan and insert the new one. Note whether the fan power is connected on the motherboard or to a power-supply connector. If your new fan has an external temperature sensor, follow the manufacturer's directions to place the sensor in the optimum location.

Upgrade to a Quiet Power Supply

Your computer's power supply may be one of the primary sources of noise. Many rely on two built-in fans to move large amounts of air. Switching to a power supply that is designed to be quiet can dramatically reduce your PC's overall noise level. Designs vary, but most units simply employ larger fans that can turn more slowly while delivering the same amount of cooling air. Some also use thermostatic controls to slow down or speed up the fans depending on the case temperature. For instructions on upgrading your power supply, see 'keep it powered, keep it cool'. Be sure to buy a power supply that has sufficient wattage to handle all of your PC's components.

Install a new CPU cooler

Today's processors run very hot—often from 145 to 175 degrees Fahrenheit. At these temperatures, effective and continuous cooling is essential. In fact, an uncooled CPU can grow hot enough to damage itself in a matter of seconds. The CPU heat sink and fan that came with your PC are probably louder than you realise. You can lower the noise by installing a specially designed CPU cooler. Most coolers include a larger, more efficient heat sink that can be paired with a much quieter fan. Buy a cooler that's designed for your processor, and follow the manufacturer's directions for installing it.

Install acoustic insulation

PC still isn't quiet enough after you've taken the preceding steps, try some more-extreme measures. Acoustic insulation kits let you add a layer of special sound-absorbing foam on the inside of your case. Installation is relatively easy: You cut the foam to size, peel off a backing, and stick it in place. The SilentDrive enclosure isolates your hard drive in a sound-deadening box. You'll need a free 5.25 inch mounting space for each SilentDrive. Follow the manufacturer's directions to install it.

Chapter 27

Digital and Webcam Imaging

INTRODUCTION

A charge-coupled device (CCD) is a device for the movement of electrical charge, usually from within the device to an area where the charge can be manipulated, for example conversion into a digital value. This is achieved by 'shifting' the signals between stages within the device one at a time. CCDs move charge between capacitive bins in the device, with the shift allowing for the transfer of charge between bins.

Often the device is integrated with an image sensor, such as a photoelectric device to produce the charge that is being read, thus making the CCD a major technology for digital imaging. Although CCDs are not the only technology to allow for light detection, CCDs are widely used in professional, medical, and scientific applications where high-quality image data are required.

WEBCAM

A webcam is a video camera which feeds its images in real time to a computer or computer network, often via universal serial bus (USB), ethernet or Wi-Fi. Their most popular use is the establishment of video links, permitting computers to act as videophones or videoconference stations. This common use as a camera for the worldwide web gives the webcam its name. Other popular uses include security surveillance and computer vision. Webcams are known for their low manufacturing costs and flexibility, making them the lowest cost form of videotelephony. They have also become the source of security and privacy issues, as some inbuilt webcams can be remotely activated via spyware. This chapter provides a brief overview of the concept image noise reduction by way of 'stacking' images and will focus on telescope CCD imaging, however the principles are applicable to many other fields (including webcam imaging). As most amateur CCD imagers know, one of the largest problems with CCD imaging is keeping the background noise in the image to a minimum. All CCD images contain some amount of noise in the image. One method used to reduce noise is to cool the temperature of the CCD chip. This can help greatly in reducing the amount of thermal noise that causes CDD images to look less than ideal. However, there are practical limits to the amount of noise reduction that this method can obtain. Another commonly used method to reduce noise is to take multiple images of the same object and then to 'stack' the images using the computer. By doing this, the noise in the image will tend to 'average out' and the object image will add, resulting in a better signal to noise ratio. For image stacking to work in an ideal fashion, a few things have to be true:

1. The noise in each image frame must be truly random.
2. The 'object' being imaged must be the same in each image.

If the above criteria are not met, the amount of image improvement obtained by stacking images will not be ideal. The noise in the image can be further reduced by stacking more images (remember that each image must be a different exposure under the same conditions). However, due to the way things work in our world, each 3 dB of noise reduction requires that double the number of images be stacked! In other words, stacking two images results in a 3 dB improvement. To get 6 dB requires that 4 images be stacked; to get 9 dB requires that 8 images be stacked. As you can see, this can really add up if you want to get something like a 20 dB improvement.

ELIMINATION OF THE CCD CAMERA NOISE IN MICROSCOPIC MEASUREMENTS OF MACHINE ELEMENTS

Camera-computer integration makes it possible to replace the observation of an object in the ocular for its more convenient observation in the visual display unit. This measurement approach has been known for many years now. This type of measuring techniques is targeted now for specified models of microscopes and towards some specified measuring tasks, and therefore lacks universality.

The Metrology and Quality Testing together with the Computer System Architecture Groups of the Wroclaw University of Technology have developed (as a joint project) a universal, teaching and research oriented test stand for the microscopic measurements of machine elements (Fig. 27.1). A large measuring microscope equipped with digital systems for the measurements of displacements of the measuring table and a CCD camera interfaced with the optical system of the microscope were employed.

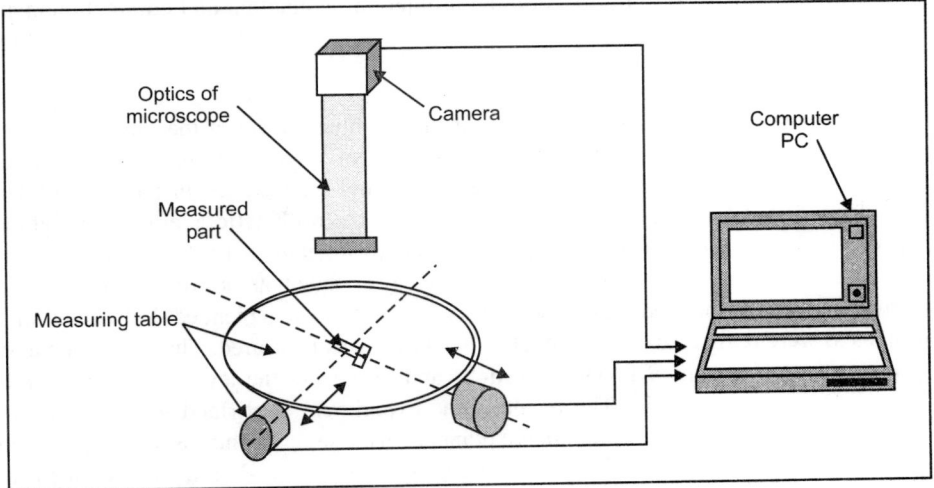

Fig. 27.1. Scheme of the measuring stand.

The measuring stand is linked to an IBM PC computer, which was provided with a video chart and a dedicated 'video measurement' software. The program establishes the matrixes of distances and angles based upon the graph of the observed image pixels, which permits to evaluate the geometry of the measured object. Some of the implemented image processing tasks are the determination of contour, of negative, contrast, zoom, and rotation.The fundamental problems of information processing, apart of the optical system faults, stem from the camera and the related noise of the vision signal. It is of crucial importance on the quality of further image processing and consequently, on the accuracy of geometrical form measurements. The elimination of noise impact and other correctional measures result in the improved image quality and make it possible to obtain an unambiguous and accurate identification of the individual features of the observed object.

Reasons of Camera Mapping Errors

Video generated images are not free of different types of defects and distortions which stem from the quality of the electronic materials used in the construction of the camera. These faults are:

1. Distortions of some fragments of the image which are due to errors in its description by individual pixels.
2. Truncated image (at the top or bottom).
3. Difference in the brightness gradients.
4. Eigen-noise of the camera electronics.

The reasons for the above listed phenomena stem from:

1. Camera assembly inaccuracies.
2. Impurities in or faults of the lenses and optical system aligning.
3. Micro-faults and inaccuracies in the process of manufacture of CCD converter matrixes.
4. Distortion of analogue signals in the camera and in the camera/video chart electronics.

Some of the technological faults may be eliminated.

For example: different sensitivity of individual pixels at constant illumination may be corrected by establishing sensitivity mapping. In this section the problem of evaluation and elimination of noise in the vision signal has been dealt with.

Experimental Work

An incentive for this research was the observation of the brightness level of the same pixels for some established conditions of image exposition. The fluctuations of brightness, which are present within one image or its sequence, stem from the non-stability of and interference in the image processing system. It is claimed that for a large number of specimens the arithmetic average of brightness for individual pixels is equal to the constant component. That also means that subsequent components of the brightness level in the series of recorded image frames may assume both positive and negative values.

To confirm the above formulated thesis, the optics of the camera was blended. Individual images (of a series of images) were registered and the level of blackness was measured. This technique allows for the control of constant elements of image distortions and to register the level of noise for individual pixels of the CCD converter matrix. The result of the experiment confirmed the base assumptions. Figure 27.2 shows the distribution of the number of points with the same brightness for a single exposition. It may be assumed as a Gauss distribution with a slight asymmetry. The dominant component with the brightness 6 and 7, which is an average brightness for the whole image, may easily be singled out. It may also be assumed that there is a symmetry with respect to the central position of points with brightness 6 and 7 which is valid within limits given by the brightness of points 2 and 11. The pattern of the noise level distribution for a single image confirms the assumption that the same pattern is valid for the consecutive images. As a result, the level of noise for a selected pixel in the bitmap in the next images will be positive or negative with respect to the constant component of the noise. The constant component of the noise is identical for each consecutive image. As a result, the brightness of a given pixel, measured as the arithmetic average, will remain constant (at a certain fixed level) for a sufficiently large number of expositions.

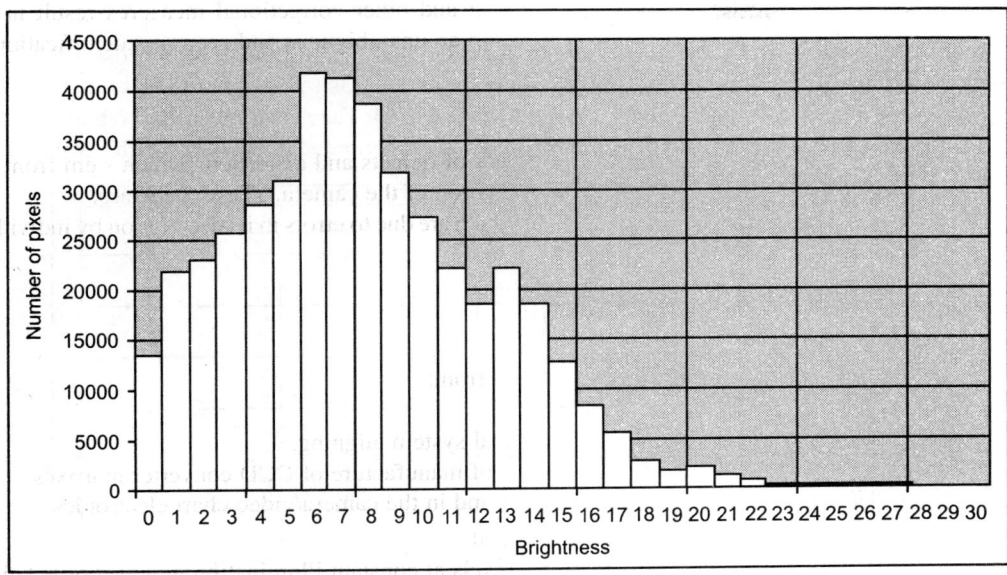

Fig. 27.2. Distribution of the number of pixels in term of pixel brightness on a bitmap made of one image.

Since the brightness value of a pixel in this process is approximately invariable and there is a possibility to include this information in an appropriate procedure, than this is the elimination of noise. The algorithm of noise elimination consists in the summation of brightness of each pixel in the image matrix (for the same exposition) and division of the results by the number of expositions. Figure 27.3 presents the distribution of noise for a bitmap made of 255 images.

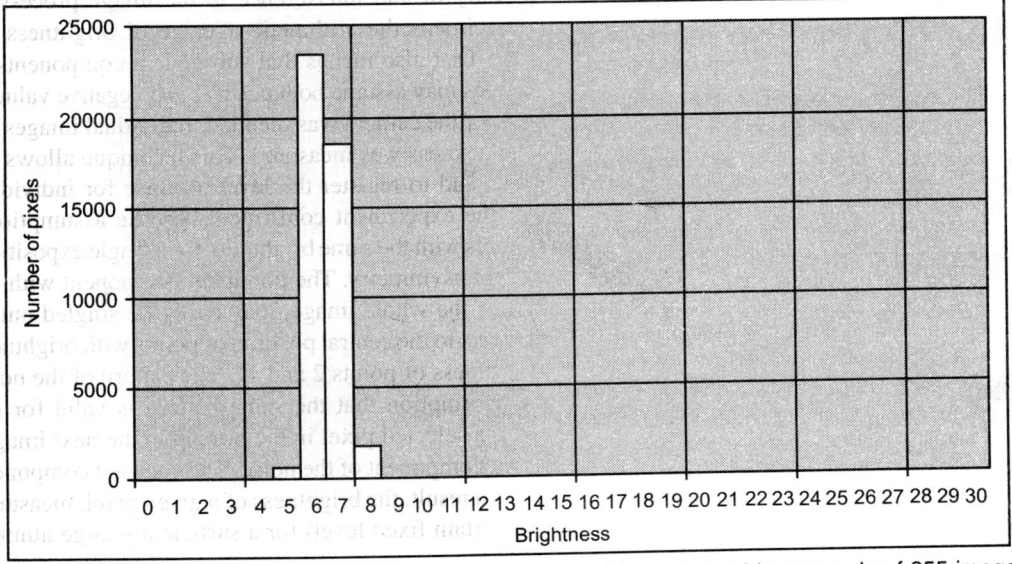

Fig. 27.3. Distribution of the number of pixels in terms of brightness in the bitmap made of 255 images.

Final Results

Calculations of the mean and standard deviations for 255 consecutive images obtained for the same expositions were made. The level of noise for individual pixels was measured and presented together with the mean and standard deviations, which is shown in Fig. 27.4. The values obtained (both in Fig. 27.4 and in the implemented algorithm of noise elimination) were determined based upon textbook formulas.

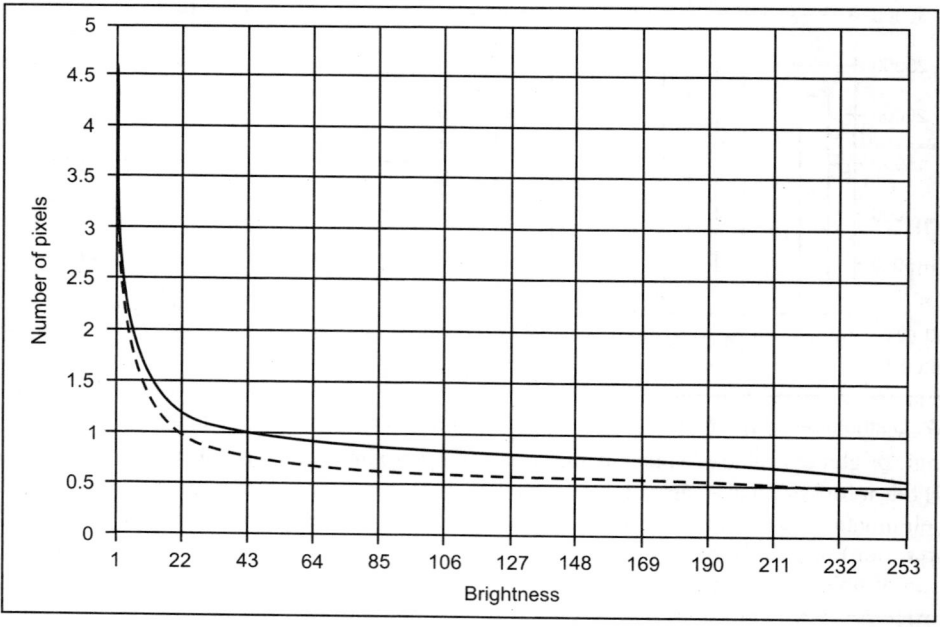

Fig. 27.4. Mean (dotted) and standard deviations (dark) for the distribution of pixel brightness in a bitmap as a function of the number of component images.

Based upon the analysis of the graph in Fig. 27.4 it is evident that with the proposed method the level of noise may be reduced 8 times (over 18 dB). This technique enhances the quality of images, which is beneficent to the comfort of operation and, consequently, to the level of accuracy in the evaluation of the geometry of the controlled objects.

Chapter 28

Mobile Phone

INTRODUCTION

It is estimated that over a third of the world's population spend time talking on mobile phones. The quality of that communication experience very much depends on the clarity of the voice that is heard at each end of the conversation. In an increasingly noisy world, it is easy to see that noise reduction and echo cancellation play a major role in helping mobile phone developers meet the expectations of this massive and demanding customer base.

Even away from this huge consumer driven market, people such as the emergency services and Formula 1 drivers depend on communication systems that can delivery clear voice communication even in the most extreme environments. This chapter discusses some of the issues faced by communication equipment developers and a few of the solutions that are applied to help them deliver a noise and echo free world.

CHALLENGE

Noise and echo can have a big influence over the performance of communication systems. For a voice activated telematics system, a recognition accuracy degraded to 90 per cent by automotive noise will render it useless to dial a 10 digit phone number. In a fire engine, poor communication with the control room due to the screaming of the siren can result in a slower or confused response, potentially resulting tragedy. For a Formula 1 driver and the pit crew, engine noise could easily wipe out their chance of a podium finish, if there was a misunderstanding regarding strategy. Although it is taken for granted by most of us, technology for handling noise and echo is essential for many of today's communication systems.

GOOD SYSTEM DESIGN—AN IMPORTANT START

Audio engineers know that many of the problems that can cause noise and echo could be minimised at the initial design stage. A good example that illustrates some of these design considerations is a hand free communication system in a car.

Simply using a uni-directional microphone pointing towards the diver, rather than an omni-directional microphone, can eliminated a great deal of road noise, sound from the music system and passenger noise (Fig. 28.1).

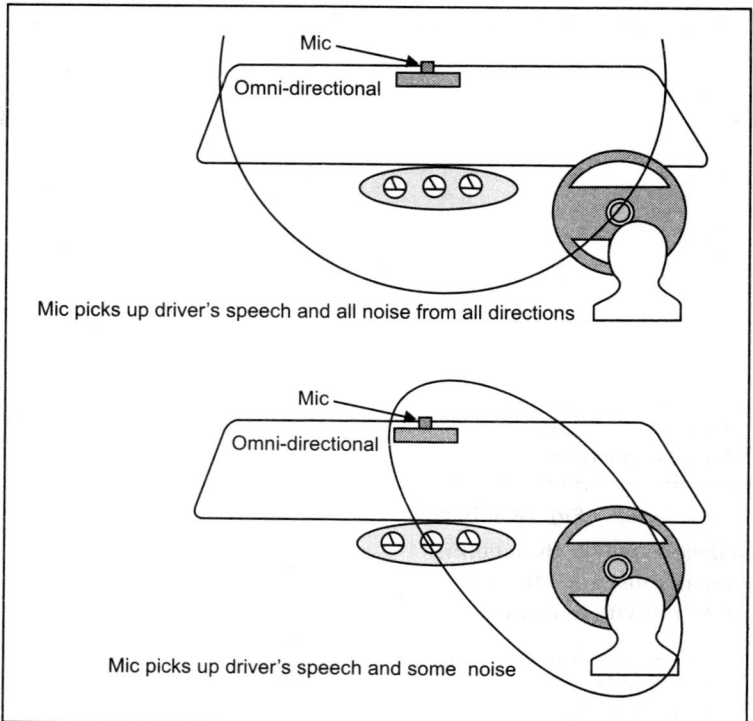

Fig. 28.1. Omni-directional microphone.

The location of the microphone relative to the loudspeaker can also have a considerable impact on the performance of the hand-free communication system, as the microphone will pick up the sound from the loudspeaker to a greater or lesser extent depending on its position and orientation relative to the loudspeaker (Fig. 28.2).

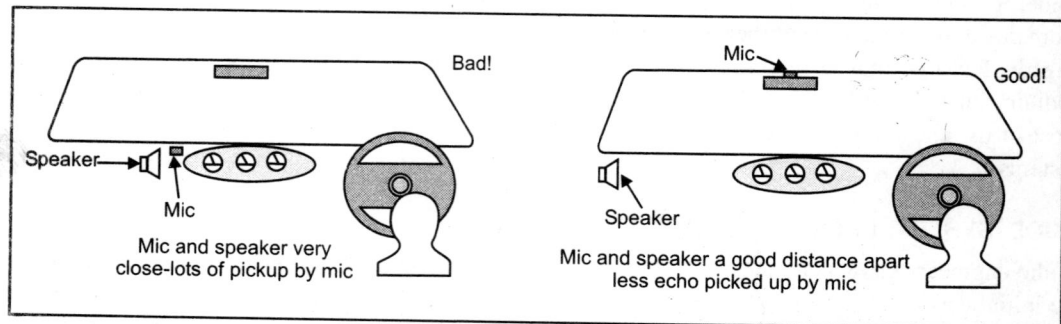

Fig. 28.2. Mic and speakers.

Developing a good analog filter for the signal input from the microphone and choosing a higher sampling rate for its conversion to the digital domain are important first steps in the electronic design. The good analog filter can reduce electrical interference and microphone buffeting, for example, while the higher sampling rate will offer a broader bandwidth, capturing more of the voice signal frequencies and therefore a better quality of voice for the system to handle (Fig. 28.3).

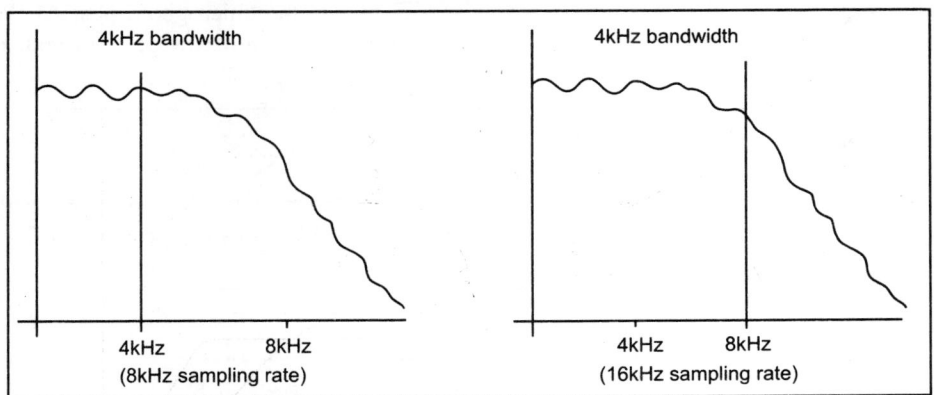

Fig. 28.3. Bandwidth and sampling rate.

A further consideration will be the amplification and management of the signal strength; too big and the signal will experience distortion through clipping, too small and it will be embedded in the noise of the system and will be difficult to extract (Fig. 28.4).

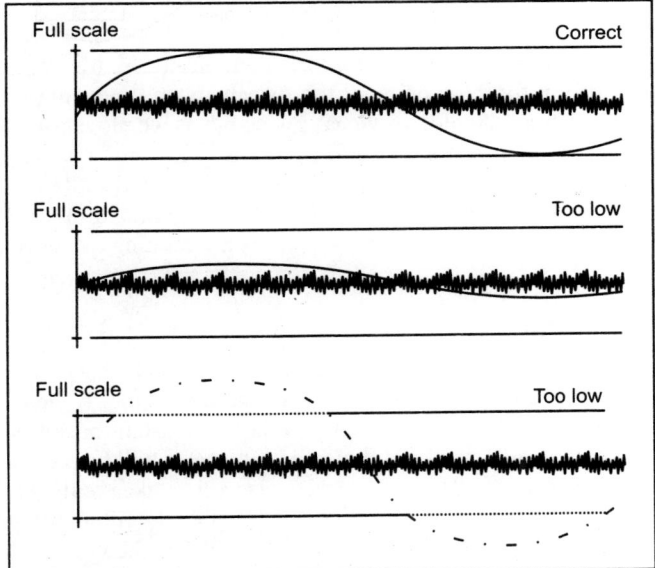

Fig. 28.4. Amplification and signal vibrations.

A strong, clear, undistorted, voice signal is an essential starting point for the achievement of good voice quality. Once this is in place, the processes of noise reduction and echo cancellation can be applied for further enhancement.

DRIVE THRU—A SIMPLE SYSTEM

The communication system at your local drive through restaurant is a great example of a voice system that could employ noise reduction software. When you speak into the 'microphone post' to place your order, the system ideally removes background traffic noise, and even the noise from your own car making order errors less likely and order taking faster (Fig. 28.5).

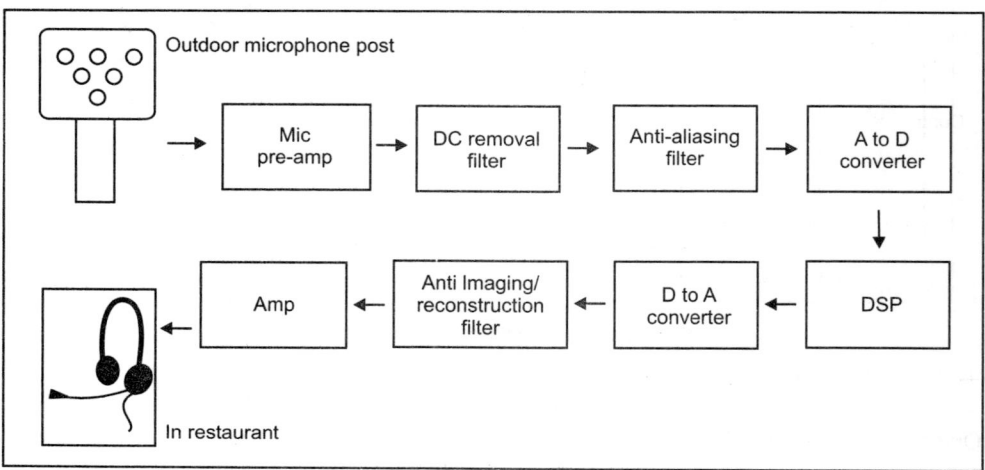

Fig. 28.5. Communication system.

You will see from the diagram above, that the system contains a DSP (digital signal processor). This processor provides the platform for the running of the speech enhancement algorithms. Most often this will be a 'Texas Instruments' or an 'Analog Devices' device but as we move forward, it is equally likely to be a Bluetooth device or some new form of WiFi processor. In some instances, these complex and highly refined voice algorithms are integrated into dedicated silicon and sold to developers as a discreet component performing all of the functions illustrated in the diagram and more. Alternatively, in more complex systems, these algorithms may be run as a task on operating systems such as OSE, Linux, Windows or QNX.

REFERENCED NOISE FILTER

The referenced noise filter is an example of a noise reduction technique that can target a specific type of problem where the noise-producing source is man-made and accessible. If we use the example of the fire engine in the introduction, the problem faced here is the constant background sound of the fire engine's siren. When the engine driver talks into his microphone, it will be picking up his voice as well as the sound of the siren (Fig. 28.6). The referenced noise filter takes a signal (direct tap) from the source of the noise (the siren) and use this as a reference to enable the filter to target the noise that need to be removed. Using this complex software algorithm on the DSP in the system, the results can be quite profound, resulting in the region of a 90 per cent reduction in the unwanted sound.

VOICE RECOGNITION ENHANCER

Voice recognition is still very much an emerging technology. One of its key applications will be for the control of in-car telematic systems. Using your voice, you'll be able to instruct your navigation system, dial up your favourite restaurant to book a table, maybe even open the sunroof and select 'Katrina and the Waves'.

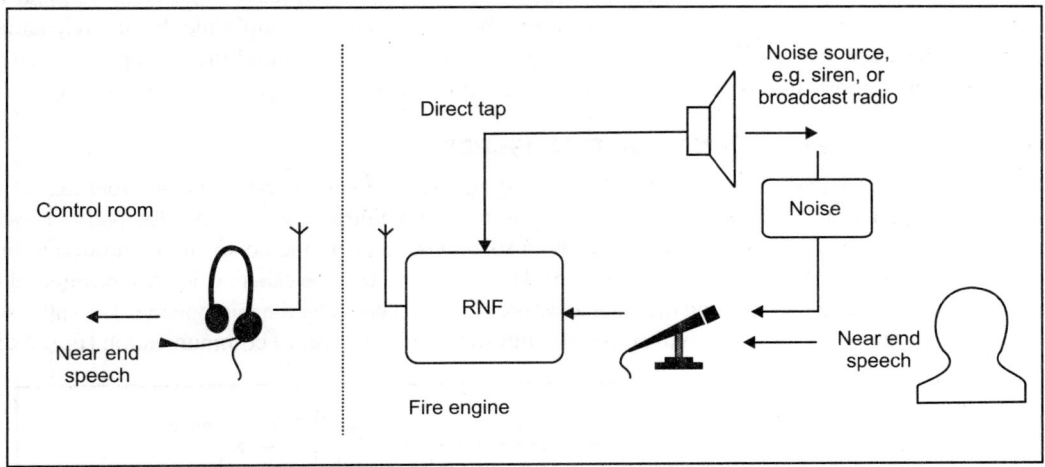

Fig. 28.6. Control room along with noise filter.

One of the biggest challenges for these systems has been the over-coming of noise. As discussed, a 90 per cent voice recognition accuracy is close to useless (one digit in a 10 digit phone number will always be wrong). The challenge is that as the car accelerates, wind and road noise starts to degrade the performance of the system (Fig. 28.7).

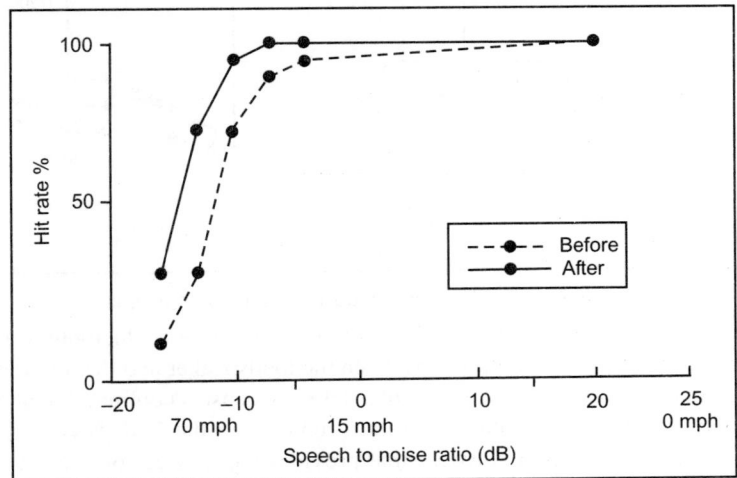

Fig. 28.7. Wind and noise graph.

The x-axis on the graph shows that as the car accelerates from 0 mph to greater than 70 mph, the 'speech to noise ratio' declines. In turn, the Hit Rate (voice recognition accuracy) also declines. Using a voice recognition enhancer software algorithm on the DSP can provide valuable improvements in the hit rate at the higher speeds. It is easy to see that a 10 per cent improvement can be the difference between a successful voice instruction and a failed one. A VRE basically works by analysing the sampled input data and making decisions on that. It decides what is speech and leaves that alone (crucial for successful speech recognition), and what is noise and then reduces that in amplitude. It can differentiate successfully between speech and noise because; speech varies rapidly in amplitude and pitch, whereas noise varies much more slowly (to the point of being what noise audio engineers term 'stationary').

HANDS-FREE SYSTEMS—A NEW SET OF ISSUES

For in vehicle communication systems, hands-free systems are essential for safety and in order to comply with driving legislation. However, a hands-free system can be complex to design effectively. As with the voice recognition system, there is concern about road, wind and engine noise but in addition to this, a hands-free system will generate echo for the caller. The problem is that the callers voice will be emitted by the loudspeaker and will travel around the vehicle before being picked up by the microphone. The caller will hear an echo of their voice and this will constantly interfere with the overall communication (Fig. 28.8).

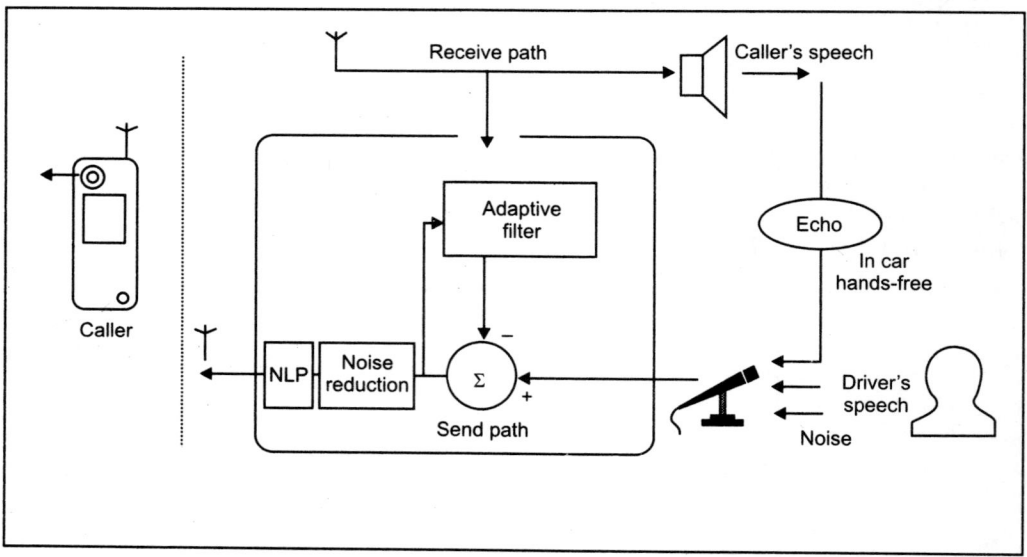

Fig. 28.8. Acoustle echo canceller and noise canceller.

Indeed this problem can exist even when using a mobile handset or a Bluetooth headset—there will be acoustic echo generated by direct coupling between the loudspeaker and the microphone to a greater or lesser degree. To make hands-free systems viable, developers use a combination of a noise reduction algorithm along with an echo cancellation algorithm running on the DSP device. Often, this is also bundled with some speech enhancing software that provides the services of some acoustic filters and gain control to boost the clarity of the voice.

HOW CAN YOU BENEFIT FROM THIS TECHNOLOGY?

The technology described is readily available as off-the-shelf software components. It is developed by expert audio and software engineers and proven in millions of devices. Developers choose to buy in these algorithms in order to ensure that they have the very best voice quality for their products. Many of these algorithms are optimised for small footprint devices and can also be tailored for specific applications, e.g. bluetooth headsets, 3G mobile phones, Tetra radios, Formula 1 racing, etc.

A range of evaluation hardware is readily available, enabling engineers to assess the impact this technology would have on their next design and to enable them to become familiar with the issues surrounding this type of software component.

CONCLUSION

When you receive the next call on your mobile phone, just listen to how clearly you can hear your caller. NCT alone has invested over 70 man-year's of mathematical, acoustic and software programming expertise to deliver the software that provides this level of clear speech. Noise reduction and echo cancellation has been a big subject for the communications sector over the last 10 years. It is expected that as voice recognition and Bluetooth systems roll out, the next 10 years will be equally as busy for voice enhancement software companies such as ourselves.

SECTION V

Noise Reduction in Construction Industries

Chapter 29

Stone Quarrying and Crushing

INTRODUCTION

A quarry is a type of open-pit mine from which rock or minerals are extracted. Quarries are generally used for extracting building materials, such as dimension stone, construction aggregate, riprap, sand, and gravel. They are often colocated with concrete and asphalt plants due to the requirement for large amounts of aggregate in those materials. The word quarry can include underground quarrying for stone, such as bath stone. Many people and municipalities consider quarries to be eyesores and require various abatement methods to address problems with noise, dust, and appearance.

STONE QUARRYING

The requirement of the stone is increasing with demands from rapidly growing infrastructural projects and housing sector. The stone is exploited from the quarries and many a times crushed at the site for getting necessary sizes. The stone is treated as minor minerals and therefore the stone quarrying do not attract the provisions of Environmental protection act.

The stone mining areas are generally identified by the mining department and collector, through the mining department regulate the operation of stone quarries. The operations of the quarries has operations including, blasting.

No quarrying operations shall be carried out without obtaining development permission of the planning authority/district collector/sub-divisional officer/tehsildar.

Noise Pollution

Blasting

1. Blast holes shall be judiciously charged.
2. No blasting shall be done when there is low cloud ceiling.
3. Millie second delayed detonation to be used.
4. All other guidelines of the explosives department and mining department regarding blasting operations shall be strictly adhered to.

Drilling

The workers shall be provided appropriate personal protective equipment viz. ear mufflers/ear plug or noise proof cabins.

Heavy earth moving machinery (HEMM)
1. The engine exhausts of HEMM to be fitted with mufflers and cabins shall be noise proof.
2. HEMM shall be properly maintained.
3. Operators shall be provided with ear mufflers/ear plugs.
4. Imposition of speed restriction of HEMM near residential area shall be enforced.
5. The haulage path of the HEMM shall be re-routed so that it is away from the residential area.

STONE CRUSHING INDUSTRY

Stone crushing activity is a significant industrial sector across the country involved in production for crushed stone of various sizes depending upon the requirement and catering requirement of raw material for various construction activities such as construction of Roads, Railway tracks, Highways, Bridges, Buildings, Canals, etc. The existing number of stone crushing units is expected to grow rapidly further in view the growing demand for development of infrastructure such as roads, canals and buildings.

The process involved in crushing large size stone boulders into different size of crushed stones depending upon the requirements in the demand sectors. The important stages involved in stone crushing activity are primary crushing, screening, secondary/tertiary crusher, screening, conveyance, storage of raw boulders and crushed stone and transportation of both ROM and crushed stones. The raw material, i.e. raw stone boulders are obtained from mining of the stone from quarries and hand picking, etc. Figure 29.1 shows typical process of stone crusher.

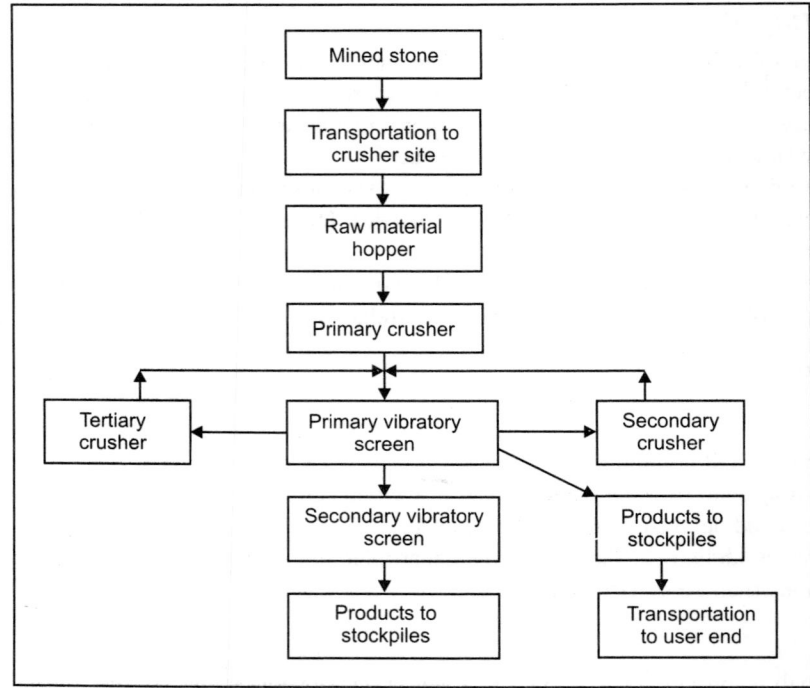

Fig. 29.1. Typical process of stone crusher.

Various types of crushers are used in the stone crushing industry such as Jaw Crushers, Roller Crushers, Cone Crushers, Impactor, Rotopoctor etc. Generally, only Jaw crushers are used as Primary crushers. For secondary and tertiary crushing application either of Jaw, cone, roller, Impactor or Rotopoctor type crushers are used.

Typical Sizes of Stone Crushers

There are large variations in the types of stone crusher setup across the country depending on geographical locations, type of demand for crushed products, closeness to urban areas, type of raw material, availability of plant and machinery locally etc. Primarily the stone crusher industry sector could be divided in three categories small, medium and large. The typical characteristics of each category of stone crushers are given in Table 29.1 briefly discussed below.

Tabe 29.1. The typical characteristics of each category of stone crushers.

Small size stone crushers	Typically the stone crushers with a production capacity ranging from 3 to 25 TPH. This category crusher have only one Jaw type crusher used as primary or secondary crusher along with one or maximum two screens
Medium size stone crushers	Typically the crushers having more than one crusher, i.e. one primary and one secondary or one/two primary and two secondary crushers along with one or more vibratory screens are categorised as medium size crushers. Medium size stone crushers will have a production capacity in the range of 25–100 TPH
Large stone crushers	Typically having two or more numbers each of primary, secondary and tertiary type crushers with at least two or more vibratory screens with mechanised loading, unloading conveying operations and producing more than 100 TPH crushed stones

Unit Operations and Technologies

The different unit operations and technologies are briefly described in the following Table 29.2.

Table 29.2. Different unit operations and technologies.

Crushing technologies	Various types of crushers are used in the stone crushing industry such as jaw crushers, roller crushers, cone crushers, impactor, rotopoctor, etc. Generally, only Jaw crushers are used as primary crushers. For secondary and tertiary crushing application either of Jaw, cone, roller, impactor or rotopoctor type crushers are used
Screening technologies	Screening is generally classified into two types: (i) coarse screening, which is achieved through grizzlies, vibratory screens, revolving screens or shaking screens, and (ii) fine screening which is achieved through vibrating screens, shaking screens
Material handling technologies	Various types of material handling technologies are used in the stone crushing industry for the purpose of moving the stones from one equipment to other, right from the point of raw material unloading up to stockpiles of products. Primarily, feeders and conveyors are used in almost all crushers

Specific noise standards applicable to DG sets

Noise limit for diesel generator sets (up to 1000 KVA) manufacture on or after the 1st July, 2003

The maximum permissible sound pressure level for new diesel generator (DG) sets with rated capacity up to 1000 KVA, manufactured on or after the 1st July, 2003 shall be 75 dB (A) at 1 meter from the enclosure surface.

The diesel generator sets should be provided with integral acoustic enclosure at the manufacturing stage itself.

Noise limit for DG sets not covered above

Noise from the DG set should be controlled by providing an acoustic enclosure on by treating the room acoustically, at the users end.

The acoustic enclosure or acoustic treatment of the room shall be designed for minimum 25 dB(A). Insertion loss or for meeting the ambient noise standards, whichever is on the higher side (if the actual ambient noise is on the higher side, it may not be possible to check the performance of the acoustic enclosure/acoustic treatment. Under such circumstances the performance may be checked for noise reduction up to actual ambient noise level, preferably, in the night-time. The measurement for insertion loss may be done at different points at 0.5 m from the acoustic enclosure/room, and then averaged. The DG set shall be provided with proper exhaust muffler with insertion loss of minimum 25 dB(A)

Requirement of certification

Every manufacturer or importer of DG set to which these regulations apply must have valid certificates of type approval and also valid certificates of conformity of production for each year, for all the product models being manufactured or imported from 1st July 2003 with the noise limit specified in the notification.

Guideline to calculate the minimum requirement of the stack of DG set

The central pollution control board (CPCB) had evolved the minimum height of the stack to be provided to the DG set and published in emission regulations part IV: Coinds/26/1986–87.

$$H = h + 0.2 \times \ddot{O}KVA$$

H Total height of stack in meter.
h Height of the building in meters where the generator set is installed.
KVA   Total generator capacity of the set in KVA.

Based on the above formula the minimum stack height to be provided with different range of generator sets may be categorised as follows:

For generator sets	Total height of stack in meter
50 KVA	Height of the building +1.5 meter
50–100 KVA	Height of the building +2.0 meter
100–150 KVA	Height of the building +2.5 meter
150–200 KVA	Height of the building +3.0 meter
200–250 KVA	Height of the building +3.5 meter
250–300 KVA	Height of the building +3.5 meter

PLASTER OF PARIS

Plaster of Paris is a type of plaster which can be used in art, architecture, fireproofing, and medical applications. When people think of 'plaster', they are often thinking specifically of Plaster of Paris, although there are a number of different types of plaster on the market including lime plaster and cement plaster. Many art and construction supply stores sell Plaster of Paris, and it can also be ordered through specialty companies.

This plaster is made by calcining gypsum, a process which involves exposing the gypsum to very high temperatures to create calcium sulphate and then grinding it into a fine white powder. When water is added to the powder to make a slurry, the slurry can be moulded in a variety of ways, and as it sets, a firm matrix is created, creating a solid shape which is also very smooth. One advantage to Plaster of Paris is that there is no volume loss, so casts made with this plaster are true to the size of the mould.

History seems to indicate that, despite the name, Plaster of Paris was invented by the Egyptians. It was used as an artistic decoration in many Egyptian tombs, and the Greeks picked up the technique, using plaster in their own homes, temples, and works of art. Paris became synonymous with this type of plaster in the 1600s, thanks to a large deposit of gypsum which made it easy to produce Plaster of Paris. The substance was also used extensively in fireproofing, giving Parisian homes a distinctive appearance.

In art, Plaster of Paris can be used to make sculptures, and test moulds for bronze and other metal castings. Plaster can also be used to make moulds which will be very dependable while withstanding high temperatures. Plaster can also be used as an architectural feature, as for example in the case of plaster mouldings mounted on doorways and window frames.

Noise Pollution

There are a number of sources from which noise arises in Plaster of Paris (POP) manufacturing like crushers, pulverises, etc. In small and medium scale units, most of these sources are intermittent, generated during the time of the breaking of raw material, product grinding, etc. Calciner does not generate high noise levels as compared to crushers. Conveyor movement is also a source of continuous noise, especially the ill-maintained conveyor systems.

During the filed studies it was observed that though the noise levels were high at shop floor levels, the noise levels at unit boundaries in ambient conditions were within the prescribed standard of 75 db(A) by CPCB for industrial zone as can be seen from Table 29.3.

Table 29.3. Noise levels in various industrial clusters.

Location	Noise levels average in dB(A)			
	Bikaner	Jammu	New Jalpaigudi	Rajapalayam
Main gate	59.0	56.3	42.1	53.4
Office building	59.4	58.8	43.4	55.85
Near pulveriser	85.5	85.2	82.8	85.9
Near calciner	72.0	71.3	82.8	80.2
Near RM crusher and conveyer	77.0	74.2	83.6	84.0
Near boundary wall	61.6	55.8	43.1	54.3
Near boundary wall	66.2	56.2	46.6	50.3
Near boundary wall	58.1	57.3	43.8	45.1

Chapter 30

Construction Process

INTRODUCTION

At first glance, it may not seem important for designers to design buildings which are safe to construct, much less to consider the particular hazards and risks associated with dust and noise. Surely this is 'the contractor's problem'? But nothing could be further from the truth. If such attitudes still prevail, they arise from the continuing belief that getting a building built is a confrontational rather than a cooperative endeavour. In fact, amongst the many hazards facing construction workers, dust and noise can be as injurious and life-threatening as falling off the edge of a roof. Designers can often greatly reduce their adverse effects by first, fully understanding how their buildings will be put together and then taking the appropriate action to avoid the hazard or mitigate the risk by redesigning or respecifying.

This chapter gives advice on various methods of eliminating, and controlling noise hazards and risks during the construction process by means of the correct specification of materials, components and assembly processes. It sets this advice in the context of the activities performed by the principal contractor, illustrating how the achievement of health and safety is a joint responsibility. Reference should be made to the construction design and management (CDM) regulations for formal definitions of the legal duties and responsibilities of the different parties.

NOISE

There is a statutory duty to control noise and to protect workers and other persons from its effects. Noise on construction sites results from the use of machinery used for demolition, piling and excavation and from plant such as, compressors, concrete mixers and dumpers. Excessive, albeit short duration, noise results from the use of riveters and cartridge operated fixing tools. The nuisance or damage caused by noise is a function of type (quality) as well as loudness. Intermittent noise may be more disruptive than a continuous pattern and high pitched sounds are more disturbing than low pitched ones. Furthermore, noise repeated day after day can build up until hearing is damaged: the effect is cumulative.

Action Levels

There is a statutory duty to control noise and to protect workers and other persons from its effects. General control is exercised through the health and safety at work act, but the noise at work regulations, 1989, apply specific controls to the amount of noise permissible within the work environment, including construction sites.

These, already referred to in section, define three action levels:

1. A First action level of 85 dB (A).
2. A Peak action level of 200 pascals.
3. A Second action level of 90 dB(A).

Although the second action level does not appear to be much louder than the first, this is not the case: the loudness scale is logarithmic, meaning, for example, that 90 dB is 10 × the intensity of 80 dB and that an increase of 3 dB doubles the sound intensity. Exposure to noise, as with dust, is measured in terms of a time weighted index: noise is averaged out over an 8 hours working day to give a personal daily noise exposure, the 'dose' being recorded as Lep, d or Leq.

Both employers and employees have a duty to control noise. At first action level, employers must:

1. Have an assessment made by a competent person and keep records.
2. Supply hearing protection to any employee who requests it.
3. Ensure that all noise producing equipment is properly used and maintained.
4. Give information and/or training on how to avoid damage to hearing, including the correct use of ear protectors.

Employees must

1. Use any protective equipment provided and report any defects to the employer.

At or above second action or peak action levels, employers must:

1. Have a noise assessment made by a competent person and keep records.
2. Reduce the noise as far as is reasonably practicable by means other than ear protectors.
3. Provide suitable ear protectors and ensure that they are worn.
4. Designate an 'ear protection zone' and identify the area with signs.
5. So far as is reasonably practicable, ensure that anyone entering an ear protection zone is wearing ear protectors.
6. Ensure that all equipment is properly used and maintained.

And employees must:

1. Properly use any ear protectors provided.

Effects of Noise

The effects of noise vary; from simple annoyance to damage to the inner ear and permanent loss of hearing. These are personal consequences, but communal as well as personal hazard results from an inability to hear shouted warnings and the fatigue which continual exposure to noise can induce.

The following examples illustrate the problem which noise can cause on construction sites, [levels given are average at the operator's position; measured in dB (A)]:

Electric hand tools	95
Hammer drill	102
Dumper	103
Circular saw bench	107
Excavator	109
Ready mix lorry	112
Batching plant	116
Compressors and compactors	120
Diesel hammer: sheet piles	136

It should be noted that, under the Noise at Work Regulations 1989, manufacturers are required to provide data about the noise levels generated by their equipment if these are likely to be above first or peak action levels. The actual amount of noise generated will also be affected by the manner in which the equipment is operated and its location on site.

Methods of Control

Before methods of control can be determined, where a noise hazard is likely to arise an assessment must be made by a competent person. The purposes of the assessment are to identify all workers likely to be exposed at or above first or peak action levels and to provide sufficient information to enable the appropriate action to be taken. As a rough guide, an assessment of daily personal exposure is likely to be needed if people have to shout to be heard by someone standing 2 metres away. Assessments of peak pressure will be needed where workers are exposed to the effects of explosive tools and equipment, such as cartridge fixers and piling hammers. Noise exposure in the construction industry is particularly difficult to assess, since exposure levels vary throughout the working day as workers move from one workplace to another. Measurement in this environment should therefore be made over a longer period and averaged out—usually over the 8 hours working day. An effective way of doing this is by means of personal dosemeters. These are small devices carried on the person with a microphone close to the wearer sear–possibly clipped to a safety helmet. Dosemeters are available which store data in a form which enables a time history graph to be printed out, thereby revealing unusually high or low levels.

Control of excessive noise often requires a combination of strategies. The first and most important is 'avoidance of the hazard in the first place. The designer, having become aware that a noise hazard may arise from a construction activity resultant from his design or specification, should review his decisions and consider whether an alternative assembly method could be adopted. For example, bored piles might be substituted for driven ones, or drilled and tapped holes for cartridge fixing. If a noisy activity is unavoidable, the following actions should be considered:

1. Carry out pre-assembly off site, where the hazard is easier to control.
2. Undersize concrete to minimise the need for scabbling (allowing for the movement of the material during setting). This will require an accurate calculation of tolerances.
3. Pre-form holes for services, especially through composite and *in situ* concrete floors.
4. Specify noise reduced plant from hire firms or manufacturers
5. Pre-plan to reduce the need to:
    (a) Drill, cut and chase concrete, including precast units and concrete blocks.
    (b) Grind steelwork to fit.
    (c) Operate noisy tools over long periods and/or in confined spaces.
6. Specify a requirement to satisfy noise standards in instructions to sub-contractors.
7. Provide visual recognition of noisy plant by marking in an obvious place.
8. Consider in detail ways of reducing noise from tools, plant and equipment. For example:
    (a) Enclose hammer and pile heads in acoustic screens.
    (b) Use vibrating rather than driven tools.
    (c) Use a resilient dolly between hammer and pile bead.
    (d) Fit efficient silencers and exhausts to diesel driven machines.
    (e) Acoustically dampen compressor casings; keep enclosure panels closed.
    (f) Keep saw blades sharp and clamp material whilst cutting.

(g) Avoid specifying, riveted fixings, for example by substituting high tensile steel bolts.

(h) Surround noisy work with acoustic screens.

(i) Seal leaks in air lines.

(j) Plan site layouts so as to segregate noisy from quiet activities.

Noise and the Environment

In addition to the effects of noisy equipment upon building site operatives themselves, nuisance can be created in the neighbourhood. Environmental noise is controlled by the Control of Pollution Act 1974, which empowers local authorities to impose limitations both on noise levels and on working hours. The principal ways of controlling noise from construction sites include the use of quieter plant, of plant located away from sensitive areas and of acoustic screens and enclosures. Screens may be simply constructed, for example of 10 mm plywood on scaffold frames so placed as to interrupt the line of sight between noise source and receiver. Finally, complaints can be much reduced if there is good communication between the site and the occupiers of nearby buildings, especially where night working is contemplated.

METHODS OF AVOIDING OR CONTROLLING NOISE DURING THE ASSEMBLY PROCESS

The following analysis lists and codifies likely noisy activities during the assembly of a typical building. All relevant activities are noted, whether or not designers can make any input to a particular action. Where they can make an input to avoid or reduce the hazard the item is asterisked. It is assumed that risk assessments ill be made where appropriate and that personal protective equipment (PPE) will always be worn as a last resort (Table 30.1).

Table 30.1. Methods of avoiding or controlling dust and noise during the assembly process.

Activity	Hazard	Methods of avoiding
1 Demolition	Localised dust clouds, noise	Inform adjoining occupiers
		Write method statement
		Erect protective screens
		Remove asbestos and other dangerous substances
2 Site clearance		
2.1 Remove topsoil	Noise from plant	Fit efficient silencer and exhaust
		Keep enclosure panels closed
		Advise neighbouring owner
2.2 Reduce levels	Noise from plant	As 2.1.
		Fix dactum levels to minimise cut and fill
2.3 Vibro-compaction	Noise from plant	As 2.1.
	Noise from pile head	Use acoustic screens
		Specify alternative consolidation method
3 Substructure:Foundations	Foundations	
3.1 Excavate strip	Noise from excavators, dumpers, tipper lorries	As 2.1.
		Correct handling of plant to avoid banding bucket, plan transport routes, etc.
3.2 Excavate raft	As 3.1.	As 3.1.

(Contd ...)

Activity	Hazard	Methods of avoiding
3.3 Piling	Noise impact on pile	Screen site
		Resilient dolly between pile and hammer head
		Use of vibrating pile heads
	Crane cables, guides, etc.	Align pile accurately on rig
	Noise from plant	As 2.1
		Use acoustic screens
	Noise affecting surrounding building, streets, etc.	Specify alternative foundation method:
		Bored not driven piles
		Diaphragm walls and ground anchors
		Shaft and kentledge
		Consolidation technique (2.3)
		Advise neighbouring owner
3.4 Basement construction	Noise and dust during excavation	Screen site
	Noise from sheet piling	As 2.1 and 3.1.
		Acoustically dampen piles
		Specify alternative methods of earth retention
		Bored not driven piles
		Diaphragm walls and ground anchors
	Noise from plant handling forms, reinforcement, concrete, etc.	Use modular forms
		Specify precast component
3.5 Placing concrete	Noise from plant handling forms, reinforcement, concrete, etc.	Locate fixed plant away from noise sensitive area
		Use electric rather than diesel powered motors on batching plant, cranes
		Advise neighbouring owner if night working necessary
4 Substructure	Work to DPC level	
4.1 Building brick/block walls	Noise from mortar mixer and high speed tools	As 2.1
		Specify as for 'dust'
4.2 *In situ* concrete walls	Noise from plant handling forms, reinforcement, concrete, etc.	Use modular forms
		Specify precast components
4.3 Holing down bolts for frames	Noise from drilling	As above
5 Ground floors		
5.1 Compacting ground and hardcore	Noise from compactor	Use roller rather than plate compactor
		As 2.1
5.2 Placing *in situ* concrete	Noise from poker vibrator and vibrating beam tamper	As 2.1 (compressor)
5.3 Power floating and trowelling	Noise from motors	Screen working areas, use PPE
		Advise L. A. and neighbouring owners if night working necessary
		Specify levelling sacreed
5.4 Forms for rafts and *in situ* slabs	Noise from assembling	Use modular forms

(Contd ...)

Activity	Hazard	Methods of avoiding
5.5 Assembling and placing reinforcement cages	Noise from cranes	Use electric powered fixed cranes
		Avoid unsocial hours when using mobile cranes
		Locate cranes where screened by existing buildings
5.6 Assembling timber ground floors	Noise from cutting timber lengths and sheets	Screen portable saw benches
		Adjust and sharpen blades clamp timber tight
		Prefabricate off site
		Specify correct sections
5.7 Assembling precast concrete floors	Noise from cranes	As 5.5
		Specify components correctly to avoid on site cutting
6 Superstructure	External walls	
6.1 *In situ* concrete walls	Noise from plant handling forms, reinforcement, concrete, etc.	As 2.1
		Use modular forms
		Specify precast component
6.2 Timber panel walls	Noise from cutting timber lengths and sheets	As 5.6
	Noise from assembling: cranes, nailing, etc.	As 5.5
		Design panels for manual assembly
		Avoid unsocial hours
7 Superstructure	Frames	
7.1 Erecting steel or precast concrete frames	Noise from cranes	As 5.5
	Noise from drilling, cutting and grinding to fit	Ensure accurate fabrication before delivery
		Assemble in correct sequence
		Plumb and align carefully
7.2 Casting *in situ* concrete frames	As 3.6	As 3.6
7.3 *In situ* casing of steelwork	Noise from assembling forms	Use modular forms
		Use steel forms
		Maximise use of forms, e.g. by designing to minimise variations in component sizes
7.4 Placing concrete in forms	Noise from crane/pump and poker vibrator	As 5.5 and 2.1
7.5 Cladding system	Noise from crane noise from cutting and grinding metal lengths, brackets and fixings	As 5.5
		Screen working area
		Specify accurate sizes
		Specify built-in fixings
8 Superstructure	Upper floors	
8.1 Assembling timber floors	Noise from cutting timber lengths, sheets and boards	As 5.6
	Noise from assembling cranes, nailing, etc.	As 5.5
		Specify modular panels
		Prefabricate off site
		Avoid unsocial hours

(Contd ...)

Activity	Hazard	Methods of avoiding
8.2 Steel/concrete composite floors	Lifting and placing steel sections	As 5.5
	Noise from cutting steel to fit round columns, for openings, etc.	Anticipate needs during design
		Prefabricate off site, code and deliver in sequence for assembly
	Noise from plant handling forms, reinforcement, concrete	Specify ready mixed concrete
		Locate fixed plant and concrete pump away from noise sensitive areas
		Operate and maintain pump correctly
8.3 *In situ* concrete floors	As 8.2 (concrete)	As 8.2 (concrete)
		Specify precast concrete components
	Noise from assembling forms	Use modular forms
	Noise from poker vibrator and vibrating beam tamper	As 2.1 (compressor)
	Power floating and trowelling	As 5.3
	Early grinding	
8.4 Precast concrete	Noise from cranes	As 5.5
		Specify components correctly to avoid on site cutting
	Noise from plant handling concrete topping	As 8.2 (concrete)
9 Superstructure	Roofs	
9.1 Assembling timber pitched	Noise from cutting timber lengths and sheets (sarking)	As 5.5
	Noise from cranes	As 5.5
	Dust from cutting tiles	Dampen work
		Fit dust bags to power saws and grinders
		Specify to tile modules
	Noise from cutting tiles	As for dust
9.2 Steel pitched, including fixing profiled metal sheeting	Noise from cranes	As 7.1
	Noise from drilling, cutting, grinding to fit	
9.3 Timber flat	Noise from cutting timber lengths, boards	As 5.6
	Noise from assembling cranes, nailing	As 5.5
		Specify modular panels
		Prefabricate off site
		Avoid unsocial hours
9.4 Concrete flat (*in situ*)	Noise from assembling forms	Use modular forms
	Noise from poker vibrator and vibrating beam tamper	As 2.1 (compressor)
		Specify precast concrete components
9.5 Concrete flat (precast)	Noise from cranes	As 5.5
		Specify components correctly to avoid on site cutting

(Contd ...)

Activity	Hazard	Methods of avoiding
	Noise from plant handling concrete topping	As 8.2 (concrete)

Building enclosed

At this stage, with roof finishes at least partially in place and doors and windows installed, the building is enclosed. Thus far, apart from activities such as handling cement and cutting tiles, noise rather than dust has been the principal hazard, especially in relation to surrounding areas. From now on, however, both can present much greater risks to workers inside the growing building.

Activity	Hazard	Methods of avoiding
10 Superstructure	Partitions	
10.1 Building brick/block partitions	Noise from cutting blocks with high speed tools	Avoid cutting in confined spaces
		Design to brick/block modules
		Specify timber or metal stud or proprietary partitions
		Specify blocks cut off site
10.2 Timber stud	Noise from cutting timber lengths, nailing	Avoid cutting in confined spaces
11 Services	First fix	
11.1 Piped services, sprinkler systems, plumbing, etc.	Noise from chasing and drilling	Design and specify as for dust
		Avoid using cartridge fixers in confined spaces
	Noise from notching and drilling timber, cutting and drilling metal pipes and sheets	Design and specify as above
11.2 Installing metal trunking	Noise from drilling for fixing	Specify as for dust
		Drill and tap rather than use cartridge fixers
12 Joinery	First fix	
12.1 Fixing wall linings, noggings, battens, ground, door frames and casings	Noise from cutting	As 10.2
		Specify as for dust
13 Joinery	Second fix	
13.1 Fixing skirtings, architraves, window boards, etc.	Noise from cutting	As 10.2
		Specify as for dust
14 Other activities		
14.1 Scabbling concrete	Noise from same	Specify as for dust
		screen areas
14.2 Maintenance	Noise from cutting out and replacing, especially in confined spaces	If reasonably practicable remove complete sections and renew faulty parts elsewhere
		Design to permit easy access

Noise

Where it is impossible to avoid a noise hazard or to control the risks arising from it sufficiently, workers are required under the Noise at Work Regulations to wear ear protectors. The trigger points are when noise levels remain at or above second action or peak action levels. As with dust masks, ear protectors are only to be used when it has proved impossible to attenuate the noise by any other means.

One reason for this is that workers may be reluctant to adopt ear protection: some forms can be intrusive, and inconvenient. Nevertheless, it is essential for the principal contractor and subcontractors to drive home to their operatives that their hearing is imperilled, possibly permanently, if they fail to use the appropriate equipment. One way of countering this reluctance, apart from enforcement, is by training, especially of new workers during their safety induction course. Another is to create a positive site attitude towards safety in general, which will then make enforcement much easier.

In those parts of the site or workshop where second or peak action levels are likely to be reached, an ear protection zone must be delineated. Special signs will be erected and all persons entering the zone must be equipped with and wearing ear protection. It should be noted that, at first action level, employees have the option to wear ear protection and employers have a duty to provide such protection if asked to do so.

There are a number of important points affecting the selection and use of ear protectors:

1. They should be provided on an individual basis.
2. They should be supplied and fitted by a trained person.
3. They should be regarded only as an interim measure, pending some better solution to the problem.
4. Workers should be trained in their use.
5. They should be suited to both the user and the type of noise.
6. They must provide a tight seal against the head.
7. They should be stored in a clean place when not in use.
8. They should be inspected regularly for deterioration or damage.
9. They should be replaced when necessary.

There are two basic types of ear protector, the disposable ear plug and the ear muff.

Ear plugs: Made of very fine mineral fibre, polythene and dense plastic foam, sometimes pre-shaped. They should be inserted correctly and, once removed, should not be reused. Cleanliness is essential for optimal performance. Reusable types of plug are available made of rubber or plastic. Different sizes may be required for each ear, they must be a good fit and be clean, since dirt can cause irritation.

Ear muffs: these completely cover the ear and are sealed to the head by means of a foam or liquid filled seal. Badly designed or fitted muffs will give little or no protection. Muffs should be matched to the different sizes and shapes of wearers heads, to differing hair styles and to whether or not spectacles are worn. Helmet mounted muffs may not fit the ear tightly, or may move as the helmet moves on the head. Muffs should be chosen to meet the particular noise risk a type good at countering low frequencies may be much less effective against high ones.

Three other points should be borne in mind with regard to ear protectors:

1. Where hazard warning signals are likely to be given, for example by cranes or reversing vehicles, either the protectors should permit the signals to be heard or the signal itself should be changed. It is not acceptable to remove protectors in order to hear signals or shouted warnings.
2. Ear protectors are only effective when they are being used: if protectors are worn for only 50 per cent of an 8 hours shift, only some 10 per cent protection is gained; if for 7½ hours out of 8, only 75 per cent.
3. Muffs are much more visible to the casual inspecting eye than plugs. Their use in an ear protection zone can therefore be more easily verified.

NOISE MANAGEMENT IN THE CONSTRUCTION INDUSTRY: A PRACTICAL APPROACH

Construction workers are among the most affected by industrial deafness. The International Labour Office encyclopedia lists the construction industry as the fourth noisiest industry sector.

The types of workers at risk include:
1. Users of impact equipment and tools (e.g. piling hammers, concrete breakers, manual hammers).
2. Users of explosives (e.g. blasting, cartridge tools).
3. Users of pneumatically powered equipment.
4. Operators of plant powered by internal combustion engines.
5. Bystanders in the vicinity of the plant.
6. Operators and bystanders in enclosed spaces where there are noisy activities or a concentration of plant.
7. Service and equipment maintenance personnel.

It is very important for the construction industry to adopt a preventive management program aimed at the reduction of workers' noise exposures. The best ways to achieve this reduction are to employ quiet work practices and use quiet construction equipment. When quieter alternatives are not available, consideration should be given to a site layout to arrange noisy processes away from workers not involved in their operation. Portable barriers can be used around static equipment like generators and concrete pumps. To achieve better results, noise control aspects should be included in all four stages of any construction project: client's specifications, tenderer's proposal, site planning and construction phase.

Stage 1

Client's specifications: A client should include noise control requirements for both occupational and environmental noise early in the planning stage for a new project. The desired noise control requirements may be included in a client specification list in the tender document. This can help to avoid unexpected and often very expensive noise control during the construction phase. It allows tenderers to plan how to overcome noise problems in advance.

The client's specifications may include the following categories:
1. Specified noise exposure levels during the construction phase, as per legislative requirements or company policy.
2. Use of quiet/silenced equipment.
3. Adoption of quiet alternative techniques.
4. Use of noise control measures like silencers, barriers, enclosures.
5. Erection of warning signs identifying noise hazard areas.
6. Time restrictions.
7. Provision of personal hearing protectors and training.

Stage 2

Tenderer's proposal: Following the client's specifications, the tenderer's proposal should cover all the specified categories and formulate a noise control policy and a noise control plan to be included in the site specific safety management plan.

The noise control plan may be a set of actions required to achieve the noise control policy and to reduce noise exposure. It may also include information on how the company is planning to meet its obligations, like:
1. List of equipment to be used—with noise levels at operator position and/or at 1 m.
2. Methods undertaken to lower noise exposure, e.g. maintenance, barriers, enclosures.
3. Restricted hours, rotation of workers in noisy places, special time arrangements like noisy work done after hours.

4. Identification of noisy equipment and processes by signs.
5. Site induction for employees and contractors to include noise levels, noise controls and correct use and maintenance of personal hearing protectors.
6. Selection and provision of appropriate personal hearing protectors.
7. Audiometric tests.

Stage 3

Planning of site activities: The main contractor should plan to coordinate subcontractors so that the activities of one do not unnecessarily expose employees of another to noise hazards. It is good practice to nominate one person as the noise coordinator for all noisy activities.

Site planning should include:

1. Preparation of guidance to workers on noise hazards and measures to be taken to reduce noise exposure.
2. Preparation of schedules of noisy plant and exposure estimates for each phase of work.
3. Laying out the site to separate noisy activities from quieter ones, e.g. concentrate compressors, pumps and generators in screened-off areas or away from the work to be carried out; workshops, stores, etc. away from noisy activities.
4. Scheduling noisy activities to take place when the minimum number of other nearby workers are present (but noise out of hours needs to be carefully planned to avoid neighbourhood annoyance).
5. Rostering workers to minimise exposure times.
6. Ensuring that workers are well trained, instructed and supervised in noise matters and responsibilities including correct use and maintenance of personal hearing protectors.

Stage 4

Construction phase: Once the construction work is in progress, it is essential to monitor the implementation of the noise control plan. This could be carried out by the client or the main contractor and could include the following:

1. Checking if equipment brought onto site complies with specifications. This could be done by obtaining information available from suppliers or by noise assessments.
2. Reducing noise from identified noise sources by exchanging equipment and/or processes for a quieter alternative or by engineering control methods to quieten the existing one.
3. Ensuring that all plant is properly maintained, e.g. all noise control measures like silencers and enclosures are intact.
4. Monitoring work schedules to check that noisy work is carried out as specified, away from other workers, outside hours, etc.
5. Monitoring if noisy areas are identified and well marked so employees and contractors can avoid entering them unnecessarily.
6. Monitoring whether training and hearing tests have been carried out and if personal hearing protectors are adequate and are being worn and maintained correctly.
7. Ensuring that the cause of any hearing loss shown up by audiometry is investigated.
8. Utilising safety toolbox meetings to provide feedback on effectiveness of noise control measures and personal hearing protectors to workers, employers and contractors.
9. Posting on safety notice boards results of noise assessments conducted and additional noise information.

Chapter 31

Cement and Concrete

INTRODUCTION

In the most general sense of the word, a cement is a binder, a substance that sets and hardens independently, and can bind other materials together. Cement used in construction is characterised as hydraulic or non-hydraulic. Hydraulic cements (e.g. portland cement) harden because of hydration chemical reactions that occur independently of the mixture's water content; they can harden even underwater or when constantly exposed to wet weather. The chemical reaction that results when the anhydrous cement powder is mixed with water produces hydrates that are not water-soluble. Non-hydraulic cements (e.g. lime and gypsum plaster) must be kept dry in order to retain their strength. Cement industry, in any country, plays a major role in the growth of the nation.

ENGINEERING CONTROLS FOR NOISE IN CEMENT INDUSTRY

Impending Mine Safety and Health Administration (MSHA) noise regulations that will require the implementation of feasible engineering and administrative controls has brought about the reassessment of current noise levels and attenuation technology available to cement plants. Cement plant noise literature concerning the area of cement plant noise is found primarily in international publications. Researchers have identified numerous sources that may contribute to high noise levels, including grinding mills, compressors, fans, blowers, coal rail car unloaders, material handlers, power transmission equipment, and crushers. Similarly, outdoor noise pollution also has been noted at some plants, thus requiring noise attenuating technologies for noise reduction out doors.

Part of the development of a successful noise control strategy with engineering controls is dependent on understanding the regulations that are currently being promulgated. In the case of the cement industry, the proposed MSHA noise rule has rules that influence the establishment of noise controls.

Tools for assessing noise levels a successful noise control program that focuses on engineering control of noise requires the institution of a hearing conservation plan and the use of proper monitoring equipment, surveys, maps, and modeling. A thorough hearing conservation plan should be established where noise exposure exceeds a 85–dBA time weighted average for eight hours. A good program consists of the following components:

1. Noise measurement and analysis.
2. Engineering control of noise sources where feasible.
3. Administrative controls and personal protection where noise control is not feasible.
4. Audiometric testing.

5. Employee training and education.
6. Record keeping.
7. Evaluation.

Sources of Noise

Noise sources are evaluated through the use of dosimeters, sound level meters, and octave band analysers. Evaluation of noise in a plant environment requires a well-constructed sound survey. Numerous sources provide useful criteria for surveys, as well as commonly used calculations. Some industrial facilities have developed sound level contour maps through noise modelling programs. These models are developed by importing noise data, noise emitter statistics, and information related to natural and artificial obstacles that can be drawn from zoning and topographical maps. Currently, a number of software programs are available for simulating noise emissions.

Causes and Control of Noise

Engineering noise controls investigation of a noise problem requires consideration of as many potential solutions as possible. Many options for control exist. The highest tier of engineering control is to guide the design of quiet machinery or to make design changes in an existing plant. The development of engineering controls at the design stages of a plant avoids retrofitting, which can be impractical on technical and economic grounds. Ultimately, if a quieter process cannot be substituted, then the noise source can be modified for attenuation. Control at the source may consist of the reduction of the driving force, reduction of radiating surface response to driving forces, or reduction in radiation efficiency of sound-generating surfaces.

One of the primary causes of noise in a manufacturing plant is the vibration produced by process machinery. By focusing on this noise source, many problems may be solved. In many cases, damping has been effectively utilised to attenuate vibration problems. Elastic materials, such as rubber and plastic, are used to dissipate energy from vibrating surfaces with a resulting reduction in amplitude and structural stress. When noise reduction cannot be achieved at the source, it may be useful to modify the noise propagation path. Measures including full or partial enclosures; sound barriers; room absorption; active control; and changes in duct geometry, fan type, and flow characteristics can be applied to a typical plant to aid in noise control.

Enclosures for noise sources or for employees in a noisy environment can significantly decrease noise levels reaching the employee. The effectiveness of an enclosure is dependent on the transmission loss of the enclosure. Acoustic performance may be affected by leaks, use of varying noise absorbent and damping materials, and rigidity of the underlying structure of the enclosure.

Barriers work effectively when placed between a noise source and a receiver. These specially designed walls reflect sound waves away from the receiver resulting in a noise shadow behind the barrier. They work most effectively when they are close to the noise source and are as high as possible within the plant. Important design elements that must be considered for effective noise attenuation include the effective height of the barrier, the wavelength of the noise, and the angle of deflection.

Placement of sound-absorbing materials on the walls of a room diminishes the reverberation of sound waves off of hard surfaces. Typically, acoustic absorbers are soft, porous materials that absorb sound waves and are described in terms of an absorption coefficient. Material composition and thickness, mounting style, and sound frequency all effect the absorbent coefficient. When noise is estimated on a subjective scale in the form of an annoyance scale, reduction of reverberation time was perceived to be

more effective than noise source reduction. Improvements in speech intelligibility, decreased risk of accidents, and improved control of machinery failure were some of the benefits of this technology.

Active noise control uses the principle of superposition of waves to attenuate noise. Active noise control can be achieved by processing the original offensive sound and subsequently injecting it back into the sound field, but at 1800 out of phase, thus generating a canceling wave form. Active noise control has been applied to one-dimensional field and duct noise, enclosed spaces and interior noise, noise in some three-dimensional spaces, and in personal hearing protection. The biggest successes have come in the area of piping and duct noise attenuation. Active noise control has had more limited applications to broad-band noise, or noise in three-dimensional spaces, because of the principles of noncausality, stability, spatial mismatch, and the infinite gain controller requirement.

Flow-related noise concerns, such as HVAC and piping noise, conveyor noise, and compressor noise, require other noise control measures, as well as some of the previously mentioned technologies. HVAC-related noise can be difficult to attenuate, especially considering the wide range of sizes of fans, ducts, and stacks.

BASIC CONSIDERATIONS FOR A QUIET DESIGN

Some basic considerations for a quiet design of HVAC systems include: minimising pressure loss within ducts, selecting of fans that operate at maximum efficiency on performance curves while providing necessary ventilation, and designing ductwork leading to fans to minimise turbulence.

Retrofitting ventilation with silencers, insulation within ducts, and active noise control also has been successful for noise abatement. Piping has been attenuated with pipe lagging modified with the addition of fiberglass layers, decoupling layers of insulation, and more efficient damping. Conveyors have been improved through substitution of quieter operating drives, damping of metal parts, and reductions in operating speeds. Compressor station noise has been reduced by various measures, including silencers, pipe lagging, and fan changes, depending on the specifics of the process.

Case studies retrofitting or redesigning a manufacturing plant for noise control can be a difficult endeavor. Proper noise control requires the teamwork of management, process engineers, mechanics, and outside vendors and acoustic consultants. Some success has been achieved by plants in a number of noisy industries, including can manufacturing, recycling, mining, and stamp manufacturing.

The area of compressor noise also has received a lot of attention in literature on case studies showing noise control through enclosures, silencers, and acoustical blankets. Blankets with thermal and noise insulation capabilities have been utilised in the power industry. Some plants with outdoor noise problems affecting residential neighborhoods have used noise barriers, absorptive panels, acoustic blankets and curtains, and silencers for noise attenuation.

CONCRETE PRODUCTS

Concrete's versatility, durability, and economy have made it the world's most used construction material. It is used in highways, streets, parking lots, parking garages, bridges, high-rise buildings, dams, homes, floors, sidewalks, driveways, and numerous other applications. Different types of hydraulic cement are manufactured to meet various normal physical and chemical requirements for purposes including low heat of hydration, high early strength, resistance to alkali-silica reactivity, and high sulphate resistance.

A properly proportioned concrete mix possesses acceptable workability of the freshly mixed concrete and durability, strength, and uniform appearance of the hardened concrete while being economical.

Noise Control in the Concrete Products Industry

This chapter gives general information for employers in the concrete products industry on common sources of noise and on how the noise may be reduced.

Noisy processes in the concrete products industry

There are many different sources of noise in the concrete products industry which can result in personal noise exposures which exceed the upper exposure action value of the control of noise at work regulations 2005 (a daily noise exposure of 85 dB) and so need to be controlled.

Examples of noise sources include:

1. Vibrating tables or conveyors for compacting concrete in flat products such as paving slabs. High levels of noise can be generated by the vibration generator itself and also from the impacts as the moulds or pallets rattle on the table surface.
2. Vibrating presses for producing blocks and tiles.
3. Impacts between pallets or moulds on conveyors or impacts due to falling waste material.

Assessing noise levels and exposures

If you think anyone may be exposed above the lower exposure action value of the noise regulations (a daily exposure of 80 dB) you should assess the risk and plan your actions to control the noise to which people are exposed in your workplace. This will often involve measuring the noise, which should be done by somebody with the necessary training and experience.

However, it is often possible to carry out a simple assessment to decide whether your employees are likely to be exposed above either of the action values. If you have to shout to be heard clearly by someone standing 2 m away, then the noise levels are likely to be more than 85 dB and a risk assessment will be needed if people are exposed for more than about 2 hours. Even if normal conversation is possible, if the noise is intrusive then it is probably 80 dB or more and if it lasts for more than about 6 hours you should have a risk assessment.

Table 31.1 shows some of the noise levels that have been measured at worker positions in the concrete products industry. This information should help you to decide whether the lower or upper exposure action value is likely to be exceeded by any of your employees.

Table 31.1. The noise levels that have been measured at worker positions in the concrete products industry.

Products	Task	Machine or process	Noise levels, dB
Flat products using vibrating tables or conveyors)	Mould filling	Steel tables	97–102
		Rubber-covered tables or conveyors	87–93
	De-moulding and stacking	Vibrating steel tables	97–99
		Tables/conveyors covered with resilient material	86–89
Concrete blocks, tiles and slabs	Operating vibratory presses	Vibratory presses (inside enclosure)	96–110
		Vibratory presses (outside enclosure)	84–93
		Unloading stations	86–88
		Inside control rooms	71–79

(Contd ...)

Products	Task	Machine or process	Noise levels, dB
	Operating secondary presses	Strapping machines, cubers, shrinkwrap, etc.	76–89
	Operating hydraulic presses	Hydraulic presses	86–97
	Production of 'Heritage' blocks	Tumblers/Rumbler	84–95
	Operating block splitters	Block splitters	86
	Operating saws	Stone saws	81–96
Reinforced concrete products	Operating vibrating tables	Vibrating tables	85–105
	Pattern makers	Woodworking equipment	85–99
Extruded tiles	Operating tile extrusion lines	Extrusion plant/processes	86–93
	Quality control inspection	Manual tasks and noise from extrusion processes	85–86
	De-palleting and racking	Manual handling tasks	84–91
	Operating secondary processing plant	Packing machines	82

There is a simple way to estimate a noise exposure from the noise level and the exposure time. If the exposure time is 8 hours, the exposure is equal to the noise level. If the exposure time is doubled (or halved) the exposure increases (or decreases) by 3 decibels.

For example, if the average noise level is 95 dB and a worker is exposed to this noise for 8 hours, then his daily exposure is also 95 dB. But if his exposure time to the same noise level is only 4 hours per day, then the daily exposure will be 95 – 3 = 92 dB. Halving the exposure time again, to 2 hours a day, would reduce the worker's exposure by a further 3 decibels, to 89 dB.

Noise control methods

Removing or reducing the noise at source, is the most effective way to control the risk of hearing damage in industry. It is more reliable, and easier to manage, than just using personal hearing protection and is often more cost effective in the longer term. You must reduce the noise exposure so far as is reasonable, if any daily noise exposures are above the upper exposure action value of 85 dB. There are several engineering approaches to noise control which are described below.

Process change and elimination of noise sources

The most efficient way to stop noise exposure is to do the job a different way, so that the noise is not created in the first place. This cannot be achieved over night, but when planning changes to a production process, or the introduction of a new process, it is often worth thinking about what will be noisy and whether this noise can be avoided.

For example, in recent years there has been a large increase in the use of self-compacting concrete (SCC). This material flows into the corners of moulds by gravity alone and requires no vibration to achieve compaction. It therefore appears to provide a long-term solution to employees' noise exposure within the concrete production industry. Use of SCC in concrete products manufacture is unusual at present in the UK, but is widely used elsewhere, such as in Japan where it was first developed. It is recommended that you consider the use of SCC when planning future production methods; this could eliminate the noise generated by vibrating tables, conveyors and presses.

Insulation

Acoustic insulation is the use of a material to block the path of the noise from its source to people's ears. The use of screens and enclosures are examples of insulation in practice. Screens may be rigid or flexible, as in the case of acoustic curtains/skirts. Whichever type is used, it is important that the screens are continuous across the face of the path of the noise transmission, with no gaps in the screen. The screen should be placed close to the source or close to the employee's position. Guards which enclose noisy plant can be turned into acoustic screens by making them solid rather than open mesh. If vision through the screens is required, they can be made from polycarbonate. Alternatively, the use of CCTV should be considered. Acoustic enclosures can involve:

1. The enclosure of the noisy process itself.
2. The provision of a noise haven for employees, such as an insulated control room.

In general, enclosures should be constructed from heavy, flexible materials which are the best for blocking sound. Enclosures will only work well if there are no openings or gaps which allow the noise to escape. Where materials pass in and out of, for example, an enclosure around a press, 'acoustic tunnels' can be provided to reduce the leakage at the openings. Another important aspect of enclosures around noisy machines or processes is that noise is trapped inside, so the levels of noise will increase within the enclosure. This can be reduced by lining the enclosure with acoustically absorbent material.

Absorption

This is the ability of a material to 'soak up' noise when the noise passes through it. Porous materials (such as mineral wool and some foams) make good absorbers. Absorbers are used to line noise enclosures and are also sometimes suspended above equipment to reduce reflected noise from hard surfaces (walls, ceiling, etc.).

Damping

Damping materials are used to reduce vibration in structures, especially flexible sheet metal components such as panels on machine casings, hoppers, chutes and skips. Reducing this vibration reduces the noise that is radiated by the vibrating surface. Damping material can be applied to the metal supports on the underside of conveyors where pallets drop, or onto any flexible sheet materials.

Isolation

The use of springs or rubber mounts to stop vibration from being transferred from a machine, such as a vibrating table or press, into the structure of the building or to another object (from which it could be radiated as noise.

Impact reduction

When two solid objects collide, the impact can be noisy. This is a common source of noise exposure in the industry: for example when pallets carrying concrete tiles collide on a conveyor or when materials fall into hoppers or skips. There are two methods to reduce the impulsive force between two objects impacting, and so reduce the noise of the impact. The first is to stop the impacts, or to reduce the speed with which the objects collide. This can often be achieved by careful control of conveyor speeds and by limiting the distance through which components or materials fall before they hit the bottom of a hopper, etc. The second method is to replace metal parts with components made from softer materials, such as

polyurethane, which absorb some of the impact energy and produce less noise. Such materials can also be used to line metal chutes, skips, hoppers, guide rails, etc. which are subjected to impacts.

Pneumatic exhaust silencing

Many processes in the industry use compressed air. A common source of noise is the exhausting of used air. Silencers are readily available and in some cases cost only a few pounds. Noise reduced air jets should be specified for all airlines and all air exhausts should be fitted with silencers. Reductions of at least 3 dB can be achieved by fitting a silencer to an air exhaust.

Other methods

One very simple method of reducing your employees' noise exposures is to move noisy equipment and auxiliary plant away from them. Where possible, noisy plant and auxiliary equipment such as should be located away from areas in which employees spend their time.

Comparatively quiet jobs should be carried out in low noise areas, rather than in a noisy area. One company reported a reduction of more than 10dB in the noise exposures of some employees by moving a routine cleaning task out of a noisy production area. Finally, many jobs can being carried out in different ways, some of which will be noisier than others. You should identify the quieter methods, where possible, and train your employees to help reduce their own exposures. This Table 31.2 identifies high risk activities or processes; and links to further information and case studies.

Table 31.2. Established noise control for high risk activities.

Product	Process	Established noise control methods
Flat products (e.g. slabs, fence posts, panels).	Mould filling, demoulding and stacking using vibrating tables or conveyors	Use self-compacting concrete Use resilient material (e.g. rubber) on tables Clamp mould to table
Reinforced concrete products (e.g. beams, steps)	Example noise levels: 95–110dB (steel tables) 86–93 (tables/conveyors with rubber covering)	Fit tunnels or enclosures over conveyors Enclose undersides of conveyors and tables Maintenance of enclosures, skirts, etc. Maintenance of vibrator motors and mountings Use wood, fibreglass or rubber moulds instead of metal to reduce impact noise
	Use of self-compacting concrete (SCC) Example noise levels: Relatively quite process: no vibration required	SCC (concrete to which chemical plasticisers are added) is increasing in popularity in the UK. Its use has the potential to eliminate the main source of noise (vibration). SCC should be discussed at visits to raise the profile and encourage innovation
Blocks, tiles, slabs	Vibratory presses	Fit enclosure (all controls outside) or provide separate control room (noise refuge)
	Example noise levels: 96–110dB (no noise reducing features) 84–93dB (outside press enclosure)	Isolate vibrating parts from floor and enclosure Maintenance of vibrator motors and mountings Silencers for compressed air exhaust Secure all parts and fittings to prevent rattling Use resilient material (e.g. rubber) for stops

(Contd ...)

Product	Process	Established noise control methods
	86–88dB (unloading stations) 71–79dB (inside control rooms	
	Rumblers/tumblers	Line barrel of tumbler with rubber lining
	Example noise levels: 84–95dB	Isolate plant from other processes and/or use plastic curtains to separate from employees
	Saws	Use noise-reduced saw blades
	Example noise levels: 81–96dB	
Extruded tiles	Extrusion plant Pallet/mould conveyors Example noise levels: 86–93dB	Extrusion plant: Use noise-reduced blow-off jets/air knives Use silencers on compressed air exhausts Conveyors: Control speed to minimise collisions between pallets (may require training) Use an impact absorbing material (e.g. polyurethane) on convenor guide rails, etc.
General	Chutes and skips	Provide chutes and skips with rubber lining Minimise dropping distances for waste material
	Mixing machines	Noise havens containing all control consoles
	Cleaning equipment Example noise levels: Up to 105 (ultra-high pressure water jetting)	Avoid or minimise the need for use of noisy equipment by washing down before the 'mix' goes off For water jetting, locate compressor in acoustic housing, restrict operating pressure
	Materials handling	Where heavy quarry type vehicles are employed, use acoustic cabs

Table 31.3 highlights the management of noise risks.

Table 31.3. Management of noise risks.

Issue	Expectation
Workplace design for reduced noise exposure	Table 31.2 deals with established technical and organisational noise control measures for a range of high noise risk activities or processes. In addition to these measures, in general there will always be benefits to be gained in considering and applying general principles of workplace design for reducing noise exposure. For example:
	Appropriate use of acoustic absorption within buildings can reduce or limit the effects of reflected sound (specialist help will be needed to put this in to effect)
	Careful planning could segregate noisy machines from other areas where quiet operations are carried out
	The number of employees working in noisy areas should be kept to a minimum
	Screens, barriers or walls can be placed between the source of the noise and the people to stop or reduce the direct sound

(Contd ...)

Issue	Expectation
	Noise refuges can be a practical solution in situations where noise control is very difficult, or where only occasional attendance in noisy areas is necessary
	Increasing the distance between a person and the noise source can reduce noise exposure considerably
Selection of tools and machinery	Employers should demonstrate a positive purchasing policy which makes sure noise is taken into account when selecting machinery
	For many types of equipment there will be models designed to be less noisy. When selecting equipment to buy or hire, besides ensuring that the tool or equipment is generally suitable for the job, employers should:
	Ask about likely noise levels for the intended use(s)
	Check that manufacturers' noise data is representative of likely noise levels for the intended use(s)
	Use the noise information to compare machines before making the final choice
	Look for warnings in the instruction book to see if particular uses of the tool or machines are likely to cause unusually high noise
	Be aware that even where manufacturers declare that their tools or machines produce less than 70 dB, levels may sometimes be much greater in your workplace
Limiting exposure duration	Restriction of the time spent in noisy areas, or doing noisy tasks, can be effective in reducing noise exposures, as can ensuring that noisy devices are only used when they are actually needed
	Where some employees do noisy jobs all day or week, and others do quieter ones, job rotation should be considered. This might need you to train employees to carry out other jobs. This system will reduce the noise exposure of some employees while increasing that of others, so care and judgement is needed. Employees will need to be rotated away from noisy jobs for a significant proportion of time to make an appreciable difference to their daily exposure
	The noise exposure ready-reckoner and exposure calculators can be used to indicate the reductions in exposure that can be achieved by reducing the duration of exposure to noise
Hearing protection	Providing personal hearing protection should be one of the first considerations on discovering a risk to the health of your employees due to noise. It should not be used as an alternative to controlling noise by technical and organisational means, but for tackling the immediate risk while other control measures are being developed. In the longer term, it should be used where there is a need to provide additional protection beyond what has been achieved through noise control
	Hearing protection use should be targeted at particular noisy jobs and activities. Personal hearing protection must be supplied by the employer to any employee whose daily personal noise exposure is likely to exceed 85 dB, or who is likely to be exposed to peak sound pressure levels above 137 dB. The employee must use the protection provided. The employer should ensure that, through the use of hearing protection, the employee's effective noise exposure is reduced to at least below the above levels
	Important factors to consider in the selection and use of hearing protection include:
	Types of protector, and suitability for the work being carried out
	Noise reduction (attenuation) offered by the protector, including taking account of 'real-world' factors, and also ensuring that not too much protection is provided

(Contd ...)

Issue	*Expectation*
	Compatibility with other safety equipment
	Pattern of the noise exposure
	The need to communicate and hear warning sounds
	Environmental factors such as heat, humidity, dust and dirt
	Cost of maintenance or replacement
	Comfort and user preference
	Medical disorders suffered by the wearer.
	The use of personal hearing protection should be managed through the provision of appropriate information, instruction and training for employees, supervision and the use of appropriately defined and demarcated Hearing Protection Zones.
Information, instruction and training	It is important that employees understand the risks they may be exposed to. Where they are at risk from noise their employer should at least tell them:
	The likely noise exposure and the risk to hearing this noise creates
	What their employer is doing to control risks and exposures
	Where and how people can obtain hearing protection
	How to report defects in hearing protection and noise-control equipment
	What their duties are under the Control of Noise Regulations 2005
	What they should do to minimise the risk, such as the proper way to use hearing protection and other noise-control equipment, how to look after it and store it, and where to use it
	The health surveillance systems
	This information should be given in a way the employee can be expected to understand (for example special arrangements might need to be made if the employee does not understand English or cannot read)
	To establish whether information, instruction and training has been carried out effectively, look for evidence that personal hearing protection is being fully and properly used, that noise control equipment is being used, and that procedures for low noise working are being followed
Health surveillance (audiometry)	Health surveillance for noise-induced hearing damage should be in place for employees whose daily personal noise exposure is frequently above 85 dB, or who are frequently exposed to peak sound pressure levels above 137 dB. Health surveillance should also be provided where exposures are lower, but where the employee may be particularly sensitive to noise. As a minimum, a program of health surveillance should include:
	Audiometric testing (baseline assessment on first entering a job involving noise exposure, annual testing for two years, then three yearly testing)
	Arrangements to receive medical advice on management of affected employees
	Arrangements to receive anonymised information to demonstrate effectiveness of controls.

Chapter 32

Construction Equipments

INTRODUCTION

Noise exposure levels of construction workers are difficult to determine due to the day-to-day variation in occupation and shift length of each worker and the itinerant and seasonal nature of the job. Nevertheless, it is clear that the construction worker is exposed to very high sound levels for considerable lengths of time. Noise control possibilities on-site are limited. Construction companies rent equipment from equipment suppliers and are not at liberty to 'improve' them. Other on-site equipment may belong to subcontactors. This too, may not have been designed for reduced noise emissions. Designing practicable, noise retrofit kits able to endure the environment of building sites is difficult. Techniques for controlling noise lie largely in the management of the site (e.g. avoiding noisy processes, scheduling noisy activities for when few workers are present) and isolating noisy processes by placing temporary sound barriers or screens. The most effective method will be realised when companies request and obtain equipment with low noise emissions, insist on 'residential' rather than 'industrial' grade combustion engine exhaust silencers, prefer electrical over pneumatic hand power tools, hydraulic over air powered rock drills, 'silenced' pile drivers over diesel pile drivers, etc.

Acoustical research has demonstrated construction tools and noise can be quieter. However, little seems to have been implemented by manufacturers except where required by law or by a few European manufacturers. Machines most successfully treated are those made the subject of legislation enacted in the 1970s in US, UK and member countries of the EC. These machines include trucks, pneumatic pavement breakers and air compressors.

Moreover the quieter saw blades referred to in the literature are not readily available, the more effective silencers are not installed on equipment as standard, and (unlike Europe) the diesel pile driver continues to remain in unrivalled and unabated use.

One conclusion is that without bylaws, legislation and enforcement, progress in noise control on construction sites will not happen. The problem can be addressed:

1. At the source by the federal government requiring noise emission labels on tools and machines.
2. At the source by Municipalities requiring construction companies use only tools and equipment which meet acceptable noise emissions.
3. At points of reception by municipal inspectors ensuring site boundary noise limits are not exceeded.
4. In the work site by WCB inspectors by ensuring tools and equipment are maintained and that workers participate in an effective hearing conservation program.

Such measures 'level the playing field' for companies interested in noise control but who, otherwise, would be put at an economic disadvantage if they alone did so.

The construction sector is of a significant size, employing as many workers as the sawmill and the heavy manufacturing sectors. Currently, 8 per cent of all hearing loss claims are attributed to this sector, comparable to the heavy manufacturing group (next sector) but far less than the sawmill group (about 30 per cent). Unlike the case with sawmills, the magnitude of construction workers' hearing losses are not noticeably falling with time. The annual average cost of hearing loss claims (excluding healthcare and rehab costs) in the construction sector is $2,95,000. This may be expected to rise with more workers being gathered under hearing conservation programs.

Noise-induced hearing losses of construction workers were probably acquired prior to the introduction, in that sector, of hearing conservation programs in 2004. One expects more workers to file claims as their awareness of their entitlement to claim for hearing loss increases as they approach retirement age. Minor reductions in the average hearing loss should be expected at least until the older workers retire and are replaced with younger workers having less hearing loss (who will have benefited from effective hearing conservation programs). Thus, the numbers and magnitudes of future claims will begin to fall over the next decades. Demographics indicate the bulk of the population will be reaching retirement age within the next decade. Thus, while older hands work their way 'through the pipe', the claim rate may initially increase for a few years. At the same time, the younger entrants into the construction workforce will have to become acclimatised to protecting their hearing.

NOISE EXPOSURE LEVELS

The hearing loss claim rate in the construction sector is lower than the forest product (FP) sector due to several factors including:
1. FP sector may have higher continuous average noise exposure levels.
2. FP sector workforce is a less transient workforce.
3. FP sector can maintain more complete control over their workers.
4. FP sector is more completely unionised to aid workers in their claim.

Only the first of the above factors is addressed in this report. West Coast Bank (WCB's) hearing conservation section has a database of noise exposure data on various trades. However, a recent report shows that we must be circumspect when referring to 'construction noise' as if noise exposures were common thoughout all types of construction; noise levels vary considerably between types of construction (Fig. 32.1). The mean L_{eq} of workers in a type of construction is determined to some degree by the power of the equipment they use or are exposed to and the environment in which the equipment is used. Earthmoving machinery used in large construction projects generate more power and noise than for a smaller project. When the noise source is inside a building or semi-enclosed space, it will create higher sound levels than if it were outside (Fig. 32.2).

Note: Noise levels are expressed in terms of the energy-equivalent continuous noise level, L_{eq}. No attempt is made to convert to L_{EX}, which normalizes the L_{eq} to an 8 hours day. This could only be accomplished given a worker's pattern of exposure to noise; workers switch to different jobs/tools/sites, and their shift length is variable and seasonal.

Extensive noise surveys conducted in Ontario provided the noise level data on various trades and activities in the construction sector. Data from a study of noise levels in UK construction are shown in Table 32.1. There seems little or no conclusive evidence from these samples that one country's construction noise is higher than anothers.

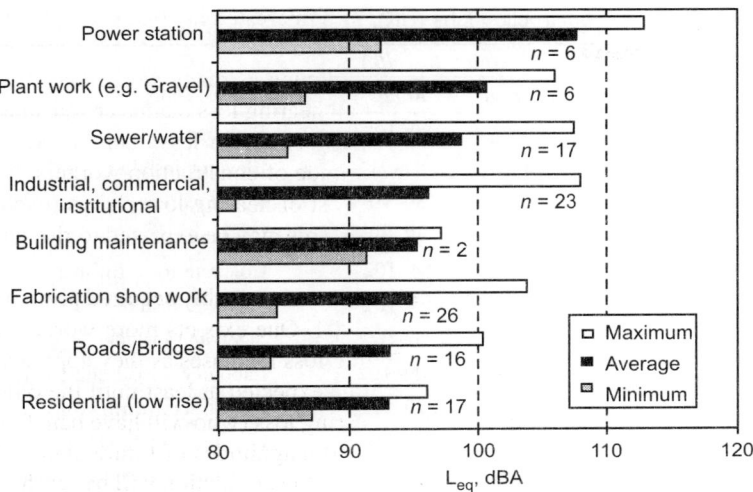

Fig. 32.1. L_{eq} by type of construction.

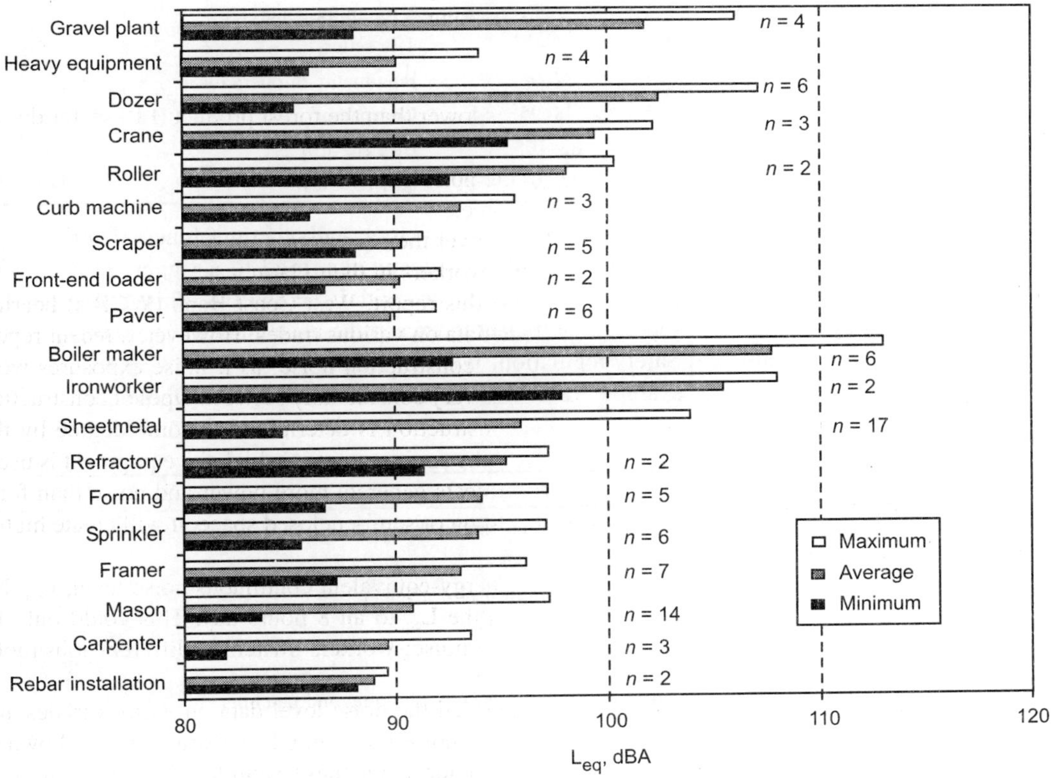

Fig. 32.2. L_{eq} in construction trades, activities or equipment.

Table 32.1. Noise levels, L_{eq}, in construction jobs (UK).

Plant/equipment	Operator, L_{EX}, dBA		Trades/Tools	L_{eq} dBA
	Average	Range		
Dozers, dumpers	96	89–103	Plumber	90
Front end loaders	88	85–91	Elevator installer	96
Excavators	87	86–90	Rebar worker	95
Backhoes	86.5	79–89	Carpenter	90
Scrapers	96	84–102	Concrete form finisher	93
Mobile cranes	100	97–102	Dry wall installer	90
Compressors	79	62–92	Steel stud installer	96
Pavers	101	100–102	Labourer—road construction	86
Rollers (compactors)	90	79–93	Labourers—formwork	88
Bar benders	95	94–96	Labourers—shovel hardcore	94
Pneumatic breakers	106	94–111	Labourers—concrete pour	97
Hydraulic breakers	95.5	90–100	Hoist operator	100
Graders, trucks, concrete pumps and mixers, generators	< 85	–	Labourers—drains and roughing concrete	100
Concrete batch plant operator	< 85	–	Tile setter	92
Poker vibrators	94.5	87–98	Pneumatic chipper/chisel	109
Saws	88.5	78–95	Compactor	108
Pile drivers (diesel and pneumatic)	98	82–105	Electric drill	102
Pile drivers (gravity, bored)	82.5	62–91	Air track drill	113

In early construction stages, we found a large variety of hand tools and other machinery contributing to a background level which is usually above 80 dBA; a dosimeter fitted to a rigger (himself an insignificant noise generator), who ranged over the site, returned with L_{eq} = 86 dBA. This 'background' sound level generated by on-site noise sources is sufficient to exceed the noise limit of section ($_{LEX}$ = 85 dBA) of the Occupational Health and Safety regulation, WCB of BC, 1998.

Individuals using power tools, for example, receive more noise than indicated in Table 32.2.

Table 32.2. L_{eq} by construction stage.

Stage of construction	Job	L_{eq}, dBA
Foundations	Sawing concrete and forms carpentry	96
Foundations	'Jumping jack' earth tamper	96.5
Foundations	Foreman, (concrete vibrator used for pour)	91.4
Foundations	Bobcat driver/First aid attendant	82
'Flying table'	Fly table carpentry–hammering	87.7
'Flying table'	Concrete finishing–electric grinder, chipping and patching	91.3
'Flying table'	Releasing, dropping flying tables	88.8
'Flying table'	Rigger working with crane	86

APPROACHES TO NOISE CONTROL

The technical literature contains research papers on a wide variety of common noisy tools and machines appearing on construction sites. Much of the research received strong impetus in the 1970s with the introduction of various environmental regulations and city noise ordinances which required noise emissions from building sites be reduced to avoid environmental noise complaints (i.e. complaints about noise generated on one site, travelling through the environment and affecting people in another location). Some of the research work on tools and machines are referenced in this review. Recent studies of construction sites showed little benefits of the research has been realised.

The control of construction noise that has been achieved is due in large part to the actions taken by governments responding to environmental noise complaints. The environmental noise law may be national statute or local bylaw and attempt to control construction noise by various means such as:
1. Prohibiting construction activity by time and date in city zones.
2. Directing trucks to less noise-sensitive routes.
3. Requiring machines conform to noise emission standards.
4. Requiring construction sites use barrier techniques to reduce noise propagation from the site.

Toronto's Noise By-law of 1987 limits noise as follows:
1. 70 L/s air compressors to 73 dBA at 7 m.
2. Leaf blowers to 70 dBA at 15 m.
3. Refuse compacting equipment to 80 dBa at 7m.
4. Pneumatic pavement breakers 80 to 92 dBA at 7m depending on weight and date of manufacture.
5. Heavy construction equipment (dozers, backhoes, mobile cranes, pile augers, trencher, etc.) to 83 to 88 dBA at 15 m depending on power output and date of manufacture.

There are in principle, several means by which to secure noise exposure reductions:
1. Hearing protection.
2. Reduction at source by retrofitting engineered controls.
3. Selection of quieter tools/machines.
4. Application of quieter processes.
5. Enclosure of the noise source.
6. Limitation of daily duration of exposure (including job rotation).

With construction tools and construction sites, some of these general possibilities encounter obvious difficulties. Recent work with construction workers has shown room for considerable improvement in the studied groups' use of hearing protection devices.

ENGINEERING NOISE CONTROLS

Mobile Equipment

Early in the site preparation stage, sound levels are dominated by heavy mobile equipment including trucks, bulldozers, front-end loaders. The operator of modern heavy equipment generally enjoys a much quieter cab (if the cab can be enclosed) than 20 years ago. Project work on a bulldozer's and a front-end loader's cab in 2001 demonstrated successful and practicable noise control techniques. Similar technology has been applied to the cabs of other vehicles, such as trucks. This was accomplished with commercially available noise control materials including absorptive headliners, floor mats, noise barrier mats and gaiters where control cables penetrated the bulkhead.

One report indicates a series of problems in attempting to comply with the OSHA noise regulation. Equipment manufacturers rejected purchase orders containing noise control specifications. Attempts to make existing equipment quieter met with several difficulties after a few months. This field testing highlighted needs for:

1. Improved quality mufflers with longer life time and more noise reduction. Muffler volumes should be about 10 times the volume of the combustion cylinder to be effective. Other work shows that when a very poor muffler is fitted, the exhaust noise is typically 10 dB above any other noise source. Addition of an improved muffler can lower noise by 6 to 15 dBA.
2. Consideration of a driver's visibility when adding cabs.
3. Floor mats and absorptive sidedoor panels to be removable for maintenance.
4. Effective adhesives for (and cleanable) sound absorbing panels.
5. Engine intake mufflers.
6. Sound absorbing hoods over engines.
7. Isolation of hydraulic reservoirs/valves from cab's control box.

Pile Drivers

A notoriously noisy item of site preparation equipment are percussive pile drivers. The diesel driver is seen and heard frequently in Richmond. One technique reported diesel drivers using a noise barrier set around the top of the pile (point of impact with the dolly and the length of the pile). A British development, named the Hush 'X' Rig, claims a noise level of 70 dBA at 15 m. with these features:

1. Rig's leaders can be adapted to accommodate any diesel hammer upto 85,000 ft-lb energy output and can accept most types of prefabricated pile.
2. Improved panelling for wide variety of hammer noise.
3. Efficient ventilation system scavenges exhaust gases, cools hammer and supplies combustion air.
4. Hydraulically operated full length doors for safety.

By comparison, a Delmag diesel pile driver seen on site locally was not fitted with any engine muffler. The L_{eq} was about 95 dBA near the site hut where the foreman reported that under the impacts experienced by the hammer, the mufflers soon fall off! Quieter, alternative pile driving techniques have been available for many years (compare levels Table 32.1) largely as a result of the introduction of environmental noise laws to minimise disturbance to residential neighbourhoods. Hydraulic and vibration pile drivers do not use the impact principle and are inherently far quieter.

Much of the incentive for the UK's development of construction site noise seems to stem from the Control of Pollution Act 1974 (UK) and British Standard 5228, 1975 'Code of practice of noise control on construction and demolition sites'.

Portable Air Compressors

Portable air compressors are used to provide the compressed air power source for pneumatic tools, e.g. the pavement breaker. The noise from compressors has been successfully reduced (10 to 20 dBA) over the years. Compressors have been silenced by a combination of enclosures fitted around the engine and compressor; additional engine exhaust muffler, lined inlet air ducts, isolating the machine vibration from the frame and possibly fitting a quieter fan. The significant noise control principles are illustrated in the sketch as the techniques could be transferred to other noisy machines (Fig. 32.3).

As with other noisy tools and equipment, quieting the portable compressor was in response to the environmental complaints of the 1970s. An enclosed compressor if properly designed should not overheat.

Enquiries were made of a road crew over letting a 'quiet' compressor run with its sidedoors open; their reply was the doors made handy shelves for the pneumatic drill's bits and chisels-overheating was not the problem.

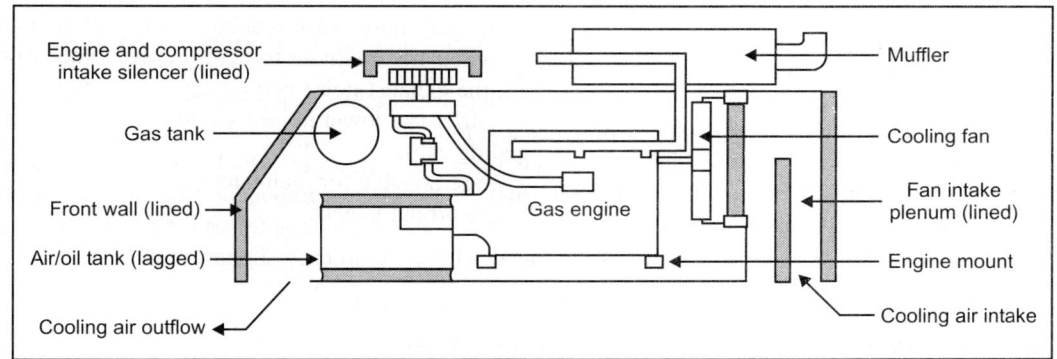

Fig. 32.3. Main noise control features of portable compressor.

Air Track Drills

Noise levels from pneumatic percussive drills are high, exceeding 110 dBA at operators' ears. Air discharge and the steel drill rod noise are roughly equal. Thus quieting one source without the other would yield a reduction of only 3 dB. US Bureau of Mines achieved a reduction of 18 dBA on a jumbo-mounted air drill 20 years ago. The technique involved development of an air exhaust muffler integrated with an acoustical enclosure for the drill. The enclosure could be easily opened for maintenance. Drilling rate was unaffected and the muffler/enclosure suffered no serious degradation after drilling 5000 ft and spending several months in an underground mine. Engineers developed a redesigned stoper drill. The redesigned chuck and steel shank corrected misalignment in the drill (causes bending waves and noise). The drill rotation was controlled by an independent air motor to avoid excess noise caused by over- or under-rotation. Other innovations included a radially symmetric main cylinder which then permitted a plenum chamber to surround the drill body over its full length, the chamber was developed into an exhaust muffler. A quieter alternative to the air drill is the hydraulic drill, whose drill steel still produces noise but the noisy pneumatic discharge is eliminated.

Joint Cutter/Stone Saw

This machine, powered by a two cycle engine, is used for cutting joints in concrete and asphalt. The circular saw blade is impregnated with diamonds around the periphery and the rotating grinding action excites the blade into vibration. In one investigation, a joint cutter was fitted with a metal hood to screen the operator from the machine and anti-vibration mounts were used to fit the engine to the frame. Reductions of 6 dBA (cutting) and 8 dBA (idling) were recorded.

Another investigation into the noise of a stone cutting saw concluded that the dominant noise was due to the excitation of the saw frame by the saw blade's vibration. The frame seemed to the major component of the noise. The work described a small pad supplied with water (normally just for cooling the blade) to serve as a viscous damping layer between pad and blade. With the water issuing from fine perforations into the clearance (about 0.2 mm) between pad and blade, a reduction of 12 dB was achieved. This device seems to be a practicable solution to the problem of stone cutting saw noise.

Hand Tools

Hand tools are difficult to quieten because:

1. The worker's ear is close to the noise source (within arm's length or less when the workpiece conducts noise much closer to the ear).
2. Any noise control measures may violate weight, size and cost restrictions required to meet tool utility making retrofit usually impracticable
3. Isolating the noise source is impracticable with most portable hand tools because a sound proof enclosure restricts access to the workpiece and tool.

Hammering nails can produce high peak levels (over 130 dB) at ear level. Fortunately, the high frequency components decay rapidly (in about 10 ms). A significant component of noise in the middle construction stage was impact noise; workers hammered to erect or release forms for re-erection on the next higher level. Others banged on frames and metal screwjacks to drop support timbers. Controlling hammering noise is impracticable.

PORTABLE PNEUMATIC DRILL/PAVEMENT BREAKER

The portable (under 40 kg) pneumatic rock drill is among the noisiest of hand tools on the construction (or demolition) site. The discharge of compressed air from the tool produces its dominant noise. Secondary noise sources include the cylinder head casing, front head and chisel.

Muffling the exhaust is usually achieved by either an integral muffler, small strap-on muffler (5 dB reduction) or by piping the exhaust air away. It is reported that strap-on mufflers large enough to enclose the cylinder case also reduce noise from this latter area at the same time to provide an extra 2 dB reduction. Strap-on mufflers enclosing the cylinder case and fronthead provide an overall reduction of 13 dB.

The noise of a typical unsilenced pavement breaker can be reduced by 16 dB. Electric pavement breakers are available in the middle to light duty categories. These include the Makita HM1800 and the Hilti TE905. The latter unit features a novel active vibration reduction system which effects a significant reduction in exposure to hand arm vibration. The electric units do not discharge pneumatic exhaust, so noise is due to the steel bit/pavement/striker interaction.

Cut-off Saw

There are several noisy portable tools powered by two-cycle combustion engines, one of which is used to cut rebar. A relative of the chain saw, Stihl TS400's spectrum featured strong peaks at 160 Hz (engine firing frequency) and at mid and high frequencies (630 and 1600Hz) when idling at full throttle. When cutting rebar even higher levels at frequencies above 1 kHz (sound level 108 dBA) were recorded. The increase is due to the rebar and saw blade responding at natural frequencies to the excitation in cutting.

Research shows that adequate exhaust silencing is achieved only when the muffler is much (ten times) larger than the engine's displacement. In the case of the TS400, engine size 64 cc, an adequately sized 'cube' shaped muffler would have sides of length of 18 cm. Rebar noise could be damped by laying sandbags along the bars' length.

Circular Saw

The carpenters' 'Skil' saw is a common noise source, the total noise output is dominated by noise in the high frequencies. Saw blade noise can be reduced (perhaps by up to 5 dB) by using sharp blades. The gear sets were a source of noticeable noise. This could be addressed by manufacturers installing better cut gears. Spiral cut gears are reputedly quieter rather than spur cut gears.

Reducing noise from saw blades is complex as several sources of noise contribute. Aerodynamic noise can be reduced by careful selection of tooth shape, gullet geometry and plate thickness, which should be selected to avoid coinciding with natural resonant frequencies of the disc. Attempts to reduce noise radiation from the vibrating sawblade using laser cuts in or near the periphery have been successful. The Laser-Q circular saw blade [manufacturered by Ernest Bennett (Sheffield) Ltd.] features a number of narrow (0.2–0.4 mm) slots made in a standard blade. The slots are elongated S-shapes, terminating in a small diameter hole, equispaced around the outer 15 per cent of the periphery. The slots are filled with a nonmetallic resin which, after curing, increases the strength of the blade reduced by the slots. The treatment can be applied to blades for non-ferrous metal and stone cutting. Examples of noise reduction are: ripping wood up to 6.4 dBA, flag stone cutting 2.8 dBA, brass >10 dBA.

Other Electrical Tools

Electrical tools found on-site can be noisy; electric drills for example, produced sound levels of 84 dBA (1/4″ bit), 92 dBA (3/8″) and 97 dBA (1/2″). Cooling fans, carbon brushes/commutator and gear box are important noise sources. Sabre saws (101 dBA), belt sanders (95 dBA), disc grinders (95 dBA), routers (95 dBA) and reciprocating saws (95 dBA) are noisy too. Quieting is practicable only by improved redesign by the OEM. The electric chainsaw (91 dBA) is notably quieter than the gas chainsaw (112 dBA). The Hitachi chipping gun's chisel rattled because it clamps in a steel chuck. The Hilti TE75 is quieter; its chuck is rubber-lined, gripping and damping over a greater surface area. Workers prefer electric over pneumatic guns as they are lighter and quieter.

PRACTICAL ON-SITE NOISE CONTROL

Planning

It is usually cheaper and more effective to plan for noise control prior to construction:
1. Plan to locate noisy plant (generators, compressors, pumps and concrete batching plant) away or screen from work areas.
2. Obtain probable noise levels of different equipment to aid in selection. Consider quieter alternative work methods (e.g. 'quiet piling' systems or rig-mounted hydraulic breaking equipment such as Atlas Copco's TEX HS).
3. Discuss noise problems which may occur with subcontractors, e.g. if subcontractors will be bringing a noisy machine on site the precautions needed to reduce noise.

Noise Control

1. Locate noisy machines away from main areas of activity. Otherwise, screen plant from work areas by using noise screens, berms or material stacked to form barriers.
2. Fit silencers to combustion engines. Ensure they are in good condition and work effectively.
3. Ensure hand-held concrete breakers are muffled.
4. Maintain machines regularly—they will be quieter.
5. Keep machinery covers and panels closed and well fitted. Bolts/fasteners done up tightly avoid rattles.
6. Check for noise problems. Do workers have to shout at arms length to converse?
7. Switch off engines or reduce to idle when not in use.

Hearing Protection Devices (HPDs)

Despite all efforts hearing protection may still be needed:
1. Ensure workers know when/why to use HPDs. Warning signs on noisy machines.
2. Offer comfortable HPDs as they are likely to be worn-workers can choose between different types.
3. Ensure workers know how to use/care for HPDs (headbands go over the head not around the neck, compress foam ear plugs well before insertion).
4. Replace damaged, hard or worn muff seals.
5. Ensure managers and foremen set good examples.

INSPECTION HINTS

Noise control
1. Is there a written noise control policy?
2. Is there evidence quieter tools/machines were selected (e.g. air compressor)?
3. Is there evidence of maintenance to minimise noise (sharp saw blades, mufflers for pneumatic tools and combustion engines adequately sized and in good condition)?
4. Are noisy machines relocated to distant places (compressor)?
5. Are sound barriers in use to reduce spread of noise (erect around noisy jobs)?
6. Are noisy processes done when other workers not present (e.g. concrete grinding)?
7. Are vibrating tools or parts isolated from other structures?

Hearing protection
1. Do all workers wear hearing protection?
2. Is there evidence of signs, instruction or supervision at noisy locations (warning signs at site entrance alone is not enough)?
3. Is some form of hearing protector available upon request on site?

Training and education
Is there evidence of training/education in the effects of noise and in noise control?

Noise assessment
1. Is there any noise measurements performed on site?
2. Can workers show their current hearing tests?
3. Is there evidence of followups when hearing loss results?

Noise Control Planning
1. Awareness of noise levels and noise management included in all trade health and safety courses.
2. Have noise control measures/quiet tools developed for specific trades included into trade guidelines.
3. Health and safety programs developed for construction trade should include a noise hazard component.
4. Tool/machine suppliers must be encouraged to offer noise emission level information on equipment labels and instruction manuals.

5. Encourage municipalities to develop noise bylaws for construction sites—companies use only tools/machines with reduced noise emission levels.
6. Articles describing successful noise management outcomes in construction should be submited to trade journals for publication.
7. Adoption of Federal government backed CSA standard for machinery noise emission levels should be supported.
8. Development of training programs for tailgate presentations on components of a hearing conservation program. Prepare another program for union safety representatives.
9. WCB inspectors to be seen doing on-site noise inspections leading to recommendations for implementation of noise management.
10. Union safety officers should be encouraged to participate in noise surveys.

SECTION VI

Noise Reduction in Roadways, Railways and Aircraft

Chapter 33

Roadways

INTRODUCTION

Roadway noise is the collective sound energy emanating from motor vehicles. In the USA it contributes more to environmental noise exposure than any other noise source, and is constituted chiefly of engine, tyre, aerodynamic and braking elements. In other Western countries as well as lesser developed countries, roadway noise is expected to contribute a proportionately large share of the total societal noise pollution.

DESCRIPTION OF BASIC VARIABLES

The intensity of roadway noise is governed by various variables such as: traffic operations (speed, truck mix, age of vehicle fleet), roadway surface type, tyre types, roadway geometrics, terrain, micrometeorology and the geometry of area structures.

Traffic operations noise is affected significantly by vehicle speeds, since sound energy roughly doubles for each increment of ten miles an hour in vehicle velocity; an exception to this rule occurs at very low speeds where braking and acceleration noise dominate over aerodynamic noise. Small reductions in vehicle noise occurred in the 1970s as states and provinces enforced unmuffled vehicle ordinances. The vehicle fleet noise has not changed very much over the last three decades; however, if the trend in hybrid vehicle use continues, substantial noise reduction will occur, especially in the regime of traffic flow below 35 miles per hour. As a pedestrian safety issue, hybrid vehicles are so quiet at low speeds that the customary warning noise may not alert the pedestrian to nearby danger, creating a potential hazard for visually-impaired people, who rely on such noise to navigate in areas of heavy traffic. Trucks contribute a disproportionate amount of noise not only because of their large engines, but also the height of the diesel stack and the aerodynamic drag. Significant interior noise is usually present inside moving motor vehicles; in fact, passengers are generally not aware that these levels are high, because experience has led motorists to expect levels commonly exceeding 65 dBA.

Roadway surface types contribute differential noise effects of up to 4 dB, with chip seal type and grooved roads being the loudest and concrete surfaces without spacers being the quietest. Asphaltic surfaces are about average.

Tyre types had considerable design changes in the 1970s, and at this juncture are probably optimised for noise control, given the of safety needs for a significant grip by the tread.

Roadway geometrics and surrounding terrain are interrelated, since the propagation of sound is sensitive to the overall geometry and must consider diffraction (bending of sound waves around obstacles), reflection, ground wave attenuation, spreading loss and refraction. A simple discussion indicates that sound will be diminished when the path of sound is blocked by terrain, or will be enhanced if the roadway is elevated so as to broadcast; however, the complexities of variable interaction are so great, that there are many exceptions to this simple argument.

Micrometeorology is significant in that sound waves can be refracted by wind gradients or thermoclines, effectively dismissing the effect of some Noise barriers or terrain intervention.

Geometry of area structures is an important input, since the presence of buildings or walls can block sound under certain circumstances, but reflective properties can augment sound energy at other locations.

TRAFFIC NOISE

Highway vehicles include automobiles, trucks, buses and maintenance vehicles. Automobiles are the primary mode of transportation and constitute the largest number of highway vehicles.

Sources of Highway Noise

Trucks

Trucks contribute to our noise problems by their exhausts, cooling fans, engines and tyres. The noise levels generated by truck exhaust systems are dependent on factors such as engine type, timing and valve duration, induction system, muffler type, muffler size and location in the exhaust system, pipe diameter, and engine back-pressure. The actual noise-generating mechanism is created by vibrating columns of gas at high pressure and amplitudes, which are produced by the opening of the exhaust valve. This noise is communicated directly to the atmosphere. Additional exhaust noise is created by the direct impingement of these released gases on the pipes and muffler shell.

In nearly all applications involving water-cooled engines, an axial flow type fan is used to draw cooling air through a forward mounted radiator. In many designs, fan noise approaches the level of exhaust system noise and is generally considered an important factor in reducing overall vehicle levels. Generally, fan noise is directly related to fan speed. It has been shown that fan noise increases at a rate of 2 dB per 100 rpm at speeds between 1500 and 2000 rpm. The noise output is also dependent upon tip speed and configuration, blade design and spacing, and proximity of accessories and other objects which affect air flow.

Engine-associated noise in internal combustion engines is produced by the compression and subsequent combustion process, which gives rise to severe gas forces on the pistons and to forces of mechanical origin, such as those produced by piston-crank operation, the valve-gear mechanism and various auxiliaries and their drives. Diesel engines are about 10 dB noisier than gasoline engines. This difference results mainly from their different mechanism of ignition. Gasoline engines initiate combustion with a spark from which the flame front gradually spreads throughout the combustion chamber until the entire fuel/air charge is burned. This yields a smooth blending with the compression. The diesel engine, however, relies on a much higher compression ratio to produce spontaneous combustion, which burns a large volume of fuel/air mixture rapidly. This yields a more severe pressure rise in the cylinder, causing more engine vibration in comparison with the gasoline engine. At constant speed, diesel engines show only slight reduction in noise, with reduction in load, due to high compression pressure. Gasoline engines, however, show a substantial decrease in noise output with decreasing load. Therefore, the change in noise level between no-load and full-load conditions is rarely more than 3 dB for a diesel engine, but can be as high as 10 dB for gasoline engines.

Truck tyre noise presents the major obstacle in limiting overall vehicle noise at speeds over 50 mph. At this speed, tyre noise often becomes the dominant noise-producing source on heavy-duty trucks. Typical noise levels from truck tyres at 50 mph range from 75 dBA for low-noise trend designs to over 90 dBA for high noise level tyres. The major offender is the crossbar design used by the vast majority of trucks on their drive wheels. These tyres may produce levels in the 80 to 85 dBA range when new, but their noise increases with wear going up by 10 dBA in the half-worn condition. This increase is attributable to a change in the tread curvature resulting from wear. Cross-bar retreads pose an even greater problem as their noise level can be as much as 95 dBA at 50 ft when operated at 55 mph in the half-worn condition. Despite their noise, cross-bar retreads are very popular for economical reasons and each tyre is recapped an average of two to three times.

Axle loading is also a major factor in the amount of noise generated by tyres. Retreaded tyres exhibit the most predominant dependence upon load. One example indicated a decrease of 15 dB resulting when load per tyre was reduced from 4500 to 1240 lb. With the tyre unloaded, the sides of the retread do not contact the road surface; hence the cups in the tread cannot seal against the road surface and compress small pockets of air. The rib-type tyre designs are generally independent of loading due to their uniform tread design across the entire cross-section. Variations in road surface also significantly affect tyre noise generation.

Another method which has been proposed is that of close fitting noise shields for engines. Noise shields can be placed around specific engine sub-systems which contribute significantly to overall engine noises. The shields are rubber mounted to isolate them from the adjoining engine surface. Under test conditions, up to 12 dB of reduction have been obtained over much of the frequency spectrum, with the exception of 250 Hz, where no change was found.

In summary, a quieter truck can be made by close cooperation between the vehicle and engine designer through:

1. Design of a vehicle giving adequate attenuation of engine noise.
2. Appropriate choice of engine design parameters.
3. Design of quieter engine structure.
4. Design of quieter tyres.

Automobiles

While not as noisy as trucks, the total contribution of automobiles to the noise environment is significant due to the huge number that are in operation. Approximately 70 per cent of automobiles on the road are over 10 years old, the average car being about 7 years old. These vehicles tend to produce higher noise levels (2 to 3 dB) under most operating conditions. This is due to deterioration of exhaust silencer performance and the response of the vehicle to pavement roughness.

For most automobiles, exhaust noise constitutes the predominant noise source for normal operation at speeds below about 35 to 45 mph, depending on the condition and design of the exhaust system. Above this speed range, in many cases, tyre noise becomes equally significant. While exhaust noise does not create a significant interior problem, certain objectionable tones may be audible inside the car. To relate back to the actual exhaust, these gases are produced by the explosions that make automobile internal combustion engines. Mufflers are designed so that the energy of these gases is spent safely and quietly in a set of baffles.

In some cases, the intensity of fan noise is almost equal to that of the exhaust. The parameters which govern fan noise generation are essentially the same as those related to trucks. There may not only be rotational imbalances, but the fan is a propeller that sucks air in from the front to flow through the radiator. More work has been done in the field of automobile fans because the objective is to reduce passenger compartment noise.

Tyre noise in passenger cars presents much less of a problem than in trucks. The principal reason for this is that standard automobile tyres do not employ the cross-bar tread design. Automobile levels at 50 to 60 mph can be as much as 25 dB less than the worst truck tyres.

Much of an automobile's noises come from its engine where there are imbalances due to acceleration of the reciprocating parts of the engine. There are also imbalances in the various rotating parts such as the crankshaft, flywheel, cooling fan, electrical generator and alternator. The generator rotates at exactly twice the engine speed, producing a very audible and disturbing beat. There are also various gas noises caused by the explosive nature of the cylinder gas pressures.

Buses

Although trucks and buses share many basic design characteristics and some common components, buses are generally quieter due to their increased packaging space. This allows larger mufflers and an enclosed engine compartment. At highway speeds, passenger buses exhibit noise levels primarily in the range of 80 to 87 dBA at 50 ft. This noise is principally due to tyre noise. The fact that buses that have been in service for long periods of time become noisier is due to damage to engine compartment seals.

Control of Highway Noises

Noise reduction at the source may be achieved through legislation and enforcement. A combination of two kinds of regulation, usually requiring staged reductions in permissible sound levels in future years, is used. Sound levels emitted by new vehicles are regulated at the point of manufacture or sale under a standard testing procedure. Although noise control at the source is the most basic approach, it is likely to be a long-term solution. Noise control at the receiver is most effective in reducing in-house noise.

Noise control of the transmission path is the third method of control. One means of controlling the transmission path is to increase the distance between noise source and receiver. This may be accomplished by re-routing of major routes and diverting traffic, particularly truck traffic at night, away from roads passing near or through residential areas and by preserving open space between major routes and adjacent residences. Zoning land adjacent to major highways to less noise-sensitive uses such as industrial or commercial use is proper abatement.

Experimentation with noise barriers has been going on for an extensive length of time. It has been found that different materials are more effective than others in deflecting noise. Some of the materials used are precast concrete, earth berms, aluminium, wood and porous concrete. Experiments show that noise barriers are fairly effective. These barriers can usually be incorporated in the existing right-of-way. Through application of architectural or landscaping techniques, these barriers can be made pleasing to highway users. Barrier heights, which are greater than the highest sound source, must be a prominent design feature.

Foliage barriers are the least effective type although they will result in a 5-dBA sound level reduction for every 100 ft depth of foliage — provided the trees or shrubs are at least 15 ft tall and sufficiently dense throughout the year. Due to the fact that little or no additional right-of-way is generally required, man-made barriers are the most promising means that the highway engineer has at his disposal to abate traffic sounds.

NOISE MITIGATION

Noise mitigation is a set of strategies to reduce noise pollution. The main areas of noise mitigation or abatement, are: transportation noise control, architectural design, and occupational noise control. Roadway noise and aircraft noise are the most pervasive sources of environmental noise worldwide, and remarkably little change has been effected in source control in these areas since the start of the problem, a possible exception being the development of hybrid and electric vehicles.

Multiple techniques have been developed to address interior sound levels, many of which are encouraged by local building codes; in the best case of project designs, planners are encouraged to work with design engineers to examine trade-offs of roadway design and architectural design. These techniques include design of exterior walls, party walls and floor/ceiling assemblies; moreover, there are a host of specialised means for dampening reverberation from special purpose rooms such as auditoria, concert halls, dining areas and meeting rooms. Many of these techniques rely upon materials science applications of constructing sound baffles or using sound absorbing liners for interior spaces. Industrial noise control is really a subset of interior architectural control of noise, with emphasis upon specific methods of sound isolation from industrial machinery and for protection of workers at their task stations.

Sound masking is the active addition of noise to reduce the annoyance of certain sounds; the opposite of soundproofing.

ROADWAY NOISE MITIGATION

Source control in roadway noise has provided little reduction in vehicle noise, except for the development of the hybrid vehicle; nevertheless, hybrid use will need to attain a market share of roughly 50 per cent to have a major impact on noise source reduction on city streets. (Highway noise is little affected by automobile type, since those effects are aerodynamic and tyre noise related.) Other contributions to reduction of noise at the source are: improved tyre tread designs for trucks in the 1970s, better shielding of diesel stacks in the 1980s, and local vehicle regulation of unmuffled vehicles.

The most fertile area for roadway noise mitigation is in urban planning decisions, roadway design, noise barrier design, speed control, surface pavement selection and truck restrictions. Speed control is effective since the lowest sound emissions arise from vehicles moving smoothly at 30 to 60 kilometres per hour. Above that range sound emissions double with each five miles per hour of speed. At the lowest speeds, braking and (engine) acceleration noise dominates. Selection of surface pavement can make a difference of a factor of two in sound levels, for the speed regime above 30 kilometres per hour. Quieter pavements are porous with a negative surface texture and use medium to small aggregates; the loudest pavements have a transversely tined/grooved surface, and/or a positive surface texture and use larger aggregates. Obviously surface friction and roadway safety are important considerations as well for pavement decisions.

When designing new urban freeways or arterials, there are numerous design decisions regarding alignment and roadway geometrics. Use of a computer model to predict future sound levels from line sources has become standard practice since the early 1970s. In this way exposure of sensitive receptors to elevated sound levels can be minimised. An analogous process exists for urban mass transit systems and other rail transportation decisions. Early examples of urban rail systems designed using this technology were: Boston MTA line expansions, San Francisco Bay Area Rapid Transit System expansion, Houston light rail system, and the Portland, Oregon Beaverton light rail line. Noise barriers can be applicable for existing or planned surface transportation projects. They are probably the single most effective weapon in retrofitting an existing roadway, and commonly can reduce adjacent land use sound levels by up to ten decibels. A computer model is required to design the barrier since terrain, micrometeorology and other locale specific factors make the endeavour a very complex undertaking. For example, a roadway in cut or strong prevailing winds can produce a setting where atmospheric sound propagation is unfavourable to any noise barrier.

Road Transportation Noise

1. Road accounts for approximately 70 per cent of total noise emissions by transportation. It must be noted that different road transportation modes have different scales of noise emissions.
2. Main sources of noise come from the engine and the friction of the wheels over the road surface. Further, travel speed and the intensity of traffic are directly linked with its intensity of noise. For instance, one truck moving at 90 km/hr makes as much noise as 28 cars moving at the same speed.
3. Ambient noise is a frequent result of road transportation in urban areas. The addition of all the noise generated by cars, trucks and buses creates a permanent ambient noise (ranging from 45 to 65 db) that impairs the quality of life in urban areas and thus the property values of residences. Nearby road arterials, ambient noise is replaced by direct noise and vibrations. The acoustics created by the surrounding environment (hills, buildings, trees, open space, etc.) alleviate or worsen local conditions.
4. Noise level grows arithmetically with speed. For instance a car travelling at 20 km/hr emits 55 db of rolling noise, at 40 km/hr 65 db, at 80 km/hr 75 db and at 100 km/hr 80 db. Available evidence underlines that around 45 per cent of the population in developed countries live in high levels of noise intensity (over 55 db) generated by road transportation. Along major highway arterials in inter-urban areas, noise emissions are likely to alter the living environment of wildlife species.

ROAD VEHICLE AND TYRE NOISE

Road Vehicle Noise

Road traffic noise is the major source of environmental noise exposure. Despite European vehicle noise standards being introduced in the early 1970s, road noise levels have not decreased. Technological progress has been largely cancelled out by increases in traffic and a trend towards heavier and more powerful vehicles and wider tyres. By combining different measures including low noise road surfaces; stricter noise standards for vehicles and tyres, and reducing traffic speed, a general reduction of road traffic noise of 10 dB(A) could be possible within 10–15 years. This would be equivalent to reducing noise to one tenth of current levels.

Road noise arises from three sources on vehicles (Fig. 33.1):

1. Propulsion noise (engine, powertrain, exhaust and intake systems).
2. Tyre/road contact noise.
3. Aerodynamic noise.

The engine noise is the dominant source at lower speeds (under 30 km/hr for passenger cars/under 50 km/hr for lorries), tyre noise dominates above that, and aerodynamic noise becomes louder as a function of the vehicle speed.

Vehicle noise standards are laid down in Directive 70/157/EEC and subsequent amendments, which regulates the technical approval of new vehicles and currently sets noise emission limits of 74 dB(A) for passenger cars and 80 dB(A) for trucks. The standards for trucks have been somewhat more effective than for cars. The noise limits include all sources of noise from the vehicle.

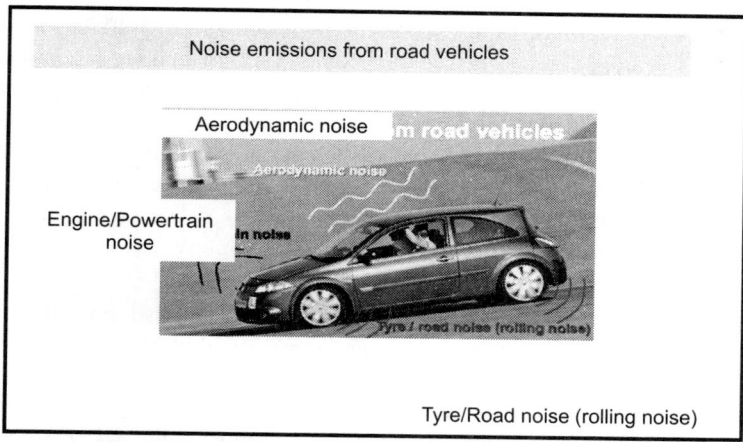

Fig. 33.1. Road noise arises from three sources on vehicles.

In practice, the EU directive follows Regulation No. 51 of the United Nations Economic Commission for Europe (UN-ECE) which harmonises rules for road vehicle sound emissions. Regulation 51 is defined at international level by the UN-ECE World Forum for Harmonisation of Vehicle Regulations. While type-approval noise limits have been gradually tightened over the years (by a reduction of more than 10 dB(A) for certain vehicles), no improvement has been made regarding overall exposure to noise generated by road vehicles, due largely to a significant increase in road traffic and a trend towards larger, heavier and more powerful vehicles. In fact, the introduction and regular tightening of these limits allowed for a harmonisation of the road vehicle fleet regarding noise emission characteristics, but did not prove to be a strong technical drive towards quieter vehicles, particularly in the case of delivery vans and trucks.

The test cycle for vehicle certification on noise levels does not accurately reflect real world driving conditions. The current test cycle does not include provision for evaluating noise performance in typical urban stop-start traffic situations at lower speeds, where engine noise is the dominant source. Another major failing is that the test parameters are set in such a way that vehicles can be designed to pass the test but are considerably louder when driven on the road. This urgently needs to change to prevent this kind of cheating, known as 'cycle beating', which is allowing dangerously loud vehicles on to our roads. The development of a new real-world test cycle is having the effect of delaying the possibility of introducing tighter limits for road vehicle noise emissions. There is a general consensus in the UN-ECE Working Party on Noise that equivalent values must be identified between the new and old test procedures, before tightening the limits can be discussed. A two year data collection period has just begun. The European Commission will collect the results. After two years at the latest (in theory, it could also be earlier), equivalent values for the new test will be identified.

Tyre Noise

Tyre rolling noise emissions have increased over time, which is predominantly due to increased use of wider tyres. As the tyre/road contact begins to dominate the noise emission above 30 km/hr for passenger cars and above 50 km/hr for lorries, it was deemed necessary to regulate tyre/road noise separately as well as its role in overall vehicle noise. Therefore Directive 2001/43/EC complements the vehicle noise standards by setting a test procedure and noise limit values for tyre rolling noise.

The directive states that the limit values should be reviewed and revised 36 months after 2001/43 enters into force. This deadline passed in summer 2004. A report making recommendations for the revision of the limit values and other aspects of the directive was presented to the European Commission in 2006. The feasibility study outlines a proposal for two phases of stricter limit values for 2008 and 2012, for tyres for passenger cars and trucks. The study recommends limit values which are equivalent to effective noise reductions of (taking into account different tyre classes and dimensions):

1. Passenger car tyres: 2.5–5.5 dB(A).
2. Commercial vehicle tyres: 5.5–6.5 dB(A).

The report also demonstrates that:

1. Quieter tyres do not compromise safety (wet grip, aquaplaning) or fuel economy (rolling resistance).
2. The proposed limit values for tyres for cars and trucks would lead to an estimated overall noise reduction of up to 3 dB(A), which is equivalent to cutting the current noise level by half, or halving the number of vehicles on the roads.
3. Enforcing these standards will not incur huge costs. The technologies and products have already been developed, and the industry is prepared.
4. Conservative scenarios estimate the societal benefit of quieter car tyres to be worth at least 48 billion, and up to 123 billion, across the EU over the period 2010–2022. Even the most cautious calculations point to benefits outweighing costs by 24:1. These figures do not even include the savings for health services, or attribute a value to annoyance or stress.
5. Benefits will be further magnified when quieter tyres are used on silent road surfaces.

Despite these convincing arguments, a proposal is not yet forthcoming. The current limit values for tyre rolling noise have had minimal impact on noise levels. 96 per cent of commercially available tyres on today's market meet the limit values.

Figure 33.2 illustrates that approximately half of the sample are over 3 dB(A) quieter than the current limits, meaning that they produce less than half the noise currently permitted. This shows that the standards are lagging behind technologies that are already available and that the current limits are not serving their intended purpose. Tyre producers currently have no incentive to produce quieter tyres. Some tyres are as much as 8 dB(A) below the limit, but as there is no noise labelling and hardly any noise information available to consumers, low noise is not exploited as a marketable attribute.

There are two principal reasons why stricter tyre rolling noise limits are urgently needed:

1. Tyre rolling noise is generally the dominant source of noise from road vehicles at medium and high speeds.
2. As the life-span of tyres is shorter than that for vehicles, addressing tyre noise standards will be one of the fastest ways to achieve road noise reductions.

The way forward should be first to achieve effective standards based on the current status of technological development. The proposed indicative limit values for 2008 are already achieved by the majority of tyres currently on the market. 76 per cent of the passenger car tyres on the market in 2004 would already meet the 2008 standards. In fact, 35 per cent already meet the proposed 2012 standards. This demonstrates that the Commission should be even more ambitious in their proposal because low-noise models have already gained type approval according to road safety requirements, sales of these models proves that they are commercially viable.

Future standards should serve as an incentive for innovation and encourage technological development. The best technology currently available (which also fulfils safety and other product approval requirements) is 8 dB(A) below the current limit values. Innovations like this should set the pace for the market. Outlining future phases of tightening in advance also provides certainty for the industry and consumers.

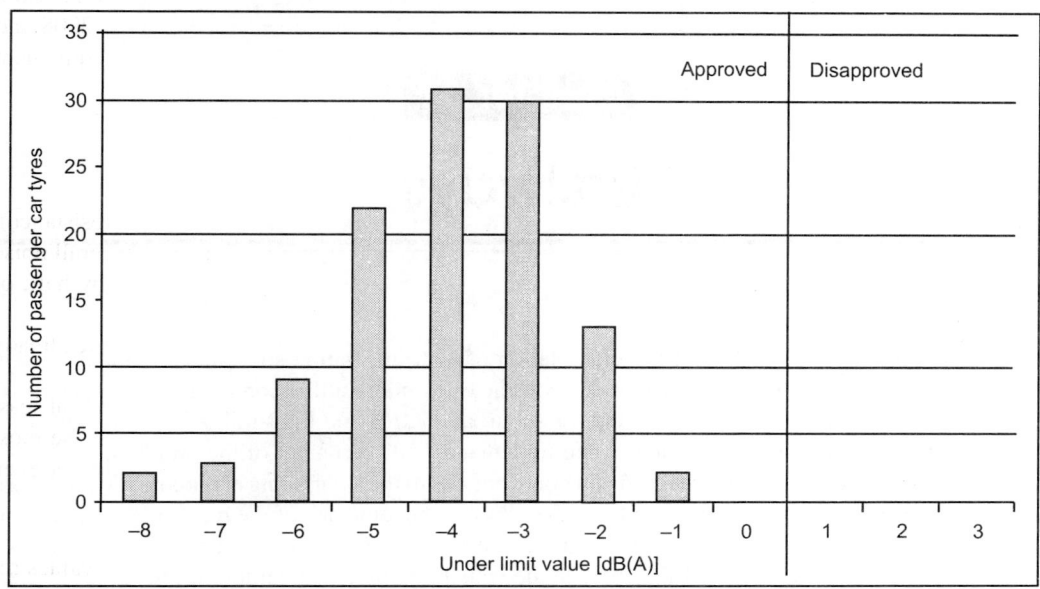

Fig. 33.2. Graph of noise level over 3 dB(A).

The report to the Commission also recommends inclusion of labelling of tyres, in the form of noise levels stamped on the sidewall or a 'low noise' label for the quietest models. At present, even if consumers request advice or information on noise levels of tyre models, sales staff are not given this information by the manufacturers. Consumer awareness of noise emissions from tyres can easily be increased by product labelling and a framework for incentives that Member States can grant to quieter tyres. Such action would clarify the range of consumer choice, similar schemes have been seen to be successful for noise ratings of electrical goods.

COMPUTER MODELS FOR ROADWAY NOISE

Because of the complexity of the variables discussed, it is necessary to create a computer model that can analyse sound levels in the vicinity of roadways. The first meaningful models arose in the late 1960s and early 1970s addressing the noise line source (e.g. roadway). Two of the leading research teams were BBN in Boston and ESL of Sunnyvale, California. Both of these groups developed complex mathematical models to allow the study of alternate roadway designs, traffic operations and noise mitigation strategies in an arbitrary setting. Later model alterations have come into widespread use among state Departments of Transportation and City Planners, but the accuracy of early models has had little change in 40 years.

Generally the models trace sound ray bundles and calculate spreading loss along with ray bundle divergence (or convergence) from refractive phenomena. Diffraction is usually addressed by establishing secondary emitters at any points of topographic or anthropomorphic 'sharpness' (such as noise barriers or building surfaces). Meteorology can be addressed in a statistical manner allowing for actual wind rose and wind speed statistics (along with thermocline data).

Chapter 34

Railways

INTRODUCTION

Railways are an environmentally cordial effectuation of instrumentation substantially suited to recent society. However, racket and ambiance are key obstacles to boost utilisation of the line networks for high-speed intercity traffic, for transport and for suburban metros and light-rail. All likewise off times racket problems are dealt with inefficiently cod to demand of discernment of the problem.

Railway transportation has made significant contributions to the flourishing economy of whole world, the environment-friendly development of the system having become one of the major challenges. Noise pollution in particular has been the subject of much attention in recent years. The primary sources of railway noise to residential communities include the whistling noise of locomotives, the noise of diesel locomotives and wheel/rail noise. However, increasing the speed on existing lines, constructing new high-speed passenger lines and developments of heavy freight transport are among the main improvement works of railways in the near future.

EXISTING NOISE PROBLEMS FOR THE RAILWAYS

Characteristics of Railway Noise

As with all traffic noise, railway noise can be described in terms of the daily average noise emission of the traffic flow, but also in more detail in terms of the noise characteristics of individual trains, vehicles and tracks. Most current national legislation is limited to reception limits for daily noise levels, which for railways is based on calculations of noise emission from the traffic flow at a given location.

Whereas the management of the traffic flow, i.e. train types, composition, timetables and speeds, is important for the daily noise emission, the noise emission characteristics of individual trains and tracks are an important factor in reducing noise at the source, as this works cumulatively.

When considering the noise emission characteristics of individual train or vehicle types, there are a number of major noise sources, which are relevant for particular situations, as illustrated in Table 34.1.

These main situations, which are relevant for the management of environmental railway noise, are the pass-by situation, which includes constant speed, acceleration and deceleration; stationary noise (in and around stations), and shunting noise, which includes a variety of noise sources.

Train speed is a major influence parameter for noise emission. The noise due to traction and auxiliary systems (diesel units, electrically driven powertrains, cooling equipment, compressors), if present, tends to be predominant at low speeds, up to around 60 km/hr. Wheel-rail rolling noise is dominant up to speeds around 200–300 km/hr, after which aerodynamic noise takes over as dominant factor. The transition speeds from traction noise to rolling noise and from rolling noise to aerodynamics noise depend entirely on the relative strength of these sources.

Table 34.1. Major noise sources relevant for particular situations.

Noise situation	Pass-by noise: Constant speed and acceleration/deceleration	Stationary noise	Shunting and other
Noise sources	Rolling	Traction/auxiliary	Squeal/Impact
	Traction/auxiliary		Traction/auxiliary
	Aerodynamic		Rolling
	(Locally: Squeal,		
	Impact, bridges)		

The predominant types of noise source can also be given per train category as indicated in Table 34.2.

Table 34.2. Main types of noise source for four train categories.

	Rolling noise	Noise from traction and auxiliary systems	Aerodynamic noise
Freight trains	++	+	
High speed trains	++	+	++
Intercity trains	++	+	
Urban trains	++	+	

+: Relevant.

++: Highly relevant.

The rolling noise, for example, depends strongly on the surface condition (roughness) of wheels and rails, whereas aerodynamic noise depends on the streamlining of the vehicle. An example of typical speed dependency is shown in Fig. 34.1.

High Importance of Maintenance

Surface roughness levels of rails and wheels even grow during normal operation. Figure 34.2 shows the roughness levels for different conditions of the rail surface and the wheel tread. Between perfectly smooth and highly corrugated rails there is a significant increase in roughness levels. In extreme situations, the variation in emission levels can be as much as +20 dB(A). Such a high noise increase will only occur with the special test vehicle with perfect wheels. In normal maintenance situations a variation of +/–3 dB(A) is found. Figure 34.3 shows the increase of the noise emission levels in Germany over a number of years after the rails have been acoustically ground.

The increase depends on the vehicle type in use on the track following grinding, for the quietest vehicles (disc braked with smoother wheels) it is with 0.9 dB(A)/year about three times as high as for cast iron block braked freight wagons where the difference is almost negligible. About 8 years after grinding the noise emission levels correspond to an average smooth rail. Figure 34.3 also shows the reduction potential of improved rail grinding which is larger, the smoother the wheels are (synergy effect). The cost/benefit consequences of additional grinding need further analysis.

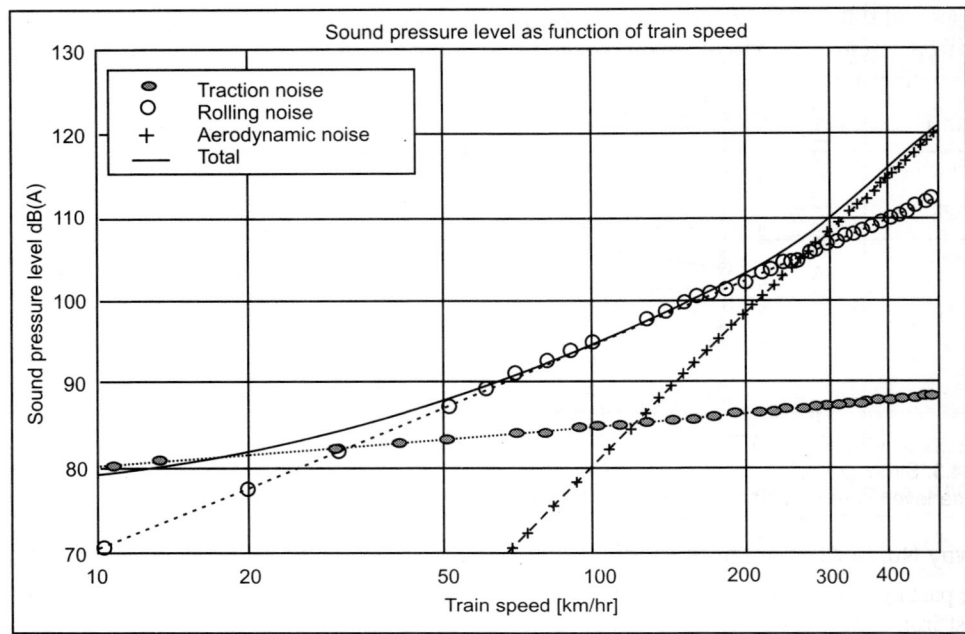

Fig. 34.1. Railway exterior sound sources and typical dependence on train speed.

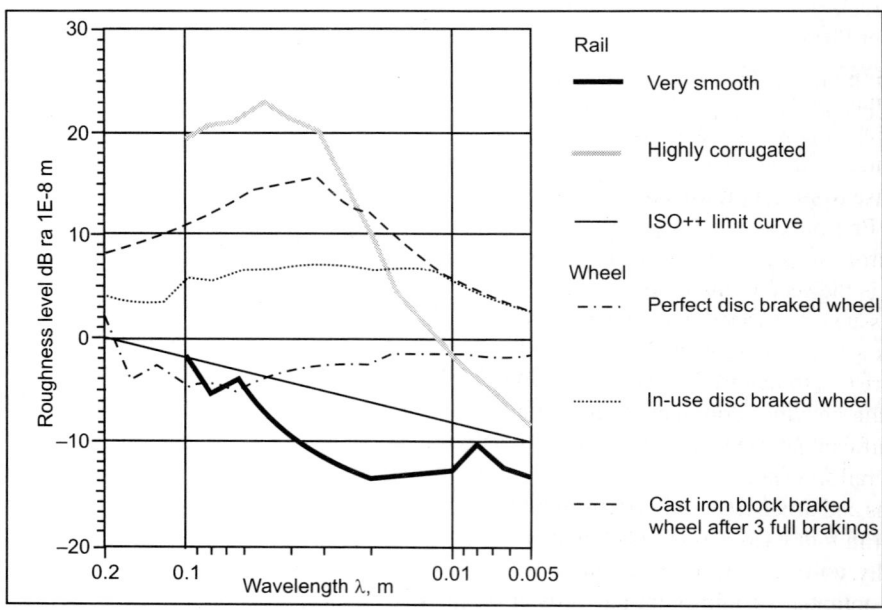

Fig. 34.2. Roughness level spectra for different rail and wheel conditions.

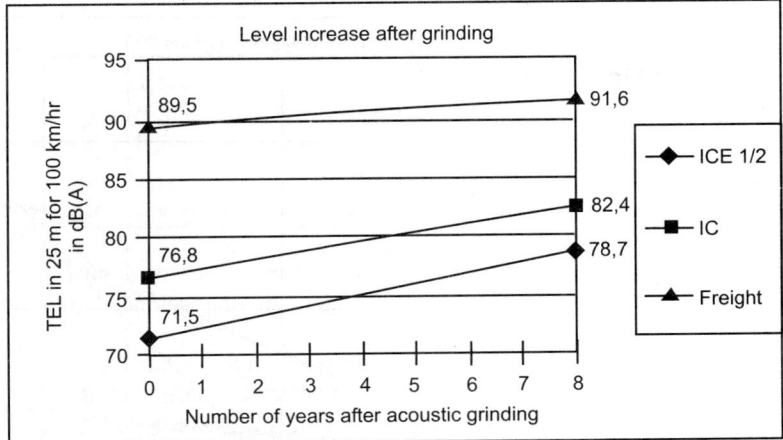

Fig. 34.3. Simulation of increase of noise emission levels after acoustic rail grinding (combining measurements of noise level increase following acoustic grinding for 3 years and for operational grinding after 3 years).

Railway Noise Abatement in the Past and Current Noise Problems

In the past railway noise has been reduced. The equipping of most new coaches with disc brakes instead of cast iron block brakes has led to a significant reduction of noise generation. Since cast iron block braked vehicles have rougher wheels than disc braked ones and wheel/rail roughness is the main driver for rolling noise, focus has to be put on the replacement of cast iron brake blocks. The replacement of jointed track with continuously welded rail across much of the European network has also lead to significant local reductions in noise creation.

However, the progress was not primarily planned as a noise reduction measure but was due to other operational requirements. Disc brakes had to be used on modern coaches to allow speeds above 140 km/hr. This was not required for freight wagons, which is why noise generation from this type of rolling stock did not change significantly during the last decades. This lack of technical progress has made noise from freight wagons the predominant railway noise issue in Europe particularly for operation at night. Present plans in Europe foresee high speed trains running at speeds up to 350 km/hr to form a *trans*-European high speed railway network. Noise from high speed lines mostly operating during the daytime is the second main noise issue.

The issue often arises at the planning stage of new high speed lines or services when noise mitigation becomes a key requirement. Noise from high speed trains (at speeds above 250 km/hr) has different characteristics to that of freight wagons. With increasing speed aerodynamic noise from the upper part of the train becomes dominant with the pantograph and recess, cab profile and gaps between carriages as a significant problem since most of the noise barriers are too low to shield this source. The third issue is urban rail transport.

Trams and urban light rail systems mainly operate in densely populated areas sometimes on a separate track but in many cases on roads together with road vehicles.

Finally, there are local railway noise issues such as curve squeal, brake screech, noise while passing railway stations, noise in shunting yards or on unballasted steel bridges which do not concern as many people as freight and high speed traffic but nevertheless can lead to a significant local annoyance.

Current Railway Noise Exposure Data

The European Environment Agency estimates in its TERM 2001 report 'that 30 per cent of Europeans are exposed to road noise levels, and around 10 per cent to rail noise levels above 55 L_{dn} dB(A) (L_{dn} day/night level over the whole day with a 10 dB(A) penalty for night time noise [22:00 to 7:00])'. The national exposure data for rail transport show that night is the critical period in countries where there is night freight. Examples of national noise levels are:

1. Germany: 3.1 per cent of the population exposed to levels at day time above 65 dB(A), but 10.3 per cent at night time above L_n 55 dB(A).
2. Switzerland during the daytime, 1.5 to 6.6 per cent are exposed to similar levels of railway noise, but 8 to 27 per cent to road noise and at night 1 to 4 per cent are exposed to rail, but 8 to 32 per cent to road noise.
3. The mean noise reception levels at 25 m from the track on the Italian lines Firenze-Bologna and Torino-Modane are 70 to 71 dB(A) at night and higher than during the day time).
4. Highest recorded levels in Germany at night are up to 79 dB(A) (L_{eq} at 25 m distance from the track centre line) caused by freight traffic. Compared with the short-term reduction targets this implies a necessary reduction, at specific problem sites such as this, of up to 19 to 24 dB(A) (this would include use of secondary measures) depending on whether or not the 'railway bonus' is applied. Further reductions will be needed if the aspired modal shift is implemented.

It would therefore seem likely that a goal of a 10–15 dB(A) reduction in exposure (focusing on the most noisy sources) is necessary across Europe in the near future to provide a significant improvement in noise exposure levels for the majority of the population affected by railway noise. Further action will obviously be required in severe situations which may include secondary measures.

DIFFERENCES IN THE MEMBER STATES

There are many differences in the member states concerning railway noise:

1. Magnitude of exposure: This varies depending on the population density, the traffic volume and characteristics (e.g. vehicle park and its emissions), geographical topology, network topology and density.
2. Importance of railway noise relative to other environmental problems.
3. Policy: The level of awareness and the priority given to environmental noise varies. Some states (NL, UK) are even taking action to protect quiet rural areas.
4. Legislation: Most of the member states have railway noise legislation for new lines, only few for existing lines and vehicles.
5. Methodology: In those states that have a national prediction scheme, those schemes show significant differences as a result of the methodology used and differences in track and vehicle characteristics. Also the fleet composition differs from country to country.
6. Population density: High population density in combination with a dense and expanding rail network increases the need to address the railway noise issue. Especially in areas where new housing closes in on existing or new lines the potential for noise problems is enlarged.
7. Investment, maintenance and public funding: The differing levels of investment and maintenance of tracks and vehicles result in differences in noise emission and exposure levels between member states, although this seems to principally relate to conventional and urban railways. The level of investment also affects the amount and type of noise abatement measures taken.

The Development of Abatement

The reduction of railway noise reception levels can be achieved by three essential types of measures: on the vehicles, on the tracks or in the sound propagation path. In the past the latter type of measures was most common. As current practice measures such as barriers (with high costs) or sound insulating windows (with limited effect) are mostly chosen instead of cost-effective source-related measures (Betuwe line in the Netherlands, Italy). The reasons for this include:

1. The sound propagation measures were normally taken due to noise reception limits which have to be observed locally whereas the vehicles are often of global origin and beyond the influence of the local authorities.
2. Vehicle emission limits which could enforce measures on the rolling stock exist only in few countries.
3. Instruments to evaluate the best solutions from a cost benefit point of view and to apportion the contributions of vehicles and tracks and the associated responsibilities have only been developed recently.
4. The application of traditional barriers and sound insulating windows does not need much innovation.
5. Lack of knowledge of viable alternatives at project management level.

In some cases vehicle-based measures were also implemented, for example:

1. In urban rail networks with propriety vehicles and limited applicability of secondary measures.
2. For completely new lines with special vehicles (high speed lines) and in countries with vehicle noise emission limits.
3. On new passenger vehicles and on a few new freight vehicles due to procurement specifications.

Recent investigations have illustrated the important contribution of measures at the source to cost-effective solutions. Therefore the principal instruments for railway noise abatement have to be assessed with respect to the enforcement or stimulation of this type of measures, and links for a common effective approach as well as instruments for the apportioning of responsibilities have to be developed.

BASIC PRINCIPLES AND INSTRUMENTS FOR THE REDUCTION OF RAILWAY NOISE

General Principles for Reduction of Negative Effects of Transport

In general the following are essential principles in reducing the negative effects of transport:

1. Avoiding transportation or making transport more efficient.
2. Lifting to modes with lower environmental impacts.
3. Reduction of the emissions (measures at the sources):
   (a) Technical measures on the vehicles and on the traffic lines.
   (b) Operational restrictions (speed, volume, night-time restrictions).

To address local problems, additional reception-related measures are available:

1. Land use planning (new lines and/or residential areas).
2. Measures applied in the propagation path.
3. Traffic regulation (bundling, use of less sensitive areas for transport).
4. Measures applied to the buildings.

This section concentrates on the technical measures at the source considering that:

1. The reduction of noise emissions is the main task of the WG.

2. Operational restrictions would counteract the transport policy target of the European Union and of the member states to shift transport volumes to rail.
3. Measures at the source generally have a favourable cost-benefit ratio.

Measures at the Source

The main railway noise sources are traction noise, rolling noise and aerodynamic noise. Noise control on these sources can be applied in new design or redesign (retrofit) and has to be retained by maintenance of vehicles and tracks. For rolling noise the following applies:

1. Smooth wheels and smooth tracks ensure minimal noise generation; this implies.
   (a) The use of braking systems that maintain smooth wheel running surface such as disc or drum brakes or composite-block brakes for block-braked vehicles.
   (b) Appropriate maintenance of the tracks and the wheels.
2. Compact, massive design incorporating vibration isolation and high damping ensures a minimum of structure-borne noise transmission in the track and the wheels. Examples are:
   (a) Smaller wheels and/or wheel dampers, optimised wheel geometry.
   (b) Fewer wheels.
   (c) Wheel-mounted disc brakes.
   (d) Optimised track design, or rail damping devices in combination with railpad selection.
3. Shielding (secondary measures) can reduce radiated sound, by applying:
   (a) Wheel-mounted, bogie-mounted or vehicle-mounted shrouds.
   (b) Low noise barriers close to the rail.

For traction noise the following applies:

1. For diesel driven locomotives or trainsets, a low noise design should be ensured for new vehicles, although retrofit may be possible. Noise control measures are:
   (a) Appropriate exhaust and intake design (high insertion loss).
   (b) Effective engine enclosure and vibration isolation.
   (c) Selection of quieter components such as turbocharger, compressors and fans.

A fundamental issue is that noise specifications are often set for unloaded pass-by, whereas in many operational conditions, locomotives pull a heavy load, producing high noise levels.

1. For electric locomotives and high speed trains, especially the noise from the cooling equipment can be a problem. This is best tackled in the design stage, although sometimes retrofit may be possible. This might include:
   (a) Elimination or smoothening of obstacles in ducts, intake and outlet.
   (b) Quieter fan design.
   (c) Increase in fan efficiency by selecting the best working point.
2. For lower speeds gear noise can be a problem. This must be dealt with in the design phase. One reduction technique is to create sufficient overall contact ratio in the gear mesh.

For aerodynamic noise the following applies:

1. For high-speed trains the aerodynamic noise can be a predominant noise source at speeds above 250 km/hr with contributions from various heights. Noise barriers lower than 4 m have insufficient effect on sources located at the top of the vehicle such as the pantographs and their recesses. Aerodynamic noise can be reduced by:
   (a) Streamlined covers for the bogies.

(b) Avoiding extruding parts or cavities along the train.

(c) Streamlining and covering of the pantograph and its recess area.

(d) Streamlined front of the vehicle.

Measures—availability of technology

At present, the following technology is available for the various noise sources:

1. Traction noise: in principle, all of the above mentioned noise control measures are available to minimise traction noise at the design stage. The remaining issues are then the cost and maintainability. Retrofitting only for the purpose of noise reduction is generally not economically feasible.

2. Rolling noise: the most effective means of control is that of wheel and rail roughness. Here the technology is available (K-blocks/disc brakes, rail grinding systems) but also depends on the cost. Add-on systems such as rail and wheel dampers are available but have limited effect; in particular the effect is not always measured properly, if wheel and track contributions are not separated. The same is true for wheel and bogie shielding. New design of wheels and tracks provides the next best option after roughness control; vehicles with smaller and less wheels, and quieter track design are longer term, but beneficial investments. Local application of low noise track has the potential to reduce noise at low and medium speeds. This can even apply for cast iron brake blocked vehicles, thereby adding to the effects of long-term retrofit programmes before all retrofitting is complete.

3. Aerodynamic noise: recent generations of high speed trains have illustrated the improvements in this field; the streamline design of new trains often benefits both noise and energy consumption. Further streamlining is possible, in particularly the covering of the bogie areas; this however has cost and maintenance consequences.

Measures Applied in the Propagation Path

Noise barriers are the most commonly applied noise abatement measure applied in the propagation path. They are applied on a wide scale both on existing and new lines. Typical noise reductions are up to 10 dB depending on the barrier height, distance to source and receiver, and barrier absorption. In many cases barrier performance is severely limited by the track layout (e.g. multiple tracks), the height of the sources and by the height of adjacent multistorey residential buildings. Barrier performance is best if the barrier is close to the source or to the receiver. Noise barriers are generally less cost effective than noise control measures at the source. Barriers also have other disadvantages such as visual intrusion and high cost. Another way of reducing sound propagation near railways is the construction of non-noise sensitive buildings between the railway and other residential buildings.

Chapter 35

Wheel and Bogies

INTRODUCTION

Excessive noise is one of the main public criticisms of rail transport in all over the world and this has an impact on the potential for growth of rail. This is not necessarily the main obstacle to a modal shift from road and air transport to rail transport, but it is important that it is tackled head on. The development of practical solutions for achieving target noise reductions is progressing and some of the newer technologies are beginning to be exploited. These will need wider adoption if the desired noise reductions are to be achieved over reasonable timescales. Considerable progress has been made in enabling the economically viable implementation of the measures.

The reduction of noise from railway sources has been the subject of significant research over the last decade and more. This paper examines the practical application of some of the options and describes their potential impact on track environmental design.

The basic principles of the different approaches to noise reduction are explained to provide an appreciation of what is possible. The detail varies according to the particular track-form being evaluated, ranging from old conventional tracks requiring retrofitting to new build on slab tracks. The paper examines measured and predicted values of noise reduction with differing approaches taking into account the different track structures' dynamic response.

Additional criteria for evaluating noise reduction measures are discussed, such as the impact on the railway from both an environmental and safety point of view.

It is often found that noise issues have a local character, which requires a sensitive appreciation of the local issues and population. The chapter attempts to give guidance on a range of possible scenarios for which different combinations of measures are useful.

WHEELS AND BOGIES

Wheel on the Rail

Railway wheels sit on the rails without guidance except for the shape of the tyre in relation to the rail head. Contrary to popular belief, the flanges should not touch the rails. Flanges are only a last resort to prevent the wheels becoming derailed—a safety feature. The wheel tyre is coned and the rail head slightly curved as shown in the diagram (Fig. 35.1). The degree of coning is set by the railway company and it varies from place to place. In the UK the angle is set at 1 in 20 (1/20 or 0.05). In France at 1/40. The angle can wear to as little as 1 in 1.25 before the wheel is reprofiled.

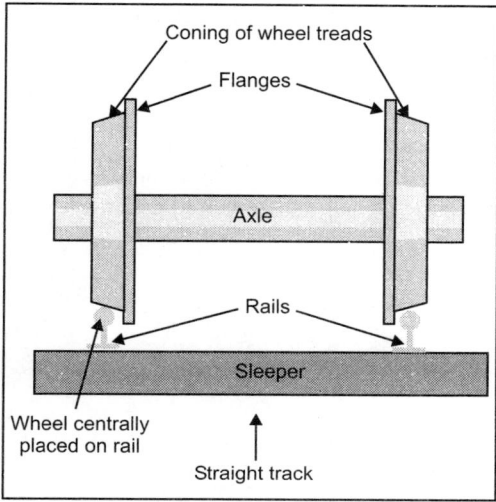

Fig. 35.1. The shape and location of wheels and rails on straight track.

This diagram is exaggerated to show the principal of the wheel/rail interface on straight track. Note that the flanges do not normally touch the rails. On curved track, the outer wheel has a greater distance to travel than the inner wheel. To compensate for this, the wheelset moves sideways in relation to the track so that the larger tyre radius on the inner edge of the wheel is used on the outer rail of the curve, as shown in Fig. 35.2.

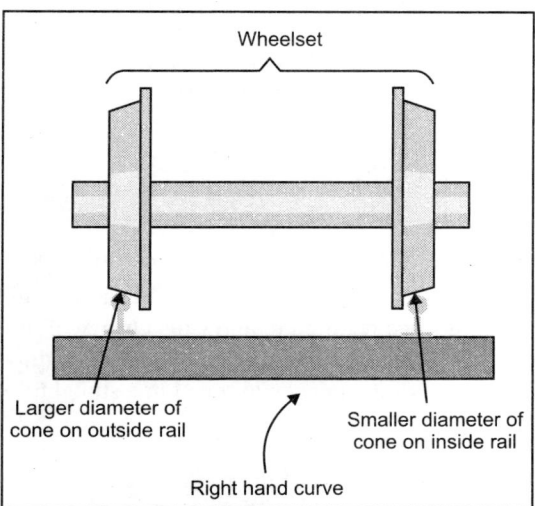

Fig. 35.2. The location of the wheels in relation to the rails on curved track.

The inner wheel uses the outer edge of its tyre to reduce the travelled distance during the passage round the curve. The flange of the outer wheel will only touch the movement of the train round the curved rail is not in exact symmetry with the geometry of the track. This can occur due to incorrect speed or poor mechanical condition of the track or train. It often causes a squealing noise. It naturally causes wear. Many operators use flange or rail greasing to ease the passage of wheels on curves. Devices can be mounted on the track or the train. It is important to ensure that the amount of lubricant applied is exactly right.

Too much will cause the tyre to become contaminated and will lead to skidding and flatted wheels. There will always be some slippage between the wheel and rail on curves but this will be minimised if the track and wheel are both constructed and maintained to the correct standards.

Bogies (Trucks)

A pair of train wheels is rigidly fixed to an axle to form a wheelset. Normally, two wheelsets are mounted in a bogie, or truck as it is called in US English. Most bogies have rigid frames as shown in Fig. 35.3.

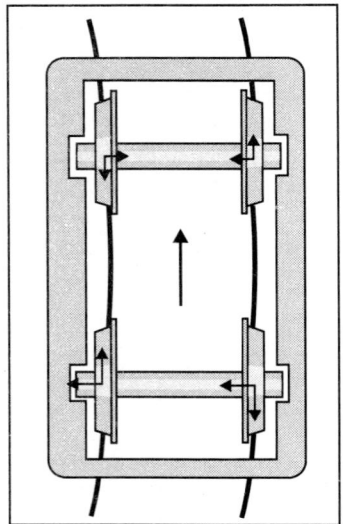

Fig. 35.3. A standard rigid bogie on curved track.

The bogie frame is turned into the curve by the leading wheelset as it is guided by the rails. However, there is a degree of slip and a lot of force required to allow the change of direction. The bogie is, after all, carrying about half the weight of the vehicle it supports. It is also guiding the vehicle, sometimes at high speed, into a curve against its natural tendency to travel in a straight line.

Steerable Bogies

To overcome some of the mechanical problems of the rigid wheelset mounted in a rigid bogie frame, some modern designs incorporate a form of radial movement in the wheelset as shown in Fig. 35.4.

In this example, the wheelset 'floats' within the rigid bogie frame. The forces wearing the tyres and flanges are reduced as are the stresses on the bogie frame itself. There are some designs where the bogie frame is not rigid and the steering is through mechanical links between the leading and trailing wheelset.

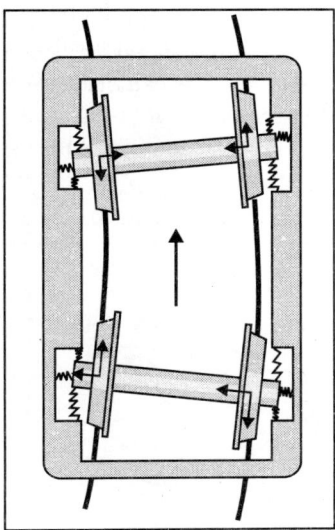

Fig. 35.4. A bogie on curved track with radially steering wheelsets.

RESILIENT WHEELS

Resilient wheels have a rubber element between the tread and the inner part of the wheel, as shown in Fig. 35.5. This has the effect of isolating the wheel web from the tread and also introducing some additional damping to the wheel. In fact, to be effective in isolation the stiffness has to be quite low, whereas to be effective in damping it should be higher. Such resilient wheels are very common in light rail and tramway applications. However, since the tragic accident involving an ICE at Eschede in Germany in 1998 due to the failure of a resilient wheel, they have not been considered for use at high speeds.

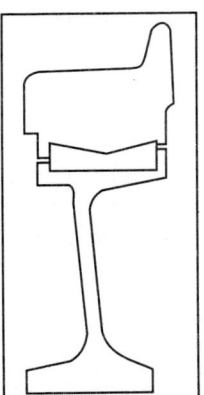

Fig. 35.5. A typical resilient wheel with 740 mm diameter from a light rail vehicle.

In a resilient wheel for 'light rail' application was analysed theoretically for a range of different values of stiffness of the resilient layer. Figure 35.6 shows the predicted components of noise from the wheel and track for various values of the stiffness of the resilient element. It was found that the noise from the resilient wheel with a national very stiff material was much less than that from the tracks, since this wheel had a small radius and straight web. The extreme case of this is where the resilient material is replaced by steel in the model, shown as 'all steel' in Fig. 35.6. However, at more typical stiffness (moderate/low) of the resilient elements, the nose from the wheel was found to be similar to that of the track.

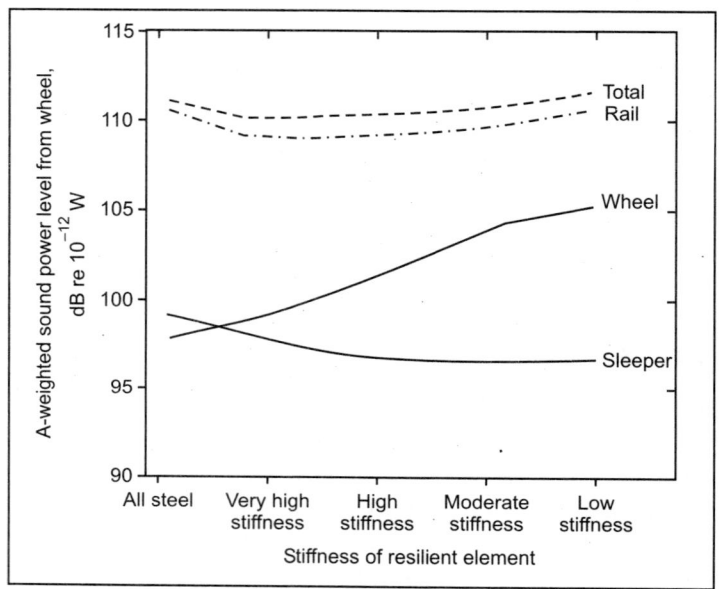

Fig. 35.6. Overall A-weighted sound power levels from resilient wheels and track for various stiffnesses of resilient element. Resilient wheel based on a 740 mm diameter light rail vehicle wheel, typical ballasted track, train speed 100 km/hr, roughness from tread-braked wheels.

A similar theoretical study was reported for a wheel intended for freight application. It was shown there, as in Fig. 35.6, that by appropriate selection of the stiffness, a resilient wheel can reduce the track component of noise. Reduction of around 2 dB were found and were attributed to a change in the balance of wheel and rail vibration amplitudes relative to the roughness, especially between 500 and 1000 Hz. The wheel component was increased, but was still sufficiently small compared with the track component at these frequencies for the rolling noise to be reduced slightly for a train speed of 100 km/hr. This has not been applied in practice, however.

Reducing the Sound Radiation from the Wheel

All the above measures are intended to reduce the vibration of the wheel and hence its radiated noise. In addition it is possible to attempt to minimise the sound that is radiated due to a particular vibration level. One means of reducing the radiation is to mount a shield on the wheel so that sound radiation from the web region is obstructed. It is the web that produces most of the noise from the wheel, particularly for a curved web design, although the tread, particularly its radial motion, should not be neglected. Such a shield must be mounted in such a way that it does not vibrate significantly, as then it would radiate sound itself; thus it must be resiliently mounted and reasonably well damped. It should also be made of a material which gives a high enough acoustic transmission loss while still being flexible. The use of thin metal plate means that bending waves excited in the panel are short compared with the wavelength of sound in air and therefore, have poor sound radiation. Good stealing is also required so that no sound can escape through gaps. In addition the damping of the wheel may also be increased somewhat by the mounting arrangement.

In the RONA project such web screens, consisting of 1 mm thick steel plates, were mounted resiliently and clipped onto the wheel. They were fitted to a wheel which had been optimised mechanically, that is in terms of mass. Its mass was only 335 kg but its initial nose radiation was greater than the reference wheel. The two screens on a wheel each added 15 kg to the mass, but their application on this particular wheel still left the mass below that of the reference wheel. The wheel component of noise was reduced relative to the reference wheel by about 6 dB. Wheels with similar shields were constructed and tested within the Silent Freight project. Compared with the reference wheel, this was estimated to reduce the wheel component of noise by about 8 dB, although it was difficult to assess because of the dominance of track noise. The concept of a perforated wheel was studied in the Silent Freight project. The idea was to reduce the radiation efficiently by introducing acoustic short-circuiting between the front and back of the wheel web. The effect depends on the size and spacing of the holes. Due to the thickness of the web and practical limitations on the distance between holes and their size, the perforated wheel was shown to be promising at low frequencies, where a 6–9 dB reduction was expected, but no appreciable effect was predicted above about 1 kHz which is where the wheel noise is usually dominant. A perforated wheel was constructed, based on the alternative optimised wheel cross-section. Unfortunately, no noise reduction was found, although in combination with the ring damper a modest reduction of around 2.5 dB was found.

AUTOMATIC DETECTION OF FLATS ON THE ROLLING STOCK WHEELS

Wheels of the railway rolling stock are never round. They have always shape errors. Their out-of-roundness causes additional dynamic load to the track, is a source of the increased noise, and may even result in damaging the rails. Wheel/rail rolling noise, dominating on straight track is produced by the changing loads generated in the wheel/rail contact zone by roughness on the running surfaces of the wheel and rail. Another cause for the noise may be the wrong wheel shape, called polygonisation, and sometimes wheel flats. Controlling the noise resulting from the mating surfaces roughness alone at the source calls for reducing this roughness or changing the interaction between wheel and rail in the contact zone, to reduce the interaction forces. The approaches include, smoothing of wheels and rails, properly treaded wheels, wheel profile modifications to suit adapt them to the actual rail profile, and the use of greases.

The development of roughness on surfaces of wheels and rails is influenced by wheel/rail dynamics as well as wear phenomena. Rail roughness value grows with the presence of multiple wheels on the rail. Overlaying of frequencies of vibrations caused by interaction of the particular wheels with the waviness of the rail leads to roughness growth at frequencies associated with peaks in the contact force that have a particular phase angle relative to the roughness excitation.

Another wear mechanism is especially serious in case of wheel flats whose effect is not only the increased rolling noise level. Wheel flats develop usually when the brakes are applied to a railway wheel. Locking the wheel results in its sliding along the rail. Usually the cause for that may be bad condition of brakes, which may be poorly adjusted, defective or frozen. Another cause may be poor adhesion at the wheel/rail interface, sometimes reduced to nil, for example due to leaves on the rail head or wrong cant of the rails in the track. In this case the wheel/rail contact takes place along a very narrow path, making breaking nearly impossible, so the wheel slides on the rail head. This sliding causes serious wear of the part of the wheel that is in contact with the rail, leading to its deformation by development of a 'wheel flat'. Such flat spaces on the wagon wheel may be typically 50 mm long but can be as big as more than 100 mm long. Discontinuities on the rotating wheels produce big impact loads between the wheel and track (Fig. 35.7).

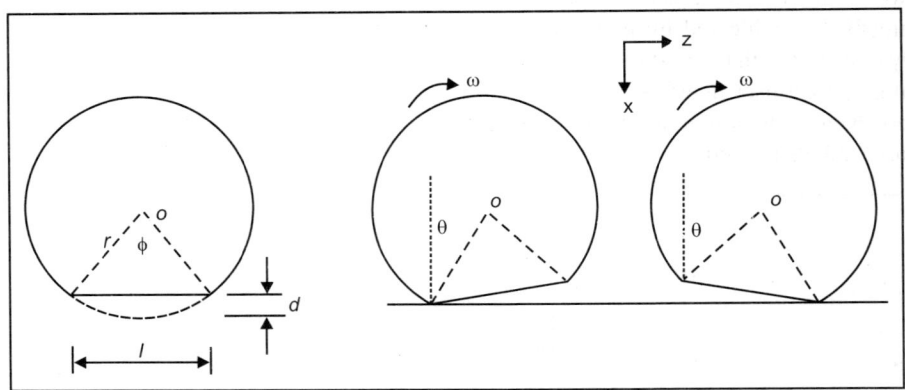

Fig. 35.7. Schema of the wheel flats interaction with track.

As a consequence, a periodic impact noise is produced, when the faulty wheel edges hammer the rail head, in addition to the usual rolling noise, which is more random in character. The dynamic forces generated by wheel flats, which may have big amplitudes, may result in damage to the track, like fatigue cracks in the rails or sleepers. Moreover, the detrimental wheel surface structure changes take also place. They heat up to high temperatures during sliding, and later cool rapidly down, which leads to development of brittle martensite within the steel beneath the wheel flat. Therefore, damage to the wheel can occur, by cracking and spalling, which lead to the additional loss of relatively large pieces of metal.

Wheel shape errors feature a serious issue when safe and efficient train operation is considered. In order to improve the efficiency of railway transportation, Chinese Railways, for instance, has decided to raise the speed of trains on the existing railway main lines. It was found that raising train speeds on the existing railway lines increases the dynamic effects of vehicles on tracks, especially in the turnout areas, in the welded rail joints, and in the sections of bridge-subgrade connections. The occurrence of wheel flats becomes more general after raising train speeds. All wheel flats traversing the track add to the premature use of all the above mentioned specific locations in the track. Unfortunately the peak of dynamic effects of wheel flats on tracks occurs in the speed range from 140 to 160 km/hr, which is just the range of line speeds for passenger trains after raising their speed. This detrimental effect of wheel flats on rolling stock operation safety is also true for metro and trams, albeit their operating speeds are much lower. In practice, a rounded flat will differ in geometry from the idealised case considered here. Rolling stock owners take countermeasures, including stricter maintenance standards for the high-speed railways, but also tram and metro monitor carefully wear rate of their wheels and carry out careful planning of their re-profiling and correcting their out-of-roundness. To this end stationary monitoring systems and portable instruments are needed as data source, as well as diagnostic databases for analysis of the wheel data.

Rolling Stock Wheels Monitoring System

The system was designed for storing information about profile and diameter of wheels and flat spaces and build up on their tread surface (Fig. 35.8). The system makes it possible to define the required information about cars, logging the distance covered by each wheel and visualisation of their wear over time as well as their repairs, inclusive generation of the wheel wear reports. The system can integrate all measurement data collected by many units of a transport company. Measurement results of wheel geometry obtained with the portable gauges and with other methods are archived in a database, with the powerful graphical user interface.

Moreover, information about bogie measurements and other car service operations defined by the system user are also saved.

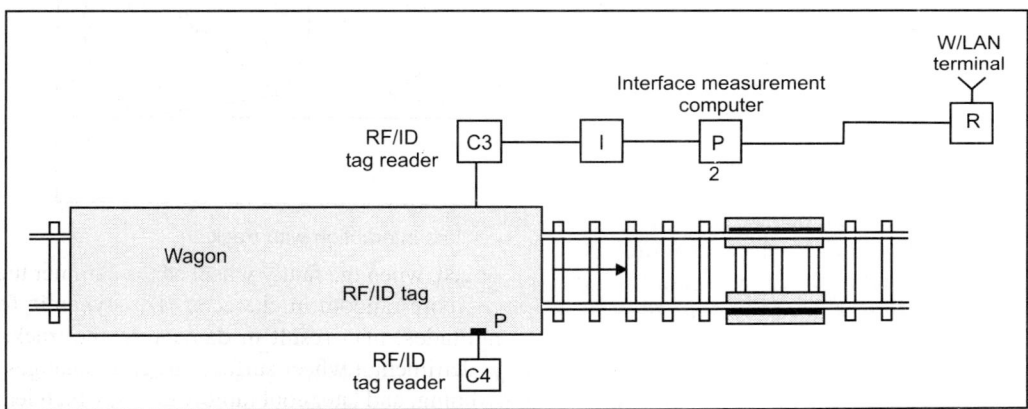

Fig. 35.8. Wheel monitoring system.

System database makes it possible to print wheel wear reports according to the user selected criteria, e.g. presenting:

1. Wheel, bogie, and cars lists sorted according to:
   (a) Wheel flange height value.
   (b) Wheel flange width value.
   (c) Wheel diameter values.
2. Differences of wheel flange widths in a bogie.
3. Differences of wheel diameters in an axle.
4. Tread surface condition.

System Operation

The presented system was designed for detecting the flats on wheels while the train passes the measurement system with speed of 10–30 km/hr. It is composed of multiple accelerometers installed on both rails over 3 m length. The sensors are installed within the rugged covers protecting them from damage. In practice, due to continued running of the wheel after formation of the flat, the profile becomes rounded at the corners of the flat, whereas the central part will remain unchanged.

The overall length of the rounded flat, will be greater than that for a new flat of the same depth. The measurement signals are sent to the measurement computer housed in a cabinet close to the system. The measurement software processes the signals from the sensors and classifies wheels into one of the four wheel flats sizes.

Wheel flats exceeding 60 mm are tagged as needing 'repair', whereas three other subranges refer to the following levels: 'warning' (45 mm), 'observation', and 'good'. Measurement results are sent by a wireless LAN to the system operator computer, where information about the wheel flats is reported and their location in the train is specified. The measurement system was installed at the entry to the depot, which makes it possible to monitor all trains. Each wagon is described in the database and all technological operations for each wheel are recorded, like turning, reprofiling, etc. Exemplary logged wheel service operations include:

1. Removing flats and out of roundness.
2. Turning the bogie.
3. Measuring wheel profile and diameter.
4. Measurement of buildup and wheel flats.
5. Reprofiling of wheel on the underfloor wheel lathe.
   (a) Current list of defects.
   (b) List of cars with no actual measurements.
   (c) Car condition report.
   (d) Train condition report.

The proven system deployment effects are as follows:

1. Effective derailment prevention.
2. Reduction of the allowed wheel diameter differences.
3. Reduction of noise level in urban areas.
4. Replacement of wheels and bogies.

Thus, in order to predict the consequent noise radiation, the wheel/rail interaction force is transformed into the frequency domain. As the train speed increases, the force spectrum and consequently the noise radiation, contains greater amplitudes at high frequencies and the overall noise level due to wheel flat excitation increases with the train. This differs from rolling noise due to roughness excitation. The noise from flats of depth 1 mm and 2 mm exceeds that due to typical roughness on tread-braked wheels and good quality track for all tram and train speeds; however the low speeds pose more problems with the reliable signal processing.

As the wheel load increases, the noise from wheel flats increases. In contrast, the rolling noise due to roughness is relatively insensitive to wheel load. Exemplary change of wheel geometry in service with visible changes after reprofiling is shown in Fig. 35.9.

COMBATING ROLLING NOISE OF TRAINS BY MEANS OF RAIL GRINDING

The main source of noise in rail traffic is rolling noise, the level of which is dependent on the travelling speed of the passing trains and the surface condition of both rail and wheel. It is, therefore, of great importance to achieve a high surface quality of both rail and wheel. A smooth rail surface can be achieved by means of grinding, whereby a distinction is made between grinding for maintenance purposes and that for noise control purposes. In order to cause less damage to the wheel surface, it is the intention of various railways to replace the cast-iron block brakes on their freight trains with those featuring K-composite bases.

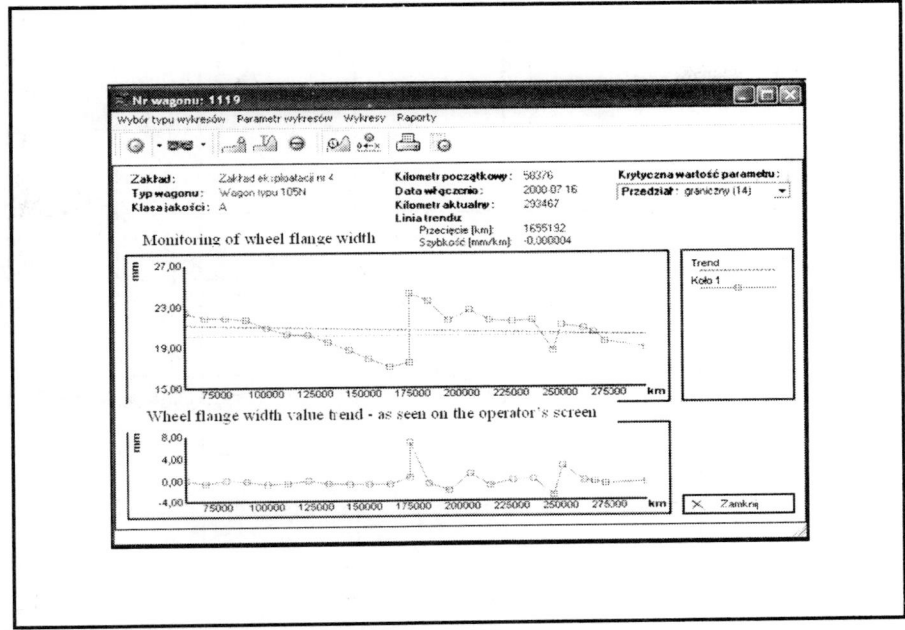

Fig. 35.9. Exemplary change of wheel geometry in service with visible changes after reprofiling.

Rolling Noise

Noise in rail traffic is mainly caused by passing trains. The type of resulting noise depends on the travelling speed of the passing trains. From Fig. 35.10, it can be observed that for speeds of:

1. Up to around 40 km/hr, the tractive noise of passing trains is dominant.
2. Between 40 and 250 km/hr, it is the rolling noise.
3. From 250 km/hr upwards, it is the aerodynamic noise.

Thus, for the main proportion of rail traffic, it is the rolling noise that is of decisive importance. In Fig. 35.11, the various causes of rolling noise are shown.

Condition of the rail surface

Under traffic, various rail surface defects can develop, such as head checks (cracks at the running edges), Belgrospis (clusters of cracks), squats (semicircular or v-shaped cracks), and periodic or irregular indentations. For rolling noise, the periodic rail defects are of particular importance. Depending on the wavelengths of these defects, they can be distinguished as follows:

1. Short waves: these mainly occur in curves, and have a wavelength range of 8–30 cm and an amplitude of 0.3–1 mm.
2. Corrugations: these typically have a wavelength range of 3–6 cm and an amplitude of 0.1–0.4 mm.

In Fig. 35.12, the level of roughness for different conditions of the rail surface are shown. A correlation exists between corrugation depth and the level of noise emitted. Although the emitted noise levels resulting from rolling noise depend essentially on the condition of the rail surface, it is important to recognise that rolling noise also depends on the condition of the wheel surface.

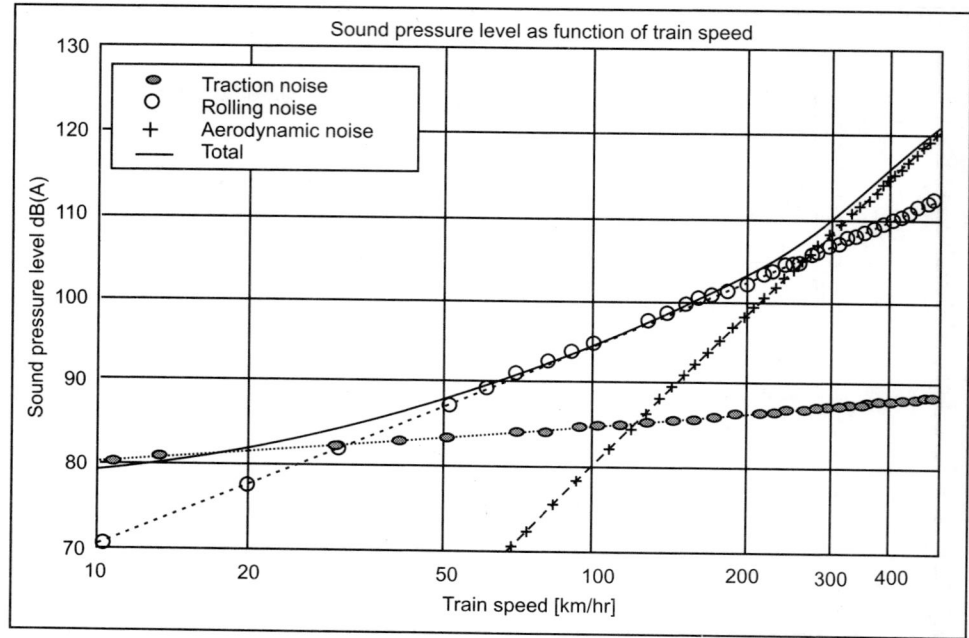

Fig. 35.10. Sources of noise in rail traffic.

Condition of the wheel surface

The condition of the wheel surface depends on the type of brake used (Fig. 35.13). The least wheel roughness occurs in the case of disc brakes and the largest in that of cast-iron block brakes, which is also reflected in the noise levels emitted. Therefore, with respect to combating rolling noise, it is not sufficient to treat only the rails, but attention should also be paid to the quality of the wheel surface. In this respect, it should be noted that it is the intention of various railways to replace the cast-iron block brakes on their freight trains with those featuring K-composite bases, which cause less damage to the wheel surface.

Treatment of Rail Surface Defects

Rail surface defects can be treated by means of rail grinding using, for instance:

1. Oscillating grinding stones: Rail grinding using this method is effected by the oscillating movement of the grinding units in longitudinal direction of the rail, in combination with the continuous forward motion of the machine. The Plasser and Theurer GWM 550 grinding machine, is equipped with ten oscillating grinding units, i.e. five per rail, each carrying six brick-shaped grinding stones. Per grinding pass, up to 0.1 mm of metal can be removed.

2. Rotating grinding stones: This method is the one most often applied to correct rail surface defects. Speno International SA is the largest and most important manufacturer of grinding machines featuring rotating grinding stones. Depending on the model, the machines feature 16, 24, 32 or 48 grinding stones. Per grinding pass, up to 0.2 mm of metal can be removed.

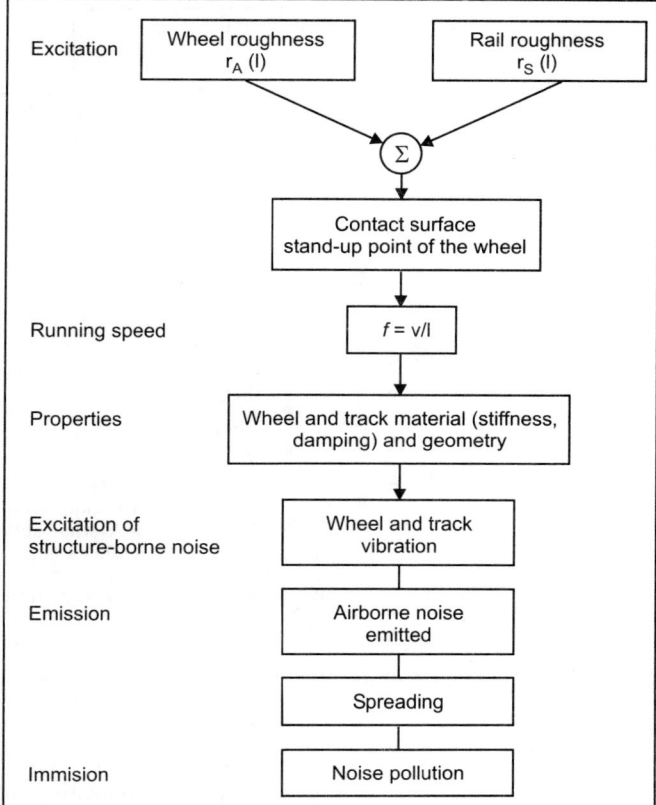

Fig. 35.11. Development of rolling noise.

Rail grinding is carried out to achieve a smooth rail surface, whereby a distinction is made between grinding for maintenance purposes and that for noise control purposes.

Grinding for Maintenance Purposes

Regardless of the noise emitted, defects on the running surface of the rail:

1. Have a negative impact on the riding comfort.
2. Cause an increase in wear of the track components and the rolling stock.
3. Are detrimental to the track geometry, due to the dynamic interaction between the wheel and the rail surface defects.

Examples described in trade literature have shown that, by means of grinding, the service life of the rail can be extended by at least 100 million tons of traffic borne. An extension of the service life of the rail is always an indicator of high cost efficiency. The elimination of rail surface defects also leads to a 30–50 per cent increase in the durability of the track geometry. For economic reasons, rail grinding should be carried out when the corrugation reaches an amplitude of 0.05 mm. As far as loosening of the rail fastenings and deterioration of the track geometry is concerned, the relevant threshold for corrugation is an amplitude of 0.1 mm.

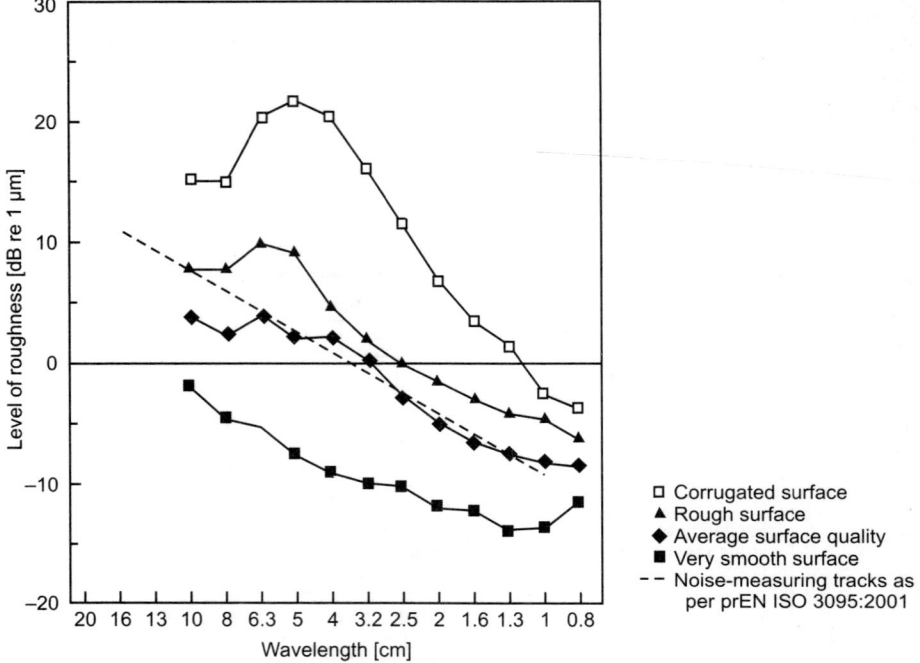

Fig. 35.12. Rail roughness for different conditions of the rail surface.

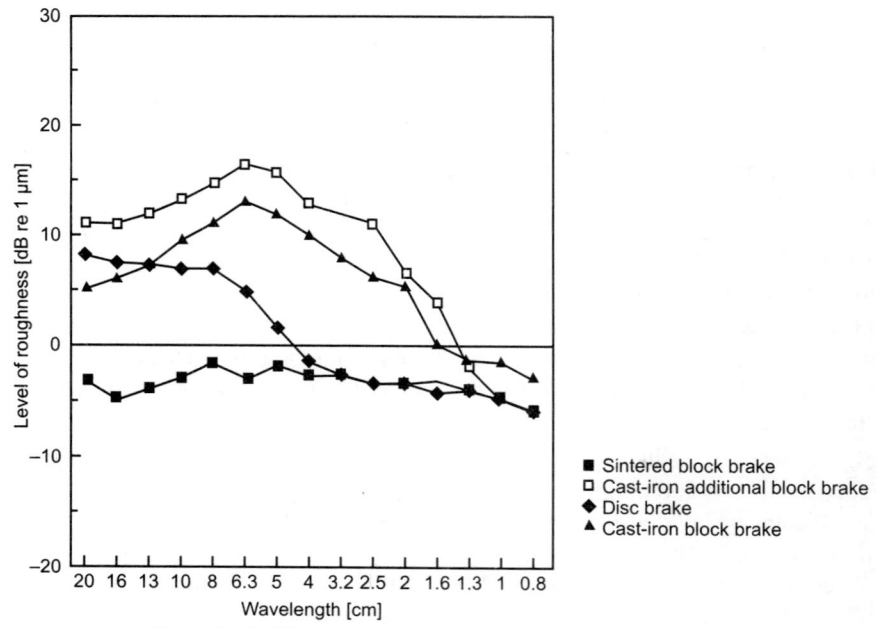

Fig. 35.13. Wheel roughness for different brake systems.

Though not its main aim, rail grinding performed for maintenance purposes also leads to a reduction in noise levels. Figure 35.14 clearly shows how, as a result of grinding for maintenance purposes (in this case by means of rail grinding using oscillating grinding stones), the third-octave roughness level of the rail surface and, consequently, the noise level were lowered by up to 15 dB.

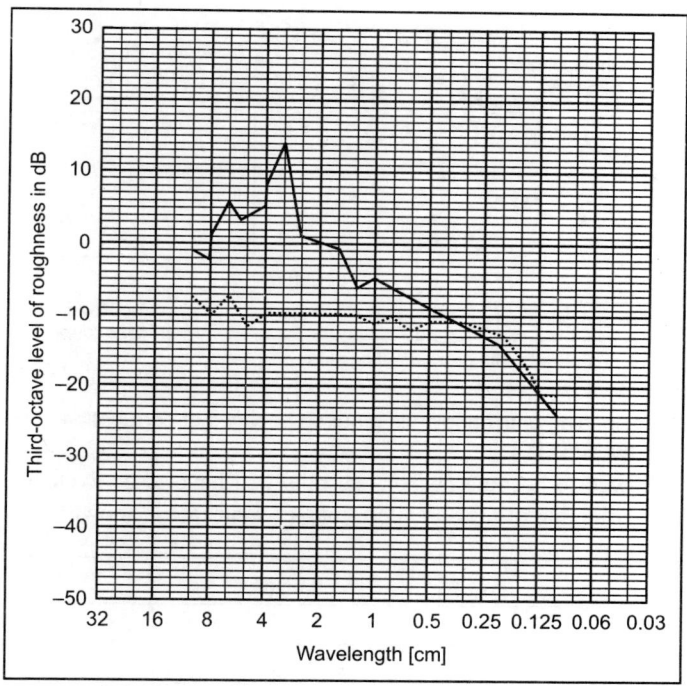

Fig. 35.14. Roughness spectra, both before (upper plot line) and after (lower plot line) rail grinding, using oscillating grinding stones.

Preventive grinding of new rails

The objective of preventively grinding new rails is to remove the low-carbon layer on the surface, i.e. the rolling skin, and any surface defects that may have arisen during their installation. As a result of preventive grinding, the occurrence of rail surface defects is delayed by more than 60 million tons of traffic borne and is, therefore, highly cost efficient, as can also be observed from Fig. 35.15.

Figure 35.15 also shows that, in the case of preventively ground rails, corrugation not only develops at a later stage but that, once it occurs, it also progresses more slowly.

Preventive grinding to combat head checks: the 'magic' wear rate

As a result of the high wear rate of the rail in the wheel/rail contact area, rolling contact fatigue (RCF) damage, or so-called head checks, occurs, especially at the running edges of the rail. As these cracks become deeper, they quickly expand into the interior of the rail. Preventive grinding should be carried out before these cracks reach a certain depth, a so-called 'magic' wear rate. As a result, a crack-free condition of the rail is achieved.

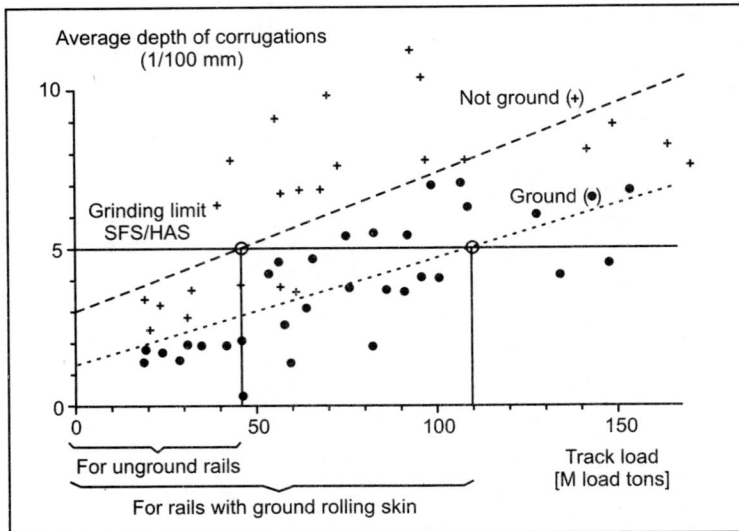

Fig. 35.15. Minimum track loading until grinding is required because of corrugation, for preventively and non-preventively ground new rail (UIC 60).

If the cracks are allowed to become deeper, the rails will have to be exchanged sooner and more often. In Fig. 35.16, the principle of the so-called 'magic' wear rate is shown.

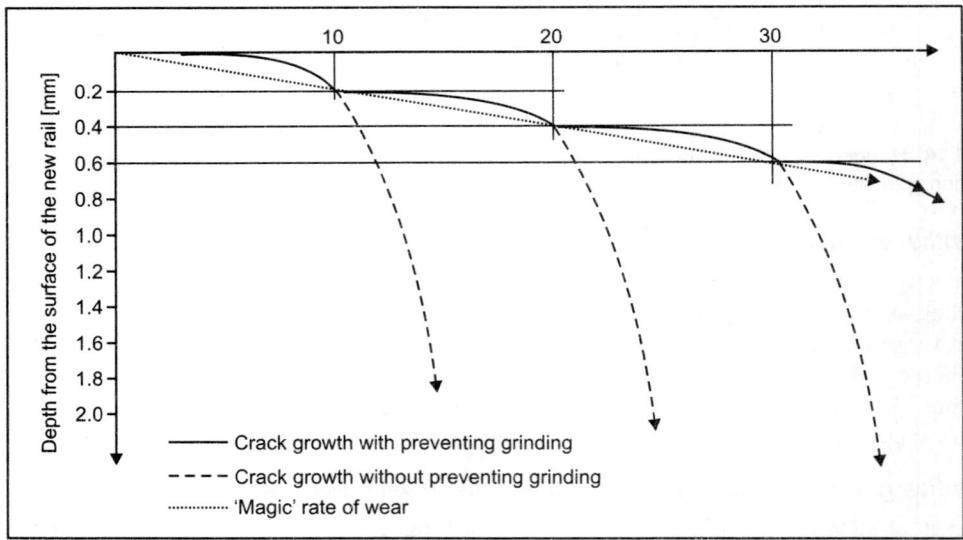

Fig. 35.16. The principle of the so-called 'magic' wear rate.

Experience gained internationally has shown that, by using this method, a service life of the rail of more than 1000 million tons of traffic borne can be achieved, even on heavily-used lines.

In curves, the service life of the rail can be expected to be as follows:

1. In tight curves: 700–1000 million tons of traffic borne.
2. In curves with average radii: 1400–2000 million tons of traffic borne.
3. In transition curves: >2000 million tons of traffic borne.

Combined tamping and grinding

Due to the high quality of the running surface that is achieved by rail grinding, far fewer dynamic forces are generated by the passing trains, thus extending the durability of the track geometry. As a result, track tamping needs to be performed less frequently and after longer periods of time. Commissioned by Austrian Federal Railways (ÖBB), a study was conducted by the Technical University of Innsbruck, Austria, whereby measurements were taken of the accelerations and oscillating speeds at the sleepers, both before and after maintenance was carried out by means of a mechanised maintenance train (MDZ), followed immediately by grinding. After six months, these measurements were repeated.

Thus, the accelerations at the sleepers directly after tamping are, in some cases, higher than before. The dramatic decrease in accelerations after grinding is striking. This condition remains unchanged, as can be observed from the measurements taken six months later. In fact, it has even improved slightly. Thus, by carrying out grinding immediately after tamping a good low-acceleration track geometry is achieved. In Fig. 35.17, the power density spectra after tamping are compared with those after grinding performed immediately after tamping.

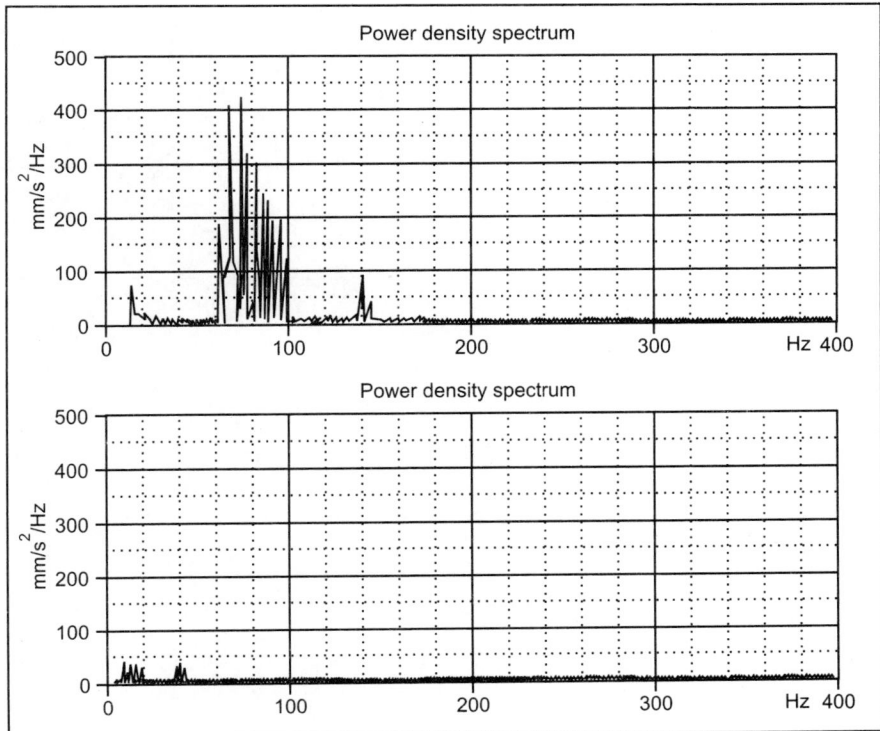

Fig. 35.17. Power density spectra after tamping (top) and immediately after grinding (bottom).

From Fig. 35.17, it can clearly be deduced that combined tamping and grinding leads to a durable geometry of the track which, as a result, requires less frequent tamping, resulting in longer maintenance intervals and cost savings.

Grinding for noise control purposes

The objective of grinding for noise control purposes is not only to eliminate rail surface defects, but also to achieve the smoothest possible rail surface. The decisive factor for grinding for noise control purposes is not to what extent the noise emission levels can be reduced, but to which level. The noise emission levels for a rail ground for noise control purposes, as compared to those for a rail ground for maintenance purposes. The grinding intervals are roughly comparable. As can be observed, over the same period of time, the rail ground for maintenance purposes shows a higher increase in noise level.

In Germany, the so-called 'specially monitored track (BüG)' was introduced. On this track, as a result of grinding for noise control purposes, a noise emission level of 48 dB(A) has been achieved, measured at a distance of 25 m from the track during the passage of a 'standard train', instead of the 51 dB(A) required in Germany. Thus, rail grinding could make the use of noise protection walls obsolete, in some cases, which could result in significant cost savings. With an assumed service life of noise protection walls of 25 years, the cost ratio would be approximately 1:10.

Chapter 36

Aircraft

INTRODUCTION

Aircraft noise is noise pollution produced by any aircraft or its components, during various phases of a flight: on the ground while parked such as auxiliary power units, while taxiing, on run-up from propeller and jet exhaust, during take off, underneath and lateral to departure and arrival paths, over-flying while en route, or during landing.

MECHANISMS OF SOUND PRODUCTION

A moving aircraft including the jet engine or propeller causes compression and rarefaction of the air, producing motion of air molecules. This movement propagates through the air as pressure waves. If these pressure waves are strong enough and within the audible frequency spectrum, a sensation of hearing is produced. Different aircraft types have different noise levels and frequencies. The noise originates from three main sources:

1. Aerodynamic noise.
2. Engine and other mechanical noise.
3. Noise from aircraft systems.

Aerodynamic Noise

Aerodynamic noise arises from the airflow around the aircraft fuselage and control surfaces. This type of noise increases with aircraft speed and also at low altitudes due to the density of the air. Jet-powered aircraft create intense noise from aerodynamics. Low-flying, high-speed military aircraft produce especially loud aerodynamic noise.

The shape of the nose, windshield or canopy of an aircraft affects the sound produced. Much of the noise of a propeller aircraft is of aerodynamic origin due to the flow of air around the blades. The helicopter main and tail rotors also give rise to aerodynamic noise. This type of aerodynamic noise is mostly low frequency determined by the rotor speed.

Typically noise is generated when flow passes an object on the aircraft, for example the wings or landing gear. There are broadly two main types of airframe noise:

1. Bluff body noise—the alternating vortex shedding from either side of a bluff body, creates low pressure regions (at the core of the shed vortices) which manifest themselves as pressure waves (or sound). The separated flow around the bluff body is quite unstable, and the flow 'rolls up' into ring vortices—which later break down into turbulence.

2. Edge noise—when turbulent flow passes the end of an object, or gaps in a structure (high lift device clearance gaps) the associated fluctuations in pressure are heard as the sound propagates from the edge of the object (radially downwards).

Engine and Other Mechanical Noise

Much of the noise in propeller aircraft comes equally from the propellers and aerodynamics. Helicopter noise is aerodynamically induced noise from the main and tail rotors and mechanically induced noise from the main gearbox and various transmission chains. The mechanical sources produce narrow band high intensity peaks relating to the rotational speed and movement of the moving parts. In computer modelling terms noise from a moving aircraft can be treated as a line source.

Aircraft gas turbine engines (jet engines) are responsible for much of the aircraft noise during takeoff and climb. However, with advances in noise reduction technologies—the airframe is typically more noisy during landing.

The majority of engine noise is due to Jet noise—although high by-pass ratio turbofans do have considerable Fan noise. The high velocity jet leaving the back of the engine has an inherent shear layer instability (if not thick enough) and rolls up into ring vortices. This of course later breaks down into turbulence. The SPL associated with engine noise is proportional to the jet speed (to a high power) therefore, even modest reductions in exhaust velocity will see a large reduction in jet noise.

Noise from Aircraft Systems

Cockpit and cabin pressurisation and conditioning systems are often a major contributor within cabins of both civilian and military aircraft. However, one of the most significant sources of cabin noise from commercial jet aircraft other than the engines is the Auxiliary Power Unit (APU). An Auxiliary Power Unit is an on-board generator used in aircraft to start the main engines, usually with compressed air, and to provide electrical power while the aircraft is on the ground. Other internal aircraft systems can also contribute, such as specialised electronic equipment in some military aircraft.

Turbofan Engine Noise Generation

There are many sources of noise from current aircraft. Turbofan engines work on the principle of sucking air into the front of the nacelle duct and pushing that same air out the back at a higher velocity. This change in momentum provides the thrust. The diameter of the engine is determined by the fan, which pulls air into the duct. This fan is a source of noise, similar to the noise caused by a propeller. The fan blades, by pushing through the air, cause noise by themselves. Once past the fan, the air is split down two different paths, the fan duct and the core duct (Fig. 36.1).

First consider the flow in the fan duct. Downstream of the fan, the flow is swirling because of the spinning fan. This swirl causes loss of momentum before the air exits the nozzle so it is straightened out with a set of vanes called stators. These stators are a large source of noise as the wakes of air from fan flow slap against the stators like waves on a beach. This regular slapping takes place at the rate of blades passing by and generates a tone at what is called the blade passage frequency, or BPF. Nonuniformities and nonlinearities result in many higher frequency tones being produced at 2 times BPF, 3 times BPF, and so on. These tones are often associated with the piercing sound generated by some engines. Fan/ Stator interaction creates more than specific tones. The unsteadiness in the fan flow (often in the form of turbulence) interacts with the stators to create broadband noise. This is often heard as a rumbling sound.

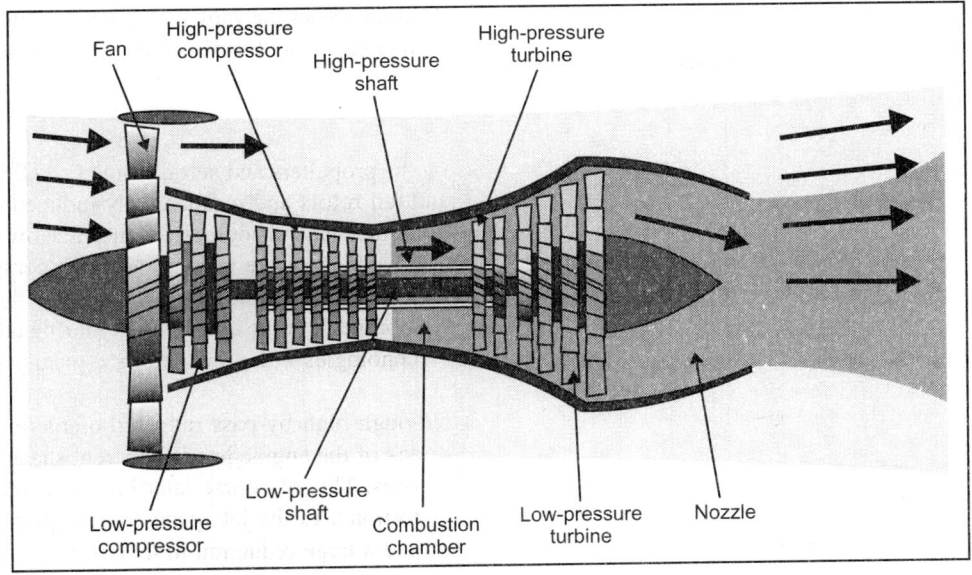

Fig. 36.1. Torbofan jet engine.

In the core duct, the air taking this path is further compressed through a series of smaller fans called rotors. Each of these rotor stages is separated by a set of stators to straighten the flow. This is another source of rotor/stator interaction noise. The compressed air is then mixed with fuel and burned. This combustion is another source of noise. The hot, high-pressure combusted air is sent downstream into a turbine which drives the fan and the compressor rotors. Since the turbine tends to look and act like a set of stators, this is another source of noise. Finally, the core duct and the fan duct flows are exhausted into the air outside the back of the aircraft. The interaction of these jet exhausts with the surrounding air generates broadband noise called jet noise.

The graph (Fig. 36.2) is representative of the noise distribution components for typical aircraft. The importance of engine noise, in particular the fan and jet exhaust noises, is clearly depicted. These two main areas are the primary focus of the research being done in the NASA Glenn AST engine noise reduction program.

HELICOPTER NOISE REDUCTION

Helicopter noise reduction is a topic of research into designing helicopters which can be operated more quietly, reducing the public-relations problems with night-flying or expanding an airport. In addition, it is useful for military applications in which stealth is required: long-range propagation of helicopter noise can alert an enemy to an incoming helicopter in time to re-orient defenses.

Sources of Helicopter Noise

1. Rotor noise.
2. Engine noise.
3. Transmission noise.

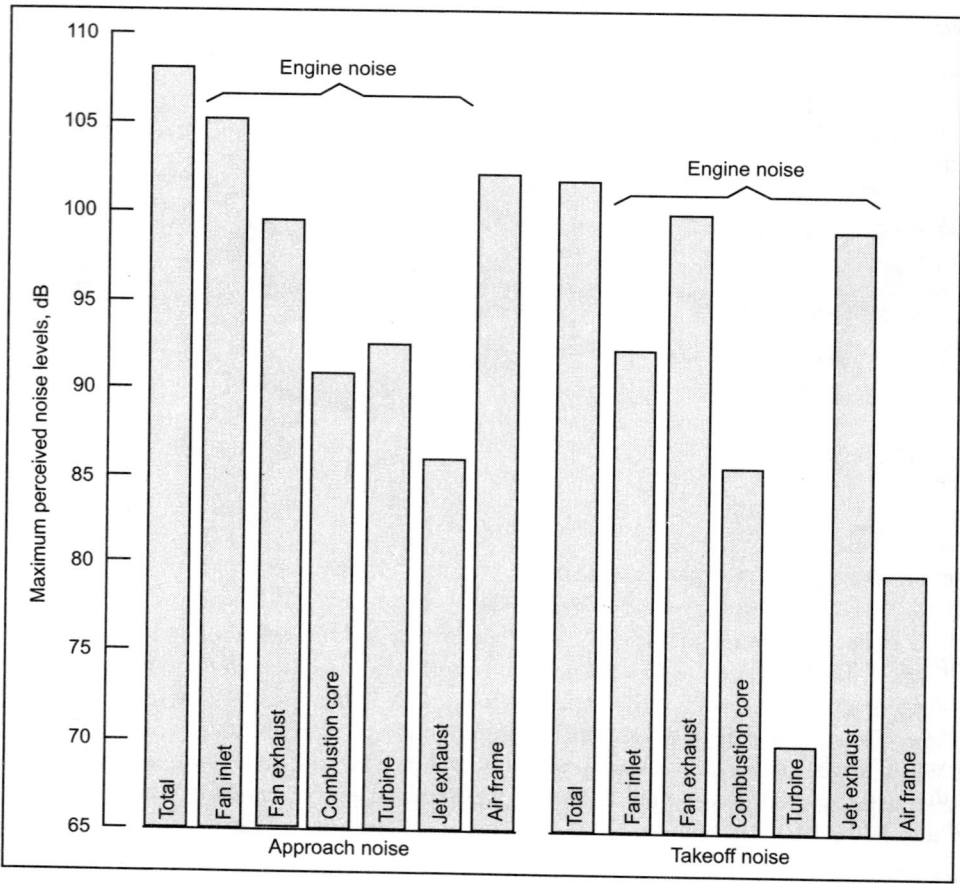

Fig. 36.2. A breakdown of the noise components of a typical engine with 1992-level technology during takeoff and approach to landing indicates the large part played by engine noise. Note: because decibels are logarithmic values, the graph's total bar is the sum of the individual bars.

The noise from a rotor can be divided into several distinct sources, which will be described as follows:

Thickness noise

Thickness noise is dependent only on the shape and motion of the blade, and can be thought of as being caused by the displacement of the air by the rotor blades. It is primarily directed in the plane of the rotor.

Loading noise

Loading noise is an aerodynamic adverse effect due to the acceleration of the force distribution on the air around the rotor blade due to the blade passing through it, and is directed primarily below the rotor. In general, loading noise can include numerous types of blade loading: some special sources of loading noise are identified separately.

Blade-vortex interaction (BVI) noise

BVI occurs when a rotor blade passes within a close proximity of the shed tip vortices from a previous blade. This causes a rapid, impulsive change in the loading on the blade resulting in the generation of highly directional impulsive loading noise. BVI noise can occur on either the advancing or retreating side of the rotor disk and its directivity is characterised by the precise orientation of the interaction. In general, advancing side BVI noise is directed down and forward while retreating-side BVIs cause noise that is directed down and rearward. It has been shown that the main parameters governing the strength of a BVI are the distance between the blade and the vortex, the vortex strength at the time of the interaction, and how parallel or oblique the interaction is.

Broadband noise

Another form of loading noise, broadband noise consists of various stochastic noise sources. Turbulence ingestion through the rotor, the rotor wake itself, and blade self-noise are each sources of broadband noise.

High-speed impulsive (HSI) noise

HSI noise is caused by transonic flow shock formation on the advancing rotor blade, and is distinct from loading noise. The source of HSI noise is the flow volume around the advancing blade tip, hence it cannot be captured by examining only the acoustic sources on the surface of the blade, HSI noise is typically directed in the rotor plane forward of the helicopter, like thickness noise.

Tail rotor noise

While most noise from a helicopter is generated by the main rotor, the tail rotor is a significant source of noise for observers relatively close to the helicopter, where the higher-frequency noise of the tail rotor has not yet been attenuated by the atmosphere. Tail rotor noise is particularly annoying to the human listener due to its higher frequency (as compared to the main rotor) which places it directly in the band in which the human ear is most sensitive.

Methods of Noise Reduction

Almost all helicopter engines are located above the aircraft, which tends to direct much of the engine-noise upwards. In addition, with the advent of the turbine engine, noise from the engine plays a much smaller role than it once did. Most research is now directed towards reducing the noise from the main and tail rotors.

A tail-rotor which is recessed into the fairing of the tail (a fenestron) reduces the noise level directly below the aircraft, which is useful in urban areas. In addition, this type of rotor typically has anywhere from 8 to 12 blades (as compared to 2 or 4 blades on a conventional tail rotor), increasing the frequency of the noise and thus its attenuation by the atmosphere. In addition, the placement of the tail rotor within a shroud can prevent the formation of tip vortices. This type of rotor is in general much quieter than its conventional counterpart: the price paid is a substantial increase in the weight of the aircraft, and the weight that must be supported by the tail boom. For example, the Eurocopter EC-135 has such a design.

For smaller helicopters it may be advantageous to use a NOTAR (from NO TAil Rotor) system. In this yaw-control method air is blown out of vents along the tail boom, producing thrust via the Coanda effect. Some designs have been done to reduce the rotor noise itself, for example the Comanche military helicopter attempted many stealth mechanisms, including attempts to quiet the rotor. Helicopter pilots can select operating modes which limits the engine torque and other parameters to ensure legal limits are respected to reduce noise. Pilots can disable the restrictions in an emergency to get extra power.

Jet Exhaust Noise Reduction

Jet exhaust consists of the fan stream and the core/combustion stream. The core flow stream is typically at a higher speed than the fan stream. As the two flow streams mix with each other, noise is created in the surrounding air. Of particular difficulty, the jet exhaust noise is actually created after the exhaust leaves the engine. This means that jet noise cannot be reduced where it is created, but must be addressed before the exhaust leaves the engine. The theory of noise generation is being studied and computer codes that can simulate the theory are being developed. The final goal of this effort is to have a computer model for jet noise that will predict the source of the noise and how it is sent into the surrounding air.

Theoretical understanding of jet noise is used to develop ideas for noise reduction concepts that are tested in model scale. Ideas that have already been tested or will be tested include mixer devices to combine the flows quickly, which reduce the noise generation area. Recently, test data have shown that a 3-dB reduction in jet noise can be achieved. The final goal is to demonstrate a 6-dB reduction.

Fan noise reduction

In order to make progress on fan noise reduction, it is necessary to understand and be able to predict that noise. Therefore, as with jet exhaust noise, effort is being put into learning the theory of fan noise generation and developing computer codes that simulate that theory. The final goal of this effort is to have a computer code for fan noise prediction that can be verified.

A second approach uses the theoretical understanding of fan noise to develop a succession of ideas for testing, with each test providing both data upon which the computer codes are verified and results upon which the next test might be built. Fortunately, the fan thrust provides many options to explore and there are many components to vary. Besides basic geometry, there are blade-wake tailoring, boundary-layer (a thin layer of air along the duct wall that moves slower than the rest of the flow) effects, fan speed, number of blades and stators, and many more. Recently, model test data showed that a 3-dB reduction in fan noise can be achieved. As with jet exhaust noise, the final goal is to demonstrate a 6-dB reduction.

Active noise control

A new approach to noise reduction is the active noise control effort. The primary principle of active noise control is to sense the noise disturbances in the engine and cancel them before they leave the engine. In effect, negative noise is made to cancel out the engine's sound waves so that no noise is heard. This is a multidisciplinary effort involving duct acoustics, controls, and actuator/sensor design.

NASA Glenn has a unique facility for this testing. The Active Noise Control Fan is a 4-ft-diameter low-speed fan designed specifically for active noise control testing. To date, several concepts have shown successful cancellation of selected acoustic modes. Because noise is the sum of all possible acoustic modes, this effort is still in its infancy, but it has potentially high pay-offs. Active noise control will contribute to the 6 dB noise reduction goal of the AST program.

SECTION VII

Noise Reduction in Hospitals

Chapter 37

Reduction and Optimisation of Hospital Noise with Six Sigma Tools

INTRODUCTION

Because hospital noise impacts patient recovery, hospitals must be quiet and calm. Hospitals must identify noise factors and discern which sources of noise are controllable. They can then use six sigma tools to measure, monitor and reduce noise. A hospital must create a quiet, calm environment for patients by providing a physical setting conducive to recovery and an organisational culture that supports patients and families through the stresses imposed by illness, hospitalisation, medical visits, healing and bereavement.

To accomplish this hospital employees must identify internal and external noise factors—is it voices, equipment or the building? The staff must also discern which noise sources are known controllable factors, known uncontrollable factors and unknown uncontrollable factors. The hospital must measure and reduce the noise in patient rooms within defined compliance levels (Fig. 37.1).

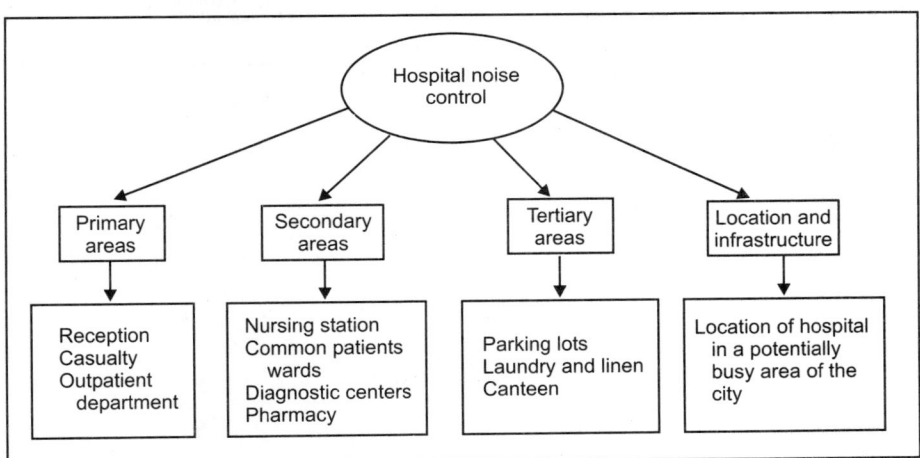

Fig. 37.1. Source of hospital noise.

IMPACT OF NOISE

Noise impacts patients in many ways including:

1. Sleep deprivation.
2. Increased anxiety.

3. Increase in noise-induced stress.
4. A 'startle reflex' resulting in physiological responses:
   (a) Facial grimacing.
   (b) Increase in blood pressure.
   (c) Higher respiratory rate.
   (d) Increased heart rate and vasoconstriction.

Continuous noise may alter a patient's memory, increase agitation, lower pain tolerance and lead to feelings of isolation.

Noise may also impact hospital employees causing:

1. Stress related symptoms.
2. Depression.
3. Irritability and decreased concentration in the work place.
4. Reduced efficiency and decreased productivity.
5. Increased medical and nursing errors.

USING THE NOISE/SOURCE MATRIX

The matrix shown in Table 37.1 displays the source of noise in various locations of the hospital and records a mean weighted average, providing a clear picture of the noisiest hospital area and the worst noise source and establishing a clear priority of what to fix first.

Table 37.1. Priority matrix–identify noise with respect to sources.

Location	Source: People (weighted score, 0.40)	Source: Equipment (weighted score, 0.50)	Source: Environment (weighted score, 0.10)	Total weighted score
Reception (A)				
Emergency department (B)				
Outpatient departments (C)				
Inpatients area (D)				
Nursing station (E)				
Diagnostic centers (F)				
Pharmacy (G)				
Operation theaters (H)				
Intensive care units (I)				
Canteen (J)				
Laundry and linen services (K)				

INTERPRETING THE STRATIFICATION TECHNIQUE

By measuring noise during peak and off peak hospital hours, analysts can detect and correct the noise. Noise meter readings established by a pollution control board offer quantitative readings that help predict, and therefore prevent, future noise problems (Table 37.2).

Table 37.2. Noise location stratification according to time of day.

Locations (noise measured in decibels)	A	B	C	D	E	F	G
Time period							
9:00 am–1:00 pm							
1:00 pm–4:00 pm							
4:00 pm–7:00 pm							
7:00 pm–10:00 pm							
10:00 pm–7:00 am							
7:00 am–10:00 am							

STRATEGIES TO SOLVE NOISE POLLUTION

After identifying the primary sources for noise pollution, the following improvement strategies can be implemented:

1. Establish stringent standards impacting patient safety.
2. Evaluate the current hospital noise through patient satisfaction surveys and by measuring the decibel levels.
3. Review the hospital's repair and maintenance policy and ensure it reflects the need for equipment to operate effectively and quietly.
4. Conduct an auditory impact query as part of every remodel and construction project, equipment purchase and staff event.
5. Change the ceiling tiles periodically from sound reflecting to sound absorbing tiles allowing patients to sleep better.
6. Convert a centralised nurse station to a decentralised nurse station.
7. Provide curtains and Plexiglas barriers in multi-bed rooms to provide both visual and auditory protection.
8. Use music therapy to replace noxious sounds with pleasant sounds—music improves restfulness and sleep, and induces relaxation.
9. Provide guidance and instruction during staff education and employee orientation sessions on the importance of maintaining appropriate noise levels.
10. Outline specific procedures regarding:
    (a) Private discussions in public areas.
    (b) Use of pagers and cell phones.
    (c) Nurse call systems.
    (d) Telephone use.
11. Place signs and slogans throughout the hospital—silent hospital help healing (SHHH)—and give patients, staff and visitors buttons that show a nurse with her finger to her lips.
12. Use sound meters to record ambient noise level at periodic times throughout the day.
13. Reduce waiting time in the outpatient departments. Schedule consulting times for the patients and fix appointments with the physicians during registration. Reducing the waiting time in turn reduces noise in the outpatient departments.
14. Display the location of offices and consultants at the reception area, and provide directories on each floor to minimise the need for visitors and patients to ask for directions.

15. Use an individual activity network diagram for each student in the hospital and ensure the faculty oversees the students in the clinical setting. This will minimise overcrowding of students in the clinical setting and streamline student activities.

16. Implement a SHHH program to recognise hospital staff and/or departments that excel at providing and maintaining a noise free environment.

Noise sources are given in Table 37.3.

Table 37.3. Noise sources.

Sources of noise	Primary area	Secondary area	Tertiary	Location
People	Unwanted movements of people (patients/ employees) Pooling of intermediate customers during consultation timing (representatives) Exchange of information between employees (human voice) Too many attenders accompanying the patients	Frequent visiting of patient's attenders in the nursing station/laboratory services for enquiries Pooling of students in the nursing station Mishandling of accessories, which creates excessive noise Load glucometer carts Staff tend to have mini-conference in the hallway creating noise	Frequent movement of people in parking facility creates unavoidable noise Overcrowding of patient attenders in the canteen Noise due to renovation/repair work done in the hospital	Noise created by people from outside the hospital (public meeting announcements)
Equipment	Ambulance noise Patient vehicles Mobility aid sounds Telephone sounds Overhead paging systems Lifts operating noise	Equipment handling Noise in laboratories Mobility aids and wheelchair sound while transfrring patients Lifts operating noise Telephones, trip alarms and intercom sound of beepers, bed rails, and ventilators Portable X-ray machine sounds, blaring TV Buzzers, beepers, multiple monitors, nurse call systems and doors	Frequent movements of vehicles Handling of equipment in the laundry and linen services Noise created during transfer of incoming essential materials in the purchase departments Utensils handling noise in canteens	Vehicle sounds

(Contd ...)

Sources of noise	Primary area	Secondary area	Tertiary	Location
Environment/ System	Improper/confused facility arrangements Lack of display boards showing facilities available in the hospital leads to unwanted enquires resulting in noise	Centralised nursing stations Facility arrangements warranting noise in inpatient wards Excessive students to patient ratio creates noise during clinical teaching	Lack of knowledge in handling the equipments by workers involved in maintenance departments Lack of space required for supportive services	Hospital floor wall and ceilings are hard and reflect sound rather than absorb it

Using quantitative and qualitative measures to identify and monitor noise levels in hospitals is critical to hospital efficiency. Reducing noise and maintaining a quiet facility will improve patient care and enhance the reputation of the hospital.

Chapter 38

Solutions to Noise Reduction

INTRODUCTION

The first step in reducing noise in hospital environments is identifying its sources. A digital decibel meter is an effective tools for measuring the sound levels of specific areas of the hospital at different times of day. Once noise sources have been identified, variety of noise abatement strategies, from sophisticated sound-masking systems to 'Quiet, Please!' signs, may be employed. Environmental design strategies for noise reduction include the maintenance and replacement of hospital equipment, the layout and acoustical treatment of patient rooms, nurses' stations, and corridors, and the implementation of emerging technologies to mask sound, reduce speech intelligibility, and introduce healing sound into the environment.

In addition to the sound emanating from all the machines and human beings working to monitor and promote patient health, a major cause of noisy hospital environments is the built environment itself. Hospital interiors and furnishing are typically made of hard, reflective materials and won't harbour infectious organisms and are easily cleaned.

All these sound-reflecting surfaces propagate noise down hallways and into patient rooms, causing sounds to echo, overlap, and linger. Rolling equipment such as procedure carts and housekeeping dollies moving across uncarpeted floors add to the din, as do pneumatic tube systems, metal chart holders, and elevator doors and alarms

EFFECTS OF NOISE

Until 2005, little work had been done to characterise and improve the acoustic condition in hospitals even though it routinely ranks among the top complaints of hospital patients, visitors and staff. Busch-Vishniac summarised many studies that indicate noise in hospital will contribute to stress in staff and suggest that noise could cause staff burnout. In addition, some researchers, such as Fife, Minckley, and Wysocki, also believe that noise affects the speed of wound healing according to their studies.

Several adverse effects have been associated with exposure to traffic noise. Studies supporting a noise 'stress' health model have suggested links between noise level and increased noradrenalin concentrations in urine, hypertension and myocardial infarction. Among the more commonly documented effects, sleep disturbances have been regarded as being the most serious.

Both noise annoyance and sleep disturbance have been proposed as important mediators of the impact of noise on health. A considerable body of research has documented the effects of noise on patient outcomes. For example, exposure to sudden, unexpected noise raises patient heart rates and has been proven to have negative influence on patient recovery times. Chronically high levels of sound on the other hand, ten to increase blood pressure levels a new study by University of Michigan researchers found a direct correlation between overall decibel levels and blood pressure levels. Higher blood pressure lead to a higher risk of cardiac problems, and a team of European researchers, in a study of 4115 patients in 32 Berlin hospitals, found that chronic noise increased the risk of heart attacks by 50 per cent for men and 75 per cent for women.

In hospital environment, where people are already ill and psychologically stressed, unnecessary noise can be very harmful.

IMPROVING NOISE LEVELS

The EPA recommended guideline values for continuous background noise are 35 decibels during the day and 40 decibels at night in patient rooms. However, a review of the literature by Ulrich and Zimring (Center for Health Design, 2007) indicates that many studies have shown that noise levels in most hospitals are much higher. And, as Ulrich and Zimring point out, there are two general reasons why hospitals are excessively noisy. First, the noise sources are numerous and loud. They include paging systems, alarms, bedrails, telephones, staff voices, ice machines, pneumatic tubes, carts, and noises generated by roommates. Second, the surfaces in hospitals—floors, walls, and ceilings—usually are hard and reflect sound rather than absorb it. They cause sounds to echo, overlap, and linger.

A third reason why hospitals are noisy might also be that there are few effective behaviour protocols in place to make staff aware of noise on a continual basis. Occasional awareness has been considered adequate when identified noises are ongoing and have direct impact on patient outcomes, as well as related costs. In defense of the staff, if the background noise level is high, then they will automatically speak louder to be heard and understood.

However, it is also true that quiet begets quiet as much as noise begets noise. The sense of appropriateness in the dynamic sound environment has yet to be established. The first step to reduce noise and improve the auditory environment is to set up and operationalise standards that are as stringent as all other environmental standards impacting patient safety. While many hospitals are committed to creating a healing environment, the auditory environment, laced with noxious inappropriate noises, is often not addressed.

A healing environment requires both a physical setting conducive to recovery in all aspects and an organisational culture that supports patients and families through the stresses imposed by illness, hospitalisation, medical visits, the process of healing, and sometimes bereavement. The sound environment, which includes communications, monitoring, paging, and technologies, must also be managed in such detail that neither patients nor staff are at risk.

When we talk about managing the sound environment, it is with full acknowledgment of the fact that there is no 'zero' noise measure. Silence does not exist, and noise is not an acoustic property. Rather it is a perception that is indefensible.

In its efforts to provide empirical standards and at the same time address the ambiguity of this issue, the EPA defines noise as 'any sound that may produce an undesired physiological or psychological effect in an individual or group'.

This definition accompanies the decibel scale. What is clear here is that there are two tests, not one, to determine whether and to what degree noise is an issue. Based on the fact that noise is evaluated by patient satisfaction measures, the EPA verbal definition supersedes the decibel level standards as volume alone does not necessarily indicate noise.

Regardless of where the patient is, equipment dominates the current hospital experience. The heavily orchestrated environment includes multiple monitors, beepers, buzzers, paging, telephones, carts, wheelchairs and gurneys, hospital beds that are electric, pillow speakers, nurse call systems, IV poles that role on tiled floors, doors that close abruptly, and carts that squeak.

SOLUTIONS FOR REDUCING AND MANAGING NOISE

Equipment Repair and Replacement

With all the rolling carts and machines in hospitals today, considerable noise reduction can be achieved by simply fixing or replacing squeaky wheels and scheduling regular maintenance to keep mobile equipment in quite working order. Other effective strategies include padding chart holders and pneumatic tube systems, and lowering volume levels on clinical and communication equipment. Making purchasing choices that are based on auditory performance.

Design of Patient Rooms and Adjacent Areas

Noise levels are obviously much lower in single-bed rooms than in shared rooms or bays. Studies consistently shows that most of the noise in a shared room is associated with the presence of another patient. In new hospital construction, there is already a trend toward standardising on single-bed private rooms. Among of these ways, MacLeod tried to improve the acoustic conditions in Vohns Hopkins Hospitals by installing sound absorbing materials. Because of the requirements for dealing with immuno-suppressed patients, the options for sound absorbing materials are very limited. They choose 2-inch fibre-glassed material for their experiment.

After they install these materials, the level of noise was dropped by 5 dB(A) and reverberation time dropped by a factor over 2. Comparing to the survey of patients, visitors and staff before and after the treatment, it also shows a very satisfied result. Here are the some images and floor plans shows the location they installed the absorbing material.

To minimise the potential for negative impacts of noise, standards or goals must be set to establish appropriate sound levels, including recommendations for modifying, maintaining, and purchasing equipment. In addition, repair and maintenance policies should be reviewed to respond to a higher quality of functionality that includes quieter operation. An auditory impact query should be part of every remodel, construction, piece of equipment, and staff event.

A recent study by Blomkvist examined the effects of poor versus good sound levels and acoustics on coronary intensive-care patients in a large university hospital in Stockholm, Sweden, by periodically changing the ceiling tiles from sound-reflecting to sound-absorbing tiles. When the sound-absorbing ceiling tiles were in place, patients slept better, were less stressed (lower sympathetic arousal), and reported that nurses gave them better care.

Moving equipment and going from centralised nurse stations (where nurses tend to congregate and talk) to decentralised nurse stations can also help reduce noise. The Karmanos Cancer Institute in Detroit, Michigan, experienced a 30 per cent reduction in medical errors on one unit after it installed acoustical panels and went to decentralised nurse stations. Methodist Hospital in Indianapolis, Indiana, attributes its improved medication error index on a redesigned coronary critical care unit to decentralised nursing and carpet in the hallways.

Judiciously using barriers, such as doors and curtains, to provide both visual and auditory protection will begin the process of controlling sounds that resonate from one area to another. Where visual access is appropriate, Plexiglas barriers can be effective. It is also well documented that single-bed patient rooms also are quieter than multi-bed rooms.

Replacing (or masking) noxious sounds with pleasant sounds moves the sound environment from being a challenge to a therapeutic modality. Enhancing the sound environment with music is a viable option if used appropriately, not in nursing stations where boom-box radios and CD players have been inappropriately broadcasting into hospital corridors.

As shown in other industries, foreground music can mask other irrelevant sounds and maintain an appropriate noise floor (i.e. the minimum noise level). In hospital settings, music combined with images of nature has been shown to reduce the amount of requested pain medication and/or improve its analgesic effect. In addition, when used appropriately, music acts as an effective audio-anxiolytic, improving restfulness and the quality of sleep, and inducing relaxation.

Another important step is staff education and new employee orientation to make staff aware of where they are and their accountability for maintaining an appropriate sound environment. While mandating staff behaviour has long been known to be the least-effective method of managing noise, behavioural standards should nevertheless be modelled and extended organisationally. This is no different than instituting dress codes and mandatory practices for infection control. Sound control must be taken seriously. This includes standards governing private or confidential discussions that take place in public areas, the use and methods of paging, and use of cell phones, nurse call systems, and the telephone.

Finally, similar to all issues of patient and staff safety, it is important to measure and notice the noise intervention on a continuous basis. Using both quantitative and qualitative measures—decibel levels and patient satisfaction surveys—a comparative analysis can be made to determine what has been achieved and which aspects of the sound environment have yet to reach the established goals.

CASE STUDY

The study was performed to measure the noise levels in specific locations of an intensive care unit and determine the disturbance levels of patients owing the noise. Studies have shown that hospital noise is a potential stressor for patients. Noise levels measured in the intensive care unit or mostly far beyond the recommended standards for hospitals, and generally measured around 60–70 dB(A). Although there are a few studies on noise levels in the intensive care unit, no study could be found that compares 24-hour intensive care unit noise measurement data at several locations of intensive care unit. The study was conducted with 35 coronary artery by-pass graft surgery patients. The intensive care unit noise level was measure by using Bruel and Kjaer 2144 Model Frequency Analyser next to the bed of each patient. A patient's disturbance owing to the intensive care unit noise was questioned.

Noise level ranged between 49 and 89 dB(A) with a mean of 65 dB(A). Peak noise levels were measured as high 89 dB(A). The noise levels measured at different locations in the intensive care unit did not differ significantly. Noise created by other patients, those who were admitted from emergency room and operating room into intensive care unit, monitor alarms, conversations among staff were the most disturbing noise sources for patients.

The patients who were located in the bed which was closer to the nurses' station were more affected by the intensive care unit noise than other patients. Having a previous intensive care unit experience also affected the patients' disturbance levels owing the noise. Relevance to clinical practice: Nurses are in key positions where they can identify physical, psychological and social stressors that affect patients during their hospital stay. Staff eduction, planned nursing activities and proper design of intensive care unit may help combat this overlooked problem.

A quasi-experimental study was performed which tested an intervention to reduce sound levels in an acute care hospital. A parallel pre- and post-test design with control group was used; patients and employees completed the Topf Adapted Sound Disturbance Scales, and environmental sound levels were recorded on Quest 2900 Sound Level Meter. Treatment interventions included an educational PowerPoint presentation for employees, minor environmental acoustical alterations, and the use of Quest 261 Sound Detector/Controller for behavioural modification. None of these interventions produced statistically significant changes is sound levels.

Patient and employees reported slightly less disturbance due to noise post intervention on the treatment unit. The findings of this study support Philbin and Gray suggestion that the use of sound-absorbing materials in the hospital's physical structure may be the most effective measure to reduce sound level in the hospital setting.

SECTION VIII

Noise Reduction in Miscellaneous Industries

Chapter 39

Plastic Industry

INTRODUCTION

Plastic industry encompasses areas of activities including manufacturing polymer materials (natural and synthetic compounds), building construction, bulk material handling, packaging or transportation. Plastic industry covers a large area ranging from aerospace to electronics industry and that is why plastic industry has a huge business potentiality over the other associated industries. This plastic industry is a vivacious and global industry that covers sectors from household commodity to polymer. Plastic industry also include fields like biochemicals, petrochemicals, ceramics.

In the early phases, plastic industry was built up with natural and synthetic materials of different forms, attributes and appearances. Over time, plastic industry started developing on the basis of organic compounds (a material that contains carbon and hydrogen and other elements), rather than natural and other synthetic materials.

FINISHED PRODUCTS OF PLASTIC INDUSTRY

Common finished products of plastic industry include industrial plastic products, plastic aircraft products, plastic bags, plastic basins, plastic bottles, plastic bins, plastic boxes, plastic cans, plastic caps, plastic clothings, plastic dies and moulds, plastic containers, plastic films, plastic foams, plastic fountains, plastic glasses, plastic lenses, plastic tubes, plastic trays, plastic toys and so many. However the machineries producing the finished products and processes produce lot of noise.

Vibrations and Noise

Sometimes the smallest cause can have a huge effect, and when it comes to buying a plastic injection moulding machine, the design in terms of service life and increasing noise emission become crucial to the decision-making process.

One of the typically small causes for production downtimes is the damage caused by vibrations, for example: Interference to the electronics, leakage from pipe joints, increased wear and tear and breakdowns in the control components of the hydraulic circuit. In addition, excessive noise emissions reduce the efficiency of staff.

For the precise alignment of electrical motors and hydraulic pumps and the effective damping of vibrations and noise on plastic injection moulding machines, offers an extensive range of suitable components and complete solutions for drive capacities of 0.55 kW to 132 kW.

Flexible bell housings

The flexible bell housing connects the electric motor to the hydraulic pump. Both flange sides are ready-machined. The torque is transmitted via a flexible coupling so that the damping element is just compressed. Noise and vibration are reduced by decoupling the rigid connection of electric motor and hydraulic pump. Compatible with mineral oil through the use of NBR rubber compound for damping element. Available for drive capacities of 0.55 to 132 kW. Cost-effective rigid version available without noise damping for drive capacities of 0.18 to 355 kW.

Damping rings

For vertical and horizontal mounting. Cost-effective noise reduction by decoupling from the tank. Compatible with mineral oil through the use of NBR rubber compound. Vulcanised sealing lip, no additional sealing required. For drive capacities of 0.55 to 200 kW.

Damping rails

For electric motors of construction type IM B35 and bell housing foot brackets to VDMA 24561. Designed for weight bearing when installed horizontally, noise reducing and oscillation damping. Compatible with mineral oil through the use of NBR rubber compound.

Couplings

The couplings are torsionally flexible and transmit the torque positively. They are fail-safe. The oscillations and shocks occurring during operation are effectively damped and reduced. Available in different hub materials: aluminium, grey cast iron (GG/GGG), steel.

Ball valves

Standard models from our range or customised solutions.

Mounting technology

Rigid and flexible elements and systems for mounting pipes, hoses and cables, reservoirs and components. Table 39.1 highlights the established noise control for high risk activities and Table 39.2 highlights management of noise risks.

Table 39.1. Established noise control for high risk activities.

Activity or process	Established noise control methods	Further information (links)
Granulators (and other size reduction machines, e.g. agglomerators, crumbers, shredders, pelletisers)	Methods include: Use feed conveyor to remove operator from higher noise areas Size reduction machines in	Generally useful: 'Noise in the plastics processing industry'
Example noise levels: 100dB (granulators) 90dB (agglomerators)	separate rooms or buildings—provide for remote or automated feeding Lag or damp the machine casing Form sound trap in feed aperture or hopper Enclose the machine	Example: Rubber granulator Example: Enclosure for rubber grinding machine Example: Strand pelletisers Example: Enclosure for pelletiser

(Contd ...)

Activity or process	Established noise control methods	Further information (links)
	Fit segmental or helical cutters Use tangential feed Fit resilient backing to knives Reduce rotor speed	

Injection moulding machines
Example noise levels:
97–100dB
	Methods include:
	Use slow speed pumps
	Control release of exhaust air
	Mount pumps and motors on
	   anti-vibration mounts and
	   incorporate flexible hoses
	   in pipe lines
	Enclose hydraulic power packs
	Convert injector guards to
	   acoustic guards
	Fit low noise nozzles to blow
	   guns, etc.
		Example: Controlling release of exhaust air

Extruders
Example noise levels:
90dB
	Methods include:
	Specify low noise design
	For hydraulic systems see
	   injection moulding machines
	   above
	Fit silencers to drive motor air
	   intakes and exhausts
	Enclose drive motor.	

Mould cleaning guns
Example noise levels:
  105dB
	Replace nozzles with low-noise
	   types (e.g. those which generate
	   an induced secondary air flow).
	Reduction of up to 10 dB
		Example: Reduced noise from mould cleaning gun

Extrusion line cut off saws
Example noise levels:
  100dB
	Methods include:
	Replace guards with solid panels
	   lined with acoustically
	   absorptive material
	Fit acoustic strip curtain at product
	   out-feed
		Example: Extrusion line cut-off saws

Ultrasonic welding machines Example noise levels: 96dB	Enclose with sound reducing material	Example: Enclosure of welding machine

Table 39.2. Management of noise risks.

Issue	Expectation	References and related guidance

Workplace design for
  reduced noise
  exposure
	Table 39.1 deals with established technical and
	organisational noise control measures for a range
	of high noise risk activities or processes. In
	addition to these measures, in general there will
		Workplace design
		Example: Coating pans
		Example: Flexible acoustic
		  screening material

(Contd ...)

Issue	Expectation	References and related guidance
	always be benefits to be gained in considering and applying general principles of workplace design for reducing noise exposure. For example:	Example: Acoustic refuges Example: Use of absorption in a noise control programme
	Appropriate use of acoustic absorption within buildings can reduce or limit the effects of reflected sound (specialist help will be needed to put this in to effect)	
	Careful planning could segregate noisy machines from other areas where quiet operations are carried out	
	The number of employees working in noisy areas should be kept to a minimum	
	Screens, barriers or walls can be placed between the source of the noise and the people to stop or reduce the direct sound	
	Noise refuges can be a practical solution in situations where noise control is very difficult, or where only occasional attendance in noisy areas is necessary	
	Increasing the distance between a person and the noise source can reduce noise exposure considerably	
Selection of tools and machinery	Employers should demonstrate a positive purchasing policy which makes sure noise is taken into account when selecting machinery	Low noise machines Noise–advice for employers [207kb]
	For many types of equipment there will be models designed to be less noisy. When selecting equipment to buy or hire, besides ensuring that the tool or equipment is generally suitable for the job, employers should:	
	Ask about likely noise levels for the intended use(s)	
	Check that manufacturers' noise data is representative of likely noise levels for the intended use(s)	
	Use the noise information to compare machines before making the final choice	
	Look for warnings in the instruction book to see if particular uses of the tool or machines are likely to cause unusually high noise	
	Be aware that even where manufacturers declare that their tools or machines produce less than	

(Contd ...)

Issue	Expectation	References and related guidance
	70 dB, levels may sometimes be much greater in your workplace	
Limiting exposure duration	Restriction of the time spent in noisy areas, or doing noisy tasks, can be effective in reducing noise exposures, as can ensuring that noisy devices are only used when they are actually needed	HSE noise exposure calculators and ready reckoners
	Where some employees do noisy jobs all day or week, and others do quieter ones, job rotation should be considered. This might need you to train employees to carry out other jobs. This system will reduce the noise exposure of some employees while increasing that of others, so care and judgement is needed. Employees will need to be rotated away from noisy jobs for a significant proportion of time to make an appreciable difference to their daily exposure	
	The noise exposure ready-reckoner and exposure calculators can be used to indicate the reductions in exposure that can be achieved by reducing the duration of exposure to noise	
Hearing protection	Providing personal hearing protection should be one of the first considerations on discovering a risk to the health of your employees due to noise. It should not be used as an alternative to controlling noise by technical and organisational means, but for tackling the immediate risk while other control measures are being developed. In the longer term, it should be used where there is a need to provide additional protection beyond what has been achieved through noise control	Hearing protection—general advice HSE hearing protection calculator Hearing protection—Over-protection Hearing protection—real-world factors Hearing protection—advice on issuing Noise—advice for employers Protect your hearing or lose it. (advice for employees)
	Hearing protection use should be targeted at particular noisy jobs and activities. Personal hearing protection must be supplied by the employer to any employee whose daily personal noise exposure is likely to exceed 85 dB, or who is likely to be exposed to peak sound pressure levels above 137 dB. The employee must use the protection provided. The employer should ensure that, through the use of hearing protection, the employee's effective noise exposure is reduced to at least below the above levels	
	Important factors to consider in the selection and use of hearing protection include: Types of protector, and suitability for the work being carried out	

(Contd ...)

Issue	*Expectation*	*References and related guidance*
	Noise reduction (attenuation) offered by the protector, including taking account of 'real-world' factors, and also ensuring that not too much protection is provided	
	Compatibility with other safety equipment	
	Pattern of the noise exposure	
	The need to communicate and hear warning sounds	
	Environmental factors such as heat, humidity, dust and dirt	
	Cost of maintenance or replacement	
	Comfort and user preference	
	Medical disorders suffered by the wearer.	
	The use of personal hearing protection should be managed through the provision of appropriate information, instruction and training for employees, supervision and the use of appropriately defined and demarcated hearing protection zones	
Information, instruction and training	It is important that employees understand the risks they may be exposed to. Where they are at risk from noise their employer should at least tell them:	What do I need to tell my employees?
		Employee and safety representatives
	The likely noise exposure and the risk to hearing this noise creates	Noise—advice for employers
	What their employer is doing to control risks and exposures	Protect your hearing or lose it (advice for employees)
	Where and how people can obtain hearing protection	
	How to report defects in hearing protection and noise-control equipment	
	What their duties are under the Control of Noise Regulations 2005	
	What they should do to minimise the risk, such as the proper way to use hearing protection and other noise-control equipment, how to look after it and store it, and where to use it	
	The health surveillance systems	
	This information should be given in a way the employee can be expected to understand (for example special arrangements might need to be made if the employee does not understand English or cannot read)	
	To establish whether information, instruction and training has been carried out effectively, look for evidence that personal hearing protection is being fully and properly used, that noise control	

(Contd ...)

Issue	Expectation	References and related guidance
	equipment is being used, and that procedures for low noise working are being followed	
Health surveillance (audiometry)	Health surveillance for noise-induced hearing damage should be in place for employees whose daily personal noise exposure is frequently above 85 dB, or who are frequently exposed to peak sound pressure levels above 137 dB. Health surveillance should also be provided where exposures are lower, but where the employee may be particularly sensitive to noise. As a minimum, a programme of health surveillance should include: Audiometric testing (baseline assessment on first entering a job involving noise exposure, annual testing for two years, then three yearly testing) Arrangements to receive medical advice on management of affected employees Arrangements to receive anonymised information to demonstrate effectiveness of controls	Health surveillance Noise — advice for employers Protect your hearing or lose it (advice for employees)

Chapter 40

Wood Working Machines

INTRODUCTION

Woodworking has some of the noisiest work places in industry. Short exposure to high noise levels can cause temporary hearing loss, but longer exposures can result in permanent damage. Sufferers often do not realise their hearing is being damaged, as hearing loss tends to be gradual. However, some effects such as tinnitus can develop more quickly. Tinnitus can be a permanent ringing or whooshing sound in the ears which can be very distressing, particularly when it's quiet, such as when you are trying to go to sleep.

People who are exposed to high noise levels, even for a short time, may experience temporary hearing loss if they continue to be exposed, they will suffer serious permanent hearing loss. This loss is gradual, and sufferers often do not realise that their hearing is being damaged. Some of the noisiest working environments are found in the woodworking industry. Noise levels can vary widely from machine to machine depending on conditions of use, but typical examples are shown in Table 40.1.

Table 40.1. Typical noise levels at woodworking machines.

Machine	Typical noise data for machines with no noise reduction measures
	Noise level (dB)
Beam panel saws and sanding machines	97
Boring machines	98
Band re-saws, panel planers and vertical spindle moulders	100
Portable woodworking tools	101
Bench saws and multiple ripsaws	102
High speed routers and moulders	103
Thicknessers	104
Edge banders and multicutter moulding machines	105
Double end tenoners	107

The Control of Noise at Work Regulations 2005 set out the legal requirements for controlling the risk of hearing damage at work. They aim to eliminate the risks from exposure to noise or where this is not possible, to reduce the risks to as low as is reasonably practicable.

Specific duties are triggered when an employee's personal noise exposure reaches certain levels, known as 'action values', which are measured in decibels (dB). These are based on daily noise exposure (the average over a working day), and maximum short duration noise (the peak sound pressure). The action values are:

1. Lower action values: daily exposure of 80 dB, and peak sound pressure of 135 dB.
2. Upper action values: daily exposure of 85 dB, and peak sound pressure of 137 dB.

You can also use weekly exposure levels when calculating personal noise exposures where noise exposure varies a lot from day to day, e.g. where people use woodworking machines and power tools on one or two days in the week but not on others.

EXPOSURE LIMITS

There are also noise exposure limits that must not be exceeded. These are a daily personal noise exposure of 87 dB, and a peak sound pressure of 140 dB. The exposure limits, but not the action values, allow the effect of hearing protection to be taken into account.

The peak sound pressure values are unlikely to apply to most woodworking machines, except perhaps nail guns. However, daily exposure for people working with woodworking machines is likely to exceed the lower action value.

What is Expected of Employers?

The best approach is to engineer out the noise at source, and to organise your workplace so that the minimum number of people are exposed to noise. When noise exposure reaches the upper action levels, the Regulations do not allow you to use personal hearing protection as an alternative to this.

You should assess the noise exposure of your employees to find out if it is above any of the action values. This means taking account of both the levels of noise and how long they are exposed to it. All noisy tasks carried out during a working day will contribute to an employee's daily noise exposure. You can use the information in Table 40.1, and knowledge of the time that employees spend working at various machines, to establish a likely daily exposure, and to decide on priorities for noise control. There are electronic spreadsheets on the HSE website (www.hse.gov.uk/noise) to calculate this for you.

Controlling Noise

Where any employee's noise exposure reaches the upper action values you must have a planned program of noise control measures in place. For woodworking machinery this is likely to include:

1. Using the best systems of work.
2. Using the most appropriate machine for the task.
3. Engineering noise control at source.
4. Effective maintenance of equipment.
5. Limiting how long people are exposed to noise.

Employees have duties to use the noise control equipment provided and to report any defects in it.

Engineering Controls to Reduce Noise Exposures

Engineering control measures are likely to be necessary to reduce noise, e.g.:

1. Change to quieter tooling.
2. Modify dust extraction systems.
3. Provide a noise enclosure.

Machine Maintenance

You have legal duties to maintain woodworking machines to reduce noise and to maintain noise by keeping machines well maintained.

What Affects Noise from Woodworking Machines?

See Table 40.2 to find out what can affect noise levels when using woodworking machines and consider these factors when planning your noise reduction measures. Make supervisors and employees aware of these issues, particularly when reducing noise exposure depends on following proper systems of work.

Table 40.2. Factors affecting noise levels when using woodworking machinery.

Variable	Relevant factor	Effect
Timber	Species	Hard, stiff timbers mean more noise (e.g. 2 dB difference when cutting oak and pine at a band re-saw) and more noise transmission.
	Width	Wide workpieces radiate noise over a greater area (e.g. a working width of 200 mm is likely to cause an increase of 2 dB over a working width of 100 mm)
	Thickness	Thinner workpieces generally vibrate more. Planing under 20 mm thicknesses can greatly increase the noise level
	Length	Long workpieces transmit noise away from the cutting area towards the operator
	Moisture content	Dry timber is brittle and a good transmitter of noise
Tooling	Width of cut	Unless helical or segmental cutters are used, the noise level immediately above the cutter increases roughly in proportion to the width of the cut (e.g. doubling the cutter length increases the noise by 3 dB)
	Cutter sharpness	Dull knives and worn blades and bands exert more force on the timber and so make more noise.
	Cutter projection	Increases in knife projection mean that more air is trapped during rotation and so more noise is produced (typically 2 to 3 dB more for each millimetre projection above 1.5 mm)
	Speed	Noise increases with tool speed (typically just under 1 dB for every m/s change in peripheral speed in the range 20 to 35 m/s)
	Balance	Out-of-balance tools means vibration and changes in cutting conditions, increasing noise
Machine setting	Timber control	The freer the timber is to vibrate, the greater the noise level (e.g. poorly set chip breakers and pressures at multi-cutter moulders lead to more transmitted noise)
	Timber support	Noise is increased if fences, bed plates, chip breakers, etc. which support the timber close to the cutting circle are not in line and as close as possible to the cutting point
Extraction	Air velocity/ system design	In a system with turbulent airflow, wood chips strike the ducting more and, without damping, this can increase noise levels

You should also follow the manufacturer's or supplier's advice to ensure that the installation and operation of woodworking machines results in the lowest noise levels possible. Noise control case studies can be found on the HSE website (www.hse.gov.uk/noise).

Buying New Machinery

Manufacturers of woodworking machines are legally obliged to design and manufacture them so they produce as little noise in use as possible. Information about the noise produced by a machine should be provided in the manufacturer's instructions as well as how to operate the machine as quietly as possible.

When buying new woodworking machinery you should include noise emission in the specification. This will allow you to make informed judgements about the likely noise exposures from a particular machine before you buy it and identify the machine that will introduce least noise into your workplace.

Noise data

When providing noise data, the manufacturer should describe the type of work being carried out by the machine when the measurements were made. This will help you to decide if the noise data is relevant to the work that you intend the machine to do. If it is not, ask the manufacturer or supplier for extra information.

Suppliers and manufacturers of machines should take into account the various ways in which a machine might be used and provide information on how certain variables affect noise levels, e.g. type of cutter, different workpieces and feed rates (see Table 40.2).

If you define your maximum acceptable noise level for new machinery, you can include this in the purchase contract. You can then check that your criteria have been met before you make the final payment.

Low-noise features

Manufacturers and suppliers of woodworking machinery should be able to describe the low-noise features of their machinery. Appropriate design measures vary depending on the machine type, but will in general include:

1. The machine structure is designed to minimise direct noise radiation, e.g. flexible panels should be avoided or treated.
2. Antivibration mountings.
3. Acoustic absorbents, shields or enclosures for control of unavoidable noise sources.
4. Use of advances in cutter design, e.g. damped or low-noise blades for sawing machines, segmental cutters for moulders and helical cutters for planers.
5. Machine tables with slotted lips to reduce noise when air gets trapped between revolving cutters and fixed surfaces.
6. Systems for minimising noise from workpieces.
7. Design of waste extraction to reduce noise generated by woodchips.
8. Silencers to reduce noise at compressed air exhausts and jets.
9. Optimised spindle speeds, tooling diameters and feed rates.

Personal Hearing Protection

While you are developing your noise control measures, your employees' hearing will be at risk. Even with all reasonably practicable noise reduction measures in place, it is likely there will still be a risk of hearing damage, so personal hearing protection will be required.

Noise reduction measures may not remove the need for employees to wear personal hearing protection, but they may benefit from being able to use a protector with a lower rating. If an employee's personal daily noise exposure is below the upper exposure action value, you should make hearing protection available if the employee requests it, and encourage them to use it until the noise-control program achieves exposure levels below the lower exposure action values. If an employee's personal noise exposure reaches or exceeds the upper action values then hearing protection is compulsory. The employer must provide it and employees must use it. You need to ensure that, by using it, their noise exposure is reduced to at least below the exposure limit value.

Avoid protectors that reduce the level at the ear to below 70 dB as this over protection may interfere with communication and hearing warning signals.

Hearing Protection Zones

Mark hearing protection zones with appropriate signs. Any woodworking machine which produces average noise levels higher than 85 dB at its operator position should be in a hearing protection zone. The zone should be extended to include any people working near the machine who might be subject to noise above this level.

Hearing Protectors

Hearing protectors vary in the degree of protection given. Select them to:
1. Protect against the noise levels in the workplace.
2. Be comfortable.
3. Be suitable for wearing with other personal protective equipment.

They should also be:
1. Issued on an individual basis.
2. Kept clean.
3. Regularly checked and maintained.
4. Replaced when necessary.

Make employees aware of the need to wear hearing protection whenever they are exposed to noise. Not wearing hearing protection for even a short period of time in a noisy environment will significantly reduce the protection they receive over the working day.

Noise Awareness

Provide information, instruction and training to managers, supervisors and employees about their likely noise exposure and the risk of hearing damage. Employees should know how to minimise the risk of hearing damage, e.g. by following systems of work, and correctly using noise-control equipment and hearing protection. Make sure they are trained to carry out regular maintenance on noise-control equipment and know when and how to report defects.

Health Surveillance

Where an employee's daily noise exposure regularly exceeds the upper action values, they should have regular hearing checks as part of a health surveillance program. These can be used to detect when employees might be suffering from the early signs of hearing damage. Where there has been damage, you should take action to prevent it getting worse. A pre-employment check will provide a baseline against which future measurements can be compared. This will show how effective your noise control strategy is.

NOISE REDUCTION AT BAND RE-SAWS

This section provides guidance for employers on how noise from band re-saws is generated and how to reduce it at source using engineering controls. It then explains how to build a noise reducing enclosure.

Band re-saws are widely used in the wood industry. Without any measures to reduce noise at source, they can produce noise levels of over 85 dB (typically 100 dB at the operator position). At this level of noise, an employee's daily personal noise exposure would reach the 85 dB upper action value after 15 minutes. The best way to deal with the problem is to reduce machinery noise at source, e.g. by providing a noise enclosure, and also to organise your workplace so that fewer employees are at risk (Fig. 40.1).

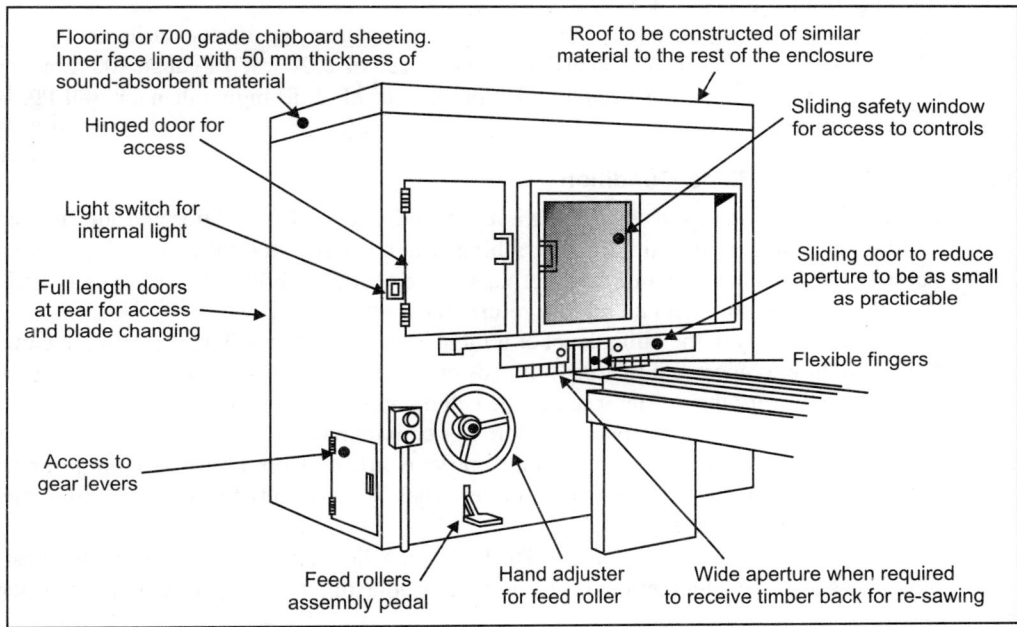

Flooring or 700 grade chipboard sheeting. Inner face lined with 50 mm thickness of sound-absorbent material

Roof to be constructed of similar material to the rest of the enclosure

Hinged door for access

Sliding safety window for access to controls

Light switch for internal light

Sliding door to reduce aperture to be as small as practicable

Full length doors at rear for access and blade changing

Flexible fingers

Access to gear levers

Feed rollers assembly pedal

Hand adjuster for feed roller

Wide aperture when required to receive timber back for re-sawing

Fig. 40.1. Band re-saw noise enclosure.

Noise from Band Re-saws

Machine noise can be reduced, particularly when the machine is idling, by maintaining the machine and blade in good condition. Well-maintained machines with pulley diameters up to 900 mm produce idling levels in the region 80 to 90 dB. Poorly maintained machines, which are otherwise identical, may idle at levels as high as 110 dB.

Cutting noise is typically between 95 and 105 dB. Machines that are cutting for any substantial part of a working day will usually need to be enclosed where practicable.

Main Source of Noise

When a band re-saw is idling, vibration of the blade is usually the main source of noise. When cutting, high vibration levels in the blade caused by sawdust trapped between the pulleys and blade, and vibration of the timber being sawn are the main noise sources. How much the blade vibrates is affected by the:

1. Gauge of the blade.
2. Condition of the saw pulley surfaces.
3. Effectiveness of the sawdust deflection and extraction systems.
4. Effectiveness of the pulley and blade scrapers/cleaners.

5. Effectiveness of the sawblade lubrication system.
6. Adjustment of the saw guides.
7. Blade tension.

The condition of the sawblade and the smoothness of the pulley faces have been found to affect idling noise levels by as much as 10 dB. How efficient the sawdust extraction and wheel scraping/cleaning systems are can have a similar effect.

Poorly adjusted saw guides can push noise levels up by 3 dB and using an unnecessarily heavy gauge sawblade produces a wider kerf (cut) and can also produce more noise. A new 19 gauge 100 mm blade running on 900 mm diameter pulleys has been found to produce levels 5 dB higher than a new 20 gauge blade on the same machine.

Indicators of Machine and Blade Condition

The difference in noise when cutting and when idling is a good indicator of the condition and adjustment of the machine and blade. On a well-maintained machine, when sawing starts the noise level should rise instantaneously by about 10 dB (sounding perhaps twice as loud) above the idling level and as the timber clears the saw, the noise should rapidly return to the idling level.

If the noise level falls slowly to the idling level or if there is no noticeable difference between cutting and idling, then the pulley scrapers, lubricating felt pads or sawdust extraction system need attention.

A high idling noise with very little difference between idling noise and sawing noise indicates that the surface of the blade or the pulleys are in poor condition or contaminated with resin and sawdust. The blade surface can have lumps, hollows or hammer marks resulting from abuse or from poor saw doctoring. Resin and sawdust (particularly from timbers such as redwood) can stick to pulley faces and the uneven surface causes the blade to vibrate.

Replace sawblades in poor condition and regrind badly-worn pulley faces. Scrape clean contaminated blades and pulley faces and maintain them in this condition by adjusting the scrapers and the lubrication system.

Experience with a Typical Machine

Tests were carried out on a typical 1200 mm re-saw in a small mill and noise levels were measured at the infeed operator's position.

Before adjustment, the idling noise level was 103 dB and the noise produced by sawing imported redwood was 103 dB. After fitting new pulley scrapers, new felt pads to the cleaner lubrication system and correctly adjusting the guides, the idling noise level was 88 dB and the noise from sawing imported redwood was 100 dB.

For a small outlay on parts and about two hours fitting time, the idling noise was reduced by 15 dB and the cutting noise by 3 dB.

Noise Enclosures

Providing a noise enclosure for band re-saws is a 'reasonably practicable' way to reduce the noise exposure of operators. As well as reducing noise, enclosures control sawdust, illuminate the cutting area (when fitted with lighting), and act as a barrier reducing the risk of operators contacting the sawblade by accident.

Effective noise enclosures for band re-saws may be constructed from a variety of materials. 'Homemade' enclosures can be as efficient as commercially supplied models and may cost much less.

A well-constructed enclosure can reduce noise by 10–15 dB. They should be built on a timber frame (minimum 50 mm × 50 mm studding). The outer skin (cladding) should have good noise reducing properties, e.g. use high-density 19 mm chipboard or plywood.

Internal surfaces

The internal surfaces of the enclosure should be lined with a sound-absorbent material, e.g mineral wool slabs with a density of 60 kg/m^3 and a thickness of 50 mm. This should be covered with thin polythene sheeting (less than 0.1 mm thick) to keep out dust. The absorbent material should be held in place by perforated metal or hardboard with at least 30 per cent open area.

Feed and delivery

Feed and delivery openings should be as small as possible. Where a wide range of timber sizes are processed, the aperture size may need to be adjustable, but in any case restricted so that the maximum cross-section of timber the machine process is not affected. Apertures should be fitted with a double row of overlapping rubber or loaded PVC flaps or fingers, each about 18 mm wide. This will allow the workpiece to pass through while creating the minimum possible gap through which noise can leak.

Access doors

Access doors should allow blade changing, machine maintenance and adjustments. Doors should be of double-skin 12 mm chipboard or plywood and built on a 25 mm timber framework. They should be rebated and self-closing (e.g. by using rising butt hinges). All gaps around the door should be fitted with soft rubber sealing strips. Robust catches or locks should be fitted to hold the door firmly closed against the sealing strips.

Viewing panels

Viewing panels should be acoustically sealed into the structure. Safety glass 6 mm thick is usually adequate, although double-glazing is preferable and sometimes necessary. Experiment with access and viewing panel positions before completing the enclosure.

Ventilation

Air-cooled equipment inside the enclosure should be adequately ventilated. You will need to acoustically lag and/or line parts of the ventilation and the dust extraction duct work as these are likely to become sources of noise breakout.

Machine controls

Some of the machine controls, such as the feed roller pedal and control wheel, should be extended so that they can be operated from outside the enclosure. The feed stop and start and emergency stop controls should also be outside the enclosure.

Joints

All joints between the enclosure walls, roof, floor, doors, extraction ducts and glazing need to be properly sealed. Expanding foam, soft pliable substrates, putties and silicone, etc. can be used.

Chapter 41

Music and Entertainment Industry

INTRODUCTION

The music and entertainment industry is unique in many ways, one of which is that high sound levels and special effects loud enough to cause noise-induced hearing loss are often regarded as essential ingredients in appealing to patrons. Permanent hearing loss from excessive noise exposure, and is some cases tinnitus (ringing in the ears), are often suffered by people who have worked in or with the industry over a number of years. Hearing damage diminishes quality of life no matter how it is caused.

The general duty of care embodied in the Occupational Safety and Health Act places responsibilities on employers, persons in control of workplaces, musicians and people who work with or near musicians, such as sound mixers and bar staff, to ensure as far as practicable that they themselves and others in the area are not exposed to hazards. While the ear can respond to sounds as loud as a raging rock band or as quiet as a distant harp, sustained high levels of sound, over a period of years, can cause permanent damage to the ear. Perhaps even worse, short sharp impulse sounds can cause instant permanent damage if loud enough. The effect is a dulling or muting of the sounds we hear, causing a loss of enjoyment of music, difficulty in understanding people talking in a crowd and general sound isolation. Hearing aids are of limited benefit in these cases and there are no transplant operations for this condition. Some people suffer long-term problems from 'ringing in the ears', or tinnitus, as a result of exposure to loud sounds. In many types of entertainment venues, music is being played at levels likely to cause damage to those present, with repeated exposure. People involved in the entertainment industry need to develop preventative strategies in relation to workplace noise whenever the 'exposure standard for noise' is exceeded (see Appendix B). Recognising the unique nature of the music entertainment industry, this code of practice forms part of an overall strategy for occupational noise control, which is contained in the Code of Practice for Managing Noise at Workplaces (hereinafter referred to as the 'Principal Code'). This code should be read in conjunction with the Principal Code, particularly in relation to the following general aspects:

1. Consultation—the Principal Code explains the need for consultation and co-operation at all stages between employers, employees and safety and health representatives.
2. Noise control strategies—the Principal Code emphasises the primary role of engineering noise controls in reducing noise levels, followed by administrative measures to reduce, the time employees are exposed to noise. Provision of personal hearing protectors is an interim measure which does not remove the obligation to reduce noise exposure as far as is practicable.
3. Information, training and education—the Principal Code lists training objectives and topics to be covered.

This code provides practical advice on developing preventative strategies to enable people in the music entertainment industry to act to minimise the risk of hearing damage resulting from the performance of live or recorded music. These strategies provide a means for employers as well as service and technical staff and others in various types of venues; to meet their legal obligations in terms of work place noise, as defined in the Occupational Safety and Health Act and Regulations.

While this code mainly emphasises the duty of the employer to employees, the strategies presented here will also benefit others in the workplace, such as performers and members of the audience.

This code does not set specific limits for music sound levels. Rather it places an emphasis on providing sufficient information on music levels and venue acoustics, so that everyone involved with the performance reaches agreement that noise exposure is as low as practicable for the style of entertainment required. Detailed explanatory information and guidance is provided in Appendices of this chapter to the code. Table 41.1 lists music venues, activities and people affected.

Table 41.1. Music venues, activities and people affected.

Type of workplace	Activities producing excessive noise	People affected
Hotels-taverns, nightclubs, and bars	Bands playing loud music Disc jockeys playing loud music through PA system Karaoke performances	Bar staff, waiting staff, kitchen staff, performers, disc jockeys, technical staff, security staff, customers
Clubs (Hospitality), Discos and Casinos	Floor shows which include loud music Bands playing loudly in entertainment area Disc jockeys playing loud music	Restaurant staff, waiting staff, bar staff, performers, technical staff, cashiers, security staff, customers
Concert halls and theatres (usually not movie theatres)	Bands, orchestras and music groups playing loudly during rehearsals and performance Recorded music for dance played loudly	Producers, directors, venue staff, performers, technical staff, audiences, security staff, first aiders, catering staff
Outdoor concert venues (stadiums, music bowls)	Bands and orchestras playing loudly Loud special effects (e.g. fireworks)	Performers, technical staff audience, security staff, first-aiders, catering staff
Cafes and restaurants	Theatre restaurant performances Bands playing loud music Loud record music	Catering staff, bar staff, kitchen staff, waiting staff, performers, technical staff, customers
Performing arts venues	High sound levels during rehearsals and performances	Performers, technical staff, house staff, audience
Education establishments	School bands playing loudly during training periods and' performances Recorded music or live bands playing loudly at school dances	Teachers, other staff, students, performers, disc jockeys
Recorded music retail establishment	Playing tapes, CDs and records loudly through store PA systems	Sales staff, managers, customers
Recreation venues	Loud recorded music during aerobics classes	Aerobics instructors, students

EXPLANATION OF TERMS

In order to be in a position to use the strategies in this code effectively, it is necessary to understand a few basic terms:

1. Music level.
2. Received noise.
3. 8 hour exposure
4. Reference position
5. Room loss.
6. Exposure standard for noise
7. Competent person
8. Practicable.

Explanations of these terms are given in Appendix B.

STRATEGIES FOR THE ENTERTAINMENT INDUSTRY

Music noise is best managed by using an appropriate strategy. Table 41.2 lists the strategies appropriate to the different people likely to be involved in music venues. It is important to know when music levels become music noise. After all you will not know when control measures are needed unless you can measure the noise.

Measuring music noise can be a complex operation, and people in entertainment venues are not normally experienced in this field. Even so, you will need to have some knowledge of how noise is measured, so that you can understand music noise control strategies. Appendices C and D have been included in this code to help you.

Table 41.2. Music noise management strategies.

Class of person	Venue owner (not operator)	Venue operator	Entertainment provider	Employer of non-music employees	Supplier and installer of sound equipment	Employee
Strategy	1	2	3	4	5	6

Appendix D of this code contains a number of examples of how the strategies can be used in different venues, both for measuring music noise and implementing controls. Select the one which is most appropriate for your situation.

Strategy No. 1

For an owner

Explanation

You are a person who owns, but does not operate an entertainment venue. Owners may lease the workplace to someone else who conducts the business, e.g. you own a suburban shopping centre and lease a shop to another person who retails recorded music. In this case you are the owner and the person running the store is the operator.

Note: An owner may be an operator. For example the owner of a casino may also operate the casino in which music noise is generated. In this case you will need to follow Strategy No. 2.

What to do

As an owner you are responsible for ensuring the practicable architectural changes which may be needed to reduce the noise exposures of people in the venue are implemented. Whilst you have no direct responsibility to provide advice or information about safety and health, it may be advisable to bring the following matters to the attention of the operator:

1. Your safety and health policy.
2. The legal requirements of the occupational.

Safety and Health Act and Regulations

1. The principal code.
2. The contents of this code.

Note: The Environmental Protection Act 1986 and the Environmental Protection (Noise) Regulations 1997 may also apply in these circumstances.

Strategy No. 2

For operators of entertainment venues

Explanation

You are the person who operates an entertainment venue listed in Table 41.1. You will normally be the person who directly engages the services of an entertainment promoter or music performer. Most likely you will be a hotel licensee, nightclub or disco proprietor, music store operator or a live theatre manager. For example, you are operating a hotel and engage the services of a rock band to play three nights a week in a part of the hotel. This also includes the situation when you engage the services of a promoter to provide the rock band. You may also employ other non-music workers such as sales people and catering staff.

What to do

General considerations

As an operator, employer or person in control of a workplace you would normally be expected to:

1. Ensure your safety and health information, policy and/or procedures include preventing risks from excessive music noise.
2. Communicate this information to employees, music promoters and/or performers.
3. Be familiar with the requirements of the Occupational Safety and Health Act and Regulations.
4. Be familiar with the contents of the Principal Code and this Code and follow the strategies in it.
5. Communicate these matters to the performers you engage and to your employees.

Noise assessment

Identify situations and areas of the venue where noise is likely to be above the exposure standard. As a rule of thumb, if a person needs to speak in a raised voice to be understood 1 metre away, the A-weighted sound level is likely to be above 85 decibels. Arrange for a noise assessment to be carried out by a competent person during a performance typical of louder performances in the venue.

Document the 'room loss' between the 'reference position' and staff locations (see Appendix C for guidance on carrying out noise measurements).

Use the following approach to account for variations in 'music levels' from one performance to the next:

1. Determine the 'music level' prior to each new performance by requesting this information from the entertainment provider. If this is not available arrange to have the 'music level' monitored.
2. Estimate the 'received noise' at each staff location by subtracting the 'room loss' from the 'music level'.
3. Estimate the '8 hour exposure' at each staff location by adjusting the 'received noise' according to the duration of the performance, using Table 41.3 (see Appendix B).

Noise reduction

If '8 hour exposures' exceed the exposure standard for noise:

1. Consider reducing the noise at source, i.e. reduction of the 'music level'. This may be approached through a process of consultation with the entertainment providers, relevant safety and health representatives and committees, if any, at the workplace. Also find out if there are any restrictions on the 'music level' needed to comply with the Environmental Protection (Noise) Regulations 1997. When a maximum 'music level' is decided on, this can be included in contractual agreements with the entertainment providers.
2. Consider reduction of noise through increasing the 'room loss'. This may be achieved through a range of architectural and other means, including:
   (a) Moving the stage and/or loudspeakers to increase the distance between the performers and staff.
   (b) Reorienting the stage and/or loudspeakers to direct less sound towards staff locations and installing sound limiters where appropriate.
   (c) Where there are multiple speaker arrays, such as discos, concerts halls or music stores, reducing the sound levels of those speakers nearest staff locations.
   (d) Increasing acoustic absorption in the room, through the addition of an acoustic ceiling, acoustic wall linings or carpet. This needs to be done with consideration for the overall environment desired, e.g. wood panelling has little absorption, so to increase-absorption-while retaining the visual effect the panels could be spaced slightly apart from each other and out from the wall with acoustic material behind.
   (e) Using local screening adjacent to the bar, kitchen or door staff locations may be feasible in some situations.
   (f) Applying sound absorption materials applied in some bar, servery or door staff areas. There are special requirements for washable, impervious facing materials in any food processing or serving area.
   (g) Ensuring adequate acoustic separation of the box office, kitchen, staff rest rooms, etc. from the entertainment area by means of doors with appropriate acoustic performance.
3. Seek professional assistance from architects, acoustic consultants and sound engineers in evaluation of the most cost-effective options.
4. In any major renovations or alterations to the layout of the venue, brief the architect or designer to consider ways of maximising 'room loss'.
5. Reduce '8 hour exposure' by reducing the amount of time staff are exposed to the noise. This may involve an agreed arrangement for rotating staff between noisy and quiet areas, if a workable system can be achieved.

6. Identify areas where peak noise levels exceed 140 decibels and instruct staff to avoid these as far as practicable.

Personal hearing protectors/education

If it is not practicable to avoid exposing persons at the workplace to noise above the exposure standard, provide appropriate personal hearing protectors to all affected people. This includes other employers employees visiting the venue as part of their work, e.g. catering staff, performers. Consideration should also be given to whether personal hearing protectors should be available for members of the audience if they request them.

Note: Providing personal hearing protectors does not remove the duty to reduce noise as far as practicable.

For each new performance with a different 'music level', advise all affected staff of their likely '8 hour exposure', and ensure the personal hearing protectors provided are adequate. Use appropriately placed signs to remind staff that the venue is a 'hearing protection area' and the audience that hearing protectors are available. Arrange training sessions on the risk of Noise-Induced Hearing Loss for regular and seasonal employees. Topics to be included in training are given in the Principal Code.

Have relevant information on the risks of noise to hearing available for new, visiting or casual employees. Arrange hearing tests as required by legislation administered by WorkCover WA.

Strategy No. 3

For entertainment providers

Explanation

You will most likely be a self-employed person who:
1. Is engaged by an operator or person in control of an entertainment venue to supply or provide music entertainment.
2. Promotes musical performances and engages the services of performers and/or technical staff.
3. Leads a band or orchestra or other musical performing group and employs the musicians who perform in that group.

Note: Performers who are employees should refer to Strategy No. 6.

What to do

General

As a promoter, or performer in this situation you should:
1. Ensure your safety and health information; policy and/or procedures include preventing risks from excessive music noise.
2. Be familiar with the requirements of the Occupational Safety and Health Act and Regulations and the Principal Code.
3. Be familiar with the contents of this code and follow the strategies in it.
4. Communicate these matters to the performers and technical staff you engage and your employees.

Music level

Identify if your performance is likely to produce exposures above the standard. As a rule of thumb, if a person needs to speak in a raised voice to be understood by another 1 metre away, the A-weighted sound level is likely to be above 85 decibels. If so, find out the 'music level' of a typical performance under typical conditions. You may wish to combine your efforts with a venue operator to have the 'music level' and 'room loss' measured on one typical occasion.

Re-measure the 'music level' when there is a significant change in the musical instrument line-up, personnel, amplifier/speaker system or musical performance.

Find out from the operator or person in control of the entertainment venue whether there is an agreed maximum 'music level' or assessment of excessive noise in the venue in which you are to perform, or the performers you have engaged are to perform. Obtain the assessment and see if it provides any useful information which will help you plan your musical performance.

Inform the venue operator of your 'music level' prior to the performance. Ensure that means are available to monitor the 'music level' during the performance. You could use direct measurement of the noise levels or refer to markings on a 'master volume' scale (see Appendix C). Adhere to the stated or agreed 'music level'.

No person should be allowed to enter an area with music noise above a peak level of 140 decibels to minimise the risk of instantaneous permanent hearing loss.

See Appendix C 'Measuring music noise' for suggestions about carrying out noise measurements.

Duty to employees

If you employ workers such as sound mixer/engineer, lighting/road crews or musicians, you will need to consider the following to prevent excessive noise damaging their hearing:

1. Increase the distance between non-performing employees and the stage area or loudspeakers.
2. Reduce the 'music level' within the workable range.
3. Reduce the 'fold back' levels on the stage to lower (but still workable) levels.
4. Reduce sound output from individual instruments, e.g. damping drums, using smaller amplifiers to reduce sound levels on stage.

Assess '8 hour exposures' and peak noise levels of performers and technical staff during a typical performance (see Appendix B—Table 41.3). Inform employees of their likely noise exposure.

Provide appropriate hearing protectors where exposures exceed the exposure standard for noise. Musicians' earplugs are now available which give more uniform reduction across the frequency spectrum.

Provide training sessions on noise as described in the Principal Code. Provide specialised training in noise/noise control for sound mixers (both front of house and backstage). Arrange for hearing tests as required by legislation administered by WorkCover WA.

Strategy No. 4

For employers of service staff visiting an entertainment venue

Explanation

Most likely you will be an employer of catering staff, security staff, promotions and media personnel, police, or first-aiders.

What to do

General

Ensure your safety and health information, policy and/or procedures include preventing risks from excessive music noise. Be familiar with the requirements of the Occupational Safety and Health Act and Regulations and the Principal Code. Be familiar with the contents of this code and follow the strategies in it. Communicate these matters to your employees.

For each venue

Consult with the venue operator. Find out if your employees are likely to be exposed above the exposure standard for noise. Instruct staff in administrative measures to reduce noise exposure such as avoiding noisy areas, rotating staff between noisy and quiet positions. Provide staff with appropriate personal hearing protectors as advised by the venue operator. Provide training sessions on noise as described in the Principal Code. Arrange hearing tests as required by legislation administered by WorkCover WA.

Strategy No. 5

For suppliers and installers of sound equipment

Explanation

Most likely you will be a supplier and/or installer of sound systems for discos, concert halls or bands, either hiring or selling the equipment. You may also operate the system at the venue. You may be self-employed and/or employ other people to do this work.

What to do

General

Be familiar with your responsibilities under section 23 of the Occupational Safety and Health Act. Be familiar with the contents of this code and follow its strategies.

Information

Provide information to customers at point of supply on potential noise hazards including:

1. Operation conditions likely to result in a noise hazard.
2. The need to monitor 'music level'.
3. Any areas where the peak noise level is likely to exceed 140 decibels.

This could take the form of:

1. Verbal advice to the receiver of the equipment.
2. Written information accompanying the equipment.
3. A hazard warning sign affixed to a prominent part of the system, e.g. the mixing desk.

Installation

Arrange the placement and orientation of the loudspeakers to minimise as far as practicable the sound directed to employee locations. Arrange the placement of loudspeakers to enable restriction of access where peak noise levels are likely to exceed 140 decibels.

Operation

Find out if there is an agreed maximum 'music level' for the venue and don't exceed it. Arrange for the 'music level' to be monitored (see Appendix C) and advise venue operator. Arrange training for employees in monitoring and methods of achieving specified levels.

Duties to employees

Provide employees with appropriate personal hearing protectors where exposures exceed the exposure standard for noise. Provide training sessions on noise as described in the Principal Code. Arrange hearing tests as required by legislation administered by WorkCover WA.

Strategy No. 6

For employees in entertainment venues (either performers or non-performers)

Explanation

Most likely you will be:
1. An employee of a catering company providing food and beverage services during a musical performance such as a rock concert.
2. An employee in an entertainment venue, working as a waiter, cashier, security officer, first-aider, police officer, chef, kitchen hand or promotions and media personnel.
3. A performer or sound engineer who is employed by someone else, such as a promoter, band leader or venue operator.
4. An employee who works in a retail store which sells recorded music.

What to do

Find out if your noise exposure is likely to be excessive. Follow your employer's or the venue operator's instructions on control strategies including:
1. Instructions relating to achieving any agreed 'music level'.
2. Abiding by any agreed arrangements for job rotation or restriction of access to noisy areas.

Do not wilfully misuse or damage any equipment provided to reduce noise in the venue. Use the personal hearing protectors provided in the manner instructed. Report any new hazardous noise situations or any hearing loss or tinnitus (ringing in ears) resulting from exposure to noise in the venue to your employer. Request for annual hearing tests.

APPENDIX A

Role in Terms of the Occupational Safety and Health Act 1984

Employers

An employer is defined by the Act as a person by whom an employee is employed under a contract of employment, apprenticeship or industrial training agreement. An employer in the entertainment industry may be:
1. The licensee of licensed premises.
2. The theatre management or drama company in a venue where a musical production is being staged.
3. A promoter.
4. The management of a venue in which concerts are given.
5. A caterer employing food service staff at a venue.
6. Any other person who employs someone under a contract of employment, apprenticeship or industrial training agreement.

Various other contractual relationships may occur within the industry, e.g. the band may employ a 'roadie' or a sound engineer, or the suppliers of sound systems may employ their own road crews. These employers owe a similar duty of care to their own employees.

Under section 19(4) of the Act, where a principal employer engages a contractor, he/she is deemed to be the employer of both the contractor and the contractor's employees in relation to the aspects of their work over which he/she has management or control. The duties of employers to employees are outlined in section 19 of the Act.

Employees

An employee is a person who works under a contract of employment, apprenticeship or industrial training agreement. In the music entertainment industry a employee may be:

1. An entertainer, such as a singer in a musical, a musician, a disc jockey or a sound engineer who may have some degree of control over the sound level produced.
2. Technical and ancillary assistants to those producing the entertainment, such as a 'roadie', a lighting person or a stage hand.
3. An employee who performs duties in conjunction with, but not related to, the production of the music, such as a bar/food service employee, glass collector or door person.
4. Supervisor/manager, e.g. bar manager, stage manager.
5. Any other person in the workplace who is employed under a contract of employment, apprenticeship or industrial training agreement.

Self-employed person

Many of the categories of persons listed under employers and employees above may at times be self-employed persons. Section 21 of the Act requires self-employed persons to take reasonable care to protect their own safety and health and to ensure that the safety and health of others is not adversely affected as a result of their work or the work of their employees. A self-employed person in the entertainment industry may be:

1. An entertainment provider whose work may result in a noise hazard to themselves and/or others at the workplace.
2. A technical or service contractor whose work does not involve the production of music.
3. A promoter.

Persons who have control of workplaces

A person who has control of a workplace may be:

1. The owner of a venue.
2. A person who is in a position to make arrangements for architectural and administrative changes to reduce noise exposure.

Section 22 of the Act sets out the duties these people have to all people at the venue.

Designers, manufacturers, importers, suppliers, erectors and installers of plant and designers and constructors of buildings

In the music entertainment industry, these may be:

1. Designers and builders of venues.
2. Manufacturers of musical instruments and sound amplification equipment.
3. Persons who provide or hire public address or other sound systems or supply and/or install permanent sound systems in entertainment venues.
4. Others, such as sound engineers, who may be regarded as suppliers in their capacity as technical support to the entertainment providers.

Section 23 of the Act sets out the duties of these people.

APPENDIX B

'Exposure standard for noise' means a noise level above which, as far as practicable, people at a workplace must not be exposed. It is defined in Regulation 3.45 as follows:

1. An $L_{AeQ,8h}$ of 85 dB(A).
2. An $L_{C,peak}$ of 140 dB(C).

Measured at the position of the person's ear without taking into account any protection which may be provided to the person by personal hearing protectors.

'Music level' means the average noise level ($L_{Aeq,T}$) of a representative portion of a typical performance, measured at a nominated 'reference position' in a venue.

'Practicable' means reasonably practicable having regard to:

1. The severity of any injury or harm to health that may occur.
2. The degree of risk of that injury or harm occurring.
3. How much is known about the hazard and ways of eliminating, reducing, or controlling it.
4. The availability, suitability and cost of the safeguards.

'Received noise' means the average noise level ($L_{Aeq,T}$) measured at the employee's ear during a representative portion of a performance. 'Reference position' means a nominated measurement position within the venue sufficiently close to the stage area that the sound level is dominated by the music.

Note: For venues which have live bands performing regularly the reference position should be 5 metres from the front of the main loudspeakers, at least 1.8 m above floor level and centrally located in front of the performers.

For other venues the reference position may be any nominated monitoring point where the music dominates the sound levels.

'Room loss' means the average drop in sound levels from the reference position back to the locations occupied by employees, measured during a typical performance with a typical crowd in attendance.

'8 hour exposure' means 'received noise' when averaged over an 8 hour period. This can be estimated by adjusting the 'received noise' according to the duration of the performance using Table 41.3.

Table 41.3. Estimation of '8 hour exposure':

Duration of performance	Decibels to be subtracted from 'received noise'
8 hrs	0
6 hrs	1
5 hrs	2
4 hrs	3
3 hrs	4
2½ hrs	5
2 hrs	6
1½ hrs	7
1 hr	9

Note: The assessment of 'received noise', '8 hour exposure' and peak noise level should not take into account any protection which may be provided by personal hearing protectors.

Measurement procedures are detailed in Appendix C of this code. The measurement and proper use of these parameters enable all parties to develop strategies for reducing exposure levels. The 'music level' defines the overall sound level of the performance.

A rock band will tend to produce a fairly constant 'music level' from performance to performance, while an orchestra or theatre company will produce different 'music levels' from performance to performance, depending on the musical work.

The 'room loss' is a characteristic of the room acoustics and room size, and is independent of 'music level'. It defines the extent to which employees are separated from the sound source and can be increased by architectural means in many cases. 'Received noise' can be estimated for any performance where the 'music level' is known, by subtracting the 'room loss' from the 'music level'. This 'received noise' value can then be converted to an '8 hour exposure' by means of a simple table, by knowing the duration of the performance (Table 41.3).

The '8 hour exposure' can be compared with the exposure standard for noise. Figure 41.1 shows example of application of terms to a hotel with a live band.

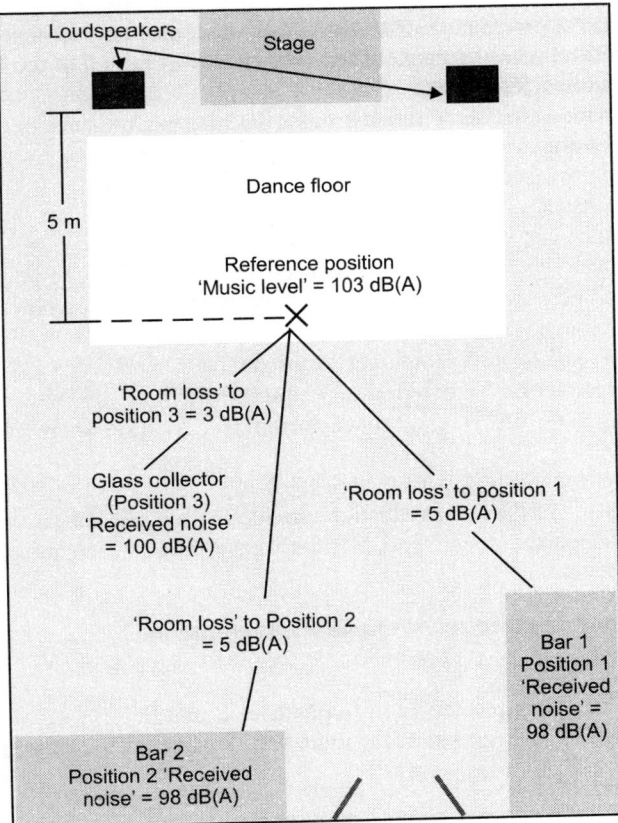

Fig. 41.1. Example of application of terms to a hotel with a live band.

APPENDIX C

Measuring Music Noise

Collecting music noise data

This section outlines the procedures to be used when conducting noise level measurements in an entertainment venue for the purpose of defining 'music level', 'room loss', 'received noise', '8 hour exposure' and levels exceeding the exposure standard, where the data is to be used to form the basis of 'noise assessment'. This data should be collected by a competent person. A list of competent persons is available in the Noise Assessors section of the Directory of Noise and Vibration Control Services, as are details on courses to give the basics of the required competency. (Available via the Internet Service on Safetyline, www.safetyline.wa.gov.au/sub30.htm).

The complexity of noise assessment may vary according to the type of entertainment venue, the number of persons at risk from noise exposure and the information on noise levels already available at the venue. If there is no prior information available, an assessment should establish if excessive noise levels are produced at the venue.

In collecting data for the preparation of a noise assessment the competent person should:

1. Conduct sound level measurements in accordance with section 4 of the Principal Code and the relevant International Standards.
2. Arrange to conduct sound level measurements during a performance or rehearsal for which the 'music level' is within 5 dB of the highest 'music level' likely to be generated within the venue, with an average crowd in attendance, under typical conditions.
3. Establish a 'reference position' within the venue from which point the 'music level' of any future performance can be measured for comparative purposes.
4. Measure the 'music level' of the performance over a representative portion of the duration of the performance and simultaneously measure 'received noise' at employee locations where this is likely to be above 85 dB(A).
5. Calculate 'room loss' for each employee location where 'received noise' is above 85 dB(A) by arithmetically subtracting 'received noise' from 'music level'.
6. Calculate '8 hour exposure' ($L_{Aeq,8h}$) for each employee location where 'received noise' is above 85 dB(A).
7. Delineate all areas where $L_{C,peak}$ exceeds 140 dB(C).
8. Assess the adequacy of the noise reduction provided by any personal hearing protectors already in use and if inadequate, recommend, alternative protectors which are adequate.
9. Document the results.

Practical considerations in conducting noise measurements

Choosing a 'reference position'

Refer to the description and examples given in Appendices B and E.

Note: A reference position selected for occupational noise purposes may also be used effectively for environmental noise control purposes.

Measuring 'music level' and 'received noise' simultaneously

The suggested procedure is to use five minute $L_{Aeq,T}$ measurements. One measurement system would be set up at the reference position, and set to print out continuous five minute $L_{Aeq,T}$ levels over a representative portion of the performance. A second measurement system would be used to conduct simultaneous five minute $L_{Aeq,T}$ sample measurements at the affected locations. Provided a representative number of five minute $L_{Aeq,T}$ samples (say three) can be taken at each affected location, and provided the five minute levels do not vary markedly at the continuous measurement station, the sampled levels can be used to estimate $L_{Aeq,T}$ levels over the performance duration.

Calculation of 'room loss'

When calculating 'room loss', use the 'music level' and 'received noise' levels determined for the full duration of the performance. The 'room loss' is the value obtained by arithmetic subtraction of 'received noise' level from the 'music level' for each affected employee location.

Measurement for moving employees

In the case of glass collectors and door staff in licensed premises, for example, the measurement of 'received noise may be carried out using a personal noise dosemeter. The start and stop times should be set to coincide with the start and stop times of the five minute samples at the continuous sampling position, unless the dosemeter is to be worn for at least a representative portion of the performance duration.

Crowd noise effects

In some locations at some venues, it may be found that crowd noise is equivalent to or predominates over music in overall sound level. In this case, while the 'room loss' should still be calculated as above, the influence of crowd noise should be noted, If this is considered to be an unusual circumstance, it may be more appropriate to repeat the measurement under typical conditions. If however, these conditions are typical, the use of a slightly higher 'music level' on another occasion, combined with the same crowd noise would provide a slightly conservative result in terms of estimation of '8 hour exposure'.

In the unlikely case where crowd noise dominates completely, a 5 dB increase in 'music level' would not cause any significant increase in 'noise exposure', in which case the $L_{Aeq,8h}$ results define the employees' noise exposure for all likely situations.

Other action

A sketch should be made of the venue showing stage position, loudspeaker positions (and orientation) and employee locations. Details of architectural finishes which may influence 'room loss' should be noted. A detailed description of the location of the 'reference position' should be provided. Any relevant settings of amplification equipment of significance in the setting of the 'music level' during the tested performance should be noted.

The name of the performing group/performance should be noted with a brief description of musical instruments used.

Conducting noise measurements for entertainment providers

When determining 'music level' for entertainment providers who move from venue to venue, such as rock bands and mobile DJs, the following procedure is recommended:

1. Arrange to conduct sound level measurements during a performance at a venue for which the 'music level' is within 5 dB of the highest likely to be produced by the performer, with an average crowd in attendance under typical conditions.
2. Where practicable, the person conducting the measurements should be a competent person.

3. The measurement procedures and instrumentation should be in accordance with section 4 of the Principal Code and the relevant International Standards.
4. The measured quantity is $L_{Aeq,T}$ over a representative portion T of the performance duration:
   (a) At the 'reference position', e.g. 5 metres from the front of the main loudspeakers, and at least 1.8 metres above floor level.
   (b) The performer's position.
5. If it is desired to relate the 'music level' to the sound level at the console or mixing desk, then simultaneous $L_{Aeq,T}$ measurements should be made at the 'reference position' and at the desk/console over a period of about 15 minutes of varied typical program.
6. Similarly, if relating 'music level' to a 'Master volume' setting on the desk or console, the 'music level' should be measured over several periods of about five minutes of typical program, repeated with the 'Master Volume' setting being changed between five minute segments and the setting noted. The steps in 'music level' between settings should be no greater than 3 dB(A).
7. Delineate all areas where $L_{C,peak}$ exceeds 140 dB(C).
8. The results should be presented to the entertainment providers on the suggested proforma of this code.

Measurement of 'room loss' in occasionally used venues

Where it is desired to determine the 'room loss' for venues which are only occasionally used for music likely to be above the exposure standard, e.g. in halls administered by local government, the following procedure could be used:

1. Set up a typical sound system in the empty hall with two large 'disco' loudspeakers on the stage.
2. Run through a typical program in the form of a series of varied records/compact discs for a period of about 15 minutes.
3. Measure $L_{Aeq,15\,min}$ at the 'reference position' 5 metres from the stage and simultaneously at the door, kitchen or any areas likely to be occupied by employees during a function. *Note*: These may be employees of a catering company, police, or security personnel.
4. Calculate 'room loss' by subtracting $L_{Aeq,5\,min}$ 'at the likely employee locations from the $L_{Aeq,15\,min}$ level at the 'reference position'.
5. Where there is an environmental noise constraint it may also be appropriate to conduct simultaneous external measurements to ascertain the appropriate internal 'music level'.
6. Provide information to hirers of the venue, (possibly in the form of an information sheet or notice in the hall), indicating the 'room loss', any limits on 'music level', or possible need for noise reduction strategies to be put in place during the performance.

APPENDIX D

Music Noise Solutions

Example 1: Hotel with live band

A hotel owner/manager engages live bands on four nights a week and arranges for a competent person to conduct a series of sound level tests on a typical night, in consultation with the relevant safety and health representatives and committee.

A 'music level' of 103 dB(A) is measured at the 'reference position' at the edge of the dance floor 5 metres from the loudspeakers. At the same time 'received noise' levels of 98 dB(A) at the bar, 97 dB(A) at the door and 100 dB(A) at the glass collector's ear are measured. The 'room loss' values are calculated: $103 - 98 = 5$ dB(A) for the bar, $103 - 97 = 6$ dB(A) for the door and $103 - 100 = 3$ dB(A) for the glass collector.

The normal performance lasts two hours (consisting of three forty-minute sets with quieter breaks in between). The adjustment to calculate '8 hour exposure' (from Table 41.3) is –6 dB(A) for a two-hour performance.

The '8 hour exposures' on the night of tests were therefore $98 - 6 = 92$ dB(A) for bar staff, $97 - 6 = 91$ dB(A) for the door staff and $100 - 6 = 94$ dB(A) for the glass collector.

The owner/manager now knows that for a 'music level' of 103 dB(A), the staff are exposed above the exposure standard.

Following consultation with the safety and health representatives and committee, the following architectural changes are initiated (Fig. 41.2):

1. Changing the location of the stage.
2. Putting acoustic absorption material on the ceiling and upper part of the walls opposite the stage and on the wall behind the bar.
3. Placing an acoustic screen at the end of the bar nearest the stage.
4. Replacing the doors to the kitchen and office with acoustic doors.

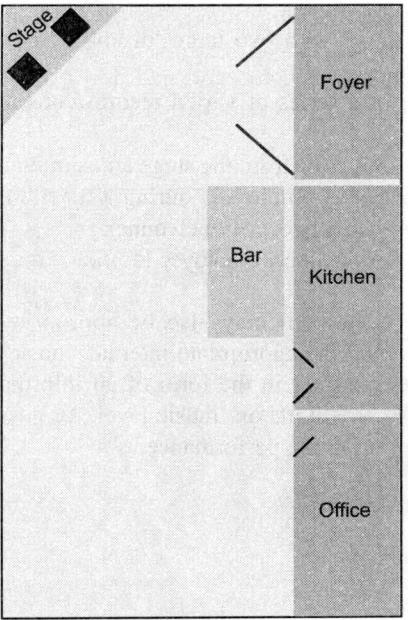 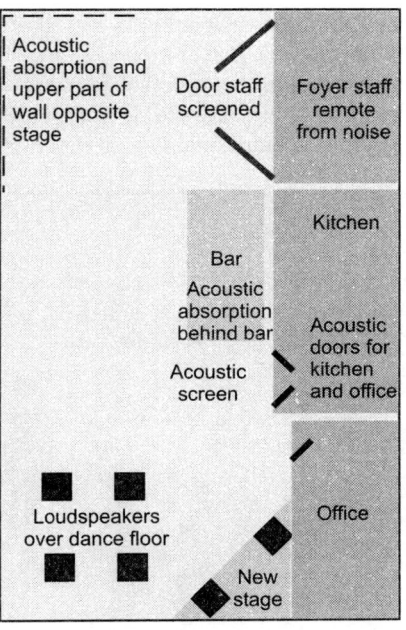

Before   After

Fig. 41.2. Possible noise control measures for increasing 'room loss' in a hotel/club with live music.

A further series of tests show the 'room loss' values are now 8 dB(A) for the bar, 10 dB(A) for the door and 4 dB(A) for the glass collector. The owner/manager introduces an agreed job rotation scheme for the glass collector to minimise his/her exposure time, thus reducing his/her '8 hour exposure' by a further 2 dB(A). A joint policy decision is made not to employ bands with a 'music level' above 103 dB(A) as a workable level for this venue.

Each new band informs the owner/manager of their expected 'music level' prior to starting the performance. The owner/manager then knows what the '8 hour exposures' of staff will be. For example, a band with a 'music level' of 100 dB(A) will cause an '8 hour exposure' of 86 dB(A) for bar staff, 84 dB(A) for the door staff, and 88 dB(A) for the glass collector.

As the glass collector, bar and door staff still receive noise above the exposure standard on nights when the louder bands play, the owner/manager needs to develop further long-term plans for noise reduction, including a review of the policy on 'music levels'. As an interim measure, the owner/manager provides suitable personal hearing protectors and an education program on the effects of exposure to excessive noise levels.

Example 2: Discotheque/Cabaret venue

A discotheque owner presents recorded music via a different disc jockey (DJ) every night. The sound system is a fixed installation with the main loudspeakers near the dance floor and others throughout the room. The disco owner, in consultation with relevant safety and health representatives and committee, selects a CD of music typically played and defines a 'reference position' at the DJ's desk. A competent person is engaged to measure the 'room loss' from that point back to the positions occupied by the bar staff, glass collector and door staff. With a reasonably practicable 'music level' of 98 dB(A) at the 'reference position', the 'received noise' at the employee locations ranges between 90 and 95 dB(A) [room loss of 3 to 8 dB(A)].

For an exposure of 5 hours, the adjustment to calculate '8 hour exposure' is −2 dB. The '8 hour exposures' therefore range between 88 and 93 dB(A). The owner now knows which staff are exposed above the exposure standard on a typical night (Fig. 41.3).

Fig. 41.3. Sound ceiling in a discotheque.

In consultation with safety and health representatives and committee, the owner plans to introduce a 'sound ceiling' above the dance floor when the discotheque is renovated in six months' time. This is a system where a ceiling with built-in speakers is suspended above the dance floor, resulting in a high sound level on the dance floor, whilst the sound propagated sidewards into the rest of the room is lowered around 10 dB(A).

These changes will increase 'room loss' with a view to reducing all staff exposures below the exposure standard:

In the interim the owner provides personal hearing protectors and an education program for affected staff. In order to maintain the 'music level' at or below 98 dB(A), she installs a sound level meter at the DJ's desk and discusses with the DJ the procedures required to keep the sound level below this level.

Example 3: Theatre for stage productions

A large theatre is used as a venue for staging performances ranging from opera to rock musicals and modern dance programs. For rock musicals the band may perform in the orchestra pit or a special platform above the stage, using a public address system brought in for the performance. Modern dance programs usually use pre-recorded music played back through the in-house PA system. For musicals, operas or ballets, the orchestra may perform from the pit, and on some occasions vocalists may be amplified slightly through the in-house PA system. The theatre management employ door staff, ushers and technical crew, some of whom work backstage.

The theatre management takes a policy decision, in consultation with the relevant safety and health representatives and committee, to prepare a noise assessment for each new show. A 'reference position' is selected at the pit conductor's position about 3 metres from the front of the stage. The 'music level' for the show is measured at this position, while sound levels are simultaneously monitored at employee locations.

The results of the tests for the first few shows indicated that it is only during rock musicals, modern dance productions and the louder operas that there is any likelihood of the exposure standard being exceeded. It is found that the 'room loss' is about the same during these types of productions. For future productions, the 'music level' is monitored during a dress rehearsal from the 'reference position'. The 'received noise' of employees is calculated from the measured 'music level' by subtracting the 'room loss' values and the '8 hour exposure' is determined by adjusting for the duration of the performance. Staffing rosters are then devised to minimise exposure and personal hearing protectors and education are provided for those still exposed above the exposure standard.

Example 4: Music retail store

An operator has two retail music stores catering to customers with differing tastes in music. Store 1 specialises in heavy metal music while Store 2 retails classical music and jazz. One of the marketing strategies is to entertain customers by playing current selections and requests via an in-house PA system, with speakers throughout the store.

The operator, in consultation with the safety and health representatives, defines a 'reference position' in the middle of each store and selects CDs of typical music' played. A competent person is engaged to conduct measurement of 'music level' and 'room loss' from those points back to the staff positions at the sales desks.

'Music levels' of 94 dB(A) and 75 dB(A) are found in Stores 1 and 2 respectively. 'Room losses' of 3 dB(A) are found in both cases. As sales staff work 8 hour shifts their '8 hour exposures' are above the exposure standard for noise in Store 1, but well below it in Store 2.

In consultation with the safety and health representatives and committees, the operator makes changes to the sound system in Store 1, enabling the volume of the speakers to be controlled independently. Those adjacent to sales desks are fixed at a low level, whilst the others are adjusted to give a 'music level' of 90 dB(A) at the central 'reference position', giving a 'received noise' of 87 dB(A) at the sales desk. Sales staff are then rostered to work 4 hours each day in each store, resulting in '8 hour exposures' of 84 dB(A).

SECTION IX

Noise Mapping

Chapter 42

Noise Mapping of Industrial Sources

INTRODUCTION

Noise is often one of the environmental variables where it seems more difficult to guarantee full compliance with legal limits in complex industrial situations and/or in other multi-source environments. This chapter describes the application of noise mapping techniques to industrial sources, such as factories, power plants, wind farms, showing the potential of its use, both at the design stage as well as for noise reduction plans of existing installations. Starting from the collection of noise sources data, reflecting and diffracting objects, cartography of the area around the site and identification of sensitive receivers, an acoustic model can be built and run according to ISO 9613 standard. This then allows one to produce noise source ranking, according to the individual contributions to the total noise, as well as comparing different noise control scenarios, thus enabling one to optimise the investment in noise control measures. Practical examples in Portugal and Spain are presented and discussed.

Noise evaluations around industrial premises have been for a long time carried out by means of short-term noise measurements, on a limited number of receivers, and by performing some simplified qualitative analysis, eventually with some simple calculations. However, experience has demonstrated that often, and above all in installations of large dimension and complexity, this approach does not enable one to get feasible results neither a clear vision of the real noise impact. Moreover, it does not, in general, produce enough information for decision-making with respect to what noise control measures should be implemented—has it does not enable one to identify and rank the noise sources—and it does not enable one to predict the noise impact of a new factory or of this or that noise control action on an existing installation. Other non-negligible aspect consists on the difficulty that 'traditional' noise assessments have in presenting results which are easily apprehended by non-specialists. This makes it difficult an effective communication of the results attained by noise control programs to the potentially interested publics, such has the neighbouring communities, governmental agencies, environmental inspectors and auditors, shareholders, insurance companies, municipalities, environmental NGOs. Thus, all the effort and investment put by the organisation on reducing its noise emissions is often not recognised by those interested parts, loosing the opportunity to enhance the image of the company.

The development of computer modelling techniques that simulate the acoustic emission and propagation, enables one, in our days, to model with good accuracy and reasonably fast, the most complex scenarios of noise generation and propagation. The results are normally presented in the form of coloured noise maps, each colour corresponding to a given interval of noise levels, typically in steps of 5 dB. Above all, such a model, if correctly developed, enables one to get a true noise monitoring and

455

management system, from which it is possible to rank noise sources, extract the individual contributions of each noise source to any given receiver, update the information whenever changes are introduced in the factory, and establish detailed noise control action plans and predicting its results. Difference maps can also be easily extracted from such a model, in order to depict before vs. after maps, total noise vs. background noise maps or scenario 1–scenario 2 maps, etc.

The need for an industrial noise map can come up to on a wide range of situations: environmental impact assessment for the installation of a new factory or for changing an existing one, environmental license such as under IPPC regulations, complaints from neighbours, certification under ISO 14000 or EMAS, where a full demonstration of compliance with noise regulations is required. In certain cases, such as industrial parks, the aim can be to control noise build up during the successive installation of new industries in the park, which can be done by establishing noise quota for each lot in the park.

Whichever the reason, the number one aim on producing a noise map — and above all the acoustic model on which it is based — should be to get a useful tool which will enable one to correctly evaluate a noisy situation, be it already existent or planned for the future, and study what the best solutions are to comply with given noise limits around the plant. By ranking the sources and predicting the practical outcome of any scenario, one can effectively optimise the investment in noise control actions.

GENERAL METHODOLOGY FOR NOISE MAPPING OF INDUSTRIAL PLANTS

The general methodology for noise mapping of industrial plants can be resumed as illustrated in Fig. 42.1.

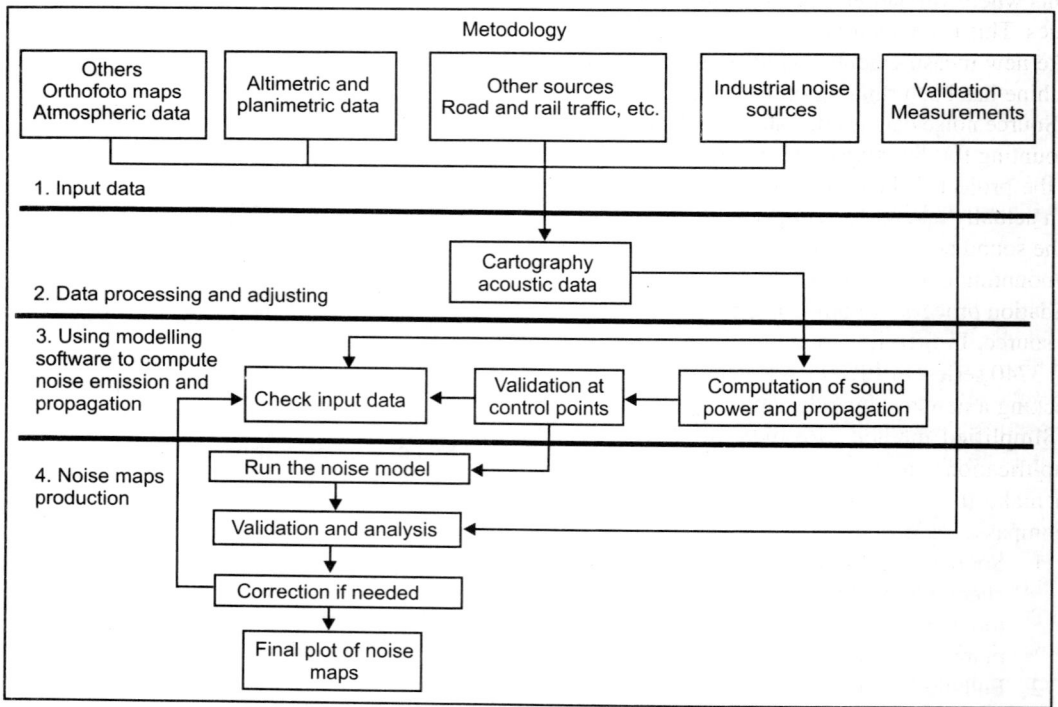

Fig. 42.1. General methodology for noise mapping of industrial plants.

Obtaining correct input data is, as in any model, the most critical part of the job: the 'garbage in, garbage out' expression fully applies here. Moreover, in industrial noise mapping projects, this is also the hardest part of the job, as very seldom one has access to accurate digital 3D drawings of the plant or, even less, to adequate acoustical data such as sound power level or directivity of the noise sources. Therefore, in most cases of existing plants, one must get all these crucial data the 'hard way', which means:

1. Actually spending several weeks on site in order to understand, as deeply as possible, how the industry actually works.
2. Getting close to all sound sources in order to measure its noise appropriately.
3. Draw by scratch all industrial buildings to insert them into de model.
4. Decide how to model each noise source.
5. Accurately position them in the drawings in order to be able to insert them at their right places in the model.

It is worth noting that the sources can easily add up to more than one hundred and, sometimes, can come close to a thousand. Also, for interior sources, one has often to estimate, or actually measure, the transmission loss of the building elements involved in the in-out propagation process.

After collecting all required data, this is introduced in the model and this usually means many adjustments, both for the geometrical data — by checking all information, taking advantage of 3D visualisation capabilities of modern noise modelling software, such as CadnaA v3.7 — and for the acoustical data — generally by running calculations at a number of control points corresponding to real points where validation measurements have been taken, and comparing the measured versus calculated values. This is an iterative process which normally means going back into the field to check out doubts, make new measurements and have meetings with engineers from the factory to verify that this or that machine has been running on its normal condition, etc.

Source noise data is normally introduced in the model in the form of octave band sound power level, accounting for directivity and for per cent working time for each reference period of the day, if relevant for the project. When data comes from actual measurements on site, either sound power levels have been actually taken (e.g. using sound intensity measurements) or, as is normally the case, an estimation of the sound power level is made from pressure measurements close to each source, taking into account its mounting conditions, the presence of reflecting surfaces or other machines nearby and performing a validation process by comparing measured versus calculated noise levels at positions further apart from the source. In fact, full sound power determination for each source according to standards such as the ISO 3740 series, or ISO 9614 is generally out of the question, except in some special situations, such as checking a new machine during set up process or start up of a new factory, and the like.

Simplified methods are therefore used, and care must be taken to ensure one makes the right simplifications and that the validation procedure is extensive enough so that your successive iterations can make the model converge into an accurate acoustic model. This validation process normally encompasses two steps:

1. Source validation: where single sources or small groups of sources are validated by means of checking measured against calculated values on a set of receivers at intermediate distances, not too close but not too far away from the sources which are being validated, normally inside the plant perimeter, but in the acoustic far field of each individual source.
2. Full model validation: where the entire plant, with all its sources, is validated by means of checking measured against calculated values on a set of receivers far away from the sources, typically outside the plant perimeter and sometimes close to sensitive receivers, such as neighbouring dwellings.

When it is possible (or it just happens) to stop some machines or groups of machines, one can take advantage of that to facilitate the process of getting and validating noise data.

Of course, when dealing with a project for a new factory, things work out differently, and one has to try hard to find acoustical data from suppliers of all types of machines or, when unavailable (as happens too much often), rely on available literature, databases of similar equipments, calculations based on machine parameters or, sometimes, try to find similar equipments running and just going there and measure it.

In any case, an important decision to be taken is on how to model each noise source. There are three basic types of sources which one can introduce in the model:

1. Point sources — adequate for small sources, such as fans, or larger sources, with well balanced dimensions, sitting away from relevant receivers.
2. Line sources — adequate for linear shaped sources, such has piping, conveyors, as well as moving sources paths.
3. Area sources — can be vertical, such as openings in a building, noise radiating façades, or very large machines, or horizontal, such as a roof, or a number of fixed or moving sources distributed on the ground.

The way calculations are made by modelling software such as CadnaA, can be set up to comply with different methods and standards. The most common for industrial sources is the method of ISO 613, which details we will not go into here. One must also configure correctly a number of software parameters, related with the calculation configuration, such as the maximum search radius, minimum distance source to receiver, maximum reflection order, minimum distance receiver-reflector, reference time, with the grid calculation (for noise maps), such as receiver spacing and receiver height, or with each source in particular, such as single band or spectrum, directivity, geometry, and type of noise levels input to the source (e.g. L_p measured at a certain distance outdoors, L_i impinging on a façade from the interior side, etc.).

The hard work involved on building and validating the acoustical model of a large industrial plant is highly compensated when one, finally, gets the final model to produce results which make sense and match the reality. It is then that one can start taking advantage of the model for practical applications, such as source ranking, calculation of individual source contributions to the total noise at any given receiving point, evaluation of different scenarios of noise control actions to propose an action plan and, of course, fully running the model to get noise maps, creating calculation grids which one can even use for further grid operations, such as arithmetic or logarithmic addition or subtraction.

PRACTICAL EXAMPLES OF APPLICATION

Chemical Plant

This noise mapping project came from the need of the company to renew its IPPC environmental license, having indications that it was not fully complying with noise limits imposed by the Portuguese regulations. Therefore, a comprehensive noise source survey was carried out, estimating its octave band sound power level from sound pressure level measurements close to each source. As happens with most process industries, most relevant sources are located outdoors and they run 24 hours a day. Input data has been appropriately validated, according to the methodology described above: source validation followed by model validation—this was not particularly difficult in this case, due to the fact that the plant is located in the countryside, away from other noise sources, except for a national road, but with sparse traffic. However helpful this may be for model validation urposes, the fact that background noise is very low does not help when it comes to noise control requirements as, according to regulations, one must reduce plant noise down to the background level. The present regulations, however, do imply that, in case background level is lower than 42 dB(A), which was the case at most sensitive receivers in this project, the limit for particular noise from the factory, at any time, is also 42' dB(A), including the correction factors for tonal components [+ 3 dB(A)] and/or for impulsive noise [+ 3 dB(A)]. From the acoustical model, a source ranking has been performed, identifying the sources which generate more noise to the sensitive receivers, which in this case are just a few isolated houses, located to the opposite side of the national road, as can be seen on Fig. 42.2.

Fig. 42.2. Location of the factory, road, houses and corresponding measurement points P1, P2, P3, P4.

Paper Mills

This information sheet has been produced by the paper and board industry advisory committee (PABIAC) to help employers and employees understand their legal duties under the Noise at Work Regulations 1989 (the Noise Regulations) to reduce the risk and injury to employees' hearing. PABIAC involves representatives from the trades unions, employer's organisations and the health and safety executive (HSE). The committee was formed in 1979 to advise the Health and Safety Commission on health and safety issues relating to the manufacture of paper and board. To successfully implement noise control measures, it is first necessary to obtain information about the noise levels to which employees are exposed, i.e. to make a noise assessment. Specific advice on this is given in the PABIAC information sheet noise assessments in paper mills, which fits into the binder guide to managing health and safety in paper mills. In this (noise mapping) information sheet, more detailed advice is given on the use of the noise mapping technique to identify the source of noise and the exposure levels at different locations within the workplace.

What is noise mapping?

In an area such as a machine hall there are a number of closely placed noise sources such as vacuum pumps, couch rolls, air and steam valves, etc. Each makes a contribution to the overall sound pressure level (noise level) at a given position. The technique of noise mapping is simply a way of taking measurements at predetermined positions identified by applying a 1 m grid to the floor plan. These measurements are then displayed on the floor plan to produce a contour map (Fig. 42.3).

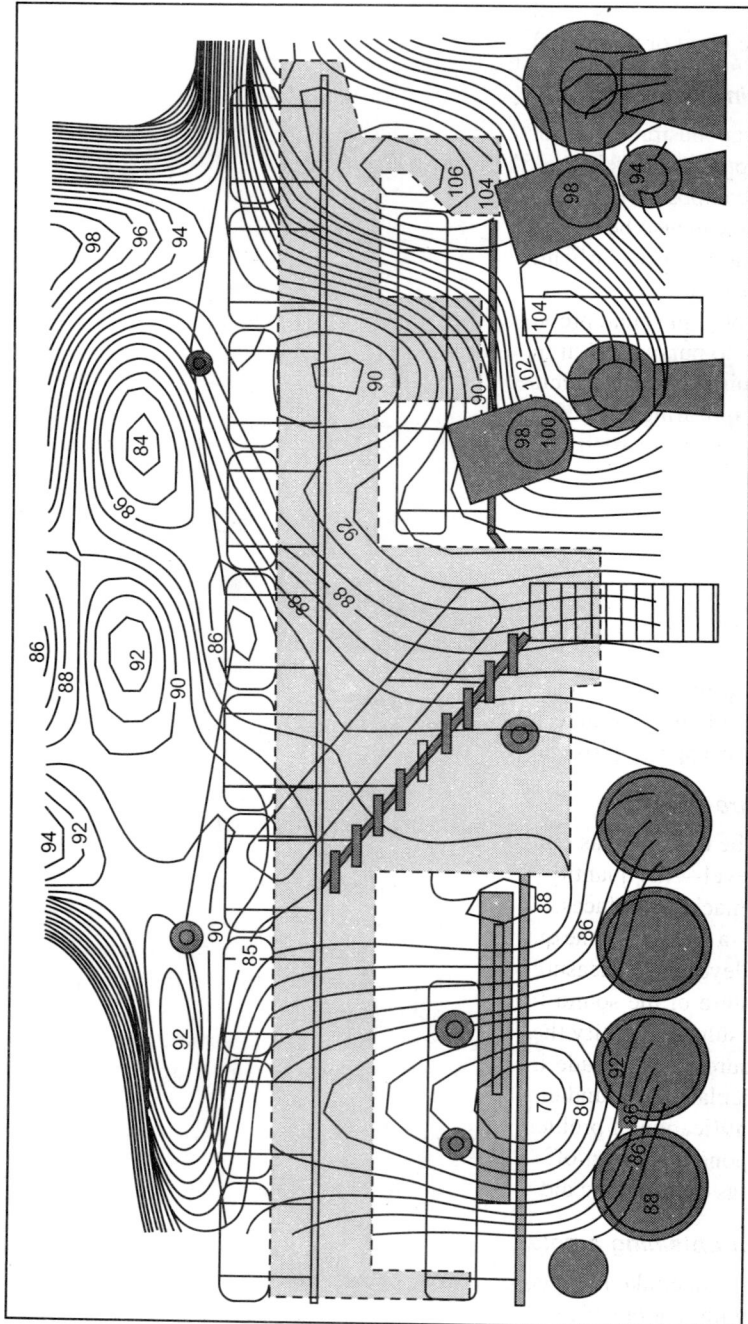

Fig. 42.3. Contour map of sound pressure superimposed onto diagram of press.

Noise mapping may be performed by capable in-house staff using readily available integrating sound level meters which can be hired or purchased. The exercise involves minimal time and cost.

How does noise mapping help?

By generating a contour map illustrating the distribution of noise, employers are provided with a graphical representation of sound pressure levels, i.e. the level of noise, within the workplace and can more readily calculate the noise exposure that employees undertaking tasks at particular locations are likely to be exposed to. The map can also be used to:

1. Identify plant creating particularly high sound pressure levels which may be amenable to noise reduction at source.
2. Locate the cause of high sound pressure levels which may be some distance from the source. It may be necessary to pursue this in more detail using a measure of sound intensity.
3. Monitor noise emission from machines or processes to highlight the need for maintenance or repair, e.g. wear in bearings of defective valves.
4. Identify low-noise corridors so that the movement of people within the building can be organised to reduce unnecessary exposure.
5. Check ear protection zones and monitor the use of ear protection.
6. For maximum benefit, the noise map can be overlaid onto existing building plans which show the location of emergency exits, fire points, etc. and can be located at the entrance for employees and visitors to see.

Do I need a noise map?

Noise maps are not necessarily needed for all parts of the mill. The more complex the workplace and the larger the number of possible noise sources, the more likely it is that you will need a map to quantify and identify sources and manage employees' exposure to noise.

What measurements are needed?

Noise contour maps can be displayed as emission profiles on a floor plan using two measures:

1. Sound pressure level—the quantity of sound energy at a particular location after being reflected by building and machine surfaces.
2. Sound intensity—a measure of the quantity and direction of acoustic energy flowing from a source.

Using sound pressure level as the measure to produce noise emission profiles is more straightforward, but due to the diffuse nature of the sound field with no obvious directional characteristics, it is more difficult to analyse the results to identify the source of high sound pressure levels.

Sound intensity measurements include information on the directional properties of noise emission which can be used to calculate the sound emission from a specific part of the machine or process. The technique allows the identification of problem areas while the machine is running; it is not necessary to sequentially switch component parts on/off. This information can then be used in calculations to design absorbing treatment and as a diagnostic aid to identify increased noise emission levels due to wear.

How should I go about obtaining a noise map?

Noise mapping should be undertaken by people who:

1. Are either from within or outside the mill organisation, although the Management of Health and Safety at Work Regulations 1999 require preference to be given to competent employees.

2. Have access to a suitable integrating sound level meter and are competent to use it.
3. Are able to develop computer graphics for display of data.
4. Have or can be provided with a good working knowledge of the mill process and relevant safe systems of work.

Where sound intensity measurements are required, specialists trained in the use of more sophisticated measurement equipment may be needed. Where it is necessary to seek professional help from a consultant to make sound intensity measurements, interpret the results and advise on the application of noise control measures, the free HSE leaflet *need help on health and safety?* provides useful guidance.

What would a typical map look like?

Figure 42.3 shows the side elevation of a paper-making machine with an overlaid contour map of the sound pressure emission profiles and illustrates the distribution of sound/noise from the machine.

Figures 42.4 and 42.5 are plan diagrams showing the distribution of sound pressure radiating into the area at the side of the press section of a paper-making machine. Figure 42.5 identifies the distribution of sound pressure for a particular problematic frequency.

Fig. 42.4. A-weighted SPL plan view of area in front of press section.

How can I use this information?

The information in these examples of mapping could be used to:
1. Control or restrict the occupancy of particular areas in the vicinity of the machine, such as the area around the couch roll identified in Figs 42.4 and 42.5.
2. Identify noise-generating components or process stages from the overall noise levels in the machine hall.
3. Design noise-control measures.
4. Provide information for the purchase of equipment emitting lower noise levels.
5. Provide a baseline measure of emission for comparison with future measurements to identify wear and the need for maintenance.

The information from these maps also illustrates how noisy components can be identified, particularly where there is no facility to operate individual parts of the paper-making machine on their own.

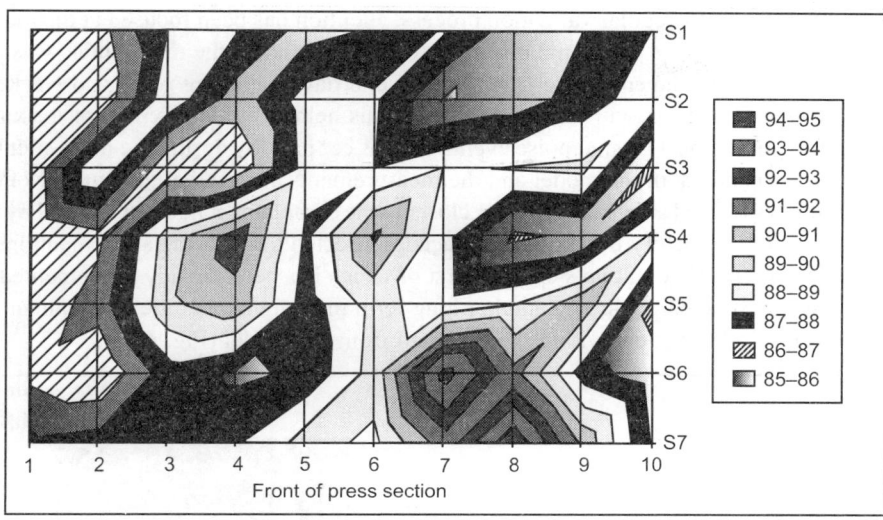

Fig. 42.5. 1600 Hz SPL plan view of area in front of press section.

For example, the area within grid references 6 to 8 and S5 to S7 in Fig. 42.4 and 42.5 illustrates the high level of noise in the vicinity of the couch roll. Figure 42.5 shows that this high noise level is characterised by a dominant frequency which in turn shows the knock-on effect the couch roll has in other areas by structure-borne transmission of vibration. This information is also necessary for the design of noise control, such as active noise control.

Noise contour maps of the whole machine hall would provide useful information on areas of lower noise exposure. These areas can often be used as preferred access routes for employees to avoid unnecessary exposure to noise.

Food Industry

This example is almost the opposite of the chemical plant, as it is a traditional old plant, located right in the middle of the city, with residential buildings all around, the background noise levels are high and most noise sources are inside buildings, except for chimneys. Although the factory does not actually stop at night, some sections of it do stop, therefore reducing its activity, and noise generation, during night time. The project started due to a complaint from a neighbour leaving near the factory. The project consisted of developing the acoustical model, produce noise maps of present situation to assess the noise impact and communicate it to the municipality, specify a noise control action plan, implement it and measure the final results, updating the noise map in the end.

Two types of sources were identified as relevant to the noise emission: interior sources, which radiate noise to the outside through the vast number of windows, and chimneys, located above the roofs. The first were not taken individually — the approach has been to measure noise impinging on the windows from the inner side, and use the CadnaA feature 'L_i from interior sources' together with the transmission loss of the windows to calculate the sound power lever per unit area radiated to the outside by vertical area sources, which were used to model the windows. The latter were measured and inserted in the model one by one as chimney sources, with a frequency dependent directivity, simulated by CadnaA from the known velocity and temperature of the gas flow at each chimney.

In this case, apart from the regular validation process, attention has been focused at the most critical receivers, namely at the house of the complainant, out of which window the microphone was mounted. Continuous long-term measurements were taken, for several days including week and weekend days, using a PC-based noise analyser, with audio recording. This helped in identifying noise sources, also enabling the filtering out of background noise events such as car pass-by during the more silent periods, such as night and weekend. Both the model and the measurements agreed that the number one sources were noisiest chimneys located above the building closer to the complainant house, followed by windows of the noisiest floors of the factory buildings with façades directed towards the afore mentioned house.

An action plan was specified in two steps, the first of which has been already implemented. Control measurements have been taken, including another long-term measurement at the complainant window, which show a clear noise reduction, within the expected from the model (Fig. 42.6).

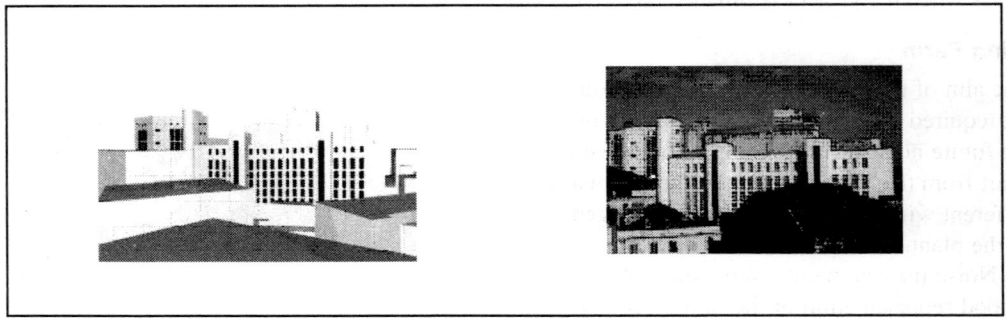

Fig. 42.6. 3D view of the CadnaA model and corresponding photo, both views taken from the window of the complainant's house.

Foundry

This project consisted on the production of noise maps in the context of an environmental impact study for the expansion of the foundry, which needed to increase its production capacity, for a number of scenarios: present situation, future situation with no noise abatement measures and future situation with noise abatement measures.

The specification of these noise abatement measures was also part of the project. The present situation noise map had already been produced in the past, although it had to be updated to the new Portuguese noise law, and it was complying to the regulations.

As the foundry was going to enlarge and new equipments were to be installed, the aim of the company was to study the problem in order to guarantee that it would still comply with the noise limits after the expansion project has been concluded.

A complete survey was carried out of all the changes which would take place, including new equipments, new layouts and new buildings. In this case, sound power levels from most new equipments were available and were introduced in the model.

Noise levels at critical points were calculated and source ranking was made in order to identify which noise sources needed to have special noise control conditioning measures prior to its installation, which could include relocation relative to the initially planned, in order to maintain full compliance with the noise limits.

Cement Industry

This example relates to a large cement factory which was starting the implementation of a large investment plan on its production lines which consisted basically on the replacement of three existing lines, which were old and ineffective, by a new production line, with a new kiln, with higher production capacity. Due to lack of space and presence of an urban agglomeration nearby, it was a complex operation and noise was one of the major issues as it should comply with ever more stringent regulations. Therefore, a noise mapping project has been carried out, which enabled the simulation of three main stages: present situation, transition situation and final situation with the new production line. It was shown that, at present, the factory is not complying with legislation and, although a noise reduction is predicted with the new layout of the factory, it has been shown that there is a high risk of not complying with the regulations. In this context, a noise control plan was studied, aiming at the full compliance with noise limits when the project is finished, i.e. at the final stage with a single production line.

Wind Farm

The aim of this study was the presentation of a noise map for the future situation, in the framework of the required permits for future installation of a wind farm, and assessment of its noise impact by comparing the future noise levels with those of the situation without the implementation of the project. Input data, apart from the cartography in vector format and orthophoto maps, were the frequency of occurrence of different wind velocities for each wind generator, and the wind rose characteristic of the surroundings of the plant.

Noise measurements were taken at different times of day and on more than one day, in order to get a good representation of the initial acoustical situation. From the acoustical model, the noise from the wind generators were calculated, at the measurement points and at all the surrounding points, by means of a noise map. In this particular case, due to the proximity of some sensitive receivers to some of the initially planned generator locations, these had to be changed to comply with noise limits and avoid future problems when the wind farm starts to operate.

To sum up computer acoustic models of industrial plants and noise maps are powerful tools for noise assessment and management, with very interesting applications both at the design stage as well as when the plant is already running. Not only it is an excellent way of making evidence of compliance to noise regulations, when this is the case, but it is even more useful when something has to be done to achieve such compliance. Enabling one to easily make a source rank, and simulate any scenario, one can be very effective in presenting noise abatement action plans and help the industries making the right noise control investments which, in many cases, are not at all negligible. Therefore, although it takes normally a large amount of work to produce the acoustical model, which of course has its own cost, it is well worth doing it whenever a new industrial plant is to be built or changed, as well as when a complex existing plant needs to produce evidence of noise compliance or is having noise complaints from neighbours.

Chapter 43

Noise Level Interpolation and Mapping

INTRODUCTION

A Noise Map is a map of an area which is coloured according to the noise levels in the area. Sometimes, the noise levels may be shown by contour lines which show the boundaries between different noise levels in an area.

The noise levels over an area will be varying all the time. For example, noise levels may rise as a vehicle approaches, and reduce again after it has passed. This would cause a short-term variations in noise level. In the slightly longer term, noise levels may be higher in peak periods when the roads are busy, and lower in off-peak periods. Then again, there is a greater volume of activity from more people and traffic in the day-time than in the evening or at night. In the longer term, wind, weather and season all affect noise levels.

This means that it is not possible to say with confidence what the noise level will be at any particular point at any instant in time, but where the noise sources are well-defined, such as road or rail traffic, or aircraft, then it is possible to say with some confidence what the long-term average noise level will be.

It may be thought that the best way of doing this is by measurement, but experience shows that this is not the case. For a start, a long-term average must be measured over a long period of time. Secondly, to obtain complete coverage of an area, measurements would have to be made on private property, where access might be difficult, and thirdly, measurements cannot distinguish the different sources of noise, so they would not be able to give information on how much noise was being made by each of the sources in an area. For these and other reasons, noise mapping is usually done by calculation based on a computerised noise model of an area, although measurements may be appropriate in some cases.

A further benefit of having a noise model is that it can be used to assess the effects of transportation and other plans. Thus the effect of a proposed new road can be assessed and suitable noise mitigation can be designed to minimise its impact. This is particularly important in noise action planning, where a cost-benefit analysis of various options can be tested before a decision is made. Industrial noise is a common cause of hearing loss among workers. In addition, industry-generated noise that exceeds community ambient noise levels is a significant annoyance and may evoke community reaction.

NOISE LEVEL AND FREQUENCY WEIGHTING

Sound is characterised by its frequency and pressure level. Sound pressure level L_p (decibels) is given by the logarithmic relationship:

$$L_p = 10 \log_{10} (p_{rms}^2/p_{ref}^2) \qquad \qquad \dots (43.1)$$

where, p^2 rms is the mean square sound pressure (Newtons/m^2) and $P_{ref} = 2 \times 10^{-5}$. Alternately, intensity level L_I (decibels) is given by the logarithmic relationship:

$$L_I = 10 \log_{10} (I/I_{ref}) \qquad \ldots (43.2)$$

where, I is the sound intensity in watts/m^2 and $I_{ref} = 10^{-12}$. For airborne sound and ordinary conditions, the difference between L_p and L_I is negligible.

Audible sound covers the frequency range from about 20 Hz to 20,000 Hz and ranges in sound pressure level from about $L_p = 0$ dB (the threshold of hearing) to $L_p = 140$ dB or higher (the threshold of pain). The A-weighting network in a sound level meter combines the effect of sound at various frequencies approximately as the human ear responds. The A-weighted sound pressure level in decibels (dBA) is a convenient and widely used measure of industrial noise. For purposes of hearing conservation and assessment of community noise, measurements in dBA are generally adequate except when pure tones are prominent.

INDUSTRIAL HEARING DAMAGE RISK

Routine hearing level tests are made with an audiometer, with hearing thresholds measured at several frequencies. Criteria for acceptable noise exposure depend on the definition of a hearing handicap. If a hearing handicap is defined as an average hearing loss of 25 dB, continuous (8 hour per day) industrial noise exposure to levels up to 80 dBA will not increase the risk of developing a handicap. This exposure level is based on the results of audiometry examinations of industrial workers and comparisons (by age group) with others who were not exposed to high industrial noise levels. Other research suggests a much lower safe exposure level on the basis of a different hearing impairment criterion.

COMMUNITY NOISE FROM INDUSTRIAL SOURCES

When industrial and residential zones abut, many problems can develop, the most pervasive often being noise. Residents may complain of sleep interference, speech interference, or simply annoyance. At a distance of for feet between talker and listener, speech communication is barely possible at normal male voice level with a background noise level of 63 dBA.

Sleep interference is more difficult to express quantitatively. In one study, it was found that the time required to fall asleep increased by 1 to 1.5 hours at noise levels between 50 and 60 dBA. Furthermore, at those noise levels, the depth of sleep was decreased. Annoyance is a very subjective effect. However, numerous studies have indicated that when intrusive noise exceeds background noise levels by 5 to 10 dBA, community action is likely.

NOISE CONTROL

Principal methods of industrial noise control include reduction of noise output at the source and control of noise transmission paths. Noise sources should be considered first. After reducing noise source output to the lowest practical levels, noise transmission paths through solids and air should be examined. For airborne noise outdoors, buildings and barriers between the noise source and receiver have a substantial effect. Finally, the distance between source and receiver may be increased, or, if the receiver is an industrial worker, he may be fitted with hearing protection.

COMPLIANCE WITH FEDERAL AND MUNICIPAL CODES AND STANDARDS

Basically, the occupational safety and health administration (OSHA) permits 8 hours per day exposure to noise levels of 90 dBA. The exposure time limit is halved for each 5 dBA increase up to the maximum permitted continuous noise level of 115 dBA. The allowable levels represent a compromise between actual existing noise levels in industry and safe levels in terms of hearing damage risk. Thus, OSHA levels may be subject to revision to reduce further the risk of hearing loss.

Community noise codes are generally based on annoyance and speech and sleep interference rather than hearing damage risk. Permitted levels for noise that intrudes on residential properties are much lower than permitted levels within factories. One city code, for example, has different requirements for light and heavy manufacturing zones. The combined output of a corporation in a restricted manufacturing zone is not to exceed 55 dBA at the boundary of a residential zone.

The limit is 61 dBA for general manufacturing zones; for general and heavy manufacturing zones, measurements may be made at the zone boundary or 125 feet from the nearest property line of a plant or operation, whichever distance is greater. One state code limits continuous noise from industrial operations to 65 dBA between 6 am and 10 pm with a night time limit of 50 dBA. For this code, measurements are made at a residential property line (even if the residence does not lie in a residential zone). The above noise limit values are given as illustrations only. The latest applicable codes and regulations must be obtained in every case.

Statistical (per cent exceeded) noise levels are sometimes used to describe noise that varies with time. The most commonly used levels are L_{10}, L_{50}, L_{90} and 40 values expressed in dBA and based on one hour measurement periods. The L_{10} level is exceeded 10 per cent of the time. For example, $L_{10} = 65$ dBA indicates that a noise level of 65 dBA was exceeded for 6 minutes during a given hour. Noise level L_{50} is the median and L_{90} is the level exceeded 90 per cent of a given time period.

NOISE LEVEL MAPPING

The distribution of resources applied to noise control should be based on careful prediction, measurement and planning. The plotting of noise contours (points of constant noise level) can be an important step in noise control. Using a few predicted or measured noise level values, it is possible to estimate and plot noise levels at other locations. This permits application of engineering controls to reduce noise transmission paths. Noise plots may also aid in siting of noise-producing operations to reduce noise impact on employees and on the neighbouring community. Other administrative controls including stationing of employees and the requirements for personal hearing protection may be indicated by noise level plots.

POINT SOURCES OF NOISE

A single piece of machinery, an exhaust stack, a fan or other relatively concentrated noise source may be considered a point source. At locations sufficiently distant from the source, noise intensity follows the inverse square law, provided reflections are not present.

INVERSE SQUARE LAW

In the absence of barriers, an ideal point source generates a spherical wave. Noise intensity (watts/m²) decreases with distance according to the inverse square law:

$$\frac{I_2}{I_1} = \frac{r_1^2}{r_2^2} \qquad \qquad \text{... (43.3)}$$

where intensity I_1 is measured at distance r_1 from the source and intensity I_2 at distance r_2.

Combining Eqs 43.2 and 43.3, we obtain the effect of distance on sound level L(dBA) for an ideal point source:

$$L_2 = L_1 - 20 \log_{10}(r_2/r_1) \qquad \qquad \dots (43.4)$$

where, L_1 and L_2 refer to sound levels at distances r_1 and r_2, respectively. When dealing with airborne sound under ordinary conditions, there is no need to distinguish between sound pressure level and sound intensity level. It can be seen that sound level due to an ideal point source decreases by about 6 dBA per doubling of distance from the source. This relationship holds for a point noise source on a hard floor as well.

NEAR AND REVERBERANT FIELD

No actual machine can be exactly represented by an ideal point source. The region close to an actual noise source is called the near field. Noise levels fluctuate within the near field, and the 6 dBA sound decrease per doubling of distance law (inverse square law) does not hold. The extent of the near field will be of the same order of magnitude as a characteristic dimension of the noise source. The far field is the region in which sound level decreases by 6 dBA per doubling of distance. Smooth walls that reflect sound waves are called 'acoustically hard'. In a room, the sound level may fluctuate near walls. The region near a wall in which the 6 dBA per doubling of distance law does not hold is called the reverberant field. The near field, far field and reverberant field are illustrated in Fig. 43.1.

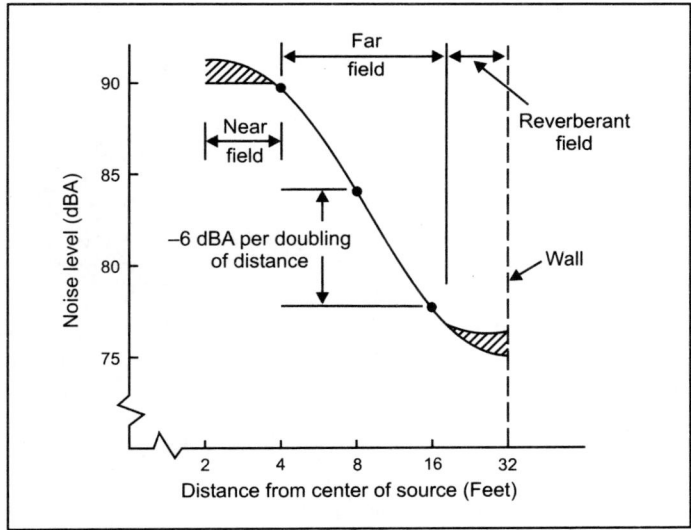

Fig. 43.1. An illustration of the variation of noise level with distance from a point source (semilog coordinates).

Noise level is also influenced by barriers, by atmospheric attenuation and by background noise. Barriers that interrupt (or nearly interrupt) the line of sight between noise source and receiver cause a substantial reduction in noise level. Atmospheric attenuation causes additional noise reduction when sound is transmitted for hundreds of feet. Background noise (traffic noise and noise from sources other than the machine in question) becomes prominent as distance from the source increases.

PLOTTING POINT SOURCE NOISE LEVELS

Using Eq. 43.4, the difference in noise level ΔL(dBA) measured at distances r_1 and r_2 from a point source is given by:

$$\Delta L = L_2 - L_1 = -20 \log_{10} (r_2 / r_1) \qquad \text{... (43.5)}$$

Conversely, for a given difference in noise level ΔL, the ratio of distances to a point source is given by:

$$r_2/r_1 = 10^{-\Delta L/20} \qquad \text{... (43.6)}$$

An increase in noise level, of course, corresponds to a decrease in distance. Using Eq. 43.6, the relationship between changes in noise level and the ratio of distances from a point source is given in Table 43.1.

Table 43.1. Change in noise level with distance from a point source.

Change in noise level ΔL (dBA)	Distance ratio r_2/r_1
0	1.000
1	0.891
2	0.794
3	0.708
4	0.631
5	0.562
6	0.501
7	0.447
8	0.398
9	0.355
10	0.316
12	0.251
20	0.100

An interpolator for point source noise may be constructed as shown in Fig. 43.2. A vertical line of any convenient length is drawn at the left of the figure. The top of the line is marked 0 dBA and the bottom of the line represents the noise source location. Measuring from the bottom of the line, 0.891 times the length of the line corresponds to $\Delta L = 1$ dBA; 0.794 times the length of the line corresponds to $\Delta L = 2$ dBA, etc. as given in Table 43.1. Lines are drawn from each of these points to a point of convergence at some convenient location at the right of the figure. The interpolator of Fig. 43.2 may be duplicated directly on a sheet of clear plastic, or a larger interpolator may be constructed from the data of Table 43.1 using transparency marking pens on clear plastic.

If the noise level is known at one point, the interpolator may be used to estimate noise level at another point in line with the first point and the noise source. Using a scale drawing of the area, the interpolator is oriented so that the 'location 1' line lies over the point farther from the source and the 'noise source' line lies over the source. The line lying over the nearer point indicates the difference in noise level between the two points.

Figure 43.3 illustrates the use of the interpolator when the noise level is unknown at a point between the source and a point of known noise level. In this example, the interpolator gives $\Delta L = 3$ dBA, which is added to the known value of 70 dBA for a noise level of 73 dBA at the point nearer to the source. If the point of unknown noise level is farther from the source than the known point, ΔL is subtracted. If one of the points is very close to the noise source, it may lie in the near field, making the interpolation invalid. If one of the points is very far from the noise source, noise from sources other than the source in question may have a predominant effect.

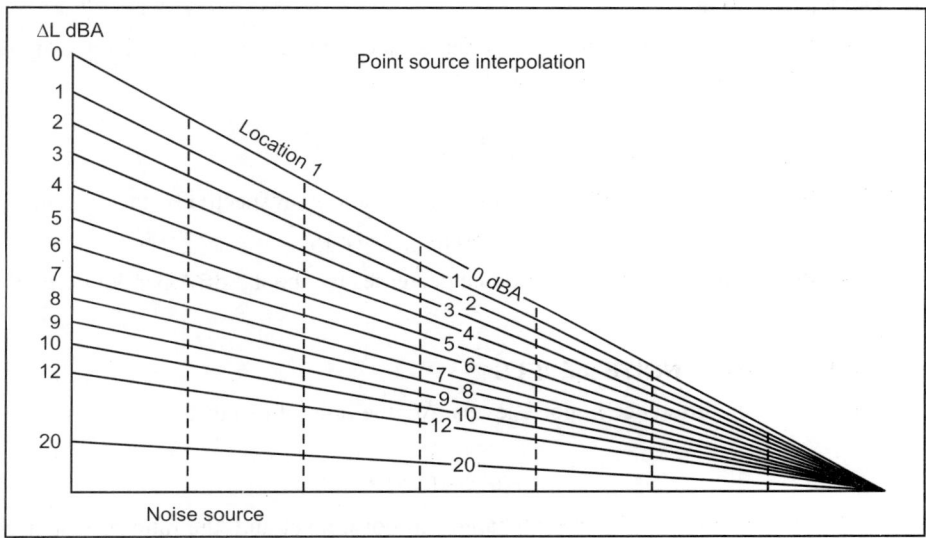

Fig. 43.2. Point source noise interpolator.

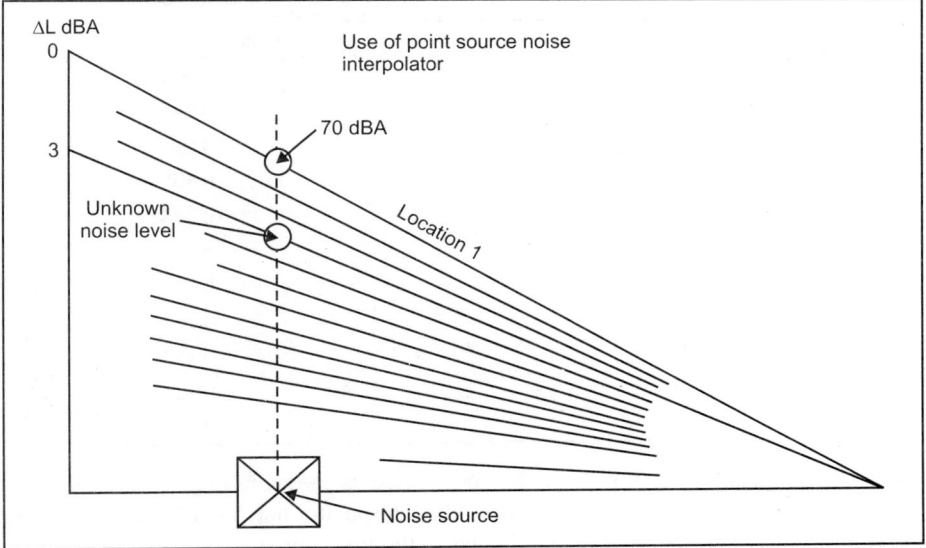

Fig. 43.3. Use of point source noise interpolator.

LINE SOURCES OF NOISE

When a number of pieces of machinery or other noise sources of equal sound power lie in a line, the situation may be approximated by a line source. At points sufficiently distant from the line source, noise intensity follows the inverse distance law, provided reflections are not present.

INVERSE DISTANCE LAW

In the absence of barriers, an ideal line source generates a cylindrical wave. Noise intensity decreases with distance according to the inverse distance law:

$$\frac{I_2}{I_1} = \frac{r_1}{r_2}$$

... (43.7)

In terms of sound levels L_1 and L_2 (dBA) at distances r_1 and r_2, respectively, this becomes:

$$L_2 = L_1 - 10 \log_{10} (r_2/r_1)$$

... (43.8)

Thus, the sound level due to an ideal line source decreases by about 3 dBA per doubling of distance from the source.

PLOTTING LINE SOURCE NOISE LEVELS

Using Eq. 43.8 for a given difference in noise levels, ΔL, the ratio of distances to a line source is given by:

$$r_2/r_1 = 10^{-\Delta L/10}$$

... (43.9)

Using Eq. 43.9, the relationship between changes in noise level and, the ratio of distances from a line source is given in Table 43.2.

Table 43.2. Change in noise level with distance from a line source.

Change in noise level ΔL (dBA)	Distance ratio r_2/r_1
0	1.000
1	0.794
2	0.631
3	0.501
4	0.398
5	0.316
6	0.251
7	0.200
8	0.158
9	0.126
10	0.100

An interpolator for line source noise may be constructed as shown in Fig. 43.4, using the data from Table 43.2. If the noise level is known at one point, the interpolator may be used to estimate noise level at another point if both points lie on a perpendicular to the line source.

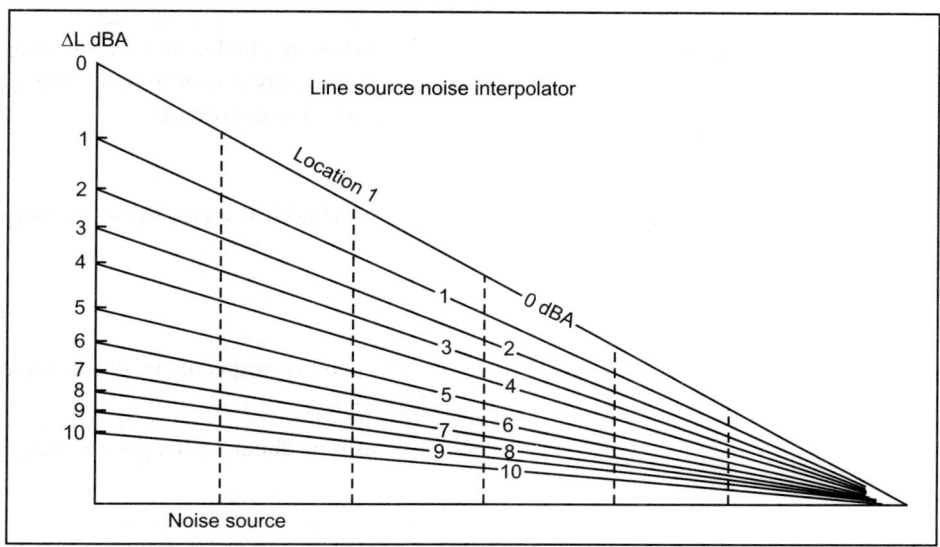

Fig. 43.4. Line source noise interpolator.

The interpolator is oriented so that the 'noise source' line lies over the source and the 'location 1' line over the farther point. The line lying over the nearer point indicates the difference in noise level between the two points.

COMBINED NOISE SOURCES

When two or more independent noise sources are present, the effect at a given point may be obtained by adding the noise intensities (watts/m^2). Most measurements and predictions, however, are given as noise levels L (dBA). For separate noise level contributions L_1 and L_2, the total noise level L_T dBA at a point is:

$$L_T = 10 \log_{10} (10^{L_1/10} + 10^{L_2/10}) \qquad \qquad \text{... (43.10)}$$

It can be seen that for $L_1 = L_2$, $L_T = L_1 + 3$ dBA (approx.). In general, if L_1 is larger than L_2 by N dBA, the combined effect of L_1 and L_2 is larger than L_1 by:

$$L_T - L_I = 10 \log_{10} (1 + 10^{-N/10}) \qquad \qquad \text{... (43.11)}$$

The difference is given in Table 43.3 for convenience in combining noise levels. The results are given in chart form in Fig. 43.5.

Using die table or chart, any number of noise contributions can be combined, two at a time. For example, let three machines produce levels of 70 dBA, 70 dBA and 67 dBA at a given point when operating singly. The effect of the three operating together is given by combining the first two levels to obtain 73 dBA, and then combining 73 dBA and 67 dBA to obtain a total noise level of 74 dBA (approximately). Due to the difficulty in making precise noise measurements and the tendency of noise levels to vary with time, final results are generally rounded off to whole decibels. It can be seen that if two noise contributions are to be combined and one is 15 or 20 dBA below the other, the lower noise level will not have a significant effect. For example, 80 dBA combined with 60 dBA produces a total noise level of approximately 80 dBA.

Table 43.3. Combined noise levels.

Difference in noise levels to be combined ($L_1 - L_2$ dBA)	Amount added to larger level to obtain total noise ($L_T - L_1$ dBA)
0	3.01
1	2.54
2	2.12
3	1.76
4	1.46
5	1.19
6	0.97
7	0.79
8	0.64
9	0.51
10	0.41
15	0.14
20	0.04

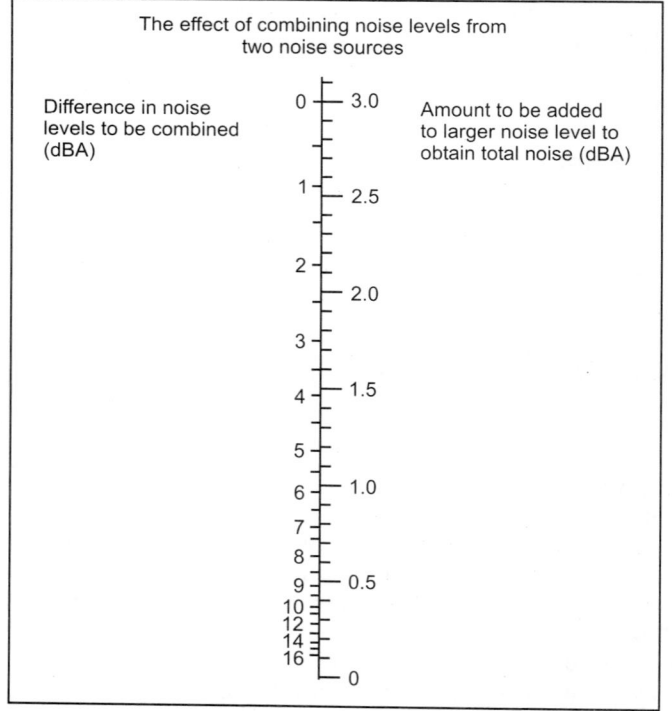

Fig. 43.5. The effect of combining noise levels from two noise sources.

NOISE CONTOUR MAPPING FOR LINE AND POINT SOURCES

It is sometimes necessary to map noise contours to estimate the impact of industrial noise on workers or on the neighboring community. The noise interpolators may be used as an aid in plotting in a free field, sufficiently for from the noise source or barriers or reflecting surfaces. For an ideal line source, constant noise level contours are simply lines parallel to the source. For an ideal point source, constant noise level contours are concentric circles. For a source that shows some directionality, the contours will be distorted accordingly.

Figure 43.6 illustrates a plan for installing a row of machines that may be approximated by a line source. Based on measurements on similar machines, it is estimated that the row of machines will produce a noise level of 78 dBA at the points indicated. The line source interpolator was used to draw lines of constant noise level parallel to the source as shown in the figure. A single machine will also be installed, and it will be approximated by a point source. Measurements indicate that it will produce a noise level of 80 dBA at the points indicated. The point source noise interpolator was used to estimate the location of curves of constant noise level due to the single machine. No curves are drawn near the machines since the point source and line source idealisations are not valid in the near field. Barriers, reflecting walls and ceilings must also be considered. In this example, it is assumed that the area of interest is free of obstructions.

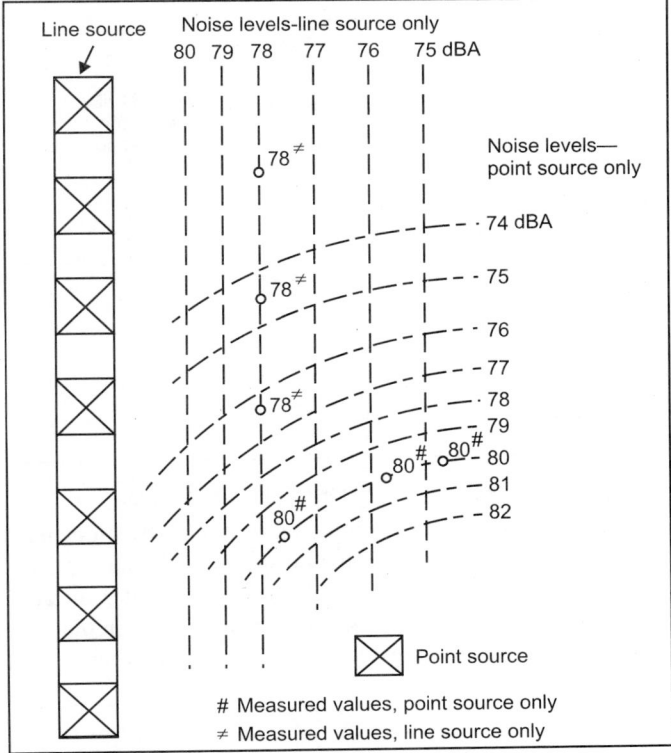

Fig. 43.6. Noise contours for a line source and a point source.

NOISE CONTOURS FOR COMBINED NOISE SOURCES

Figure 43.6 illustrates the approximate effect of two noise sources operating separately. If both sources operate together, the combined effect will be of interest. This may be obtained as in Fig. 43.7, considering the intersection points of the noise contours and using Table 43.3 (or Fig. 43.5) to sum the noise contributions. For example, the intersection of two 77 dBA contours represents a combined effect of 77 + 3 = 80 dBA, and that point is so marked. The intersection of a 74 dBA contour and an 80 dBA contour represents a combined effect of 80 + 1 = 81 dBA (approximately) and it is so marked. As indicated previously, precision of better than 1 dBA may not be warranted.

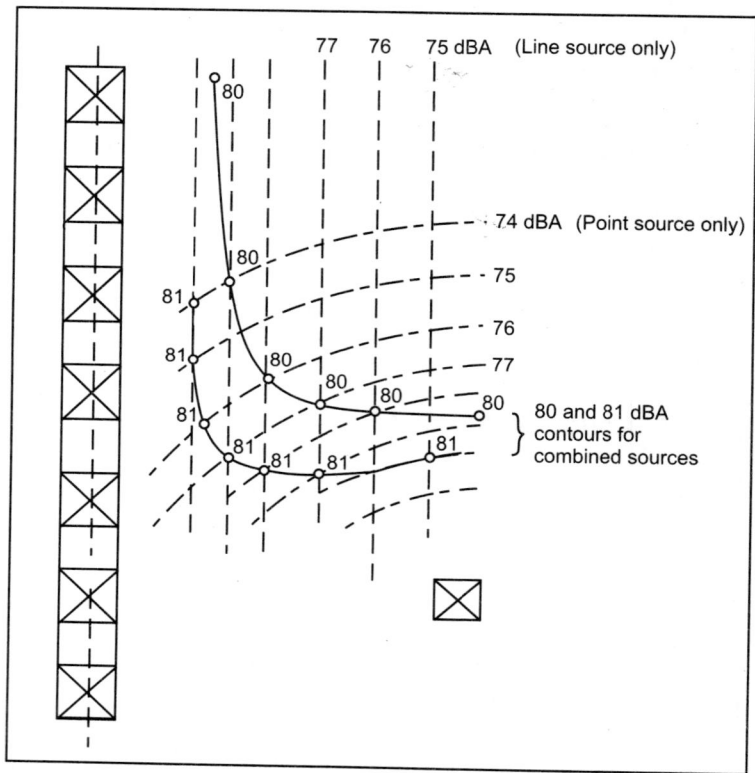

Fig. 43.7. Noise contours for combined sources.

The combined noise contours are obtained by joining all of the intersections marked 80 dBA, then all of the intersections marked 81 dBA, and so on. In the illustration, only the 80 and 81 dBA combined noise contours have been plotted. If noise levels are of interest at greater distances from the noise sources shown, the plotted values may approach background noise levels. In that case, the background noise levels would be combined with the noise levels due to the point and line sources.

SECTION X

Case Studies

Chapter 44

Road, Rail and Aircraft Noise

INTRODUCTION

The natural starting point for the reduction of environmental noise is to control the noise at source. The need to apply secondary measures, such as road design or land use, often depends on how well the control of the sources, including their number, has succeeded. However, the costs usually form a factor of prime importance. At best, noise control may be very inexpensive, if applied at the right time. But at a later stage, especially the secondary forms of noise control tend to be expensive, with the costs loaded on society or on the receiver rather than onto the user or producer of noise sources.

There are many approaches to source control. The most obvious is to control the vehicles or machines themselves through technical improvements. However, as progress is made in reducing the component sources of noise, it is also necessary to consider the number of sources and the environment where the sources operate. For instance, the road surface design can influence the noise emission, particularly at moderate and high speeds. In addition, source control can be extended to the operating conditions, e.g. driving style of road vehicles, or type of in-flight operations for aircraft.

In environmental policy, it is generally agreed that controlling noise at source is the preferred abatement method. However, additional measures which attempt to limit the spread of noise, or to give consideration to the land use near major noise sources are often needed as well. Other approaches are the improved design and insulation of buildings and non-technical (i.e. economic or social) forms of abatement.

NOISE REDUCTION AT THE SOURCE

Road Traffic: External Factors of Source Emission

Components of emission

The most important noise generating factors in motor vehicles are:

1. The power train (including the engine and transmission).
2. The tyre/road interaction (also termed as rolling noise).

The relative importance of the mechanisms depends on the vehicle type and driving conditions. Typically, the tyre/road noise is logarithmically related to speed: there is approximately a 12 dB level increase per doubling of speed. Power-train noise is only slightly influenced by speed. Therefore, there is a crossover speed above which tyre/road noise dominates the overall noise.

Rolling noise has a negligible contribution to the overall noise from heavy vehicles at low speeds, but above about 30 km/hr for cars and 40 km/hr for heavy vehicles, rolling noise becomes a significant part of the noise. Above 50 km/hr, rolling noise is the dominant noise source for cars, and above 70 km/hr for heavy vehicles. Since the range of traffic speeds in cities is spread both below and above the crossover speed, it is obvious that both power-train and tyre/road noise must be reduced in order to obtain a better environment. In highway traffic, almost no reduction of overall noise is possible, unless tyre/road noise is reduced. Significant reductions in the noise produced by the power train are still considered feasible. However, these effects will not be fully realised for a wide range of vehicle operating conditions, unless the rolling noise components are also reduced. Currently, and with the new generation vehicles, tyre/road noise plays a bigger role in urban traffic noise than expected before. It has been found that tyre/road noise may determine much of the overall noise even in the standard acceleration tests of new vehicles (Fig. 44.1).

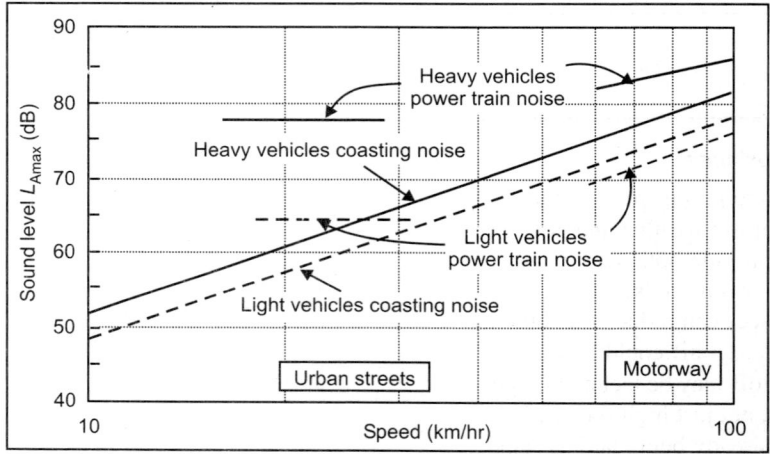

Fig. 44.1. Relative significance of main noise sources in motor vehicles as a function of speed.

The most important component of rolling noise is generated by the tyres rolling over the road surface. The mechanism of noise generation is complex. At present the many factors involved are qualitatively understood. Quantitative knowledge of the interaction between the different mechanisms is, however, still incomplete. The main factors affecting tyre noise are the speed of rotation of the tyre, the type of tread pattern and material, and the texture applied to the road surface. Earlier, it was a general belief that changes to the tyre construction and tread pattern have a smaller effect on rolling noise than changes made to the road surface material and texture. But today it is considered that there are potentially equal possibilities for controlling rolling noise by improving the road surface or by changing the tyre design.

The European Commission has invested in research with the aim to reduce the rolling noise caused by tyres. Following theses studies, the Commission has made a proposal for a European Parliament and Council Directive amending Directive 92/23/EC relating to tyres for motor vehicles and their trailers. This proposal aims to reduce rolling noise by setting limit values for tyre and road noise emission and by defining their level test and measurement conditions. The final adoption of this Directive is expected during 2014. One confusing factor in designing low-noise tyres is that today most ordinary car tyres are optimised for driving at very high speeds, up to 190–300 km/hr, resulting in poor optimisation in normal conditions. A general maximum speed limit would obviously be needed on the way to quieter tyres in ordinary traffic. However, in this inventory, the tyre part of tyre/road noise will not be considered further.

Emission limits

The noise emission limit values in use today are linked to type testing in standard conditions which emphasise the noise generated by the power train during acceleration. The limits have been rather ineffective in decreasing the emission as a whole in typical driving situations. However, key operators of motor vehicles, such as cities and municipalities, need not adhere to common emission limits when purchasing new equipment. Stricter 'custom' limits for buses, garbage, cleaning and other working vehicles, as well as for trams and local trains may be applied as one means in noise abatement programs (Table 44.1).

Table 44.1. The evolution of the European emission limits over time, for selected motor vehicles, L_{AFmax} [dB].

Vehicle category	1982	1992	1998/2000	2005/06
Passenger car	82	80	77	74
Urban bus	89	82	80	78
Heavy lorry	91	88	84	80

Low-noise road surfaces

Recently, open textured porous road surfaces have been developed, which offer the advantages of good skidding resistance in wet weather and good noise reduction and sound absorption characteristics. Thus these surface types provide both safety and considerable reductions of not only tyre noise but also to some extent the power train noise.

Porous road surfaces can reduce the total noise emitted by vehicles by, typically, 2–4 dB, when the surfaces are new. Up to 5 dB could be reached on high-speed roads. With suitable refinement, greater reductions up to 6–7 dB may be technically possible. This improvement seems to apply to most vehicle operating conditions, not just high speeds. In some countries (for instance the Netherlands and France), porous surfaces are already being laid as standard on a large proportion of the major road network.

However, the long-term durability is crucial, and not yet proved to be sufficient. Also, clogging can close the pores and the cleaning of the surfaces is expensive. Winter conditions are particularly rough. Especially in the Nordic Countries, icing of the pores may easily break the structure, and the use of studded winter tyres forms an additional wear factor. At present, considerable research effort is being devoted to improving the durability of this type of road surface (Fig. 44.2).

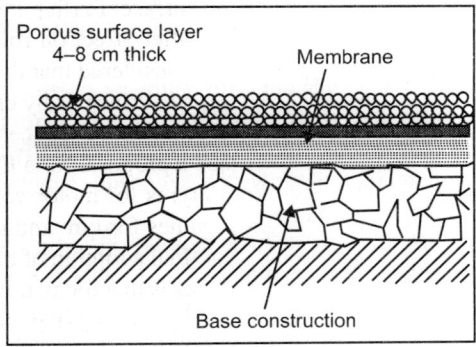

Fig. 44.2. Typical structure of a porous low-noise road surface.

Road maintenance

Noises from the vehicle body and suspension and from movement of the load carried can be very objectionable, but these aspects are not easily dealt with. It has been suggested that some form of legislative action is needed to control this form of noise. Good road maintenance is a prime requirement for reduced load and body noise. Also, poor maintenance of roads and streets may to a large extent increase traffic noise as a whole via the increase of tyre/road noise. On the contrary, frequent resurfacing may reduce the noise.

Road Traffic: Operational Conditions

The most important factors affecting the noise generated by the traffic stream are, besides the traffic speed, the traffic volume and the proportion of heavy commercial vehicles in the traffic flow. Another significant factor is the traffic flow, described as free flowing or interrupted as at traffic lights and junctions.

The foremost practical techniques to control the noise from traffic stream include rerouting, restricting access both in terms of type of vehicle, (e.g. lorry bans or preferred lorry routes), and in terms of time of day. For instance, night bans have been imposed in some areas. Control methods dealing with traffic volume and proportion of heavy vehicles form the primary steps for local authorities in reducing noise. The measures affecting speed and smooth flow are secondary steps, eventually needed to abate the noise of the remaining traffic.

Road traffic management

Traffic volume and redirection

The most obvious way to reduce traffic noise is to move the traffic away from the noise-sensitive section of the road. Concentrating urban traffic on a few main routes including bypasses, thus reducing noise in the remaining area can provide considerable benefits to many people. For example, halving the traffic flow on a residential street with light traffic may reduce noise by 3 dB. Yet the number of redirected vehicles could be quite small and easily absorbed in neighbouring roads built purposely to take higher traffic flows.

Reductions in journey times and accidents are usually additional benefits. However, closing road sections from all traffic can present problems of access, and exemptions can reduce substantially the effectiveness of the ban. On the other hand, bypasses themselves may bring along negative aspects connected to the visual landscape and access.

The effect of traffic volume controls depends not only on the proportion of traffic removed but also on the volume of traffic before and after the restrictions. Halving the traffic flow reduces noise, if other parameters do not change. However, traffic volume and speed are generally correlated and a reduction in volume is normally associated with an increase in speed. The result may be that the optimum benefits from the reduced flow are not achieved (Fig. 44.3).

Removing traffic from one road produces an increase in noise on other roads. However, the fact that noise is logarithmically related to traffic volume can be used to good effect. For example, taking traffic from a lightly used road and placing it on an already heavily used road gives little increase in the noise from the latter, but the improvement on the lightly used road can be substantial.

Bypasses, specifically designed to take high traffic flows to relieve residential streets from traffic, can produce considerable noise benefits. For instance, the noise reduction found in a typical case in

Vienna varied between 0–4 dB. In areas where bypasses do not already exist it may be possible to use shopping streets for rerouted traffic during the night.

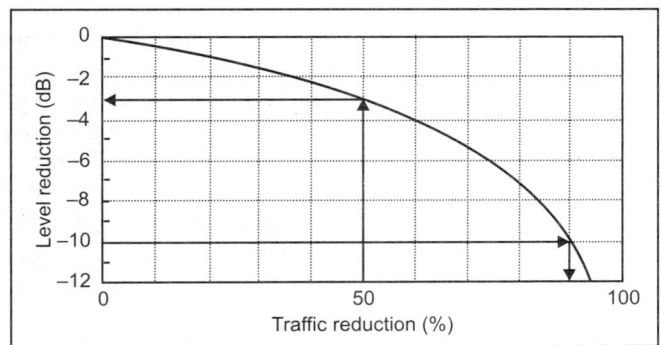

Fig. 44.3. Effect of traffic volume reduction by redirection. Halving of traffic volume produces a noticeable noise level reduction of 3 dB. The subjective halving of noise (–10 dB) requires a 90 per cent reduction in traffic.

Restrictions in time and area

Restricting the number of heavy vehicles using 'sensitive' routes is one possible control measure. Generally, the noise of heavy trucks easily becomes the dominating factor in streets with low traffic speed. Bans on heavy vehicles entering a prescribed district may take the form of a total ban on all commercial vehicles above a certain capacity or weight, or restrictions at certain times, usually at night and over the weekend.

In Switzerland, heavy trucks are not allowed to run anywhere in the country during the night with the exception of buses, fire engines and trucks carrying certain perishable goods. Zürich provides an example of a combination of traffic measures to combat noise impact, involving a ban on trucks, vehicle-free zones and very quiet public transport. Quiet zones with access limited to 'low noise vehicles' can, in principle, provide incentive both to manufacturers and operators of vehicles. It offers an attractive solution to control noise impact on a limited scale. However, potential problems may rise if standards are not harmonised between regulatory authorities. Harmonisation is similarly important internationally both for manufacturers and users. Several Central Europe cities and regions have also introduced weekend restrictions and bans on heavy goods vehicles in order to reduce noise and gas exhaust emissions. In March 1998 the European Commission made a proposal for a Council Directive on a transparent system of harmonised rules for driving restrictions on designated roads. In general, roundabouts produce fewer noise problems than signalised intersections. Compared to the noise of equivalent free flow traffic stream, the noise near roundabouts typically increases by about 1–2 dB.

Traffic calming, speed reduction and smooth traffic flow

Speed limits

The reduction of traffic speed is in principle an effective control measure for traffic noise. On high speed roads, halving the average vehicle speed could reduce the noise by 5–6 dB. However, speed reductions cannot always be achieved easily in practice. Speed limits are one of the most commonly used traffic restraint measure. They are generally introduced for reasons of safety. The reduction in speed caused by this restraint, if effective, will generally also lead to reduction of noise. Local speed limits are effective from a noise point of view only if they can be enforced without measures that increase the acceleration of vehicles. Another problem is that a considerable proportion of the motorists may exceed the limits. However, the limits generally tend to lower the highest speeds.

The design of the traffic speed restriction methods is important. The measures should introduce sufficient restraint on the motorist to cause speed lowering, without affecting gear changing which could result in a net increase in noise levels. The methods adopted should also ensure that traffic flows freely to encourage a non-aggressive style of driving. Speed control measures can have other positive advantages, such as reducing accidents.

The limits can also positively influence driving style and behaviour. The drivers might accelerate and decelerate less aggressively than when driving in a street with a higher speed limit. It is estimated that the noise reduction caused by driver behaviour changes may range between 2–4 dB depending upon the speed actually achieved.

Speed monitoring

Automatic speed monitoring and even control could be one means of ensuring that designated speed limits are adhered to. Generally, there is no conflict between the goals of safety and noise abatement, when limiting the speed of road traffic is considered. The noise aspects of speeding are emphasised during the evening and night hours of low traffic, because of the eventual health effects related to sleep. The technology for automatic speed monitoring is largely available today, including electronically enforced automatic speed control. One could argue that pressure from the environmental side might be one factor, eventually leading towards the introduction of the method within, say, one or two decades (Fig. 44.4).

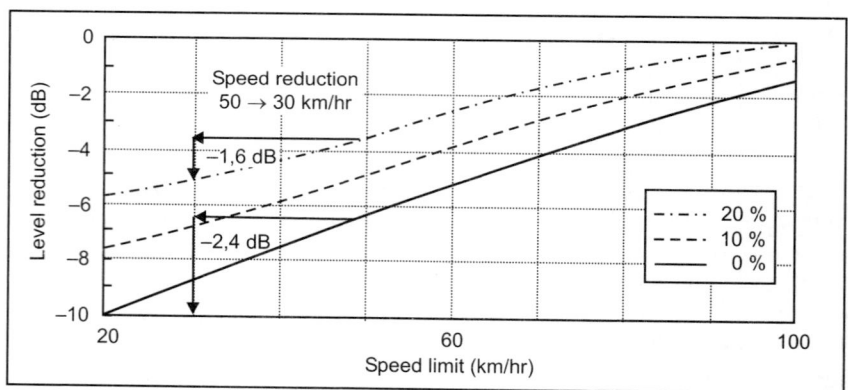

Fig. 44.4. Effect of speed on noise from road traffic; parameter: proportion of heavy vehicles.

Traffic restraints

The speed can often be reduced more efficiently by using physical restraints in addition to just speed limits. Speed can be reduced by using, for example,

1. Humps laid across the road surface.
2. Striping of the road to give the motorist a greater awareness of speed.
3. Road narrowing and road bending.

Road narrowing can be established for instance by introducing coned or widened areas for pedestrian use or dedicated bicycle lanes. Other means of narrowing can be implemented by introducing car parking bays perpendicular to the traffic stream, and road bending by varying the orientation of parking bays. These control measures are capable of substantially reducing the speeds and also the number of noisy events. Noise reductions of typically 2–3 dB can be achieved. A potential disadvantage is that too prominent restraints, such as high humps, may cause excessive braking and acceleration, and thus an increase of noise emission.

Road junction design

Noise can often be higher near road junctions than alongside roads with smooth traffic flow. Vehicle noise increases substantially during acceleration, particularly when the initial speed is low. At junctions, vehicles accelerate and decelerate, stop and start over. In junction design, it is important to consider how to smooth the traffic flow to minimise vehicle accelerations and thus to provide noise benefits. Also traffic management plans which aim at reducing journey times and accidents have the same objective. For example, some noise reduction can be expected from the use of linked or demand-controlled traffic light systems. The disadvantage is that often the improved flow leads to an increase in capacity which may result in more traffic. Or an increase in traffic speed may bring increase in noise. The overall noise reductions are generally small, usually less than 2 dB (Fig. 44.5).

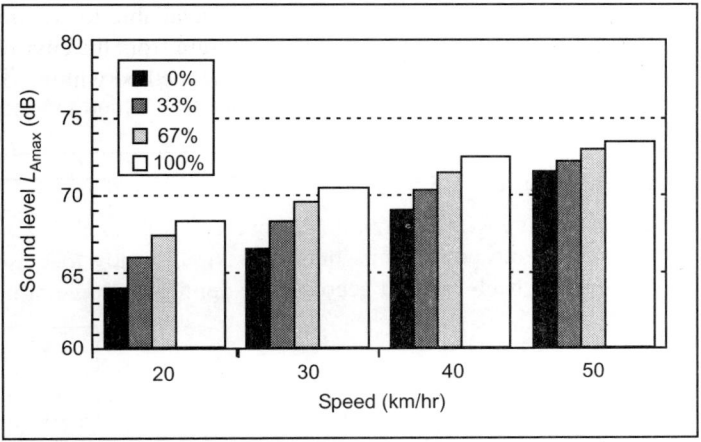

Fig. 44.5. Effect of uniformity of traffic flow; parameter: unevenness.

Another measure to smooth the flow through junctions is to switch off the traffic lights at low-density junctions during the night. However, usually no systematic improvement occurs in noise since vehicle speeds are generally increased which offsets the advantages of fewer vehicles accelerating from rest.

Road design and alignment

The noise radiated by traffic is influenced by both the vertical and horizontal alignment of the road. Generally, the steeper the longitudinal gradient the greater the resulting noise. When designing for less steep gradients, cuttings and tunnels should reserve fair attention as an effective noise control measure and the solution to be preferred over elevated embankments and viaducts.

Driver behaviour

The noise generated by an individual vehicle depends not only on the vehicle speed, but also the gear selected and whether the vehicle is accelerating or decelerating. These features may vary constantly as drivers attempt to cope with the traffic and road conditions. Driving styles may differ substantially, some drivers will drive less aggressively than others. The influence of the driver on noise generation can be considerable. Therefore, the variations in driver behaviour form a potentially useful means of controlling noise. The behaviour can be affected especially when coupled with traffic restraint measures such as speed limits. Publicity campaigns and providing information can help very much both in accepting the restraints and to adopting a driving style which is effective as to noise.

In general, driving style which reduces noise also improves fuel efficiency, reduces gas exhaust emissions and improves traffic safety. Therefore, aiming at noise control supports other goals in the areas of traffic and environmental policy. Educating drivers to be more aware of the possibility of conserving fuel can thus also promote noise control. Instrumentation for fuel consumption could lead to more economical driving and less noise. Also, measures may be introduced to influence behaviour to save fuel. For example, the use of cruise control and speed limiting devices in vehicles could be useful.

Passive driving style can result in considerable fuel saving, only small increase in journey time, and substantial noise reduction. The average reduction can be approximately 5 dB for cars, 7 dB for motorcycles, and 5 dB for commercial vehicles.

Tampering

Another form of driver behaviour that can increase noise is tampering with the vehicle. This is mainly a problem associated with motorcycles. The noise increase can be as much as 20 dB. Such actions may be difficult to control via in-use enforcement. A more appropriate method is to introduce regulations which ban the sale of poor quality replacement silencers.

Rail Traffic: External Conditions at Source

A similar situation than that existing with the emission from road traffic applies generally to noise from railways as well. At low speeds the power unit contributes significantly to noise generation, but at higher speeds the interaction of wheels and rail becomes the dominant source of noise.

Rail wheel noise

Generating mechanisms

The major component of train noise is caused by the interaction of the steel wheels and the steel rails, which generates sound by the vibration of wheels, rails and vehicle structure, track support and ground. The following vibration generating mechanisms may occur as a result of rail/wheel interactions.

1. The impact of the wheel on a rail joint; a mechanism that is not present in the case of continuously welded rail (reduction 3–5 dB).
2. The impact of wheel flanges against the rail.
3. The motions caused by track and wheel irregularities: corrugations in the rail and flats on the wheel as well as smaller scale roughness in both rail and wheel. Can raise the noise level by 10–20 dB.
4. Sliding contact of wheel flanges resulting in flange squeal (control measure is to avoid tight radii of curvature).
5. Vibration of the supporting structure.

Control methods

Methods of controlling rail/wheel noise generally follow two directions. Besides actions on the train itself, it may be possible to reduce or control the roughness of both wheels and rails and to reduce the formation of wheel flats and rail corrugations. For example, it has been found that both wheel flats and rail corrugations may be substantially reduced by employing disc brakes rather than the more conventional wheel tread braking. Also, rail corrugations can be controlled by routine grinding of the rails.

Other methods of control include the use of bogie skirts, rail-side noise barriers, reducing the number of wheels and the employment of rail isolation techniques. These include the use of resilient rail fasteners to aid damping in the rail and to decouple the rail from the support structure, and the use of ballast mats on bridge decks to limit vibration coupling to the bridge structure. In addition, enhancing the ground absorption between and beside the rails may further reduce the noise by a few decibels.

Rail maintenance

The conditions of the surface of the rail and the tread of the wheel have a significant effect on noise from a train. Defects in the wheel tread such as flats (due to wheels sliding during braking), loss of portions of the wheel tread due to thermal or mechanical fatigue, various rail running surface defects, and rail joints are all major causes of train noise.

Air Traffic: Operational Controls

Over the last decades the problem of noise from aircraft has received substantial attention and some major successes in controlling it has been achieved. The methods dealing with the abatement of the noise emitted can again be divided into two:

1. Development of quieter power units.
2. Regulation and control of aircraft operation in the vicinity of airports.

Also the number of sources (volume of air traffic) affect the overall noise reduction at the receiver. For instance, the noise emission from aircraft has decreased during the past 20 years more than 10 dB, but the increase in traffic volume results in roughly no reduction as a net effect.

The noise emitted by the aircraft themselves have been subject to continuously tightening emission limits for a number of years. The tightest limits were declared some 25 years ago, and no further lowering of the emission limits have been agreed upon.

The European Union has been concerned with the lack of further progress on a new noise stringency standard. Consequently, The EU Council has adopted Regulation 925/99 on the registration and operation of modified and recertificated aircraft (i.e. equipped with the so-called 'hush-kits').

In-flight operations

The noise of an aircraft is closely related to the way it is operated. For instance, the noise can vary considerably depending on the climb-out or landing approach procedures used. In addition, ground operations can influence airport noise impact.

Takeoff/landing restrictions

Aircraft are at their loudest during takeoff when full power is used. If residential or other sensitive areas are situated close to the airport, the aircraft have not climbed to a sufficient height when crossing them and noise reduction measures are needed. For example, engine power is reduced when a safe height is reached and the climb is continued more gradually. Under an approach route to an airport the noise may be comparable although the engine power is lower than at takeoff. The rate of descent must be low to keep it within safe limits. Typical angle of descent for the last phase of approach is 3° from an initial height of 600 m. Both landing and take-off restrictions are able to reduce the size of an airport noise contour. Many air carriers have adopted a noise abatement procedure as a policy wherever they operate. At many airports the reverse thrust is used when landing to reduce aircraft speed. However, this produces a loud noise event for a short period, and at airports with sensitive areas nearby can give rise to noise problems. Some airports apply thrust reversal controls as part of their noise abatement plan and several airports do not permit reverse thrust procedures due to noise.

Noise monitoring

At most major airports climb-out is controlled using noise monitoring stations under the departure routes. They indicate whether the prescribed climb profile and engine power settings are followed.

Noise abatement flight tracks

Flight-track procedures with unique flight tracks prescribed for departures or arrivals are commonly used for avoiding flight over noise-sensitive areas. The flight track is the projection onto the ground of the flight path of aircraft. For noise abatement purposes, the flight tracks can be useful for both approach and departure in positioning the aircraft relative to ground or land uses.

Many airports assign headings that place aircraft over non-populated land including water; or agricultural and wilderness areas. However, for airports with a dense surrounding population, this kind of flight track method may not be applicable. Then a rotation of operating runways may be used, with flight tracks distributed in a more or less equal pattern, in order to attempt to spread the noise geographically more evenly to surrounding communities.

Time of day restriction (curfews)

Air controls of this type generally apply to the time aircraft are permitted to operate, typically limiting the hours in which an airport may permit flight operations. Many major airports have some form of restriction during the night. No jet operations are allowed at night in certain specified areas (Washington), or the number of take-offs during the night is substantially restricted. In a region with more than one jet airport, a night curfew can be a feasible proposition. Switzerland imposes a night-time operational curfew for all traffic. Examples of the noise zones of an airport as a result of noise control actions are shown in Fig. 44.6.

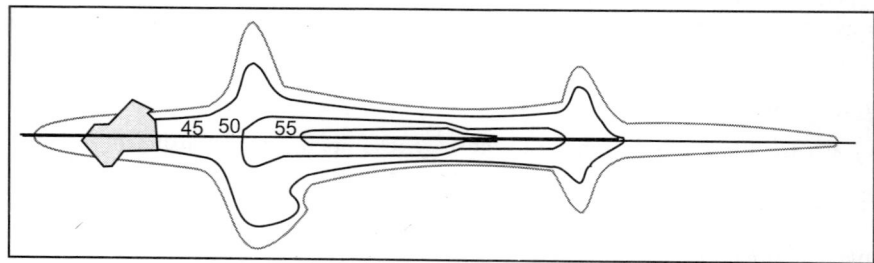

Fig. 44.6. Example of the noise zones of an airport as a result of noise control actions.

A partial curfew is also common where the airport permits certain operations at night, based upon the type or class of aircraft. For example, scheduled departures of noisier aircraft may be prohibited. Some airports place a restriction on the total number of operations over a selected long time period, for instance summer. Restrictions in many countries are not limited to only civil aircraft. In rare instances complete curfews are in effect restricting any aircraft, usually at night. In the European Union, discussions have taken place on the possibility of banning the whole of air traffic during the night (hours 23–06). One disadvantage from this action is a difficulty of scheduling flights with longer distances, perhaps involving multiple time zones.

Perimeter rule

This rule is used to limit the stage length of flights departing from and arriving at the airport. It is sometimes applied when there are nearby airports that can operate without such a restriction. This can influence noise in several different ways:

1. It can impact the capacity of a airport. In general, the fewer operations limit the overall noise.
2. With restricted lengths the maximum take-off weight, heavily influenced by fuel, is less. This permits more lift and can reduce the ground noise contour of the aircraft.
3. The aircraft type needed for a reduced stage length may be quieter than aircraft used for longer flights.

Ground operations and traffic management

Aircraft ground operations can also cause noise problems close to airports. The sources of ground operation noise include engine testing and run-up prior to taxiing, and standing aircraft noise on apron and terminal stands. Methods of controlling the noise from these operations include, for run-up noise, reorientating or relocating the aircraft away from noise-sensitive areas or the use of suppressors and barriers. Other ground operations are controlled using space to separate noisy operations, such as start of roll, from sensitive areas, and buildings and screens to shield the noise.

The management of the airport can have a significant influence on noise control. Airport design and its operational runways, run-up areas, buildings and noise barriers can influence community impact. Administrative controls and remedies, such as operationally based charges for noise, are increasingly common. The foremost ground-based control methods include:

1. Slots, limiting the number of operations within a specified time period. This can involve restrictions on how many air carrier movements are allowed during a 24-hour period.
2. Capacity, generally refers to the number of flights or passengers permitted over a defined period, e.g. a year. A major reason is to limit the noise.
3. Preferential runways, the most common operational technique in use.
4. Displaced threshold: A point along the runway is used as a landing threshold for arriving aircraft or a take-off point for departing aircraft. If displaced, the approach or departure takes place higher above the ground. Increasing the altitude reduces the noise exposure contour, thereby affecting fewer people.
5. Ground run-up: Static run-up tests of engines, associated with maintenance and repair, can generate noise impact, depending upon location, time of day, aircraft type, and facility. Run-up noise control is usually achieved applying the measures of noise emission limit, time of dayrestrictions, location, site design or abatement equipment.
6. Aircraft towing: The towing of aircraft for noise control purposes is not common, but the technique may reappear, depending upon safety, energy, and noise cost-benefits in the future.
7. Noise based charges, the idea being that the aircraft operators should pay a fee proportional to the generated noise. The operators of noisier aircraft are financially penalised while the operators of quieter aircraft are rewarded by reduced landing charges.

Training restrictions

Aircraft training activities are an important subject for regulation in many airports, and training restrictions are a common type of operational noise control method.

Industry: Control at Source

Industrial noise is generated by a wide and mixed collection of various types and forms of noise sources. Therefore, only a very general outline of their abatement measures can be drawn in a short inventory. One typical class of sources are rotating machinery: fans, pumps, compressors as well as gas turbines, diesel and electric motors, and gears connected to these. Another is formed by moving or flowing gas, liquid or solid particles, in ducts, pipes, transmission lines, through valves, at openings into open air etc. A third class may be working machinery, often impulsive as to noise.

Specific examples of strong concentrated sources are stone crushing plants, stone quarries, pithead installations and asphalt stations. Power stations are also often considerable noise sources. Wide area-type outdoor sources may be found in petrochemical complexes and saw mills. A general 'factory' may radiate noise from machinery located outside or on the roof of the building, or through walls and openings. Moving machinery, fork-lift trucks, excavators or cranes may operate outdoors, etc. A rough list of typical control measures at source may include the following:

1. Silencers, attenuators or mufflers, in connection with rotating machinery and ducts/pipes leading to/from these.
2. Screens, enclosures, cladding or even dedicated separate huts or buildings.
3. Improved sound insulation of buildings, for walls, windows, doors, openings, ventilation, etc.
4. Alternate, inherently less noisy solutions or devices in processes or production procedures (such as lower velocity, screwing or cutting instead of striking, etc.).

Permission policy adopted by the (usually local) authorities is a general means for abating or controlling industrial noise.

Summary: Control at Source

Table 44.2 gives summary of control at source.

Table 44.2. Summary of control at source.

Measure	General effect, dB	Local effect, dB
Road traffic		
Power train	0–3–5	–
Tyres	0–7	–
Low-noise road surfaces	0–2	0–5–10
Traffic management		
Volume, redirection, by-passes	–	0–5
Restrictions in time and area	–	0–15
Speed reduction, traffic flow		
Speed limits*	0–2	0–2–4
Speed monitoring	–	–
Traffic restraints	–	0–3

(Contd ...)

Measure	General effect, dB	Local effect, dB
Optimal flow, accessibility, green wave	–	0–2
Road junction design	–	0–2
Road alignment	–	0–2
Driver behaviour	0–3	0–5
Rail traffic		
Power unit	–	–
Rail-wheel interaction	0–3	0–6
Air traffic		
Power unit	–	–
In-flight operational controls	–	–
Landing/takeoff restrictions	–	–
Noise monitoring	–	–
Noise abatement flight tracks	–	–
Time of day restrictions	–	–
Perimeter rule	–	–
Ground operations, traffic management	–	–
Training restrictions	–	–

*Requires efficient control.

LIMITING NOISE PROPAGATION

Controlling noise at source is usually regarded as the principal method. However, measures which attempt to limit the spread of noise, once generated, are often of considerable value as well. For instance, for road traffic the techniques include road design and alignment, the management of traffic flows, the use of screens and barriers. A more general method is the consideration given to the land use near major transport routes.

Also, even if significant reductions in aircraft noise exposure have been accomplished through operational procedures, additional methods of control are often required. Non-operational controls primarily include the management of land use in the surroundings of an airport.

Land Use Planning and Management

A major road, motorway or railway line generates high noise levels in its vicinity. When a new traffic route is planned through an existing urban area much of the existing built environment will remain. The layout and design of the route becomes crucial in minimising the noise impact from traffic. If, however, a road or railway passes through an area that is undeveloped or scheduled for re-development, noise impact control by appropriate management of the adjoining land use can also be considered.

Opportunities for successful acoustical site planning are determined by the size of the available space, the terrain and the zoning policy applied. Suitable techniques include:

1. Placing as much distance as possible between the noise source and the noise-sensitive activity.
2. Placing noise-compatible activities such as parking bays, open spaces, shopping areas and commercial facilities between the noise source and the noise-sensitive areas.

3. Adopting cluster development concepts for housing as opposed to more conventional grid patterns or 'ribbon' development where the first row of houses tends to take the full impact of the noise.
4. Using the natural land and built form, and plantings as barriers to screen sensitive areas.

Location of noise sources, spatial separation

Road and rail traffic

One of the main methods of control is to use distance of the traffic route from the area to be protected. For instance, doubling the distance will generally lead to a 3–5 dB level reduction from a given traffic flow depending on the attenuation of the ground between the road and the receiver point. Dwellings can be protected from traffic noise by setting them well back from the transport route or line. However, this approach is often uneconomic. For example, on a level site next to a motorway the noise seldom falls below 70 dB at distances less than 100 m from the road. It has become a tradition to always consider spatial separation as one essential control measure. Nevertheless, in some cases it may be the only possible solution. For instance in mixed developments the high-rise blocks cannot be easily screened by barriers. Instead, these should be located as far from the road as possible. The rest dwellings of low-rise type can often be protected by roadside barriers or by ground attenuation.

During the last decades this approach has led to rather wide noise protection zones, say, 100–200 m on each side of a motorway. The result has been an increase in the total number of vehicle kilometres per year (the total traffic work), due to a sprawled city. In the future, these zones could rather be considered for sites of new buildings, protecting the older buildings. The results could be a good noise environment at the original buildings, and a negative cost of noise control. In general, an extensive assessment has shown that compact cities are not more problematic than more sprawled ones as to their noise climate.

Air traffic

Generally, airport noise needs good land use planning, providing adequate spatial separation of noise sources and noise-sensitive area. However, suitable land for a possible site of a new airport is becoming relatively scarce. It is usually found only at some distance from the cities. Montreal airport, at a distance of 85 km from the centre, is one of the first examples of the siting of a new airport where the total area of the land purchased includes the area to be impacted by noise; in this case approximately 300 km^2.

Where noise problems occur as a result of an existing airport or where space cannot be used to effect a satisfactory solution, other land use management options can be employed. These include:
1. Land zoning, to provide separate areas for commercial, recreational and residential development which are compatible with the noise exposure.
2. Negotiated and compulsory acquisition of land and property by the airport authority. This permits the proprietor to develop the land with a more noise compatible use.
3. Financial compensation (for, e.g. sound insulation improvement) where the property owner receives payment in return for permitting the over-flight of aircraft. (This option is not a noise reduction measure, because the noise and its effects remain.)

Noise zoning

One way of ensuring that spatial separation is given full consideration is for the local administration to impose a zoning policy. Land adjoining a major traffic noise source has development restricted to activities which are not noise-sensitive (e.g. commerce, agriculture, industry).

Such a technique offers the advantages of clearly defined development policy. But there is usually not enough demand for noise-compatible land use to afford adequate protection for every community. Also, this type of strip zoning may not be compatible with other plans for the growth and development of the community. Or it could be in conflict with the development of adjacent communities. Where areas blighted by traffic noise are not taken into noise-compatible use they can become useless waste land, often too expensive to maintain.

Defining a general zoning policy for an area may work reasonably well when all sectors of the community combine to form an agreeable plan. For example, in a typical system of zoning for land use alongside motorways, no housing is allowed within certain distance (e.g. 30 m) from the road and apartments within an adjacent zone (e.g. 30–80 m) have to have additional sound insulation. Along other roads, house construction is not permitted within a short distance from the road. A typical arrangement for air traffic is used in Germany: Around certain airports the protection zones are fixed (zone 1 with noise level below 75 dB, zone 2 below 67 dB), and land use for building is restricted. In zone 1 housing is not allowed, and in zone 2 only allowed if the buildings are insulated. Schools and hospitals are not allowed in either zone.

Use of insensitive premises

A conventional grid subdivision of land affords no real noise protection from an adjacent transport route. The first row of houses receives the full impact of the noise. In contrast, cluster developments enable the whole area to be planned as a single unit taking into account the required density of housing, depending on noise exposure and the use of both space and noise-compatible development as buffers. Figures 44.7 and 44.8 show examples of cluster development close to a major road. For instance, the placement of light industrial units near the road provides some screening to the dwellings located closest to the road. The dwellings further from the road are not as well screened but benefit from the increased distance. In this way an acceptable acoustical environment can be achieved for all residents in the area. Similarly the cluster development concepts in Fig. 44.8 show how the use of space near the road can provide protection for dwellings, aesthetic appeal and community value.

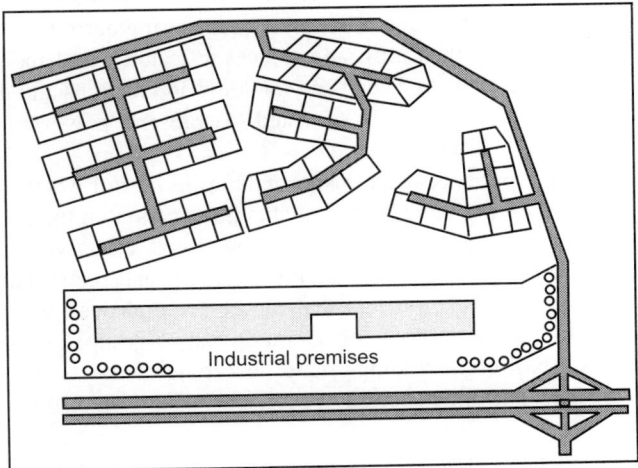

Industrial premises

Fig. 44.7. Example of noise compatible land uses.

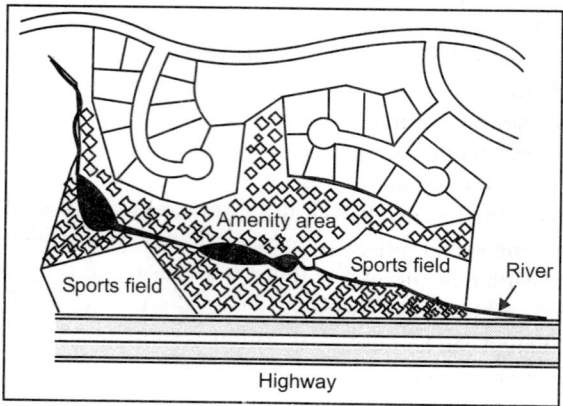

Fig. 44.8. Example of using open space for reducing noise impact.

Noise Screening

Barriers

A cost effective control method for traffic noise is to erect a barrier or screen alongside the road (or railway line). The main requirement is that the barrier should be sufficiently high and long enough to provide a reasonable vertical and horizontal overlap with the line of sight of the road from the receiver point. Barriers can reduce the noise level by up to 15 dB. When the buildings to be screened are close to roads with heavy traffic, the practically achievable noise reduction is usually of the order of 5–10 dB. However, at greater distances the screening potential may be substantially lower. In some extreme cases the sound level at a large distance may be higher with the screen than without the screen. This may happen if the noise source is situated low in comparison to the surrounding terrain and the screen is relatively low.

A wide range of materials have been used for barriers, including earth mounds, wood, steel, aluminium, concrete, masonry block, acrylic sheeting and rubber mats. Absorbing barriers of various constructions are widely used. Absorptive facing on the traffic side reduces reflected sound and is claimed to improve screening. Barriers over 8 m in height have been used for some applications and novel capped barriers, angled barriers and vegetative barriers have been tested. At present, however, only relatively small amounts of vegetative barrier have been constructed.

There is now a considerable experience in the design and location of noise barriers and comprehensive prediction and design methods are available. The available information on the acoustic performance of various types of barrier is for most cases sufficient to allow a cost effective application. Barriers do have also adverse effects: for instance, the degradiation of the visual scene and impression, and the increased difficulty of crossing a road. Barriers that may offer improved performance over simple reflecting barriers and therefore are worthy of consideration can be grouped as follows:

1. Absorbing barriers—barriers including sound absorbing elements that absorb a significant proportion of incident sound and hence reduce reflected sound which could contribute to overall noise.

2. Capped barriers—barriers that have a specially shaped top section which is claimed to further attenuate waves diffracted over the top of the barrier.

3. Double barriers—two parallel simple barriers built along one side of a road so that the sound from the traffic stream is diffracted over the top edges of both barriers.
4. Longitudinal profiled barriers—barriers that vary regularly in height along their length and are designed to reduce noise by creating destructive interference effects behind the barrier.
5. Angled barriers—barriers that are tilted away from the vertical in order to reflect traffic noise upwards and away from residential areas.
6. Dispersive barriers—barriers that have contoured surfaces angled so as to disperse the noise, in order to prevent strong sound reflections into most sensitive directions.
7. Embankments and earth mounds which may be used in combination with a conventional barrier.
8. Vegetative barriers—barriers made partly or wholly from vegetation which is rooted in a retained soil mound. The mound can be retained by various means, e.g. woven willow branches.
9. Covers—can take many forms, e.g. a grid or set of louvres set horizontally over a cutting or a complete cove screening both sides and the area above the road.

The following conclusions are stated in state-of-the-art reviews of noise barriers.

1. Absorptive barriers are commonly used. However, because they are relatively expensive compared with simple reflective barriers their acoustic performance should be determined more precisely by full-scale testing for a range of highway screening situations.
2. Double barriers are very seldom used although theoretical models have shown their potential effectiveness. It is desirable that these barrier configurations should also be tested in full scale since they could prove effective and they may widen the range of options available.
3. Multiple edge barriers and other barrier shapes are worthy of further tests since cost effectiveness could be high.
4. Tilted and dispersive barriers should be considered as an alternative to absorbing barriers, particularly where barriers are constructed on both sides of the road.
5. Vegetative barriers have attracted interest because of their environmental and aesthetic appeal, but little is known of their acoustic performance. These barriers could be tested in full scale to determine whether foliage has a significant effect on acoustic performance.
6. Covers which surround the highway are relatively expensive but because noise reductions can be very large the costs might be justified in environmentally sensitive areas.
7. The cost of noise barrier construction can be high. Computer programs have been developed to aid the barrier design so that the desired degree of protection is provided at minimum cost.
8. The opportunity to check the performance of barriers alongside highways should be sought where full scale tests have shown the potential for relatively high cost effectiveness.
9. Consideration should be given in further research to the screening efficiency of barriers at a wide range of distances and under different meteorological conditions.

Depressed roads

The noise radiated by traffic is influenced by the vertical position of the road. In areas where distance cannot be utilised to produce a desired effect the road may be placed low relative to the surrounding terrain, or depressed in an actual cutting, where the sides of the cutting act to screen the sensitive area. Or, where a higher degree of noise attenuation is required, the use of covers, enclosures or tunnels can be considered. However, the high cost of the latter constructions generally precludes their use in most situations.

It is always advantageous from a noise point of view to place the road low in the landscape. Even if the road is not depressed in a cutting, a low placement maximises the ground attenuation caused by acoustically soft natural soils. Roads in cuttings are generally well screened by the edge of the cutting wall. Reflections from the opposite wall, however, can reduce the screening performance unless absorption is used. Where the space is not limited, the cutting can be formed by sloped embankments. The noise is not greatly affected by the depth of cutting because the improved screening provided by the increased depth of cut is offset by the increase in reflected noise from the opposite wall of the cut area. Improvements in the screening can be obtained using absorbing wall surfaces or by sloping the reflecting wall.

Buildings as noise barriers

Additional noise protection can be achieved by arranging the site plan to use buildings as noise barriers. A long building, or a row of buildings parallel to a highway can shield other more distant houses or open areas from noise. A two-storey building can reduce noise levels on the side of the building away from the noise source by about 13 dB. Further rows of buildings may only produce a small additional benefit, e.g. 1–2 dB beyond the second row only (Fig. 44.9).

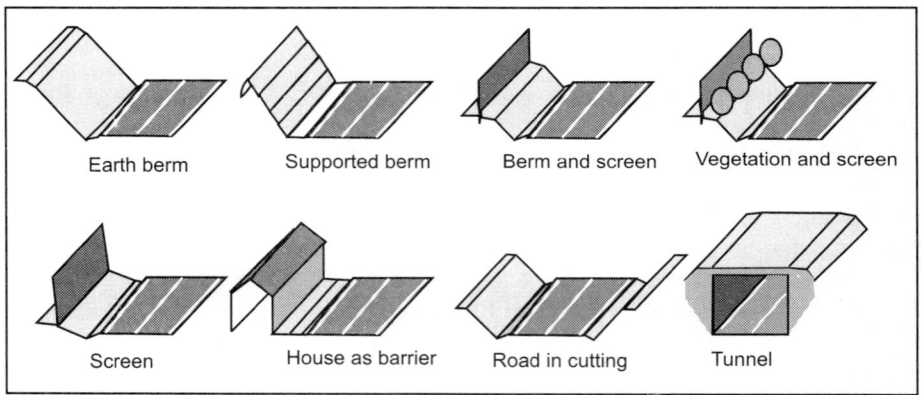

Fig. 44.9. Examples of noise barriers.

When sound propagates through areas with detached houses, the losses by screening and scattering are small. A typical value is 5 dB/100 m, but not more than 10 dB overall. Single buildings should always be situated parallel to the street or railway. Then at least the windows on the backside are in the sound shadow of the house, where the sound level is much lower than on the front side.

It is not recommended to situate houses perpendicular to the street, because then both sides are almost fully exposed to the traffic noise. The traditional arrangement of closed rows of buildings along the streets is the best way to create quiet zones. These serve as screens for houses or areas and backyards behind them. The higher the buildings alongside the street are, the better the zones are quiet. Existing gaps should be closed, if possible.

Combination of buildings and barriers

Screening can also be realised with a combination of a building and a barrier:

1. A row of single-family houses combined with a mound or a dike; earth against the totally closed facades on the side of the road (dwellings built into a dike); sound reduction up to 13 dB, depending on the height of the dwellings.

2. A noise screen (of glass) combined with a gallery-type of block of flats (the screen forms the outer facade of the building on the side of the road); sound reduction up to 20 dB.

Tunnels

A tunnel is the most effective means of noise screening, but very expensive and seldom possible because of noise abatement reasons only. Tunnels are built in urban centers where land is very expensive, and especially when they can be covered with buildings. Construction costs and the costs for maintenance, illumination and ventilation of tunnels are high (Fig. 44.10).

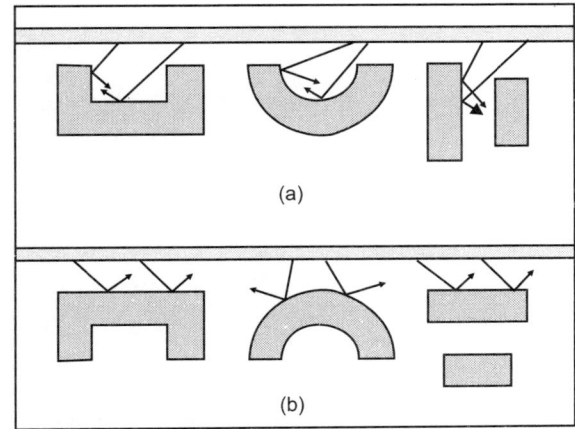

Fig. 44.10. Examples of the use of buildings as noise barriers: (a) to be avoided, (b) preferred.

The ramps are usually clad with sound absorbers in order to reduce the sound reflected by the walls. The absorbing claddings can reduce the noise by about 10 dB. Though this is mostly not needed elsewhere that at tunnel mouths, sound absorbing claddings are often used because they are less expensive than tiles. Sound insulation measures may also be needed. For instance in Berlin a motorway is led through an apartment building. Big efforts were made to reach an efficient insulation against airborne and structureborne sound.

Vegetation as noise shield

The attenuation by trees, hedges and woods is mostly overestimated by architects and urban planners. People often believe to hear less when they see less. Some general conclusions can be drawn:

1. Plantings which are high and dense enough to obscure the traffic visually provide mere attenuation than that due to the mere distance provided by the buffer strip. An attenuation of, at most, 1 dB per 10 m depth of planting can be expected. Shrubs or other ground cover are necessary to provide the required density near the ground.
2. The psychological effect of planting should not be ignored. By removing the noise source from view, plantings can reduce human annoyance to noise. The fact that people cannot see the highway can reduce their awareness of it even though the noise remains.

Summary: Limiting Propagation

Table 44.3 gives summary of limiting propagation.

Table 44.3. Summary of limiting propagation.

Measure	General effect, dB	Local effect, dB
Land use planning and management		
Location of noise sources, spatial separation		
Road and rail traffic	0–2	0–6
Air traffic	–	–
Noise zoning	–	0–20
Use of insensitive premises	–	0–20
Noise screening		
Barriers	–	0–15
Depressed roads	–	0–5
Buildings as noise barriers	–	0–20
Combination buildings-barriers	–	0–20
Tunnels*	–	0–30
Vegetation†	0–2	0–1

*Effective but expensive.
†Generally overestimated.

PROTECTIVE MEASURES AT THE RECEIVER

In most practical situations the overall effect of controlling noise at source and limiting its propagation are not sufficient methods of control. A further technique, which applies to all transport modes, as well as other environmental noise sources, is the improved design and insulation of property to minimise disturbance within buildings.

Especially, aircraft noise requires often additional methods of control, even if significant reductions in exposure have been accomplished through operational procedures. Besides methods related to land use, nonoperational aircraft noise controls include the sound insulation of property as well as other forms of subsidy and compensation.

At the planning stage of a new building, the shape, orientation and location of the building and the arrangement of the internal spaces should be chosen to reduce potential noise problems. In existing buildings the acoustic environment can sometimes be made more acceptable by altering the use of the rooms, but in general it is necessary to improve the noise insulation of the building.

When the noise reduction is to be provided by the building enclosure it is essential that all possible paths for the transmission of sound are considered. The less isolated parts on the outside of the building have to be paid special attention. There may be restrictions and limitations on the lifestyle of the people within the building. For example, windows cannot be opened to provide natural ventilation without reducing the sound insulation.

Sound Insulation

Today, good sound insulation of the outer shell of the building is widely considered to be more or less a must as soon as the outdoor level exceeds 55 dB in the day or 45 dB at night. However, sound insulation can only provide a satisfactory acoustic environment within the building, but not outside. The balconies, gardens and other spaces outside, on the noisy side of a building will still be subjected to noise.

Therefore, reduction of noise by the building itself should only be considered as a last resort; that is when insufficient reduction can be made at the source or between the source and the building. In one estimate, the additional cost of construction for noise-insulated dwellings is stated to be about 15 per cent (of construction costs) when the external sound level is about 77 dB. By careful design this additional cost is said to be reduced to only 2 or 3 per cent, but these numbers refer to countries of low standard in thermal insulation. As a whole, the costs of sound insulation is largely a question which combines good modern building technology versus old building fabric, thermal insulation and economical way of heating. In countries with colder climate, no additional cost is needed with new housing.

Sound insulation of outer walls and windows

The sound insulation of new buildings, including means provided by layout, is usually taken care of as parts of the overall design process. Improvements to the sound insulation of existing buildings usually involve some form of retrofit. Often it is not possible to avoid living rooms and bedrooms having windows exposed to high traffic noise. Then the rooms have to be protected sufficiently by sound insulating outer walls and windows. When the outer walls are massive and heavy (concrete or brick walls), the difference between outside and inside level is determined solely by the sound insulation of the windows.

Doors and windows provide the most obvious components of low sound insulation in a building. Generally the quality of these components dictate the degree of insulation achieved by the building as a whole. For instance, if an external wall has an opening or gap of about 10 per cent of its area (a value typical for windows), the overall noise reduction will only be about 10 dB even if the rest of the wall provides high insulation.

Normal modern windows with double panes, with at least a 100 mm separation between panes and good sealings have weighted sound reduction indices around 30 dB and are sufficient in most cases. Solid well-fitting doors, with good quality gaskets and rebated sills, can achieve a sound insulation of 25–30 dB in typical residential installations.

Typical new triple or four-pane windows used in the Nordic countries for good thermal insulation can achieve a sound insulation of approximately 35 dB. But near busy streets or railway lines sound reduction indices above 40 dB may be needed. Then special windows are required. Sometimes however, even new windows may be erroneously designed from a noise point of view.

Most rooms can be ventilated by opening the windows for short periods from time to time. But bedrooms should be sufficiently ventilated without the need to open a window at the noisy front of the house. However, tightly closed or sealed windows cannot be used for natural ventilation. A mechanical ventilation system or air conditioning system must be provided. The associated air vents and inlets should be located away from the noisy facade or should have silencers so that they do not provide paths for noise transmission.

One special means to protect rooms with windows on the noisy facade and at the same time to provide for ventilation is to place them behind conservatories (wintergardens). Another version of this is a glazed balcony. A conservatory or closed balcony acts as a buffer for noise. If it is somewhat damped, it sufficiently reduces the sound level in most cases.

Vibrations and structureborne sound near railways

At distances below about 50 m from railways, vibrations (structureborne sound) transmitted through the ground into adjacent buildings may become a problem—especially with floating floors. Long before these vibrations are felt, they are heard as 'secondary sound' which is radiated by the floor and ceiling of the rooms. This rather low-frequency noise can sometimes be amplified by the resonance of floating floors. The primary control method is vibration isolation under the track or under the house or both. Especially the latter means is extremely difficult and expensive to implement afterwards, in already existing instances.

Insulation from air traffic noise

Around many of the larger airports, domestic dwellings that are seriously affected by noise are eligible for a grant for improving sound insulation to reduce the internal noise. The methods of sound insulation primarily involve improvements to the insulation of windows and doors in a manner similar to traffic noise insulation requirements.

However, roof insulation and the attenuation of roof ventilation and blocking of chimney flues may also be required for aircraft noise. As with traffic noise, insulation against the provision of fixed or multiple windows will often require additional ventilation by fans.

Building Design

Room plan

In some rooms of dwellings, people are less annoyed by noise from outside than in others. As traffic noise is usually only a problem for the rooms facing towards the source, the noise-sensitive rooms should be identified and located on the other side of the building. The less noise-sensitive rooms can then provide a barrier to the penetration of the noise into the other rooms of the building (Fig. 44.11).

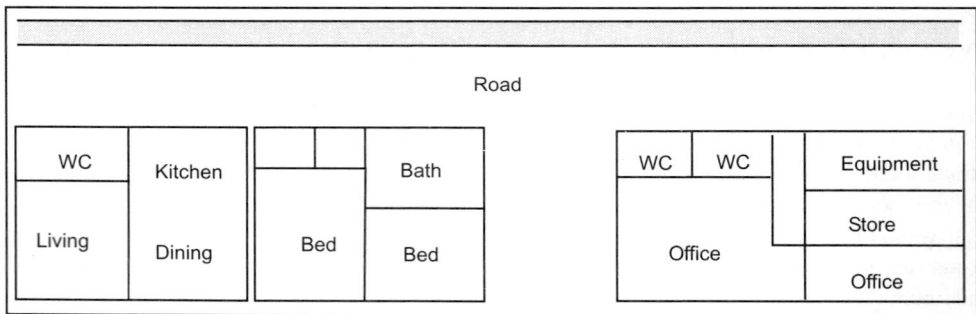

Fig. 44.11. Examples of noise compatible arrangements of rooms.

The use of the housing arrangement with the living rooms and bedrooms facing in the opposite direction to the noise source, and stairwells, kitchens, bathrooms, etc. toward the street, can provide noise benefits to sensitive areas indoors. Other building layouts can also be used to minimise noise disturbance.

Shape and orientation (quiet sides)

The need for expensive construction with high sound insulation can be minimised if the shape and orientation of the building is planned with due regard to the noise sources. The aim is to avoid reflecting sound from surfaces so as to direct it towards the noise-sensitive rooms of the building itself or any nearby building. The simplest case of this principle is found in streets with closed rows of houses on both sides. Especially when the street is narrow, the noise at the facades of the houses is higher than in streets with detached houses.

Reflections between the fronts of houses increase the sound level by up to 4–5 dB. Therefore it is recommended to use such houses preferably for commercial purposes. If they are used for apartments, living rooms and bedrooms should be oriented to the quiet side of the house if possible. In a more dedicated design, the shape of the building can be utilised to provide a self-protecting building where some parts, such as wing walls and balconies, provide shielding from road traffic noise.

Some examples of building shapes which can diminish noise intrusion are shown in Fig. 44.12. These same building shapes can be positioned so that noise is not reflected back into the building. A bonus of the design shown is that the building itself can be orientated to act as a barrier to the road traffic noise and thereby provide shielding for the spaces behind it.

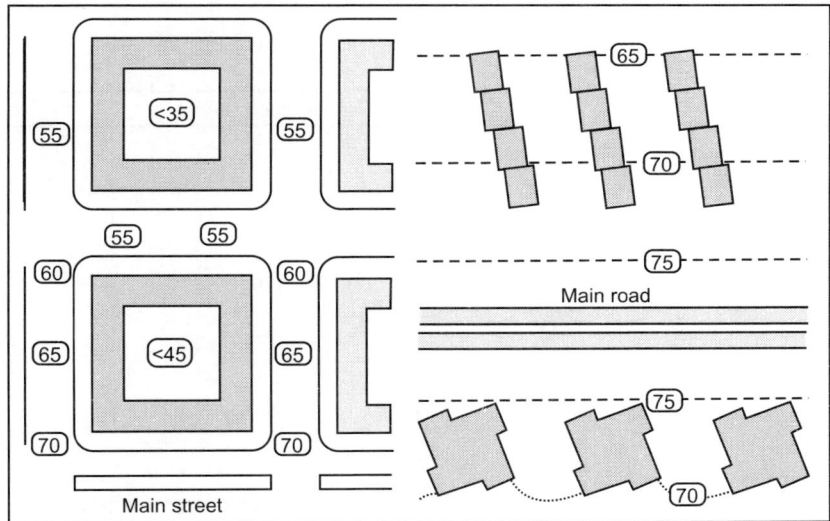

Fig. 44.12. Examples of building shapes and orientations: (a) main street 50 km/hr, 15,000 vehicles/day, 5 per cent heavy, noise situation good; (b) main road 70 km/hr, 50,000 vehicles/day, 10 per cent heavy, noise situation bad.

Two other examples of self-protecting buildings are shown in Fig. 44.13. A high-rise building is set well back on the site so that only the facade of the low-rise podium is exposed directly to the road traffic noise. Glass areas, which generally have low sound insulation, are facing away from the road on one side of the building and shielding is provided by the shape of the building. The glass areas on the other side are protected by a solid wing wall which also provides a quiet courtyard.

A thin-walled courtyard or a row of sheds or garages between a house and the traffic has been shown to provide an attenuation of the order of 12 dB. The reduction by balconies can range from 5 to 14 dB depending upon the width of the window, the angle between the road and the window, the depth of the balcony and the height of the boundary wall. For balconies located well above the level of the traffic, the underside should be designed so as to reflect sound away from windows at lower levels or it should be covered with a sound absorbing material.

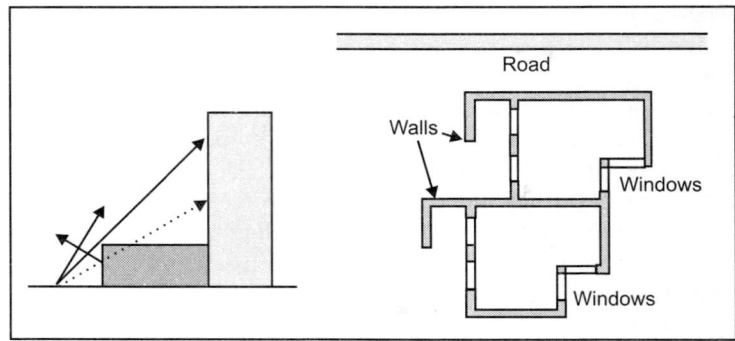

Fig. 44.13. Examples of self-protecting buildings.

Table 44.4 gives summary of control at receiver.

Table 44.4. Summary of control at receiver.

Measure	General effect, dB	Local effect, dB
Sound insulation		
Outer walls		
Windows	–	0–10
Roofs (air traffic)	–	0–3
Noise induced by railway vibrations	–	0–6
Building design		
Room plan	–	0–10
Shape and orientation (quiet sides)		
Orientation	–	0–20
Self-protection	–	0–20
Assisted protection	–	0–10

ECONOMIC AND SOCIAL REGULATING METHODS

The impact of emission limit legislation on overall noise levels is likely to remain limited. Effective noise abatement action will require increased use of other instruments, such as land-use planning and economic instruments in combination with stricter standards. In order to put greater emphasis on the polluter pays principle, economic instruments in particular should have a greater role.

Costs and Benefits of Noise Abatement

The technical solutions to noise problems described in the previous sections have a potentially strong influence in traffic planning decisions, particularly if the costs can be kept low in relation to the total investment and to the benefits gained. However, noise abatement measures have a finite cost, and this can often be a key issue in the decisions. The decision maker needs to know whether the costs of the noise control methods are worth the benefits in noise reduction. So far, proper cost-benefit analysis is not always done in reality, thus this tool for the decision process may not be available. Many of the most important benefits, such as better health and better sleep or social benefits, are difficult to evaluate in economic terms. However, environmental economy has been greatly developed in the last few decades. There are several different methods available that could be applied to noise questions, also a large variety of evaluation methods for the assessment economic costs and noise benefits in financial terms.

Still costs and benefits are often not necessarily assessed on the same scale and, in the absence of further guidance, the balancirg of these requires some form of judgement by the decision maker. The planning decision is often further complicated by the fact that noise, while being a major pollutant associated with transport, is rarely the only environmental factor of concern.

A new road may introduce noise into a quiet area but it may also affect other factors such as severance, visual intrusion, air pollution and perceived danger, which are not necessarily related to noise impact. In addition, some noise control measures may adversely affect other environmental considerations. For example, a noise barrier may reduce the quality of the visual scene and make crossing the road more difficult. The decision maker should consider all the environmental impacts of transport and be able to 'weight' the importance of impacts to arrive at an assessment of the impacts and benefits.

Economic Instruments

Economic measures form an important type of noise abatement methods. They include economic incentives to encourage noise abatement, special taxes and charges to raise funds for the abatement, and the payment of compensation to people who are affected by noise impacts.

Polluter pays principle

The guiding principle used for allocating costs of pollution prevention and abatement measures is the polluter pays principle. This means that the polluter should bear the expenses of carrying out the measures decided by public authorities to ensure that the environment is in an acceptable state. The cost of the measures should be reflected in the cost of goods and services which cause pollution in production or consumption. Environmental appraisal methods and particularly cost-benefit analysis are examples of decision making tools which, at least in principle, are available for responsible authorities in controlling noise and other resulting environmental impacts. In this case the responsibility for control is with the authority and not with the transport user. Invariably the transport user only 'sees' the benefits provided by the transport facility and does not 'feel' the environmental 'costs' created by its use (the case of externality of costs).

An alternative approach is the internalisation of costs, i.e. to leave the decision making with the users or operators of the transport system, but, in addition, to force them to see the impacts they create as costs. In practice this can be achieved with a tax. Ideally, it would be equal to the cost of the caused impact to the society. This is one interpretation of the polluter pays principle; it is intended to cause the producer to reduce environmental impacts to an optimum level. At the same time the producer compensates society for the produced impacts.

Possibilities, approaches and limitations

Vehicle noise limits provide a convenient means of enforcing the adoption of new technologies, but the process is rather slow in reacting to available changes. Therefore, it is not a particularly good instrument to encourage innovation. In some cases governments initially finance the development work for technically and economically feasible quiet vehicles, before appropriate and more stringent regulations can be introduced. The manufacturers are naturally reticent in providing new technology without an obvious market for change with resulting economic benefits.

Another complementary approach is to create market or demand for quiet vehicles by the use of economic incentives. Some economic instruments are intended to encourage the manufacturers and users to realise the social costs of the noise. These methods, which form one interpretation of the polluter pays principle, impose a noise-related tax or charge, thus encouraging the producer or user to abate the noise in order to reduce the costs. Therefore, the demand for quieter vehicles is increased. At the same time revenues raised from the charge compensate society for the nuisance caused. The process of charging the polluter has both an incentive and a redistributive or financing role.

Some cases where this approach is applied are:

1. Aircraft take-off and landing charges. A noise charge is included in the take-off or landing fee to encourage operators to use quieter aircraft with reduced operational costs.
2. Motor vehicle noise charges. The incentive is to encourage manufacturers to design quieter vehicles and therefore to make their products more competitive, and buyers to choose guided by benefits from the reduced charges of quieter vehicles.

Additionally, economic incentives can take the form of a tax concession, grant or subsidy to industry or to research organisations to promote the development and use of quiet vehicles and low noise vehicle components. These methods used in association with regulations are also seen by some regulatory parties as a promising approach to encourage innovation. Others are, however, less supportive and point to the disadvantages of additional costs of implementation. Financial aid is not consistent with the polluter pays principle. Economic instruments designed to assess or reduce noise emission are partly hampered by the lack of relevant methods to describe the emission itself. Type approval of motor vehicles as it is done today is the foremost example. This is considered to be a severe disadvantage which may make economic instruments ineffective and difficult to introduce.

Noise taxes and charges

Imposing taxes or charges on a polluter can be an incentive to reduce emissions or a way of raising funds for use in abatement. In principle, charges can be set to persuade polluters to comply with a pre-set standard. If the charges are equal to the true social cost of the pollution, polluters will, given certain assumptions, adjust their level of pollution to the socially optimal level.

Theoretically it is cheaper to reach a given overall level of pollution by means of charges than by regulatory standards. The latter are inefficient because all polluters have to comply, irrespective of cost. Charges allow the necessary overall reductions in pollution to be achieved mainly by those polluters who can reduce their emissions most cheaply. Charges are best used in combination with direct controls such as regulatory standards which reflect international obligations. Despite their advantages, noise charges are not widely used expect for aircraft noise. This is partly due to lack of confidence in their effectiveness, and partly because they tend to be opposed by some parties involved. There are also fears about transition costs, and concerns about the difficulty in deciding on the level of charges. The latter concern is largely misplaced if the objective is to reduce noise rather than to achieve a theoretically optimal level of pollution.

Aircraft noise charges

For many years noise charges have been applied successfully to civil aircraft operations whereby a noise charge is incorporated into the take-off or landing fee. The intention is to encourage operators to reduce operational costs by using quieter aircraft and to provide a source of financing to pay for alternative noise abatement measures such as sound insulation of buildings.

Aircraft are subject to take-off/landing fees at airports. A noise charge can therefore be applied as an additional fee. Such a charge acts as a complement to existing ICAO standards which limit aircraft noise emissions and to existing local antinoise procedures (take-off and landing procedures, restrictions on airport and aircraft use, etc.). Several European countries use this method whereby the fee is divided into categories according to the noise emission of the aircraft type. An aircraft noise charge has three main functions, the first two being incentive functions and the third a financing function:

1. To encourage airlines to retrofit or replace their noisiest aircraft, and to assign their noisiest aircraft to long-haul journeys (minimising therefore the landings giving rise to a noise charge) or to locations where no noise charges are collected.
2. To encourage manufacturers to develop and produce quieter aircraft.
3. To allow the financing of sound insulation, rehousing and various measures aimed at protecting airport neighbours from noise.

The impact of take-off/landing charges for aircraft to reduce noise remains somewhat unclear. The OECD 1990 evaluation argued that the efficiency had been low and did not influence the airlines choice of aircraft, whereas a report on the situation in Germany pointed to considerable success.

Motor vehicle taxes and charges

The possibilities for taxing noisy vehicles include:

1. A tax on new vehicles dependent on their noise category (which may depend on noise emission and type of use/average annual mileage).
2. An annual tax dependent on noise category. Such taxes may be used in conjunction with in-service checks that a vehicle is still within its designated noise category (this would also open the possibility of operators being able to reduce their annual tax by fitting noise suppression equipment).
3. A charge on noisy vehicles when they are used in an environmentally sensitive area. This could be achieved by selling permits to operate inside a designated area, the price of the permit depending on the noise category of the vehicle.

A problem with charges connected to use or area is that they are difficult to control. Therefore, coupling the charge with the noise emission of vehicles, and thus promoting a switch towards quieter vehicles, has been considered to be an appropriate way to internalise the noise. Whether charges do act as an effective incentive depends on the level of charge in comparison with other operating costs, and the feasibility of penalising only noisy activities. The imposition of a noise charge system for road vehicles has met considerable opposition.

Noise taxes or charges for road transport have been used in Europe even less than economic incentives and, where used, have generally been set too low to encourage noise reduction. Their main function has been to raise funds for noise control measures such as the insulation of buildings.

One of the reasons is the difficulty in finding a link between a scale of taxation and the produced noise impact. Although the noise potential of a vehicle is (loosely) assessed in a type approval test, the noise impact produced depends on the driving style and usage by the owner. Also, the driver of a car cannot be identified and controlled as easily as an aircraft operator. Therefore it is commonly considered that an impact related tax does not provide the desired benefits. In addition, noise taxes or charges are sometimes seen as a way of paying for the right to pollute, while it would be better to prohibit excessive noise instead of compensating it. If the charge is set too low the noise producer finds it cheaper to pay the charge rather than to abate the noise. The difficulty in finding the 'right price' to charge for noise continues to generate both controversy and indecision. This is obviously one reason for the lack of progress in vehicle noise charging.

Noise taxes or charges paid by manufacturers have the advantage of encouraging them to produce quieter vehicles. However, if users pay, they have an incentive to reduce noise by maintaining the vehicle, fitting better noise suppression equipment, and using the vehicle less (assuming that the taxes or charges are made dependent on in-service noise and distance travelled). For example, Austria planned to introduce a road user charge that differentiates according to the noise (and also emissions) of vehicles in 1996.

Fuel pricing

Fuel taxes are an incentive to reduce noise in as much as they reduce distance travelled. This form of taxation represents a charge on the actual use of the vehicle and is not really linked to the intrinsic noise output of each vehicle type. Consequently, a driver of a car with high fuel consumption producing low noise emission will pay a higher charge than a driver of a car with good fuel economy producing more noise. The noise charge relates, however, to the total distance travelled, placing a higher burden on the users of vehicles which impact more people. Fuel taxes may also be used to encourage:

1. The use of 'quieter' fuels, e.g. petrol as opposed to diesel.
2. The use of more fuel-efficient vehicles. These will tend to be newer and better maintained, and hence quieter.
3. Operators to favour smaller vehicles, though the noise effects of this are not straightforward to predict.
4. Fuel-efficient, and quieter, driving styles.

In addition, the funds raised by fuel taxes can be used for noise abatement or compensation payments. An addition to fuel price has been used in the Netherlands for mainly financing the Dutch noise abatement program. The charge is about 0.9 per cent of the price of gasoline and 1.2 per cent of the price of diesel fuel. Thus there is also some price differentiation with noisier diesel powered vehicles.

Road pricing

Strictly, road pricing is the charging of road users for the full marginal cost of trips as they affect other road users. A consequence of road pricing should be optimal use of the road network and this may lead to both increases (e.g. if speeds increase) and decreases in noise nuisance. However, inclusion of a noise cost element in the charge should be an incentive to reduce noise on these routes. Road pricing is implemented, for instance, in Singapore and Norway.

Financial aids

One noise abatement policy is to pay polluters for reducing their level of pollution to a socially optimal level. Setting a subsidy to achieve this optimal level requires a knowledge of the true social costs of the pollution, and there are complications in that subsidies can actually increase total pollution by increasing the number of polluters (e.g. number of vehicles). Choosing subsidies that will achieve predefined (though non-optimal) levels of pollution is also problematical.

Financial aid has an important practical advantage over other economic measures: it seldom has direct adverse effects on any interest group and is therefore easier to introduce. For this reason, financial aid has been widely used in environmental protection, despite its shortcomings.

Grants

An example of financial aid is government funding of for the development of quieter vehicles or components. This can be seen as encouraging manufacturers or research organisations to undertake research and development that they would not otherwise do, or as demonstrating the feasibility of producing quiet vehicles as a complement to imposing tighter emission regulations. Several countries have provided grants to help finance the development of low noise vehicles:

1. In Germany, a research program was launched in 1978, sponsored by the Environment Agency (UBA) which was aimed at producing production prototype vehicles with low noise emission characteristics. Government sponsorship amounted to approximately DEM 8 million over a 5 year period.
2. In France, research on quietening vehicles was carried out through the thematic Action Program (ATP) which was active between 1971–1982. Public funding was administered through the INRETS organisation with government sponsorship in the region of FRF 20 million.
3. In the United Kingdom a program known as the QHV90 was initiated. It was managed by the Transport and Road Research Laboratory and the Department of Trade and Industry and provided support to develop a range of production prototype quieter vehicles. The total costs of the program were approximately GBP 10 million over a period of 6 years and they were shared equally between the industry and the government. The project helped manufacturers meet the 1989 noise limits.

Other subsidies

Another type of incentive is the granting of subsidies or tax relief to stimulate investment in quiet vehicle development. In the Netherlands a tax relief scheme has been in operation since 1980 which provides financial incentives to private sector companies investing in low noise vehicles. The amount of relief is in the range 3.0–7.5 per cent of the price of the vehicle depending upon the weight, power and noise level of the vehicle. It is estimated that 70 per cent of trucks and buses registered since 1980 have benefited from this policy measure. In the first six years a total of NLG 36 million were saved by private sector companies as reduced taxes. The potential disadvantages of such schemes are that there is an additional cost in implementing and administering such a complex taxation system, and that there is no certainty that the quieter vehicles remain quiet in service.

Another form of subsidy involves allocating grants to operators to purchase soundproofing kits for vehicles in service. In France, for example, the government gives a 50 per cent subsidy on the extra cost of purchasing retrofitted 'hush kits' for urban buses.

Although of limited scope, this type of initiative is likely to become more widespread in the future and could be extended to include incentives for tyres and road surfaces producing lower noise: the tyre/road noise problem will have to become an important part of abatement policy in the future.

Other forms of economic incentive

Compensation

Instead of subsidising a reduction in pollution, compensation could be given to those it affects. Compensation is defined as a payment, in cash or in kind, designed to restore as far as possible a person to his state of welfare before the damage occurred. The need to compensate for noise is particularly acute in existing built-up areas around airports and along highways.

Compensation for house price depreciation caused by noise or other environmental impacts is a well established policy in some countries, e.g. the USA. House price depreciation does not reflect the full social welfare cost of pollution but compensation may in principle also be paid to cover this. There are however difficulties in deciding what level of compensation this would require. It is debatable whether adopting compensation as a policy instead of noise abatement is socially desirable, although social justice may require people suffering from noise to be compensated. It may not be good for society as a whole for people to live in environmentally degraded conditions even if the individuals themselves are satisfied by the compensation payments. In principle, noise compensation is not an incentive as such but rather a kind of financial penalty on those responsible for the noise, and a redress for those suffering from noise. In practice however, it may have some important characteristics of an economic incentive. If the polluter, or the authority making decisions about pollution, has to pay the compensation (thus another form of the polluter pays principle), they may try to minimise the payment. This is particularly so in the planning of new airports, highways and railways where projected compensation payments may act as an economic incentive to the authority to reduce or mitigate environmental impacts.

Tradeable permits

Tradeable permits to pollute are an attempt to combine the advantages of regulations and charges. They avoid the need to decide what level of charge will be sufficient to achieve a desired standard. Authorities decide on a required pollution standard, and issue permits that allow pollution only up to that point. Polluters who wish to produce more pollution can only do so if they can purchase sufficient permits from others who are willing to reduce their own pollution. Therefore, tradeable permits act as an incentive because polluters who reduce their pollution can sell permits. Also polluters with low pollution need less permits initially. Permits have been applied in the control of air pollution from static sources. Suggested application to noise abatement is based on noise emission certificates which producers would have to purchase for each built vehicle. The price would depend on the vehicle noise category. Owners who modify their vehicles to reduce noise would benefit by being able to sell their certificate and buy a cheaper one. It is assumed that the permit system would have the advantage of offering flexibility to manufacturers to introduce quieter vehicles in a shorter time, as compared to a system with strict standards and type approval certificates. Difficulties in controlling the system are thought to be one disadvantage.

Promotion of quiet vehicles

Restrictions on the use of noisy vehicles at certain times of the day and area have been put into operation in many countries. Although this can be regarded as mainly a traffic control measure it does also provide an economic incentive because it promotes the production and purchase of quiet vehicles. Germany has introduced the concept of 'low noise vehicles' which are authorised to enter a protected area covered by traffic restrictions. Vehicles qualify as 'quiet vehicles' if they comply with the following sound levels.

Engine power, kW	Sound level, dB
<75	77
76–150	78
>150	80

A severe drawback of this approach is that the sound levels stem from the standard type testing which cannot properly identify or classify low-noise vehicles. The actual noisiness should be defined in terms of noise level in ordinary traffic, for which, however, approved standard methods are still lacking.

Other Instruments

Favour of low-noise means of transport

A noise-wise meaningful shift in the mode split, i.e. in the division of use between various means of transport, is a strategic long-term goal in one the most influential areas of noise abatement.

Low-noise public transport

Public transport is not 'low noise' by definition. The key issue here is connected with the introduction of new equipment for public transport. Local authorities are in a highly influential position to affect noise abatement in the community, by setting up ambitious noise specifications when vehicles and systems are planned and purchased. It has been shown that if trams or trains are occupied to some 20–30 per cent of their capacity it makes no difference as to noise whether the travellers use railbound public transport or private cars. Thus higher occupancy and low-noise equipment have a favourable impact on community noise. From the 1970s onwards, substantially quiet city buses have been on the market, which have a marked influence on the noise situation, already with typical figures of average occupancy.

Light transport

Walking and bicycling are truly low-noise means of transport. Successful promotion and wider use of light transport are connected to two development trends. Locally, the actions needed are improvements in practical possibilities: safe and logistically usable lanes, routes, areas and zones. In a wider scope, the possibilities of promoting these forms of traffic depends on the development in land use and community structure. The foremost precondition is shortening distances between homes and workplaces, commercial premises, schools, etc.

Freight transport

Potential possibilities for noise reduction may also be found with the transport of goods. Ways to accomplish this may be more efficient combination of transfer tasks, control of traffic flows, and use of logistics and information technology.

Community development

A general short- or medium-term goal in community development, traffic and land-use planning concerning noise issues is to introduce and widen noise-free zones and areas. Typical examples of such are pedestrian areas, residential precincts and recreation areas. More long-term measures might be the development in the community macrostructure which is capable of reducing the amount of overall transport, and the evolution towards replacing the need of physical transport with telecommunications and information technology.

Increasing the density of community macrostructure

One theoretical study has arrived at the conclusion that, in general, compact cities are not noisier than more sprawled ones. The noise level (at least in cities with no less than half a million inhabitants) is roughly constant regardless of population density, leading to no conflict between noise control and energy consumption. Thus these two considerations go together.

Reduction of need for transport

One possible, and partly ongoing, development trend in the society is an increase in the transfer of information, as opposed to the transfer of people—actually physically moving from one place to another, using (inherently noisy) transport. A practical example is the trend of increasingly working distantly from the 'original' workplace, for instance at home, with the aid of telecommunications. Another is teleconferencing. This evolution has potentially beneficial effects as to noise. The trend is also more generally easy to accept and promote, as its effects on energy consumption and air pollution are also beneficial.

Education and information; noise labelling, public awareness

These activities are important in promoting acceptance of and compliance with noise regulations. They can also be used in their own right to encourage noise abatement. Possibilities include:

1. Educating technical staff, decision-makers and elected representatives in the application of noise abatement policy and the importance of noise as a problem.
2. Educating the public to gain acceptance of noise abatement policies and promote low-noise behaviour such as choosing quiet vehicles or adopting a low-noise driving style.
3. Demonstrating the benefits that improved noise abatement can bring.

Consumer pressure is a potentially effective force leading towards noise reduction, if successfully provoked. Noise labelling on vehicles and products is one prerequisite for this, and arousing public awareness of such a system is also needed in implementing the measure.

Furthermore, there are many kinds of related regulatory incentives, such as noise abatement campaigns, consumer information and education, noise surveys, noise labelling and certificates, noise-related advertising, public purchases, bans of noisy equipment, and information of sound insulation of buildings.

Summary: Economic Instruments in Noise Abatement

Table 44.5 gives summary of economic instruments in noise abatement.

Table 44.5. Summary of economic instruments in noise abatement.

Instrument	Objective	Comments
Taxes and charges	Incentive to:	
Higher take-off/landing fee for noisy aircraft	Reduce emissions (if charge depends on emission level)	Compared to standards alone, charges are a low-cost method of achieving a given level of noise pollution
Charge refunds for quiet aircraft or road vehicles		
Taxes on noisy vehicles	Change to a less environmentally damaging mode of transport	Not widely used in noise abatement because of:
	Fund raising for noise abatement	Lack of political confidence
		Difficulty in choosing the optimum tax level
		Possibility that the polluter may be taxed for optimal or suboptimal pollution
		Fears about compatibility with legal system
		Fears about transition costs

(Contd ...)

Instrument	Objective	Comments
		Existing noise charges too low for much incentive effect, and used instead to raise funds for noise abatement
		Charges may be used to encourage a market for quieter vehicles, thus complementing other economic noise abatement policies such as R&D subsidies
Fuel taxes	Reduce impact of noisy vehicles by restricting speed or distance travelled	
	Promote use of more fuel efficient (quieter) vehicles	
Road pricing	Reduce congestion, with concomitant effects on the environment	Strict definition of road pricing is to charge road user for the full marginal costs of trips as they affect other road users, but can include other external costs including noise
		Road pricing can redistribute traffic onto other roads and increase or reduce environmental impacts
Financial aid	Incentive to:	
R&D grants	Assist and encourage development of low noise vehicles	Widely used in environmental protection because, unlike charges, seldom adversely affect any interest group
	Demonstrate feasibility of reducing emission limits	
Subsidised purchase of quiet vehicles	Encourage purchasing of quiet vehicles, thereby creating a market for them	Subsidies can risk increasing the total level of pollution by increasing total number of vehicles or amount of travel
Tax reductions for quiet equipment or modes of operation	Subsidise a less environmentally harmful mode of transport	
Compensation	Incentive to:	
For house price depreciation	Reducing or removing the financial impact of noise on individuals	Methods of determining the correct amounts of compensation for welfare loss problematical
For loss of social welfare	Counteracting effects of noise on welfare	Removes incentive for people to move away, install insulation, etc.
	Providing incentive to the payer of compensation to reduce noise impacts instead	Not desirable as a noise abatement policy on its own
Tradeable permits	Incentive to:	So far mainly used for air pollution from static sources
Vehicle noise emission certificates	Reduce noise, i.e. polluter can reduce permit costs	Advocated as a way of overcoming lack of political faith in efficacy of emission charges

(Contd ...)

Instrument	Objective	Comments
		Advantages:
		Minimises total cost of pollution abatement
		Authorities can vary standards by buying and selling
		Groups could buy-up permits to reduce pollution (but governments could issue more)
		Avoids difficulties of choosing a noise charge that will achieve desired standards because actual standard is set by the issuer of permits, not by the unpredictable influence of a given charge on the behaviour of polluters

RUBBER RESILIENT WHEELS CUT URBAN RAIL NOISE

The good noise and vibration characteristics of rubber resilient wheels and the long service life of their tyres make them ideal for use on intensively-used vehicles such as trains, metro cars, and suburban trains. As static loads as well as wheel speeds are higher for metro and suburban trains compared with trains, the development of such designs is more demanding in both the design and test stages.

The search for solutions to traffic congestion problems in urban areas is high on the agenda of local authorities in most countries. Environmental friendliness is one of the most important reasons for the growing importance of rail transport. But even rail transport has environmental problems that have to be tackled. In urban areas, where the track is often very close to densely populated areas, noise and vibration is of increasing concern to residents. Noise is mainly caused by wheel-to-rail contact, especially in curves, switches, and loops. An effective way to resolve this problem is the application of resilient wheels, equipped with rubber elements between the wheel centre and the tyre. Resilient wheels not only reduce noise but also, due to their resilient characteristics, reduce wear of the tyre and track, thereby extending considerably the life of the tyre. The principle of a resilient railway wheel has been known and applied for some time. Nevertheless, Bonatrans has taken a different approach from other designs of resilient wheels that improves the quality, ease of maintenance, and strength of resilient wheels, while at the same time reducing their weight. In addition, Bonatrans does not use any linking elements such as bolts or pins in its resilient wheels. This increases the area of the wheel web available for other accessories like brake discs. It also considerably simplifies maintenance, as a worn tyre can be exchanged directly under the vehicle in service without the need to release the bogie. Development of resilient wheels by Bonatrans came to fruition in 1997 when the first resilient wheels designed and manufactured by them were fitted to a type T3 tram originally manufactured by CKD, Czech Republic, and operated by Ostrava Transport Authority. The resilient wheels replaced the original Tatra-type resilient wheels.

Following trials lasting about one month, first without and then with passengers on board, the tram was put into normal passenger operation. Up to now the tram has completed almost 4,00,000 km and passed the first tyre reprofiling that took place after 2,24,000 km. The wheel has a nominal diameter of 700 mm and was designed for a maximum vertical load of 55 kN. Prior to its introduction into service, the wheel was subjected to numerous tests of components, including rubber elements. These tests were followed first by static tests of a complete wheel for the verification of wheel stiffness characteristics in both radial and axial directions, then by dynamic tests for the verification of wheel behaviour under long-term loading simulating life-cycle tests. Dynamic tests were carried out either on a roller test rig or on hydraulic resonance equipment.

Another design of 610 mm wheel was developed for low-floor Astra-Inekon trains manufactured by Skoda Dopravni Technika in Pilsen. Bonatrans resilient wheels in those vehicles have been in operation since 1998, mainly in the Czech Republic. Considerable attention was given to noise analyses. The wheel design has lower levels of emitted noise from wheel/rail contact, which is evidenced by results of external and internal noise measurements, compared with a tram that was equipped with an original Tatra rubber resilient wheel. For example, in the T3 trains in Ostrava, attenuation of external noise between 1 and 4 dB, depending on frequency, was achieved in a straight line at a speed of 40 km/hr, while attenuation in a curve with a diameter of 30 m at a speed of 25 km/hr was between 5 and 18 dB, depending on frequency.

The wheel can also be equipped with a disc noise and vibration damper, provided the bogie or wheel design permits it. The contribution of the damper is not negligible. This was demonstrated by laboratory noise tests, during which the tyre was kick bounce excited at the tread simulating rolling noise for travel along straight track and rail/flange noise for travel through a curve. The main components, such as the wheel centre and tyre, are manufactured from proven materials that have long been used in the railway industry in compliance with UIC standards.

To produce a good wheel means not only having a sound wheel design but also using good materials. That is why R&D department of Bonatrans has been working on the development of materials with extended service life and operational safety. Besides the standard middle-carbon refined steel, new tyre materials are tested in operation: low-carbon micro-alloyed steel as well as higher-carbon steel with high strength and good plastic characteristics. Most new materials are operated together with standard steel grades and their characteristics and material life can be compared under identical operational conditions.

Chapter 45

Bhilai Steel Plant, Diesel Locomotives, TISCO and Hindustan Zinc Ltd.

INTRODUCTION

This chapter discusses case studies of Bhilai Steel Plant (Bhilai, MP), Diesel Locomotives Ltd. (Varanasi), Tisco–Jamshedpur, and Hindustant Zinc Limited–Udaipur.

BHILAI STEEL PLANT

Oxygen Plant Compressor Noise Control

BOC, Mannessman Demag – Germany and Demag Kirloskar were engaged in setting up a large oxygen plant for Bhilai steel plant. One of the major equipment for the oxygen plant is 2 Nos. large air compressor of capacity 1,47,800 kg/hr at 11.73 kg/cm² discharge pressure driven by a 9500 KW motor supplied by M/s. Mannessman Demag. M/s.AIPL was approached to design, manufacture, supply and commission a suitable noise control scheme to control the noise level within 85 dBA at a distance of 1 m around the compressor. The compressor package consists of a large multi stage compressor, 9500 KW motor, after coolers, intercoolers and moisture separator. The coolers are having a diameter of 2700 mm and length of 9.7 m. The compressor with motor was installed at an elevation of 4.5 m. The large after coolers and intercoolers are installed right below the compressor. It was also observed that the noise emanated from the coolers was much on the higher side. The noise level was observed from the compressor are shown in Table 45.1.

Table 45.1. Noise level from the compressor.

Freq. (Hz)	63	125	250	500	1k	2k	4k	8k
Suction SWL dBA	113	117	116	119	121	121	118	107
Discharge SWL dBA	126	130	129	132	134	134	131	120
Blow off SWL dBA	109	113	112	115	117	117	114	103
Casing SWL dBA	114	108	107	110	112	112	109	98

AIPL's responsibility was to design a suitable noise control system to reduce the noise emanated from the compressor well within 85 dBA limit. The temperature rise inside the acoustic enclosure is to be maintained below 5°C above ambient.

On careful analysis of the whole system involved in this project, it was observed that the coolers are posing problems not only from the point of temperature rise, but also emits very high noise from the casing. This necessitated suitable acoustic treatment by providing a brick walled enclosure for the cooler assembly package. If the cooler assembly package was to be covered inside an enclosure, then the temperature would shoot up inside the acoustic enclosure affecting the performance of the cooler. After a very careful analysis the following scheme was worked out.

1. Acoustic enclosure for the blower with motor.
2. Suction silencer for the compressor.
3. Discharge silencer for the compressor.
4. Blow off vent silencer in the bypass line.
5. An enclosure for the cooler assembly below 4.5 m elevation.
6. A ventilation package to remove the heat dissipated from the compressors, motors, after coolers, intercoolers and moisture separators.

Acoustic insulation of the suction, discharge piping and acoustic treatment of various other piping in the lubrication circuit was done. The acoustic enclosure designed was having a dimension of 9.7 m long × 9.2 m wide × 3.4 m height for the compressor/motor and the brick walled chamber of dimension 16 m length × 10 m wide × 6 m height for the cooler assembly with adequate number ventilation inlet silencers and soundproof door assembly.

The ventilation air sucked through the intake silencers takes away the heat dissipated by all the coolers and moves up through the floor grills and gets inside the compressor acoustic enclosure to take away the heat from the compressor and motor and escape through the exhaust blowers mounted on the roof of the enclosure with ventilation exhaust silencers.

After installing the system, the noise level was found to be within 85 dBA around the enclosure. The temperature was measured inside the enclosure and found to be within 50°C above ambient.

DIESEL LOCOMOTIVE WORKS—VARANASI (UP)

2600 HP Engine Test Bed Noise Control

M/s. Diesel Locomotive Works (DLW) is the manufacturer of 2600 HP and 1350 HP Loco Engines. These engines are performance tested in a Test Bed, subjecting the engines up to full load conditions prior to assembling on locomotive. DLW proposed to establish 4 Nos. of Engine Test Bed and awarded the contract to M/s. Acoustics India Private Limited for design, engineering, manufacture, supply, erection, testing and commissioning. The scope of the project included complete civil work for engine foundation, construction of control room, fuel oil/lube oil/cooling water circuits, electrical work, control panel, MCC, cables, piping, water rheostat type load boxes, cooling tower, heat exchangers, sound proof enclosures for the test beds, test bed ventilation system, sound proofing the control room, air conditioning, engine exhaust silencers, etc. on turnkey basis.

In this section, we would cover only the noise control part of the test bed, control room, ventilation and air conditioning.

Noise sources

As 4 Nos. test beds for testing 1350 HP and 2600 HP engines are located side by side, the engines were releasing very high acoustic energy which was transmitted as fluid borne noise through engine exhaust ducting and turbocharger running at 25,000 rpm. The casing radiated noise from the engine coupled with the generator was also predominant. The noise levels measured around the engines are given in Table 45.2.

Table 45.2. Noise levels measured around engine.

Freq (Hz)	63	125	250	500	1k	2k	4k	8k
Engine noise (at 1.5 m distance)								
SPL dBA	91	97	98	99	97	95	94	91
Engine exhaust (at 3.0 m distance)								
SPL dBA	93	102	98	91	90	90	84	72

The noise control system was designed to reduce the noise level within 85 dBA at 1 m distance from the enclosure and within 55 dBA inside the control room.

Complexity of the project

Being test beds, engine will have to be loaded/unloaded everyday on the test bed. Hence loco engines exhaust ducting and fresh air intake ducting should not be a fixed installation. Performance of the acoustic enclosure system/silencer designed should not get deteriorated due to flexibility of connections. Dismantling the acoustic panels of the test bed acoustic enclosures would not be possible everyday as the engines had to be loaded on the test bed almost on daily basis.

Enormous amount of heat rejected by the engine would have to be exhausted, limiting the temperature raise inside the enclosures. The broadband noise generated by the engine which is fluid borne in the exhaust line will have to be controlled within the limit (Fig. 45.1).

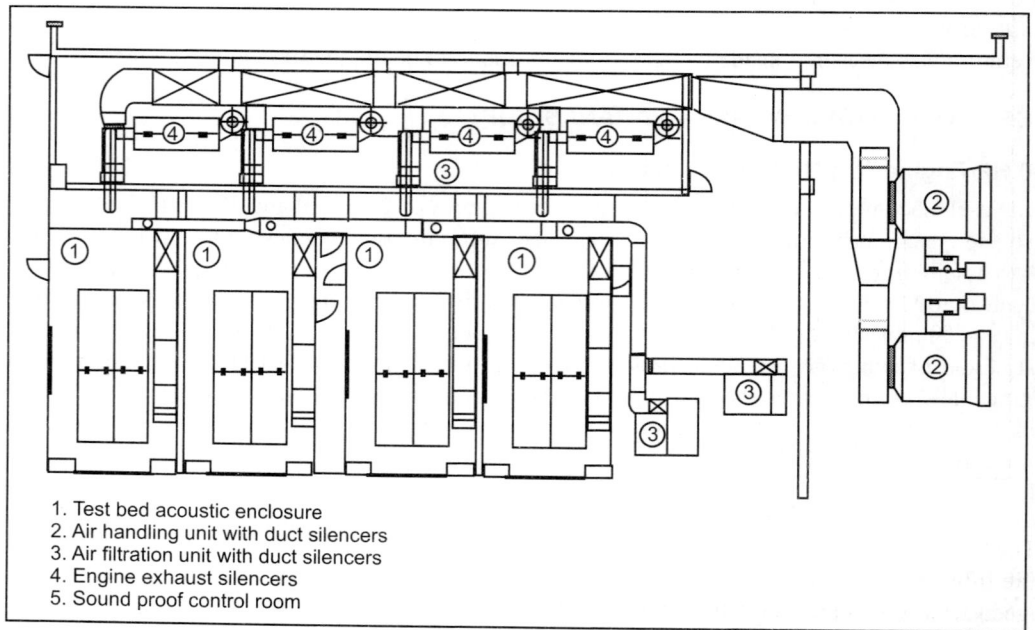

1. Test bed acoustic enclosure
2. Air handling unit with duct silencers
3. Air filtration unit with duct silencers
4. Engine exhaust silencers
5. Sound proof control room

Fig. 45.1. Engine test bed project layout.

The turbocharger noise evidently with peak amplitude in the higher frequency required a very careful design to limit the noise emission.

The scheme

The noise control package was divided into the following:

1. Test bed acoustic enclosures.
2. Engine exhaust silencers.
3. Turbocharger intake silencer.
4. Ventilation system with duct silencers with air washer based air-handling unit (AHU).
5. Turbocharger air intake filtration system with intake silencer.
6. Sound proofing of the control room with sound proof glass windows including air conditioning of the large control room.

The major requirement that was considered while designing the acoustic enclosure was the provision for loading the engine inside the enclosure from the front and from the rooftop. Each of the 4 Nos. acoustic enclosures was designed to have a dimension of 5500 mm width × 12,900 mm length × 4040 mm height. The acoustic enclosure was constructed with acoustic panels and brick wall with acoustic panel lining. Figure 45.2 shows performance of engine acoustic enclosure.

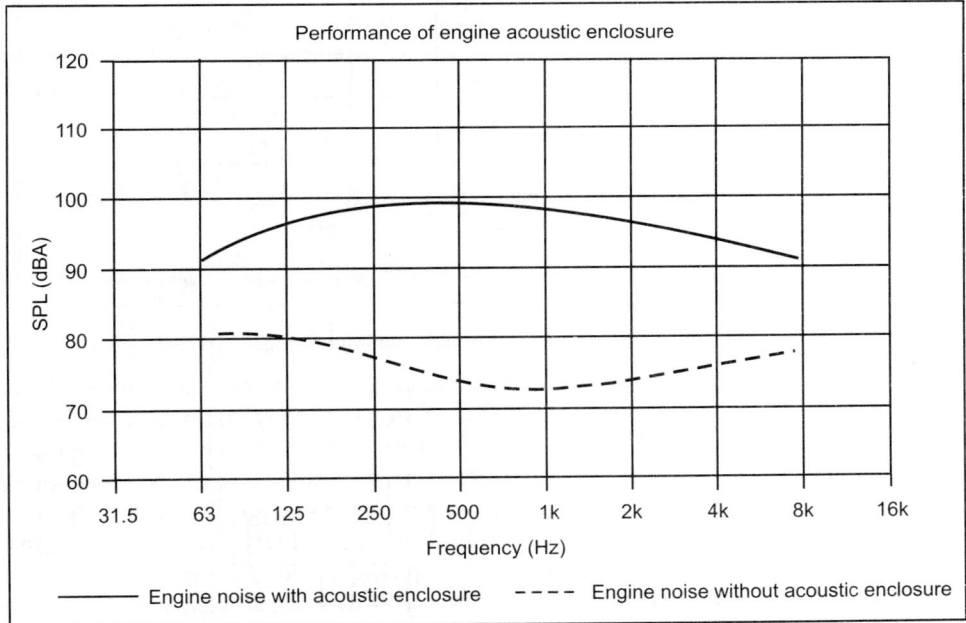

Fig. 45.2. Performance of engine acoustic enclosure.

The front side of the acoustic enclosure was designed with motorised sliding twin doors enabling the engine front-loading trolley to move inside the test bed enclosure. The roof loading arrangement was designed with a motorised sliding panel on the rooftop to have a clear opening of 3400 mm width × 6500 mm length, which enabled top loading of the engine from the rooftop using EOT crane.

The operations of the motorised mechanism of the sliding doors are controlled from the control desk in the control room. The engine exhaust system consists of a reactive and absorptive silencer with the associated ductwork and acoustic insulation. A flexible connection was used for coupling the engine exhaust connector to the fixed exhaust piping. The turbocharger intake air was ducted from an air filtration package, ducted to each of the 4 Nos. test bed, and connected to the inlet of turbocharger through flexible coupling. The noise emission from the turbocharger was attenuated through the duct silencers. Figure 45.3 shows performance of engine exhaust silencer.

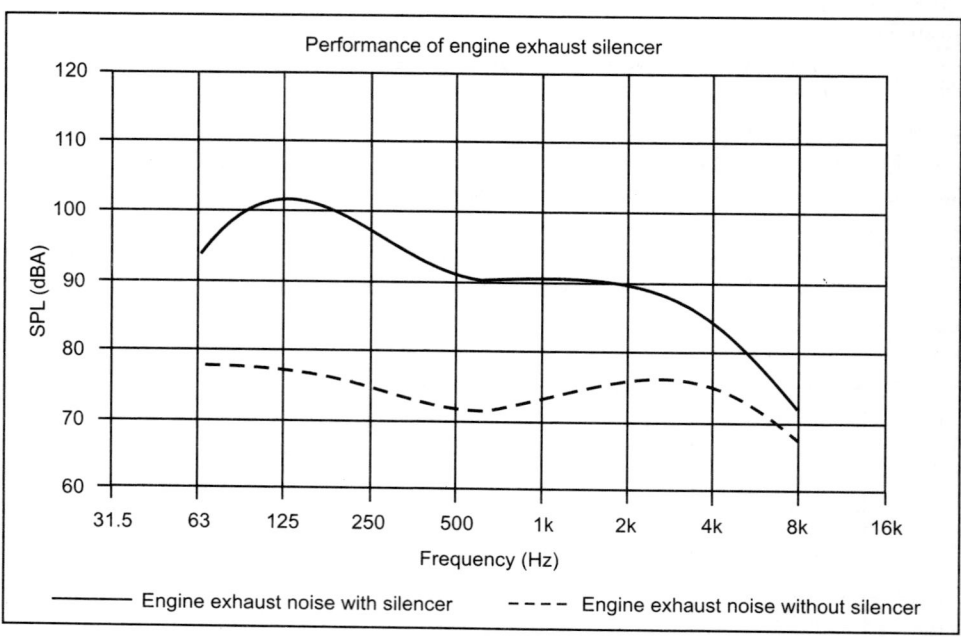

Fig. 45.3. Performance of engine exhaust silencer.

Being a test bed a very effective cooling system was designed to ensure evacuation of heat dissipated from the engine. This comprises of separate air washer based air handling unit with supply air ducting with duct silencers to each of the enclosure ensuring a cool environment inside the acoustic enclosure.

The control room sidewalls and roof were acoustically treated with acoustic panels and soundproof doors. The dimension of the control room is 26.5 m length × 4.5 m width. The control room was air conditioned with centralised unit and duct silencer. Each of the 4 acoustic enclosures dividing wall between the control room and the acoustic enclosure was provided with triple glazed industrial sound reducing glass windows to ensure visibility of test bed from control room and at the same time attenuating the test bed noise to restrict within 55 dBA inside the control room.

Thus, the whole system was tested and achieved a noise level of within 85 dBA around the test bed enclosure and within 55 dBA inside the control room.

TATA IRON AND STEEL COMPANY (TISCO)—JAMSHEDPUR

Sinter Exhaust Gas Fan Noise Control

Tata Iron and Steel Company Limited awarded the contract to Acoustics India Private Limited for the Noise Control of Fan handling 9,80,000 m³/hr hot dust laden sinter gases at 1650°C driven by a 6500 KW motor. The noise control scheme is designed to bring down the noise level within 85 dBA at a distance of 1 m around the blower and the silencer is designed to offer minimum pressure drop.

AIPL carried out the design analysis and designed a system, which comprises of acoustic enclosure, Sinter fan discharge silencer, acoustic insulation of fan intake ducting and acoustic insulation of fan discharge transition ducting. The silencer was designed to offer the insertion loss shown in Table 45.3.

Table 45.3. Insertion loss.

Frequency (Hz)	125	250	500	1000	2000	4000
DIL dB	15	24	35	43	26	16

The silencer is of absorptive type and rectangular in construction and installed in the discharge side of the Sinter fan. The internals of the silencer comprised of splitter elements arranged in parallel, inside the rectangular steel casing. The internal splitter was designed for better aerodynamic performance and for minimum pressure drop. The inlet dampers connected to the hydraulic actuators posed a serious problem due to complex arrangement. The acoustic treatment scheme for the damper actuators thereby offers noise reduction without compromising the functional requirement.

The whole system was installed at site and the noise level was measured as 83.6 dBA as against the guaranteed noise level of 85 dBA.

HINDUSTAN ZINC LIMITED—UDAIPUR (RAJASTHAN)

Control of Noise Problems

M/s. Hindustan Zinc Limited (HZL) has been operating the Zawar Group of Mines, consisting of four underground lead-zinc mines, viz. Mochia, Balaria, Zawarmala and Baroi in Block-1 with beneficiation facilities in Udaipur district of Rajasthan. HZL's Zawar Group of Mines has been certified for Quality Management System (ISO: 9001), Environment Management System (ISO:14001), Occupational Health and Safety (OHS: 18001) and Social Accountability (SA: 8000).

Impact on noise levels and ground vibrations

With the mining operations, deployment of machinery, drilling and blasting of mine development, excavation and transportation of ore and men are expected to generate the noise levels. However, almost all the mining activities are installed more than 50 m below the ground level. The noise levels due to operation of mining equipment are being confined to underground only and attenuated due to the depth of the operation. As a precautionary measure, blasting is carried out at the end of the shift.

The noise levels and vibration induced by blasting is instantaneous in nature and is attenuated due to depth of the mine below ground. Hence the impact of noise and vibration may not be significant.

The main noise generating sources are compressors, crusher house, ball mills, floatation cell 6 MW DG set. The noise levels at the source for these units will be in the range of 80–90 dB(A). The noise dispersion from the mine has been estimated through computer based mathematical model results reveals that the incremental noise levels will be well within the CPCB standards.

Almost all the mining activities take place more than 50 m below ground level. The noise levels due to operation of mining equipment is confined to underground only and attenuated due to the depth of the operation.

The blasting is carried out at the end of the shift at a depth of 50 m or more, below ground. The noise levels and vibration induced by blasting are instantaneous in nature and attenuated due to depth of the mine below ground. Impact on noise and ground vibrations are not likely to affect the structures within the vicinity of the mine lease area and hence not significant. Blast vibration is kept below the limit prescribed by Director General of Mines and Safety (DGMS).

Impact of off-site traffic on air quality

The air quality predictions have been carried out by using the air quality model CALINE-4 developed by California Department of Transportation. The model is based on Gaussian dispersal equation and uses a mixing zone concept to characterise pollutant dispersion over the roadway.

Mitigation measures for noise control

The noise level monitoring carried out in the area has indicated that the present noise levels are in the range of 38.6–63.8 dB(A). The nose at the surface is generated by movement of heavy machinery and in underground due to drilling, blasting, excavation and crushing.

Mitigation measures to reduce ambient noise levels

The following control measures are being adopted to keep the ambient noise levels well below the limits:
1. The prime movers/diesel engines are of proper design and properly maintained.
2. The operator's chamber is safe guarded with proper enclosures to reduce the noise levels.
3. A thick greenbelt provided in phased manner around the periphery of the mine to attenuate noise.
4. All the guidelines and precautions specified by GDMS are strictly adhered to in order to control the noise levels.

Measures to protect workers from high noise levels

The following measures are taken to protect the workers from exposure to higher noise levels:
1. Provision of protective devices like ear muffs/ear plugs.
2. Provision of sound insulated chambers for the workers deployed on machines producing higher levels of noise.
3. Reducing the exposure time of workers to the higher noise levels.

Mitigation measures for blast induced noise and vibration

Blasting contributes to noise pollution and vibration. The intensity of the vibration and noise depends upon the maximum charge per delay during blasting. Blasting causes ground vibrations that travel through the ground in the form of energy waves away from the blast or from the point of initiation of the charge. The measures undertaken are:
1. Latest technology of the blasting undertaken to reduce the blast-induced noise and vibrations.
2. Drilling parameters like burden, spacing are properly designed to give the proper blast.
3. More number of delays are used in the round to reduce charge per delay.
4. Effective stemming.
5. Workers are provided with personnel protective equipment like ear muffs/plugs.
6. Secondary blasting for breaking of big boulders is being done underground ground.

Mitigation measures for ground vibrations

1. Use of latest electro-hydraulic drills along with Simba and DTH drills.
2. Use of electronic along with non-electric ignition system.

3. Regular vibration monitoring through seismographs.
4. Determination of predictor equation for group vibration based on maximum charge per delay (MCPD).
5. Ground vibrations maintained below stationary limits.

Subsidence—surveillance and monitoring

1. Extraction of ore in planned by leaving a minimum 30-m of solid cap rock.
2. Ensuring global stability of the mines by:
   (a) Characterisation of rock masses-RMR and Q-system.
   (b) Determination of geotechnical and physico-mechanical properties of rocks.
3. Determination of stope and pillar dimensions;
   (a) Using empirical Mathew's stability graph method.
   (b) Authentication using numerical modelling (FLAC-3D software).
4. Stress monitoring of mine pillars.
5. Ground movement monitoring with multi-point bore hole extensometers.
6. Regular subsidence measurement and monitoring of cap rock.
7. Inhouse rock mechanics department for designing and monitoring and the results are being analysed and validated by CIMFR and NIRM.

Subsidence: Remedical measures

1. Erection of isolating stopping walls, etc.
2. Fencing-off the area at the surface.
3. Isolating the problem area from rest of the time.
4. Erection of isolating stopping wall, etc.
5. Close interval monitoring of the area in particular and other adjoining areas for assessing the ground conditions.
6. Surface cap is monitored periodically for subsidence.

Noise Control from Loudspeakers (Mumbai)

INTRODUCTION

The same sound may be experienced differently when heard by different persons or by the same person at different times. Noise is unwanted sound. But unwanted can change from time to time depending on the frame of mind of the listener. I may enjoy listening to music when out with my friends and in a relaxed mood, and resent the imposition of the very same music at a time when I am studying for an important examination. Various factors including age, general health level, and participation in the activity responsible for causing the sound count. The same level of sound at different times may be enjoyable, or may be 'noise'.

In a city like Mumbai, where people in many different situations of health and social conditions live in close proximity, it is impossible to keep the sounds associated with the fun or the need of one section apart from pain caused to some other section of society. For example, Rang Bhavan is situated next to 2 hospitals, GT Hospital and Cama Hospital. While those enjoying rock music at Rang Bhavan experience the sound in a pleasurable way, the very same sound may cause trauma and become noise to those others who are in critical health conditions in the adjoining hospitals. While extremely loud volumes for political rallies held at Shivaji Park may be imperative to candidates contesting an election and wishing to reach out to the maximum number of people, these may be the cause of distress to a new born infant in the adjoining maternity home.

In the congested residential areas of Mumbai, old persons, infants, students, those requiring rest and quiet conditions live side by side with others wanting to enjoy noisy celebrations. It is impossible to ascertain the exact condition of all your neighbours prior to conducting a social or religious function in such a city, where population is one of the most densely gathered in the world, and it is the prevailing practice to brush aside such considerations when planning an important function.

Under the circumstances, it is imperative to statutorily protect the health and well being of whole sections of society by restricting imposition of unwanted and dangerous levels of sound, or 'noise' at any time. While respecting the need to hold cultural, social and other functions, citizens of Mumbai also need to learn a respect for the rights of those whose situation demands a noise free environment.

NOISE CREATORS: CAUSES, EFFECTS, EXPERIENCES AND SOLUTIONS

Noise may take many forms, yet all types of noise share one basic: they cause injury to somebody. Noise is, in fact, a form of assault upon the senses. Who then, are these people who would thus assault their neighbours? The burden is shared by each and every one of us. In varying forms and for various

522

good reasons, we have all contributed to and continue to contribute to increase in noise levels until such levels have reached dangerous proportions. Traffic, construction, recreation—the sounds of all of these at the same time and in a limited space combine to produce this heightened effect. In the shared responsibility, some of the activities we undertake to contribute are loudspeakers and firecrackers used for social, religious and political functions and for recreational and commercial activities: Loudspeakers are used for all and every type of function, whether social, political, religious, commercial or recreational. Noise from loudspeakers in addition to other types of noises which already exist, adds to the already high overall decibel levels. Noisy firecrackers are often used, either alongside or on their own.

Traditional and Social Aspects

Indian festivals are traditionally celebrated with song and dance in large groups, using musical instruments, drums and fireworks. Unfortunately in recent years, it has become a trend to amplify the celebrations using loudspeakers and the festivals have become very noisy. Noise levels have greatly increased in recent years by the use of loudspeakers and other sound amplification systems used indiscriminately in crowded cities during festivals, marriages and other functions.

The firecrackers also traditionally used for celebration are another major source of excessive noise. They are used indiscriminately in residential areas, next to hospitals, schools, with little consideration for the effect on the well being of persons unable for a variety of reasons to bear the high level of noise created. Atom bombs, chain bombs, etc. are permitted for manufacture provided they do not exceed 125 dB—the level of a jet engine taking off at 25 metres. These are then used in areas where quiet may be essential to the health and well being of individuals. In cities like Mumbai where, as discussed earlier, it is impossible to ensure that noise is contained in spaces where it does not cause distress to others, use of such excessive sources of noise may not be appropriate.

Health

The 2006 WHO report on community noise says it all. It speaks of noise as a hazard to physical and psychological health. Millions of city based Indians have experienced these hazards at first hand and know the trauma that can be caused. They are suffering the adverse health effects detailed in the report due to noise exposure on a regular basis but have never spoken out. The reason for their silence is due to their fear of repercussions from deliberate noise makers, who are typically politically supported or have underworld sponsors, and belief that some types of noise are unavoidable, and that the suffering has to be endured in the absence of a cure. Although laws restricting noise levels exist, these laws are only just beginning to be taken seriously and implemented.

Road Blocks

Many traditional Indian celebrations call for high volumes of sound, which exceed levels acceptable for healthy living in crowded cities. Noise and inconvenience to others is a factor which is discounted when planning such celebrations. Political and administrative will to implement noise control laws is absent, since the politician is often the noise maker himself. A display of power and wealth in functions which are ostentatious and extremely noisy is common. These glittering functions are sought to be emulated by other sections of society, who also use the loudspeakers at high volumes, chain type firecrackers and atom bombs used by the rich and powerful.

In the first stages of implementation of noise rules, it is perhaps natural that the rules are enforced only for the weaker sections of society, while use of loudspeakers and firecrackers becomes more of a statement of power than ever.

It will require a collective will to practically enforce noise rules equally for all and on a long-term basis. This is only just beginning to happen. The process requires constant monitoring in its early stages since obstruction can be expected from all interested parties: festival organisers, firecrackers manufacturers and dealers, politicians, builders who would be required to spend money on equipment to reduce noise levels and who may have to limit their hours of work — the list goes on.

Sound Map

Mumbai also has another side and the conflict between these two sides has become a subject of public debate. There is a real need for continuing cultural, religious and political activities within the very limited space available. It would be desirable to invest in a thorough and professional research of the overall situation and careful planning is required to ensure that these two separate needs of our city may be balanced. Without such planning, it is only too likely that we will slide back into a situation where suffering citizens are forced to bear noise levels far in excess of those permitting good health or that all public functions in the city come to a complete standstill.

A 'sound map' of Mumbai would provide a base line for this and also indicate locations where it would be desirable to allow cultural and other activities to proceed without endangering the health of citizens. Simultaneously, strategies to contain sound levels within the audience for which the sound is intended could be worked out with the help of professional sound/acoustic engineers and/or agencies of the government who would be professionally qualified to undertake such a job. It would be possible to strike a balance between the need to keep sound levels low and the need to hold various public functions and activities.

Modern Technology

Technology can make the impossible bearable, sound that is enjoyable for the protagonists and comfortable for their neighbours. Professional sound engineering methodologies have been designed and new technologies built that serve this purpose.

For deliberately amplified sound (music, speeches, religious prayers and songs, and drama) the purpose of so doing has been to ensure that the source sound can be heard by a greater number of people, the audience. Unfortunately, the cost of doing so in the past has been that inadvertently others, who have no wish to — or may actually be injured in — listening to the sound, are forced to suffer. Clearly, this is a form of antisocial behaviour, and laws have been promulgated to control it.

There are at least two approaches to sculpting amplified sound such that it becomes what the audience wants to hear, without inflicting discomfort on others. Each has its merits and disadvantages, which can be discussed at length.

The first is a technology that 'shapes' the travel of sound from the speakers to the audience. The total sound energy is delivered in packets to the audience. Each packet of sound is carried in an envelope to a particular location, with very little leakage. Each packet is characterised by a delay signature, equivalent to the time taken for the sound to reach from the speaker elements to the section of audience for whom it is intended. Thus all sections of the audience receive the sound signal at exactly the same time, and do not hear any signal that is not intended for their ears. Aside from the very high quality of audience experience, a significant result is the low spillover of any sound outside the audience listening area.

The second technology wraps the audience in a blanket of sound experience, by dividing the sound emission to many speaker sets placed throughout the audience listening area. Members of the audience directly hear the sound emerging from close by, and also get an additional small spillover of sound from further speakers. The latter sound, while almost inaudible, adds a very pleasant ambient listening quality to the overall experience. Since each speaker delivers a very low sound output, the spillover of sound outside the audience area is easily controlled. Delivery of sound to each speaker can be accomplished either through wires or wirelessly. If the audience area is large, or the location is only temporary (i.e. any multi-purpose grounds), then wireless is a far superior choice. Fixed locations such as special amphitheaters can gain by having the speaker installation carried out with concealed wiring during construction. For any other location, the cost of temporary wiring will be very high.

Another relevant factor is the energy consumption levels. Any solution that depends on the delivery of sound energy over large distances will consume more energy. Alternatively, if the signal content is delivered wirelessly using the electromagnetic spectrum, the energy cost is extremely low. A typical amphitheater like Rang Bhavan will need about 50,000 watts for a good sound experience throughout the main audience area using traditional technology, about a tenth of that if sound is delivered through wires to distributed speakers, and less than a tenth of that (under 500 watts, if about 100 radio speakers each delivering 5 watts are distributed around the area) using wireless (the transmitter will consume less than 100 milliwatts, which barely affects the energy calculation at all).

Available Legal Framework

The laws governing noise levels are contained in the Environment Protection Act 1986, The Noise (Regulation and Control) Rules 2000 and in various other Acts and Rules depending on the source of noise. Firecracker manufacture is controlled by The Explosives Act 1884 and The Explosives Rules 1983, where maximum noise levels are stipulated.

Decibel level control and zoning are the key elements of the Noise Rules. All areas are designated into appropriate zones. Areas within 100 metres of hospitals, educational institutes, courts or religious places are designated 'Silence zones' where decibel levels cannot exceed 50 dB in the daytime and 40 dB at night. Other areas are Residential Zones, Commercial Zones and Industrial Zones. Maximum decibel levels for each of these are laid down in the Rules. Implementation is delegated to the Commissioner of Police.

The Environment Protection Act (in section 15) provides for penal and criminal action in case of violation of Rules. Any violation carries a jail term upto 5 years and penalty upto Rs 1 lakh. Government servants are also not exempted from these penalties. Unfortunately while these laws exist, there has never been any serious attempt to implement them until recently. It has been the accepted contention that enforcement of noise laws is impractical and unnecessary, since it might lead to a 'law and order problem' when the persons (typically powerful or moneyed) making the noise were thwarted, and that those who suffered as a result of the noise should somehow just put up with it.

It was necessary to obtain a court order from the Mumbai High Court as the first step towards implementing the noise laws in Mumbai, to motivate the police to take action. The first court order, obtained just prior to Ganpati 2003, was half heartedly implemented. We measured and recorded noise levels during Ganpati and produced the findings before the Chief Justice of the High Court of Mumbai. At that time, Mr. Chhagan Bhujbal, then Deputy Chief Minister of Maharashtra, had announced relaxation of noise laws until midnight for 4 days out of the 9-day Navratri Festival, when loudspeakers would be allowed for use anywhere in Mumbai until midnight. We succeeded in obtaining an Order from the High Court banning the issue of loudspeaker licences in Silence Zones and cancellation of any licences already issued for all the days of the Festival, traditionally one of the noisiest festivals of all. Navratri 2003 was the quietest ever.

We solicited complaints from citizens and set up a 'hotline' to advise them about making complaints in case noise laws were violated in their areas. During Navratri alone more than 200 complaints were received, in spite of the visible decrease in noise. The Police rose to a difficult challenge, in the face of opposition from their political bosses, by setting up a Control Room to register telephonic complaints from citizens, enforced the Silence Zone restrictions and the 10 pm statutory deadline, and took action in most cases by issuing a warning or confiscating equipment when warnings were not heeded.

After Navratri, no loudspeaker licences have been issued in Silence Zones for any other functions, including marriage functions, cultural functions and political functions and loudspeakers used at other locations have been at lower volumes and meticulously shut down at 10 pm. The Police have confiscated equipment in some instances for violating the statutory 10 pm deadline. Predictably, the exceptions to this have been politically motivated (such as a rally at Shivaji Park with loudspeakers placed in a Silence Zone and addressed by Shiv Sena chief Bal Thackeray, which was permitted through a direct order of the Deputy Chief Minister, and another at the same location inaugurating the Congress electoral campaign and addressed by Congress President Sonia Gandhi).

During the run up to the recent national elections, the Nagpada Police Station registered a criminal case against noted underworld don turned politician, Arun Gawli, for violating the 10 pm deadline while campaigning—a first Deputy Prime Minister L.K. Advani was also shut down at 10 pm while campaigning at Patna.

Another first—an FIR (First Information Report), the first step towards criminal prosecution of a noise maker was recently filed by a citizen, Mr. Kiran Shukla, against a builder constructing in a Silence Zone. This builder was granted written permission by MHADA (Maharashtra Housing and Development Authority), to whom the land belongs, to continue active construction until 10 pm and loading and unloading of materials through the night. In fact, active building at the site often continues until midnight and begins before 6 am. We await the result of Mr. Shukla's action, and would encourage other citizens and groups of citizens to take civil and criminal action in other neighbourhoods when they face distress from noise from any source.

Role of Implementing Agencies

Government

A section of the government is responsible for policy making, another for monitoring, and yet another for implementation. The policy maker, in this case the Ministry of Environment and Forests, has, after due research and after inviting comments from citizens and NGOs, notified the Noise Pollution (Regulation and Control) Rules in February 2000. Monitoring is the responsibility of a statutory body, The Maharashtra Pollution Control Board, and implementation is delegated to the Commissioner of Police or the District Magistrate as applicable, depending on the location.

Commissioner of Police

In Mumbai, the Commissioner of Police is responsible to protect citizens from noise, as from other forms of assault, by ensuring that no unnecessary or unwanted sounds exceed statutory decibel limits at any time. A formidable task, given the ground reality of Mumbai, the traditional use of drums and music for street celebrations, the insensitivity of sections of citizens to the discomfort caused to others by their recreational activities, by the indiscriminate use of horns, construction activity at night, etc. We cannot and should not expect the Police to take on the role of anything but what it is: a law enforcement agency. The responsibility of ensuring sensitivity to the situation of others, for their right to health and well being and for respect of the law falls on every citizen.

The role of the Police, then, in its capacity of a law enforcement agency and of it's Commissioner as the Head, is to maintain law and order. Its role is to ensure compliance with the law in situations where noise is wilfully imposed on unwilling hearers. In such instances (of which there are many) the Police is bound to treat noise, an assault on the senses, as any other physical assault would be treated, and to take penal and criminal action against the offender. This action of the Police is not a substitute for public awareness or sensitivity. It is a deterrent to the breach of law and order caused by neglect or by wilful disregard of laws and the consequent assault to the physical well being of others in situations where they are helpless to stop this assault without aid from the Police.

Maharashtra pollution control board

The MPCB is a statutory body to check and control environmental emissions of various types including noise. It is mandatory for the MPCB to monitor and prevent noise emissions beyond those permitted under the Act and Rules.

NGOs

The Police cannot substitute for the sensitivity that we, the citizens need to imbibe. NGOs working in the fields of health, education or welfare of children would be able to further awareness and to help ensure that adverse health effects of noise are minimised. It is not until noise is commonly recognised for the danger to health that it is, and until education about the benefits of minimising noise levels is imbibed by children of school age and younger, that long-term solutions to excessive noise levels can be practically sustained.

Organised groups

Citizens and organised groups of people have an important role to play. Some have already volunteered to work long-term on curbing noise pollution by monitoring noise levels in their areas and taking appropriate action to curb excessive noise levels. Many have acquired noise meters and are being effective. One has filed an FIR with the Police as the first step towards criminal prosecution for breach of noise laws. We hope to encourage such citizens and continue our effort to reduce noise pollution in India by building a network capable of handling the challenge.

Laws and Statutes to be Implemented by the Police

Laws and statutes to be implemented by the police are:
1. Under the Noise Pollution (Regulation and Control) Rules 2000, maximum permissible decibel levels for different zones are given in Table 46.1.
2. Loudspeaker permissions are granted by the Police. Such permissions are granted after taking into consideration the location for use of loudspeakers. Zones are designated according to their nature — silence zones, residential zones, etc. No loudspeakers are permitted in silence zones at any time, day or night, or in residential zones at night and decibel levels in residential zones are restricted during the day.
3. Violation of the Noise Rules is a cognizable offence, carrying a jail term of upto 5 years and penalty upto Rs 1 lakh under the Environment Protection Act.

Table 46.1. Ambient air quality standards in respect of noise.

Area code	Category of area/zone	Limits in dB(A) L_{eq}	
		Day time	Night time
(A)	Industrial area	75	70
(B)	Commercial area	65	55
(C)	Residential area	55	45
(D)	Silence zone	50	40

Note: 1. Day time shall mean from 6.00 am to 10.00 pm.

2. Night time shall mean from 10.00 pm to 6.00 am.

3. Silence zone is defined as an area comprising not less than 100 metres around hospitals, educational institutions, religious places and courts. The silence zones are zones which are declared as such by the competent authority.

SILENCE WILL BE GOLDEN AT SHIVAJI PARK—MUMBAI

The High Court ruled in an interim order on May 5, 2010 that Shivaji Park would be a silence zone. DNA asks people about effectiveness of the ban that came in after a six year long battle and use of loudspeakers by political parties.

Expert View

An inspiration for others affected by the problem: Declaring Shivaji Park as a silence zone is really appreciable. After six years of battling frequent noise and din from political rallies, residents are able to get the High Court order. Now when there is court order residents will be relaxed. It is very important to realise that more than any rally or party function, peace and health are more important for all residents. This order can be an inspiration for many people who are suffering from noise pollution. We must accept that the menace of noise is growing everywhere in the city. Citizens must come together and oppose such hazard. Moreover the government must ensure that silence zone rules must be followed by politicians. They should not remain applicable to common people only. Rules must be same for everyone. For keeping the environs silent, people should report any violation promptly— *Sumaira Abdulali, founder, Awaaz Foundation.*

Activist View

People should know the rules to battle ahead: The High Court's order is a great relief for the residents of Shivaji Park. Firstly, one of the residents checked the level of noise pollution of loudspeakers used in political and other rallies, with an instrument. It revealed that the decibel level was quite high and this was the reason why many people residing near Shivaji Park felt some discomfort. So the residents approached the civic corporation but they did nothing to resolve the problem of noise pollution. We then moved to the Maharashtra State Pollution Control Board and they declared that there was grave noise pollution despite many residential colonies, schools and religious places. They sent a recommendation to state government too. People also got a copy of the Noise Pollution Rules 2004. But it didn't yield any results. So in the end people filed a public interest litigation and finally won the battle. You must get the list of silence zones from your ward office. If the list says that your area comes under the jurisdiction of silence zones and no rules are being followed then you can approach the authority—*Ashok Ravat, founder, member Shivaji Park ALM.*

Motorists violate norms as well: Its a good move and should help in ensuring that one of the most iconic landmarks becomes a peaceful place. Similar bans should be instituted at other places as well. It comes across as a huge menace especially for students. Spending quality time with friends or family almost becomes impossible because of blaring loudspeakers at festivals and during elections. The ban acknowledges the growing problems that people are facing. However, motorists blare horns in front of various silence zones like shrines, schools and hospitals and clinics. During festivals and elections these rules are completely forgotten. These silence zones need to be followed with utmost sincerity — *Preeti Doiphode, associate consultant.*

More realisation about hazard: This is a big relief for all residents, especially senior citizens who are in a majority in this area. People are becoming conscious about the environment and realise that noise pollution is a growing menace and silent hazard in the city. Other steps which can be taken to curb the menace are shifting rallies by political parties and other functions to non-residential areas. Also, nobody adheres to the 10 pm deadline so the culprits should be dealt seriously. This will deter people from blaring loudspeakers for public or private functions. More awareness should be created about silence zones across the city and ensure that everyone respects them— *Devlina Dutt, management associate.*

Loud revelry is not celebration: The move is welcome though, I'm not sure of the application of the ruling. I hope political parties do not appeal against the ruling. In a sense this is an acknowledgment that the problem of noise pollution needs to be addressed. But importantly, it will be good if we look at the attitude of our citizens. Celebrating any occasion with loud music, without having any consideration for other residents is not correct. The attitude of most citizens needs to change, only then will legal steps be helpful. Unless we change our attitude, curbing the menace drastically will not be possible. Political parties also need to show concern. The educated class of the city needs to alter their cultural activities — *Harshal Shah, finance consultant.*

Gatherings are to bind people: It is completely justified to ban loudspeakers in public places. Noise pollution is a great nuisance, especially for senior citizens, infants and young children appearing for exams. Noise pollution can be curbed by using loudspeakers only in private forums. Political and cultural activities are meant to bind people and hence should be conducted in a conducive manner. Shivaji Park has several residential colonies and it will be selfish to not keep their well-being in mind. Closed private auditoriums can be made use of for these occasions. If it is feasible for us to create huge sound-proof auditoriums for these purposes, it will be of great use— *Divya Hinge, PR professional.*

Apply the law to politicians too: Declaring the area as a silence zone will be effective and a welcome move. However, the government should ban loudspeakers all over Mumbai, not only in Shivaji Park. There should be a rule prohibiting use of loudspeakers especially on roads. The law must be applicable to everyone including politicians. This is a good step taken towards curbing noise pollution, but other measures must be taken to curb the problem in Mumbai immediately. There are some more initiatives to discourage noise pollution. Firstly, politicians shouldn't be allowed to speak on the roads and canvas for candidates during the election. Secondly, during festivals people should not be allowed to put loudspeakers on the road— *Jaya Mistry, admin executive.*

Need time limit on other days: I'm happy that finally Shivaji Park is declared as silence zone. I have been residing here for almost 25 years and the misuse of the playground for political and religious rallies has led to the menace of noise pollution throughout the year. However I'm unsure about what will happen during Ganesh Chaturthi. During the festivities, there is chaos and loud music for eleven days. If they can resolve and tone down the festivities, it will be appreciated. Also they need to put some restriction and cut-off time on the use of loudspeakers for dates like Republic Day and Ambedkar Jayanti. Also there is a senior citizens' enclosure, it will benefit them to a large extent — *Jimit Shah, account executive.*

Appendices

This appendix highlights the noise control measures related to various industries such as plastics, woodworking, agriculture, air transport, ceramics, construction, docks, engineering, food and drink, foundries, glass (flat and container), motor vehicle repair, paper and printing, quarries, and rubber.

Table 1. Generic noise control measures.

Activity	Noise control measures (where reasonably practicable)
Air movement	Relocate/segregate static plant, e.g. compressors, vacuum pumps, blowers, etc. to lesser or non occupied rooms. The process could be acoustically enclosed within an accessible and adequately ventilated noise reducing enclosure. Use of low noise emission portable generator sets and compressors, e.g. 'hush packs'. Fans inlet/discharge fitted with flexible connections and silencers to reduce duct borne noise. Reduce excessive line air pressure or fit low velocity (quiet) nozzles to 'open ended' fixed position or portable blow-off pipes for removing swarf cuttings, wood chips, lubricants, water cooling, components ejection or segregation. Reduce impulsive noise emission from exhaust ports of pneumatic actuator/manifolds using porous metal or plastic port silencers; and maintain good connector seals to avoid noisy air leaks.
Conveying/ Transporting	Use damped/composite materials for rollers; use of component guide/sequencing release mechanism to reduce component impact noise; maintain adequate lubrication of bearings/rollers; reduce the speed of conveying; suspend conveying ductwork using anti-vibration hangers to reduce structure borne noise; external damping of ducting conveying materials; fit sectional acoustic tunnel hoods over open conveyor lines; add external damping compounds or rivet plates of sheet metal to lightweight flat surfaces, e.g. non-critical machine panel work, chutes, trolley tables, conveyor sides, etc. to reduce vibration and noise emission; Internally line material stock feeder tubes, e.g. auto lathe bar stock.
Forming	Relocate or segregate machinery, e.g. presses, moulding machines, corrugating machines, bowl polishers, blast chillers or freezers, block making machines, granulators, static compressors, blowers to lesser or non-occupied rooms; Machinery could be acoustically enclosed within an accessible and adequately ventilated enclosure; Use hydraulic rams to realign distorted fabrications after welding, forming or alternatively, use magnetic damping mats or sandbags if realigning by hammering; Reduce noise emission cutting thin sheet metal, e.g. motor vehicle panel work, using magnetic damping mats, sandbags, etc. also where feasible, eliminate noise by laser type profiling, etc.
Processing	Relocate/segregate noisy machinery and/or ancillary equipment, e.g. compressors, presses, fans, saws, cutting-off moulding, fabrication, grinding, fettling, etc. to lesser and or non-occupied rooms; Machinery could be acoustically enclosed within an accessible and adequately ventilated

(Contd...)

noise reducing enclosure; Where not reasonably practicable to remove or enclose, e.g. long process lines, local noise refuges can be installed for operators to control/oversee processes; Where possible minimise the use of handheld grinders by improved component design, e.g. machine weld preparation and removal, or using 'low noise' discs fitted to both portable and (possible pedestal grinders); Maintain sharpness of cutting tools, and/or reducing speeds with increased number of cutting teeth or blades; Avoid cut materials falling from excessive heights into un-damped collection bins, use of damped or deadened steel chutes, hoppers, bins. Clamp materials being cut along their length to minimise vibration, i.e. 'bouncing' on supporting surfaces to reduce noise emission infeed/discharge chutes, hoppers; Use of damped percussive and rotary percussive tools, e.g. chisels, (in chipping hammers, rock drills and breakers); Breaking materials using quieter hydraulic crushing or bursting rather than percussive methods, e.g. crushing concrete instead of using pneumatic or hydraulic breakers, or cutting using damped wall saws or diamond wire to profile area and using bursting methods to remove materials.

Table 2. Plastics.

High risk activity/process	Example noise levels, dB	Established noise control methods
Granulators (and other size reduction machines e.g. agglomerators, crumbers, shredders, pelletisers)	100 (granulators) 90 (agglomerators, etc.)	Methods include: Use feed conveyor to remove operator from higher noise areas. Situate size reduction machines in separate rooms or buildings —provide for remote or automated feeding. Lag or damp the machine casing. Form sound trap in feed aperture or hopper. Enclose the machine. Fit segmental or helical cutters. Use tangential feed. Fit resilient backing to knives. Reduce rotor speed.
Injection moulding machines	97–100	Methods include: Use slow speed pumps. Control release of exhaust air. Mount pumps and motors on anti-vibration mounts and incorporate flexible hoses in pipe lines. Enclose hydraulic power packs. Convert injector guards to acoustic guards. Fit low noise nozzles to blow guns, etc.
Extruders	90	Methods include: Specify low noise design. For hydraulic systems see injection moulding machines above. Fit silencers to drive motor air intakes and exhausts. Enclose drive motor.
Mould cleaning guns	105	Replace nozzles with low-noise types (e.g. those which generate an induced secondary air flow). Reduction of upto 10 dB.

(Contd...)

High risk activity/process	Example noise levels, dB	Established noise control methods
Extrusion line cut off saws	100	Methods include:
		Replace guards with solid panels lined with acoustically absorptive material.
		Fit acoustic strip curtain at product out-feed.
Ultrasonic welding machines	96 (typical)	Enclose with sound reducing material.

Table 3. Woodworking.

High risk activity/process	Example noise levels, dB	Established noise control methods
Circular saws	97–102	When purchasing new blades obtain 'low noise blades'.
Vertical spindle moulders	95–100	The use of limited cutter projection tooling will reduce noise levels and should have been in place since 2003 under PUWER.
Multi-spindle planer moulders	upto 105	Segmented blocks (widely available) can reduce in-feed noise levels.
		Properly designed and maintained chip extraction systems (where not part of integral enclosure) will reduce idling noise levels.
		Use smoother profile blocks with low blade projection
		Slotted or perforated table lips can reduce idling noise levels.
		Reductions in noise can be made by reducing the cutter's rotational speed, and increasing the number of knives on the cutter.
		There should be a noise enclosure, either as an integral part of the machine or retrofitted. As with all noise enclosures it should be of suitable design, form as complete an enclosure as possible, and be properly maintained and used.
Band resaws	95–105	Maintenance of machine (e.g. pulley scrapers, lubricating felt pads or sawdust extraction system) and blade, combined with blade adjustment, are extremely important for noise levels.
		Noise enclosure of band-resaws is considered to be reasonably practicable.
Planer thicknesser	97–101	Reductions of 7 to 13dB have been achieved during thicknessing only by adjustment of the table to slightly increase gap between cutter and table. Not to be used when the machine is used for planning when the timber is fed across the top of the cutter.
Small hand fed thicknesser	104	Enclosure (can be as simple as a 15 mm lined chipboard box)
Chipper/hoggers		Segregation of machine from work areas, or enclosure of machine.

Table 4. Concrete and cement products.

Product	Process	Example noise levels, dB	Established noise control methods
Flat products (e.g. slabs, fence posts, panels)	Mould filling, demoulding and stacking using vibrating tables or conveyors	Steel tables: 95–110 Tables/conveyors with rubber covering: 86–93	Use self-compacting concrete (see below) Use resilient material (e.g. rubber) on tables. Clamp mould to table. Fit tunnels or enclosures over conveyors.
Reinforced concrete products (e.g. beams, steps)			Enclose undersides of conveyors and tables. Maintenance of enclosures, skirts, etc. Maintenance of vibrator motors and mountings. Use wood, fibreglass or rubber moulds instead of metal to reduce impact noise.
	Use of self-compacting concrete (SCC)	Relatively quiet process: no vibration required	SCC (concrete to which chemical plasticisers are added) is increasing in popularity in the UK. Its use has the potential to eliminate the main source of noise (vibration). SCC should be discussed at visits to raise the profile and encourage innovation.
Blocks, tiles, slabs	Vibratory presses	No noise reducing features: 96–110	Fit enclosure (all controls outside) or provide separate control room (noise refuge).
		Outside press enclosure: 84–93	Isolate vibrating parts from floor and enclosure.
		Unloading stations: 86–88	Maintenance of vibrator motors and mountings.
		Inside control rooms: 71–79	Silencers for compressed air exhaust.
			Secure all parts and fittings to prevent rattling. Use resilient material (e.g. rubber) for stops
	Rumblers/tumblers,	84–95	Line barrel of tumbler with rubber lining. Isolate plant from other processes and/or use plastic curtains to separate from employees.
	Saws	81–96	Use noise-reduced saw blades.
Extruded tiles	Extrusion plant Pallet/mould conveyors	86–93	Extrusion plant: Use noise-reduced blow-off jets/air knives. Use silencers on compressed air exhausts. Conveyors: Control speed to minimise collisions between pallets (may require training).

(Contd...)

Product	Process	Example noise levels, dB	Established noise control methods
			Use an impact absorbing material (e.g. polyurethane) on convenor guide rails, etc.
General	Chutes and skips		Provide chutes and skips with rubber lining .
			Minimise dropping distances for waste material.
	Mixing machines		Noise havens containing all control consoles.
	Cleaning equipment	Chipping hammers: can be > 120 dB	Avoid or minimise the need for use of noisy equipment by washing down before the 'mix' goes off.
		Ultra high pressure water jetting: up to 105 dB	For water jetting, locate compressor in acoustic housing, restrict operating pressure.
	Materials handling		Where heavy quarry type vehicles are employed, use acoustic cabs.

Table 5. Agriculture.

High risk processes/equipment	Established measures for noise control	Sector and other guidance
Use of tractor without a 'Q' cab	Consider replacing with 'Q' cab tractor where used for field work on arable enterprises >100 ha. Where not reasonably practicable or being used for other operations (yard, road, etc.) hearing protection must be worn.	AS8rev3 noise in agriculture.
Use of a 'Q' cab tractor with with significant breaches of the cab by additional services	Hearing protection must be worn until items are replaced or the openings blocked missing doors or windows or to restore noise protection to original level.	
Pig feeding	Consider timed/automatic valves on swill outlets to troughs and other automated feed systems. Where entry is required during feeding time mark area as hearing protection zone and ensure hearing protection is worn.	
Barn machinery	Consider operator station noise enclosures on large installations. Fit automatic cut-offs to roller mills, etc. Where not reasonable practicable and entry is required during feeding, mark area as hearing protection zone and ensure hearing protection is worn.	
Forestry and arboriculture equipment (e.g. chainsaws, chippers/shredders, brush cutters) and amenity/landscape equipment (e.g. mowers, strimmers)	Selection of low or lower noise equipment. Wearing of suitable hearing protection.	

Table 6. Air transport.

High risk processes/equipment	Established measures for noise control	Sector and other guidance
Loading aircraft, etc. with aircraft engine operating or auxiliary power unit (APU) or ground power unit (GPU) in use	General risk assessment based approach, i.e. limit persons/time spent, etc. Manage control use of APU/GPU. Use stand alone generator fitted with enclosure (i.e. HUSH pack generator set). Otherwise use ear protectors.	HSG209 Aircraft turnround.

Table 7. Ceramics.

High risk processes/equipment, typical noise level (sample L_{Aeq})	Established measures for noise control
Clay stockpile raw material handling above 90 dB	Where heavy quarry type vehicles are employed, use acoustic cabs noise havens containing all control consoles.
Body preparation (high speed blungers, ball mills, pug mills, vibrating screens)—above 90 dB	Where possible, separate body preparation from other activities. In large works, consider using video cameras or other means of remote viewing/monitoring of conveyors, etc. Provide noise refuges, etc. Options which may be reasonable practicable include: enclose drives for high speed blungers, use rubber linings for ball mills, enclose screens and relocate vacuum pumps on pug mills.
Making machine pumps and motors presses (dust, ram) auto towing machines— above 90 dB	Variety of pumps used. Noise levels can be reduced significantly by simple enclosures lined with sound absorbent materials. Replace vacuum pumps by rotary pumps where possible.
Glaze spraying (sanitary ware and tableware — above 90 dB	Pressures selected for sanitary ware spraying should be the minimum commensurate with satisfactory performance. Automatic tableware spraying machines can be sited away from other work areas. In both cases, careful selection of spray nozzles can reduce noise levels.
Sanitary ware reclaim —above 90 dB	Segregate from other work areas. Use noise absorbent lined hoods.
Kiln fans— above 90 dB	Locate away from occupied areas, provide partial enclosures or screening.
Vibromills— above 90 dB	Segregate from other areas. Fit acoustic enclosures and noise dampening to the bowls.
Grinding and polishing of imperfections —c. 90 dB	Noise levels vary considerably depending on amount of grinding. Traditional methods use small grinding wheels. Using abrasive belts or single arm linishing belts greatly reduce noise levels.

Table 8. Construction.

High risk processes/equipment, typical noise level (sample L_{Aeq})	Established measures for noise control
Tunnelling by hand with clay spade or jigger pick 95–117 dB	Use mechanised tunnelling techniques in all but the smallest tunnelling jobs; if hand digging is used then use lower noise emission tools: silenced body/damped picks; silence pneumatic tool exhaust; maintain equipment/air lines; operate in accordance with manufacturers' instructions.

(Contd...)

High risk processes/equipment, typical noise level (sample L_{Aeq})	Established measures for noise control
Scabbling 94–105 dB	Scabbling purely for architectural aesthetic effect is not acceptable. Specify finishes that do not require scabbling. Some finishes can be designed into shuttering using special moulds); design to allow larger concrete pours/consider work sequencing; specify/use non-mechanical scabbling methods; use lower noise emission tools. Surface preparation to ensure good concrete bond can be achieved by other methods, e.g. cast in proprietary joint formers, or chemical retardants and water jetting.
Breaking concrete, asphalt, etc. with hand operated breakers 96–105 dB	Breaking in new concrete/masonry and other breaking work. Plan cast in ducts, detail box-outs to minimise the breaking of new concrete. Use boom-mounted hydraulic breaker on construction plant with noise-protected cabs; use lower noise emission tools: silenced body/damped chisel; maintain equipment/air lines; operate in accordance with manufacturers' instructions. Pile cap removal: Pile cap removal using hand-operated breakers is not acceptable. Consider alternative solutions, e.g. pile head removal using bursters/crushers; Elliot method, Recipieux method, or use hydraulic pile croppers and design pile spacing and pile re-bar for mechanised cropping. A limited amount of dressing of the pile cap with hand held breakers may still be required.
Abrasive disc cutters/angle grinders 98–104 dB	Consider elimination of need for on-site cutting by design/prefabrication; sharpen cutters/replace discs regularly; maintain equipment and operate in accordance with manufacturers' instructions; minimise numbers exposed.
Striking (dismantling) proprietary falsework (using metal hammers to free collars) 107 dB (L_{Cpeak} 136 dB at 2m from activity)	Maintain the falsework legs properly-follow manufacturers instructions on cleaning and lubrication to reduce effort required to release legs. Use a purpose made spanner whenever possible. Minimise use of hammers, if hammers must be used use plastic/rubber hammers and wear hearing protection, shield others from the noise.
Dump trucks /site dumpers 93–95 dB	Purchase/hire lower noise emission plant; maintain plant; damp vibrating panels; consider lining load section; use noise reduction techniques for cab, etc. (dump trucks); consider retrofitting silencers/diffusers to exhaust (may also be applicable to other construction plant).
Cartridge tools C-weighted peak noise level 143–157dB	Minimise numbers exposed and shield others (e.g. with portable enclosure); where used on steel plates, use damping (e.g. sandbags).
Concrete pumping 91–93 dB	For independent pump, enclose pump/motor (consider need for ventilation!) silence exhaust; use quieter plant; maintain plant and operate in accordance with manufacturers' instructions.
Powerpacks/compressors 85–91 dB	Locate away from occupied areas; provide acoustic enclosure (consider need for ventilation!); use low noise emission equipment (e.g. 'hush packs'); keep access panels closed; maintain equipment/air lines; operate in accordance with manufacturers' instructions.

(Contd...)

High risk processes/equipment, typical noise level (sample L_{Aeq})	Established measures for noise control
Grit blasting 96–100 dB	Minimise numbers exposed; provide local enclosure, maintain equipment/air lines; operate in accordance with manufacturers' instructions.
Driven piling 115–132 dB	Consider alternative design solutions to minimise noise; consider using quieter methods: (e.g. vibration methods instead of drop hammer); enclose noise source; use damping on sheet piles; minimise numbers exposed.

Table 9. Docks.

High risk processes/equipment	Established measures for noise control
Loading car delivery vessel car decks, i.e. unaccompanied freight	Risk based approach to manage time spent/persons exposed/vehicle running times, etc.

Table 10. Engineering.

High risk processes/equipment, typical noise level (sample L_{Aeq})	Established measures for noise control
General	Engineering noise task group guidance
	Top 10 controls: (http://www.hse.gov.uk/pubns/top10noise.pdf) Determining the best option for control: (http://www.hse.gov.uk/pubns/noisesources.pdf).
Air carbon arc gouging, 105–120 dB	Eliminate need for it, i.e. use single sided welding, non-welded restraining aids, etc. consider process substitution, i.e. oxy-fuel gouging, gouge during break times to minimise risk to others.
Use of chipping tools, 122–128 dB	As above, also increase accuracy of cut, cut angled edges in preparation for welding etc. NB When used on ship structure noise transmitted via the vessel can be significant source of exposure to others.
Abrasive blasting, up to 110 dB	Fit silencer to compressed air exhaust port, enclose compressor or site from work area, prevent compressed air leaks, use 'quiet blasting nozzles' or enclosed (mechanically propelled) blasting equipment NB Current standard for blasting helmets does not consider ear protection, exclude any protection they provide.
Grinding, 85–109 dB	Where possible eliminate cosmetic dressing, use low noise, flexible or laminated grinding discs; high frequency grinders (as opposed to grinders with universal motors); silenced pneumatic grinders; the lowest spindle rating needed, place magnetic mats on external surfaces of workpiece.
Hammering steel, 95–100 dB	Eliminate their use, i.e. correct weld distortion using hydraulically actuated straightening devices, use magnetic, hydraulic or screw fairing aids instead of welded lugs and wedges. Use soft faced, recoilless hammers; vibration damping, i.e. damping sheets or magnetic mats.
Metal cutting saws, 100 dB <	Purchase quieter machines when replacing machinery. Keep the blade sharp, use damped saw blades; noise/vibration absorbing material on feed table surface, damp the machine subframe, enclose the cutting area, locate the saw in a separate room.

(Contd...)

High risk processes/equipment, typical noise level (sample L_{Aeq})	Established measures for noise control
Power presses and CNC punch presses, 95 dB	Purchase quieter machines when replacing machinery. Use anti-vibration mountings; quiet tooling; damped machine panels; acoustically treated discharge chutes; noise enclosures.
Riveting, 100–110 dB	Consider use of radial and orbital riveting machines instead of conventional cold impact riveting machines. Fit a silencer to air exhaust on pneumatic machines; cushion impact noise by using a damping compound between actuator ram and tool ram. Replace 'percussion' riveting with 'squeeze' riveting.
Shears (high-speed continuous), 100 dB	Unattended machines: fit with a noise hood together with hold-down rollers to reduce vibration of the feed stock.
	Manually fed machines: fit wear resistant rubber material to the clamp base, reduce the clamp descent rate.
	Distance scrap metal falls should be reduced to the minimum; chute can be lined with rubber material.
	Designers may be able to improve the noise performance of these machines by setting the blade at a slight angle to the vertical.
Ultra high pressure water jetting, 105 dB	Limited scope for reducing noise levels other than locating the compressor in acoustic housing. Correct use of ear protection essential, should be compatible with waterproof clothing worn, i.e. fit below hood of jacket. Often ear protection will be worn by the operator but not others in immediate are who are also at risk.

Table 11. Food and drink.

High risk processes/equipment, typical noise level (sample L_{Aeq})	Established measures for noise control
Glass bottling	Replace glass bottles with plastic ones.
90–95 dB (dairy)	Design out noise at source: specify acceptable noise level when purchasing machinery.
85–95 dB (brewing and soft drinks)	Reduce inter-bottle impact: slow down speed of line and increase spacing of bottles.
100 dB (high speed bottling, 400–800 bottles per minute)	Dampening of impact surfaces: fit dampening material at impact points.
	Fit acoustic enclosure over bottle conveyor.
	Provide acoustic barrier around cap feeder bowl and fit noise reducing mountings.
	Limit worker exposure time: job rotation.
Product impact on hoppers	Design out noise at source: specify acceptable noise level when purchasing machinery.
95 dB (confectionery)	Reduce product-hopper impact: reduce drop height of product.
>90 dB (frozen food)	Reduce or fill in gaps at feed and take-off of pelletisers.
>100 dB (animal feed)	Reduce impact noise:
	Use hopper made of sound-deadened steel.

(Contd...)

High risk processes/equipment, typical noise level (sample L_{Aeq})	Established measures for noise control
	Line inside of hopper with impact deadening material.
	Line outside of hopper with noise dampening material.
	Line guards/panels with noise dampening material (can produce 5dB noise reduction).
Wrapping, cutting wrap, bagging, etc. (e.g. sweets) 85–95 dB	Design out noise at source: specify acceptable noise level when purchasing machinery.
	Reduce drop height of product.
	Enclosure:
	Line cover panels with noise dampening material.
	Fill any gaps in cover panels with noise absorbing material.
	Fit full acoustic enclosure over bagging line.
	Regularly maintain machinery.
	Limit worker exposure time: job rotation.
	Provide noise refuges for workers.
Bowl choppers (meat) >90 dB	Design out noise at source: specify acceptable noise level when purchasing machinery.
	Maintenance: regularly maintain rotating parts, machine mountings and sharpen blades.
	Fit acoustic hood/enclosure over bowl chopper.
	Fit noise-dampening material to bowl or panels.
	Segregate bowl choppers from quieter machinery/areas exposure time: job rotation.
	Limit worker exposure time: job rotation.
	Provide noise refuges for workers.
Pneumatic noise and compressed air 85–95 dB	Design out noise at source: specify acceptable noise level when purchasing machinery.
	Use low-noise air nozzles.
	Fit manifolds/silencers on exhausts.
	Move compressor outside or to a people-free area or enclose compressor (but ensure no overheating).
	Regularly maintain potentially noisy equipment.
Milling operations 85–100 dB	Design out noise at source: specify acceptable noise level when purchasing machinery.
	Locate mill in a separate room away from workers.
	Enclose hammer mills, roller mills and mixers with acoustic enclosures.
	Fit noise dampening material to panels.
	Reduce drop height of pellets and line hoppers with impact absorbing material.
	Enclose outside of pipes carrying particulate product (e.g. with half cylinder sheet steel lined with 50 mm mineral wool slabs which can provide 10–15 dB noise reduction).
	Limit worker exposure time: job rotation.

(Contd...)

High risk processes/equipment, typical noise level (sample L_{Aeq})	Established measures for noise control
	Provide noise refuges for workers.
Saws/cutting machinery 85–107 dB (meat)	Design out noise at source: specify acceptable noise level when purchasing machinery.
	Ensure preventative maintenance/inspection is carried out on blade alignment, blade sharpening, lubrication, floor mountings, etc.
	Use noise dampening on saw blades.
	Limit worker exposure time: job rotation.
Blast chillers/freezers 85–107 dB	Design out noise at source: specify acceptable noise level when purchasing machinery.
	Replace plant with a less-noisy model.
	Enclose plant with acoustic panelling (e.g. sheet steel outer skin, perforated steel inner skin, 75 mm mineral wool slabs can provide >20 dB noise reduction).
	Limit worker exposure time: job rotation.
	Noise refuges for workers.
Manually pushing wheeled trolleys/racks	Design out noise at source: specify good quality wheels/bearings when purchasing trolleys.
Up to 107 dB (from wheels/wheel bearings especially those subject to high/low temperatures in ovens/freezers)	Regularly maintain wheels/bearings.
	Improve flooring to reduce damage to wheels/bearings and cut down noise.
	Use conveyors to move product where possible.
	Improve layout to minimise movement of product.
Packaging machinery 85–95 dB	Design out noise at source: specify acceptable noise level when purchasing machinery.
	Install noise reducing enclosures.
	Fit silencers to noisy exhausts.
	Limit worker exposure time: job rotation.

Table 12. Foundries.

High risk processes/equipment	Established measures for noise control
Compressed air lines (various processes)	Segregation/enclosure of compressors, provision of low noise nozzles and exhaust silencers; regular maintenance; rectification of leaks.
Induction furnaces	Current control to prevent resonance; isolation/segregation of process.
Arc air gouging in fettling	Avoid use of this process where practicable; segregate if possible; hearing protection.
Moulding machines	Various noise control measures, e.g. local enclosures.
Rumbling machines	Elimination by better casting quality; segregation and other measures depending on specific machine.
Mechanical shake-out	Enclosures; noise damping materials on machine; hearing protection.
Fettling	Elimination by better casting design; better design of tool; avoidance of use of chipping hammers.
Knock-out	Segregate this process from others; mechanisation.

Table 13. Glass (flat and container).

High risk processes/equipment, typical noise level (sample L_{Aeq})	Established measures for noise control
Most noise in glass container manufacturing is generated by pneumatic noise and/or glass to glass contact (cullet and product).	
Batching/mixing plant, 96 dB Storage hopper vibrators and vibratory conveyors	Noise haven containing all control consoles.
Basement, 90 to 100 dB Fans, cullet transport and tipping, dumper trucks	Inlet and outlet silencing on fans, enclosure of fans and drive motors. Provision of cabs on dumpers and other vehicles.
Furnace area, 92 to 105 dB Furnace combustion air fans, furnace cooling fans	Silencing of fans. Noise haven containing all control consoles.
IS Machine area, 90–105 dB Pneumatic noise, cullet, mechanical noise, cooling fans	Pneumatic noise from blanks and moulds cooling, air exhausts and exhaust of forming air. Minimum air pressures, inlet and outlet silencers, wide bore pipe for ducting air exhaust from occupied area. Proper timing of forming air. Cushion cullet chutes and maintain machinery. Automatic spraying or permanent coatings reduce manual lubrication at machines.
Line reject container chutes at delivery end of Lehr 85 to 95 dB	Chutes for reject containers can be lined to eliminate glass to glass and metal contact and reduce reverberation, e.g. use old conveyor belting. Enclosure may also be necessary.
Inspection/Packing 88 to 92 dB	Line reject container chutes (e.g. old conveyor belting), to eliminate glass to metal contact and reduce reverberation. Conveyors designed to regulate bottle flow reduce glass to glass contact. Covering of conveyors has been attempted without great success for quality inspection reasons.
Palletiser 85 to 95 dB	Fit silencers to pneumatic exhaust.

Table 14. Motor vehicle repair.

High risk processes/equipment, typical noise level (sample L_{Aeq})	Established measures for noise control
Vehicle body repair 85–107 dB	Get suppliers of machinery and equipment to specify noise levels at operators' position and choose quiet machines or equipment (especially air saws and chisels which can generate noise levels up to 107 dB and air grinders and orbital sanders up to 97 dB). Isolate bodywork in separate rooms or fix ceiling high partitions.

Table 15. Paper and printing.

High risk processes/equipment	Established measures for noise control
Buckle Folders	Enclosure at all buckle plates.
Paper making machines	Provision of hood (acts for both noise and heat control).
Corrugators	Enclosure.
Vacuum pumps and compressors	Site away from work rooms; shield or enclose.
Sheet-fed printing machines	Ensure adequate spacing and housing; ensure vacuum pumps and compressors are dealt with as above.

Table 16. Quarries.

High risk processes/equipment	Established measures for noise control
Blasting	Blast design, adequate covering of detonating cord, in-hole initiation, shock tube initiation.
Drilling	On hand-operated machines: fitting of mufflers, hearing protection. On drilling rigs: hydraulically driven motors, mufflers and exhaust silencers (and remote positioning of exhaust), control cabins for operators.
Compressors	Remote positioning of compressor units; provision of silencers.
Excavators and draglines	The cabins of new machines offer good noise protection. On older machines soundproofing may be required, and maintenance.
Wheel loaders, dump trucks, etc.	Insulation and covers around engines and fans; good soundproofing of driver's cab, keeping windows and doors closed (air conditioning may then be required in hot weather), silencers on intake silencer.
Crushing/milling	Resilient mountings, chute linings, acoustic curtains, lagging, covers, etc. can bring about useful reductions in noise levels. May need separate soundproofed cabin for operator.
Screening	Use of synthetic screen mats to replace traditional metal plate or woven wire; chute linings and enclosures are usually practicable.
Conveying/feeding	Noise problem possible at the feed or discharge end. Efficient maintenance helps the problem, also reducing the drop height and preventing material hitting empty bins and hoppers. Also use of spiral chutes or lined cascade towers.
Heating/drying	Fitting enclosures to burners and fans. Silencers on inlet and outlet sides of fans. Anti-vibration mountings can prevent reverberations around structure. Remote operation may be practicable.
Saws	Use of dampened saw blades with enclosures; reducing speed of the blade; remote and automatic control of the machines.

Table 17. Rubber.

High risk processes/equipment, typical noise level (sample L_{Aeq})	Established measures for noise control
Grinders/granulators, 96–115 dB	Specify low noise design, special segmental or helical cutters, etc. use tangential feed, fit resilient backing to knives, reduce rotor speed, lag or damp the machine casing, form sound trap in feed aperture or hopper, use feed conveyor in an acoustic tunnel, enclose the machine.
Two roll mills, 90–104 dB and Internal mixers, 84–100 dB	Specify low noise gearboxes, fit helical gears, lag and/or damp the gearbox casing, enclose gearbox and drive, use individual rather than line shaft drives, fume control systems should be designed and installed to reduce noise, isolate and damp guards and other vibrating parts. Fit suitable silencers, pipe exhaust away from operator position, specify low noise gearboxes, lag or damp gearbox casing, isolate and damp thin metal panels, isolate, i.e. use anti-vibration mounts, enclose the gearbox, use belt conveyors instead of vibratory feeders.

(Contd...)

High risk processes/equipment, typical noise level (sample L_{Aeq})	*Established measures for noise control*
Injection moulding machines, 97–100 dB	Specify low noise design, use slow speed pumps, provide damping for control valves, insert hydraulic silencers, mount pumps and motors on anti-vibration mounts and incorporate flexible hoses in pipe lines, enclose hydraulic power packs, convert injector guards to acoustic guards, fit low noise nozzles to blow guns, etc.
Wire twisting machines, 91–97 dB	Use resin bonded fibre gears, damp and acoustically lag machine panels and guards, enclose the machine, specify low noise design.
Tyre curing presses, 83–97 dB	Use low noise nozzles, link blow off nozzle operation to machine work cycle and control by on/off switches, fit suitable silencers to pneumatic system exhausts, duct air away from operator, eliminate steam leaks.
Tyre buffing machines, 85–92 dB	Change process — use peeling to remove bulk of rubber, use buffing brushes rather than rasps, silence air exhausts, silence extraction system and choppers, enclose the buffing machine.
Tyre skiving, 85–92 dB	Specify low noise tools and select carefully, use electrically powered tools, fit silencers to the exhaust ports of pneumatic tools, cable driven tools are more difficult to manoeuvre than pneumatic tools so particular care is needed in locating and supporting them.

Table 18. Stone Masonry.

High risk processes/equipment	*Established measures for noise control*
Chipping hammers	Segregate the process where possible, hearing protection.
Saws	Controls include segregation, enclosures, dampened saw blades, reduced blade speed, remote/automatic machine operation.

Table 19. Textiles.

High risk processes/equipment	*Established measures for noise control*
Worsted and cotton preparation and spinning (especially gill boxes and double twisting): Weaving, Textile finishing (especially crimping), Woven carpet and rug manufacturers, Rope/twine manufacturers, Narrow fabrics (especially braiding), Knitting (some processes, e.g. sock knitting	For all textile machinery some reductions in noise can be achieved by preventive maintenance programmes. However for many processes control of exposure will rely on an effective programme of personal hearing protection.

Glossary

Absorption coefficient	:	A measure of the quantity of sound lost on impinging on a surface. It is a property of the material on which the sound impinges and is dependant on thickness of material and frequency of the sound.
Absorptive silencer	:	A silencer making use of the absorptive properties of materials incorporated in it to reduce the sound passing through it.
Ambient noise	:	The existing background noise in an area can be sounds from many sources, near and far.
Anechoic room	:	A specially constructed room in which as much sound as possible is absorbed at its boundaries. It is typically achieved by using sound absorbing wedges.
Attenuation	:	Noise reduction.
Attenuator	:	It is a noise-reducing device - often colloquially known as a 'silencer'.
Background noise	:	The existing noise associated with a given environment, can be sounds from many sources, near and far.
Breakout	:	It is the escape of sound from any source enclosing structure, such as ductwork and metal casings.
Cross talk	:	It is the transfer of airborne noise from one area to another via secondary air paths, such as ventilation ductwork or ceiling voids.
Cut off frequency	:	It is the frequency at which performance of an acoustic item or material starts to fall below normal or below criterion. Applied to anechoic wedge treatment it refers to the frequency below which the absorption coefficient is worse than 0.99.
Decibel (dB)	:	One tenth of a Bel, a Bel being a unit of amplification corresponding to a tenfold increase.
Directivity factor	:	When sound radiates from any source sound levels can be higher in certain directions than others. This is called 'Directivity'. Directivity Factor is the ratio of the increased level to the average value.
Directivity index	:	Is directivity factor expressed in decibels (dB). It is usually designated by DIO where, O is the angle between the axis of the source and the direction of the measuring point.
Discrete frequency	:	It is a single frequency signal, or a single frequency noise sufficiently dominant over other frequencies to be distinctly audible.
Dynamic insertion loss (DIL)	:	It is a measure of the acoustic performance of an attenuator when handling the rated flow. Not necessarily the same as Static Insertion Loss because it may include regeneration and/or other velocity effects and will account for the effects of the actual fluid and fluid conditions for which the silencer is designed.
End reflection	:	End reflection occurs when sound energy radiates from a hole. The sudden expansion to atmosphere causes some low frequency noise to be reflected back

towards the source. Expressed in decibels (dB), the effect is dependent on hole size and frequency. Maximum at lowest frequency from smallest hole.

Flanking transmission : It is the transfer of sound between any two areas by any indirect path, usually structural. It can also apply to noise transmitted along the casing of a silencer.

Free field : It is a sound field, which is free from all reflective surfaces. A simulated free field can be produced inside a anechoic room. It is a sound field, which is based on a flat reflective surface with no other reflective surfaces present is known as a hemispherical free field.

Frequency (Hz) – sound : It is the number of sound waves to pass a point in one second.

Frequency - vibration : It is the number of complete vibrations in one second.

Helmholz resonance : A resonance created by the mass of a 'plug' of fluid acting on the resilience of 'spring' of a volume of fluid, e.g. a 'plug' of air in a bottleneck resonates on the volume when one blows across the neck. This principle can be used in silencers, etc.

Hertz (Hz) : It is the unit of frequency equivalent to one cycle per second.

Insertion loss : The reduction of noise level by the introduction of noise control device established by the substitution method of test or by 'before and after' testing. The term can be applied to all forms of treatment including silencers and enclosures.

Insulation (sound) : It is the property of a material or partition to oppose sound transfer through its thickness.

Inverse square law : The reduction of noise with distance in terms of decibels, it means a decrease of 6 dB for each doubling of distance from a point source when no reflective surfaces are apparent. This is only applied in free field conditions where the source is small in comparison with the distance.

Laminar flow : It is colloquially used to describe the preferred state of airflow. Strictly means undisturbed flow at very low flow-rates where the air moves in parallel paths.

Level difference : The difference in sound pressure levels between two positions, e.g. inside and outside an enclosure. (This is not the same as Insertion Loss, Transmission Loss or Sound Reduction Index although in some circumstances they may be similar.)

Masking (sound) : Extra sound introduced into an area to reduce the variability of fluctuating noise levels or the intelligibility of speech.

Mass law : Heavy materials stop more noise passing through them than light materials. For any airtight material there will be an increase in its 'noise stopping' ability of approximately 6 dB for every doubling of mass per unit area. It applies only over a certain frequency range and with this range sound reduction index (SRI) is given approximately by:

$$SRI = 20 \log_{10} Mf - 43$$

where, M is the surface density of the material in kg/m^2 and f is frequency in Hz.

Natural frequency : The frequency of a system or material at which it freely vibrates when a force is applied and removed (e.g. kicked).

Near field : It is the area close to a large noise source where the inverse square law does not apply.

Noise : Unwanted sound.

Noise criterion curves (NC) : An American set of curves based on the sensitivity of the human ear. They give a single figure for broadband noise. It is used for indoor design criteria.

Noise rating curves (NR) : A set of curves based on the sensitivity of the human ear. They are used to give a single figure rating for a broad band of frequencies. It is used in Europe for interior and exterior design criteria levels. They have a greater decibel range than NC curves.

Noise reduction : It is used to define the performance of a noise barrier. Established by measuring the difference in sound pressure levels adjacent to each surface.

Octave bands : It is a convenient division of the frequency scale. Identified by their center frequency, typically 63 125 500 1000 2000 4000 8000 Hz.

Pink noise : It is noise of a statistically random nature, having an equal energy per octave bandwidth throughout the audible range.

Pressure drop : The difference between the pressure upstream and down stream of a silencer at given flow conditions. If the silencer is to be installed other than in a duct system of constant cross-section care must be taken with regard to measuring positions and methods to allow for difference in velocity head.

Pure tone : It is a single frequency signal.

Random noise : A confused noise comprised from large number of sound waves, all with unrelated frequencies and magnitudes.

Reactive attenuator or : An attenuator in which the noise reduction is brought about typically by changes resonant attenuator in cross section, chambers and baffle sections.

Regeneration : The noise generated by airflow turbulence. The noise level usually increases with flow speed.

Resonance : This is the build up of excessive vibration in a resilient system. It occurs when the machine speed (disturbing frequency) coincides with the mounted machine natural frequency or support system. Similar effects can occur in acoustic systems (i.e. sound energy in a gaseous fluid).

Resonant frequency (Hz) : It is the frequency at which resonance occurs in the resilient system.

Reverberation : Reflected sound in a room, that decays after the sound source has stopped.

Reverberation room : A calibrated room specially constructed with sound reflective walls, e.g. plastered concrete. The result is a room with a 'long smooth echo', in which a sound takes a long time to die away. The sound pressure levels in this room are very even.

Reverberation time : The time taken in seconds for the average sound energy level in a room to decrease to one millionth of its originally steady level after the source has stopped, i.e. time taken for a 60 dB decrease to take place. It is usually related to frequency bands as it varies with frequency.

Room constant : It is the sound absorbing capacity of a room, usually expressed in m^2.

Sabines : It is a unit of absorption comprising the sum of the products of absorption coefficients and areas of the materials of a room. It must be qualified by the units of area used, e.g. square meter sabines.

Sabine's formula : Predicts the reverberation time of a room or enclosure from know room volume and absorption characteristics. Becomes inaccurate when absorption is high.

Silencer : It is Colloquialism for attenuator, spoken by optimists.

Sound insulation : It is the property of a material or partition to oppose sound transfer through its thickness.

Sound level meter : It is an instrument for measuring sound pressure levels. It can be fitted with (noise meter) electrically weighting networks for direct read-off in dBA, dBB, dBC, dBD and octave or third octave bands.

Sound power : It is a measure of sound energy in watts. It is a fixed property of a machine, irrespective of environment.

Sound power level (SWL)	:	It is the amount of sound output from a machine, etc. cannot be measured directly. It is expressed in decibels of SWL.
Sound pressure level (SPL)	:	It is a measurable sound level that depends upon environment. It is a measure of the sound pressure at a point in N/m^2. It is expressed in decibels of SPL at a specified distance and position. It can also be considered as a measure of intensity of terms of sound energy per unit area at the point considered, but is not a vector (i.e. directional) as sound Intensity strictly is.
SPL direct field	:	It is the sound radiating directly from the source(s) to the receiver without reflection. The direct components of a sound level field are calculated from a given SWL by using inverse square law and directivity, etc.
SPL reverberant field	:	It is the sound reaching the receiver after one or more reflections. The reverberant component of a sound level calculation from a given SWL by using room constant values from reverberation time and volume.
Sound reduction index (SRI)	:	A set of values measured by a specific test method to establish the actual amount of sound that will be stopped by the material, partition or panel, when located between two reverberation rooms. Average SRI can be calculated by averaging the set of values in the sixteen third octave bands from 100 Hz to 3150 Hz. It is a property of the material(s) or construction, not directly measurable in the field.
Sound spectrum	:	It is the separation of sound into its frequency components across the audible range of the human ear.
Standing waves	:	These occur due to room geometry. Sound levels at some locations in the room at certain frequencies will be intensified by additive interference of successive waves, and in other locations reduced by cancellation.
Static insertion loss (SIL)	:	The insertion loss of an attenuator under static (no flow) conditions (cf insertion loss, dynamic insertion loss).
Transmission loss	:	A set of values measured by a specific test method to establish the actual amount of noise that will be stopped by the material, partition or panel when placed between two reverberation rooms.
Threshold of audibility	:	The minimum sound levels at each frequency that a person can just hear.
Threshold of pain	:	The sound level at which a person experiences physical pain (typically 120 dB).
Threshold shift	:	A partial loss of hearing caused by excessive noise, either temporary or permanent, in a person's threshold of audibility.
Third octave bands	:	It is a small division of the frequency scale, three to each octave. It enables more accurate noise analysis.
Turbulent flow	:	A confused state of airflow that may cause noise to be generated inside, for example, a ductwork system.
Velocity head or velocity pressure (Pv)	:	It is a measure of the inertia of a flowing fluid used in assessing pressure losses in duct systems and/or silencers for air at atmospheric conditions.
White noise	:	It is noise of a statistically random nature having an equal energy level per Hertz throughout the audible range.
Wavelength	:	The distance between two like points on a wave shape, e.g. distance from crest to crest.

References

Allen, T., *Noise and Effects on Man*, Edward Arnold, London.

Budyko, P., *Noise Abatement and Control*, Wiley Inter Science, New York.

Downe, A.L., *Noise Reduction and Control*, Blackwell Scientific Publications, Oxford.

Evans, L., *Technology for Reducing Industrial Noise*, Reinhold, UK.

Fishlock, N., *Abatement and Measurement of Control Valve Noise*, Open University Press, Buckingham.

Gouldman, S.M., *Noise and Vibration Control*, McGraw Hill, New York.

Gross, Z., *Handbook of Noise Measurement*, Reinhold, USA.

Harding, C.M., *Handbook of Noise Control*, McGraw Hill, New York.

Irishik, B.J., *Instruments for Noise Analysis,* Elsevier, USA.

Jylleus, K., *Industrial Noise Problems and Their Solution*, Noyes, UK.

Kryter, S., *Noise Abatement and Control*, Henser Publishers, USA.

Kushale, J., *Sound, Vibration and Noise,* Van Nostrand Reinhold, USA.

Luwis, F., *Impairment of Hearing from Noise*, John Wiley and Sons, New York.

Morton L., *Acoustic Design and Noise Control*, Maclaren, London.

Paul N. Cheremisinoff, *Industrial Noise Control Handbook*, Ann Arbor Science Publishers Inc., Michigan.

Pearsons, K.S., *Handbook of Noise Ratings*, Butterworth, Oxford.

Peter, K.J., *Occupational Safety and Health Standards*, Noyes, UK.

Powell, M., *Silencers, Shields and Barriers-Designing with Lead*, Van Nostrand Reinhold, Amsterdam.

Rettinger, M., *Noise Vibration Control*, Sterling Publications Ltd., Hong Kong.

Riviere, E., *Personal Safety Devices,* Elsevier Applied Science Publishers, London.

Rustom, K.L., *Handbook of Industrial Pollution*, Reinhold, USA.

Smith, B., *Acoustic Design and Noise Control*, Elsevier, New York.

Straughan, L., *Sound and Noise*, Surrey University Press, Guildford.

Taylor, L.S., *Sound Control Materials*, Academic Press, UK.

Van Vlack L.H., *Surface Transportation and Traffic Noise Reduction*, Addison-Wesley, UK.

Wells, J.K., *Silencers, Mufflers and Active Noise Control*, Academic Press, New York.

Wynmer, K., *Instrumentation for Noise Analysis*, Pergamon Press, Oxford.

Yulleus, K.T., *Handbook of Noise Control*, Maclaren, London.

Index